Lecture Notes in Physics

Editorial Board

H. Araki, Kyoto, Japan
E. Brézin, Paris, France
J. Ehlers, Potsdam, Germany
U. Frisch, Nice, France
K. Hepp, Zürich, Switzerland
R. L. Jaffe, Cambridge, MA, USA
R. Kippenhahn, Göttingen, Germany
H. A. Weidenmüller, Heidelberg, Germany
J. Wess, München, Germany
J. Zittartz, Köln, Germany

Managing Editor

W. Beiglböck
Assisted by Mrs. Sabine Landgraf
c/o Springer-Verlag, Physics Editorial Department II
Tiergartenstrasse 17, D-69121 Heidelberg, Germany

Springer
Berlin
Heidelberg
New York
Barcelona
Budapest
Hong Kong
London
Milan
Paris
Tokyo

The Editorial Policy for Proceedings

The series Lecture Notes in Physics reports new developments in physical research and teaching – quickly, informally, and at a high level. The proceedings to be considered for publication in this series should be limited to only a few areas of research, and these should be closely related to each other. The contributions should be of a high standard and should avoid lengthy redraftings of papers already published or about to be published elsewhere. As a whole, the proceedings should aim for a balanced presentation of the theme of the conference including a description of the techniques used and enough motivation for a broad readership. It should not be assumed that the published proceedings must reflect the conference in its entirety. (A listing or abstracts of papers presented at the meeting but not included in the proceedings could be added as an appendix.)

When applying for publication in the series Lecture Notes in Physics the volume's editor(s) should submit sufficient material to enable the series editors and their referees to make a fairly accurate evaluation (e.g. a complete list of speakers and titles of papers to be presented and abstracts). If, based on this information, the proceedings are (tentatively) accepted, the volume's editor(s), whose name(s) will appear on the title pages, should select the papers suitable for publication and have them refereed (as for a journal) when appropriate. As a rule discussions will not be accepted. The series editors and Springer-Verlag will normally not interfere with the detailed editing except in fairly obvious cases or on technical matters.

Final acceptance is expressed by the series editor in charge, in consultation with Springer-Verlag only after receiving the complete manuscript. It might help to send a copy of the authors' manuscripts in advance to the editor in charge to discuss possible revisions with him. As a general rule, the series editor will confirm his tentative acceptance if the final manuscript corresponds to the original concept discussed, if the quality of the contribution meets the requirements of the series, and if the final size of the manuscript does not greatly exceed the number of pages originally agreed upon. The manuscript should be forwarded to Springer-Verlag shortly after the meeting. In cases of extreme delay (more than six months after the conference) the series editors will check once more the timeliness of the papers. Therefore, the volume's editor(s) should establish strict deadlines, or collect the articles during the conference and have them revised on the spot. If a delay is unavoidable, one should encourage the authors to update their contributions if appropriate. The editors of proceedings are strongly advised to inform contributors about these points at an early stage.

The final manuscript should contain a table of contents and an informative introduction accessible also to readers not particularly familiar with the topic of the conference. The contributions should be in English. The volume's editor(s) should check the contributions for the correct use of language. At Springer-Verlag only the prefaces will be checked by a copy-editor for language and style. Grave linguistic or technical shortcomings may lead to the rejection of contributions by the series editors. A conference report should not exceed a total of 500 pages. Keeping the size within this bound should be achieved by a stricter selection of articles and not by imposing an upper limit to the length of the individual papers. Editors receive jointly 30 complimentary copies of their book. They are entitled to purchase further copies of their book at a reduced rate. As a rule no reprints of individual contributions can be supplied. No royalty is paid on Lecture Notes in Physics volumes. Commitment to publish is made by letter of interest rather than by signing a formal contract. Springer-Verlag secures the copyright for each volume.

The Production Process

The books are hardbound, and the publisher will select quality paper appropriate to the needs of the author(s). Publication time is about ten weeks. More than twenty years of experience guarantee authors the best possible service. To reach the goal of rapid publication at a low price the technique of photographic reproduction from a camera-ready manuscript was chosen. This process shifts the main responsibility for the technical quality considerably from the publisher to the authors. We therefore urge all authors and editors of proceedings to observe very carefully the essentials for the preparation of camera-ready manuscripts, which we will supply on request. This applies especially to the quality of figures and halftones submitted for publication. In addition, it might be useful to look at some of the volumes already published. As a special service, we offer free of charge LaTeX and TeX macro packages to format the text according to Springer-Verlag's quality requirements. We strongly recommend that you make use of this offer, since the result will be a book of considerably improved technical quality. To avoid mistakes and time-consuming correspondence during the production period the conference editors should request special instructions from the publisher well before the beginning of the conference. Manuscripts not meeting the technical standard of the series will have to be returned for improvement.

For further information please contact Springer-Verlag, Physics Editorial Department II, Tiergartenstrasse 17, D-69121 Heidelberg, Germany

S. M. Deshpande S. S. Desai R. Narasimha (Eds.)

Fourteenth International Conference on Numerical Methods in Fluid Dynamics

Proceedings of the Conference
Held in Bangalore, India, 11-15 July 1994

 Springer

Editors

Suresh M. Deshpande
Department of Aerospace Engineering
Indian Institute of Science
Bangalore 560 012, India

Shivaraj S. Desai
Computational & Theoretical Fluid Dynamics Division
National Aerospace Laboratories
Bangalore 560 017, India

Roddam Narasimha
Jawaharlal Nehru Centre for Advanced Scientific Research
Indian Institute of Science
Bangalore 560 012 , India

Cataloging-in-Publication Data applied for.

Die Deutsche Bibliothek - CIP-Einheitsaufnahme

International Conference on Numerical Methods in Fluid Dynamics <14, 1994, Bangalore>:
Fourteenth International Conference on Numerical Methods in Fluid Dynamics : proceedings of the conference held at Bangalore, India, 11 - 15 July 1994 / S. M. Deshpande ... (ed.). - Berlin ; Heidelberg ; New York : Springer, 1995
  (Lecture notes in physics ; Vol. 453)
  ISBN 3-540-59280-6
NE: Deshpande, Suresh M. [Hrsg.]; HST; GT

ISBN 3-540-59280-6 Springer-Verlag Berlin Heidelberg New York

This work is subject to copyright. All rights are reserved, whether the whole or part of the material is concerned, specifically the rights of translation, reprinting, re-use of illustrations, recitation, broadcasting, reproduction on microfilms or in any other way, and storage in data banks. Duplication of this publication or parts thereof is permitted only under the provisions of the German Copyright Law of September 9, 1965, in its current version, and permission for use must always be obtained from Springer-Verlag. Violations are liable for prosecution under the German Copyright Law.

© Springer-Verlag Berlin Heidelberg 1995
Printed in Germany

Typesetting: Camera-ready by the editors
SPIN: 10501414   55/3142-543210 - Printed on acid-free paper

Editors' Preface

The present volume constitutes the Proceedings of the Fourteenth International Conference on Numerical Methods in Fluid Dynamics, held at Bangalore, India, during 11–15 July 1994. The Conference had the following format. Each day of the conference began with a plenary session at which an invited lecture was delivered. During the rest of the day there were two sessions in parallel in which oral presentations were made. The Conference concluded with a valedictory lecture on the last day. A total of 224 manuscripts were submitted to the Conference, from all over the world, and were reviewed by three regional committees chaired by Prof. J.-J. Chattot for papers from the Americas, by Prof. F. Sabetta for papers from Europe, and by Prof. K. Oshima and Dr. K. Srinivas (on behalf of C.A.J. Fletcher) for papers from Asia and Australia.

Out of the 112 manuscripts accepted, 19 were withdrawn before the conference and 5 could not be presented at the Conference. In this volume the invited lectures and the valedictory lecture are grouped together separately from the contributed papers. The rest of the papers, all presented at Bangalore, are included under two broad headings: Numerical Methods and Algorithms, and Applications.

The Conference was attended by 216 participants from 18 countries across the globe. To those of us who organised the Conference in Bangalore, it was a memorable and rewarding occasion; and we hope that all participants carried with them memories of a fruitful, friendly, and pleasant meeting. If the Conference was a success, we owe it to the support of the International Organising Committee, in particular Dr. M.Y. Hussaini and Prof. K.W. Morton who made many valuable suggestions about its organisation, to the many Indian agencies (listed elsewhere) who generously provided financial assistance, and to innumerable friends and colleagues in India who worked very hard for the professional and social success of the meeting.

Bangalore
June 1995

S.M. Deshpande
S.S. Desai
R. Narasimha

Acknowledgements

The Organising Committee wishes to express its gratitude to the following Indian organisations for their generous financial support to the conference and this volume.

- Council of Scientific and Industrial Research, New Delhi
- National Aerospace Laboratories, Bangalore
- Jawaharlal Nehru Centre for Advanced Scientific Research, Bangalore
- Indian Institute of Science, Bangalore
- Indian Institute of Technology, New Delhi
- Aeronautical Development Agency, Bangalore
- Tata Information Systems Limited, Bangalore
- Aeronautical Research and Development Board, New Delhi
- Department of Science and Technology, New Delhi
- Centre for Development of Advanced Computing, Pune

The National Organising Committee acknowledges with gratitude the support received from members of the local committees and the spontaneous and painstaking effort from many friends and well-wishers who contributed to the success of the Conference, and in particular from Dr. Dinesh K. Prabhu and Mr. Satyaprasad of the National Aerospace Laboratories (NAL). Special mention is made of the careful reformatting of many papers by Mr. V.S. Narasimhan and Ms. D. Shobha and the excellent infrastructural and administrative support given by colleagues from NAL. Last but not least, the Committee acknowledges with pleasure the friendly care and generous help received from the staff and management of Hotel Ashok, where the meeting was held.

CONTENTS

1. Invited Lectures

D.E. Keyes
Aerodynamic Applications of Newton-Krylov-Schwarz Solvers   1

A. Leonard, P. Koumoutsakos, G. Winckelmans
Vortex Methods for Three-Dimensional Separated Flows   21

B. Mohammadi, O. Pironneau
On Wall Laws in CFD   31

T.S. Prahlad
Status of CFD in India   39

N.P. Weatherill, D.L. Marcum, A.J. Gaither, O. Hassan, M.J. Marchant
The Development of High-Quality Unstructured Grids on Parallel Computers   54

2. Valedictory Lecture

A. Jameson
Computational Methods for Aerodynamic Design   71

3. Numerical Methods and Algorithms

a) Kinetic/Boltzmann Schemes

J.-P. Croisille, R. Khanfir, G. Chanteur
A Kinetic Flux-Splitting Scheme for the MHD Equations   86

S.S. Deshpande
A Boltzmann Taylor Galerkin FEM for Compressible Euler Equations   91

J.-L. Estivalezes, P. Villedieu
A New Second-Order Positivity Preserving Kinetic Scheme for the Compressible Euler Equations   96

A.K. Ghosh, S.M. Deshpande
A Robust Least Squares Kinetic Upwind Scheme for Euler Equations   101

K. Xu, L. Martinelli, A. Jameson
Gas-Kinetic Finite Volume Methods   106

S.V. Raghurama Rao
Peculiar Velocity Based Upwind Method for Inviscid Compressible Flows   112

F. Rogier, J. Schneider
Discrete Boltzmann Equation for Solving Rarefied Gas Problems   117

b) Grids/Acceleration Techniques

J.G. Andrews, K.W. Morton
Spurious Entropy Generation as a Mesh Quality Indicator 122

R. Biswas, R.C. Strawn
Dynamic Mesh Adaptation for Tetrahedral Grids 127

F. Chalot, C. Kasbarian, M.-P. Leclercq, M. Mallet, M. Ravachol, B. Stoufflet
Extensive Analysis and Cross-Comparisons of Past and Present Numerical Formulations for Flow Problems on Unstructured Meshes 133

A. Chatterjee, G.R. Shevare
A Time-Accurate Multigrid Algorithm for Euler Equations 139

M. Gazaix
Hypersonic Viscous Flow Computations with Solver and Grid Sensitivity Analysis 144

J.P. Gregoire, G. Pot
Speed-Up of CFD Codes Using Analytical FE Calculations 149

D. Hänel, R. Vilsmeier
A Field Method for 3-D Tetrahedral Mesh Generation and Adaption 155

J.F. Lynn, B. van Leer
Multigrid Euler Solutions with Semi-coarsening and Local Preconditioning 161

B.K. Soni, H. Thornburg
CAGD Techniques in Structured Grid Generation and Adaption 166

S. Ta'asan
Canonical-Variables Multigrid Method for Euler Equations 173

V. Venkatakrishnan, D.J. Mavriplis
Agglomeration Multigrid for the Euler Equations 178

Zi-Niu Wu
Steady Transonic Flow Computations Using Overlapping Grids 183

c) Boundary Conditions

A. Dadone, B. Grossman
Surface Boundary Conditions for the Numerical Solution of the Euler Equations in Three Dimensions 188

W. Jia, Y. Nakamura
The Ψ-q Formulation and Boundary Condition for Incompressible Flows in Multi-connected Region 195

C.P. Mellen, K. Srinivas
Treatment of Branch Cut Lines in the Computation of Incompressible
Flows with C-Grids 201

Yo Mizuta
Concepts on Boundary Conditions in Numerical Fluid Dynamics 206

D. Schulze
Far Field Boundary Conditions Based on Characteristic and
Bicharacteristic Theory Applied to Transonic Flows 211

D. Sen
An Accurate Cubic-Spline Boundary-Integral Formulation for Two-
Dimensional Free-Surface Flow Problems 216

H.S. Udaykumar, W. Shyy
ELAFINT: A Computational Method for Fluid Flows with Free and
Moving Boundaries 221

d) Euler/Navier-Stokes Equations

R. Arina, F. Ramella
An Implicit Spectral Solver for the Compressible Navier-Stokes
Equations in Complex Domains 228

F. Bassi, S. Rebay
Accurate 2D Euler Computations by Means of a High-Order
Discontinuous Finite Element Method 234

L.A. Catalano, P. De Palma, G. Pascazio, M. Napolitano
A Higher-Order Multidimensional Upwind Solution-Adaptive
Multigrid Solver for Compressible Flows 241

V.V. Dedesh, I.V. Yegorov
On Some Advances in Fluid Flow Modeling Through Coupling of
Newton Method and New Implicit PPV Schemes 246

L. Flandrin, P. Charrier
An Improved Approximate Riemann Solver for Hypersonic
Bidimensional Flows 251

B. Fortunato, V. Magi
An Implicit Lambda Method for 2-D Viscous Compressible Flows 259

J. Groenner, E. von Lavante, M. Hilgenstock
Numerical Methods for Simulating Supersonic Combustion 265

A. Iollo, M.D. Salas, S. Ta'asan
Shape Optimization Governed by the Quasi 1D Euler Equations Using
an Adjoint Method 274

V. Karamyshev, V. Kovenya, S. Cherny
Relaxation Method for 3-D Problems in Supersonic Aerodynamics 280

G.H. Klopfer, S. Obayashi
Virtual Zone Navier-Stokes Computations for Oscillating Control
Surfaces … 285

J. Rokicki, J.M. Floryan
The Domain Decomposition Method and Compact Discretization for
the Navier-Stokes Equations … 287

J. Szmelter, A. Pagano
Viscous Flow Modelling Using Unstructured Meshes for Aeronautical
Applications … 293

I. Toumi, P. Raymond
Upwind Numerical Scheme for a Two-Fluid Two-Phase Flow Model … 299

M. Valorani, B. Favini
A Compact Formalism to Design Numerical Techniques for Three
Dimensional Internal Flows … 307

S. Yamamoto, H. Nagatomo, H. Daiguji
An Implicit–Explicit Flux Vector Splitting Scheme for Hypersonic
Thermochemical Nonequilibrium Flows … 314

e) Viscous and Turbulent Flows

Ch.-H. Bruneau, P. Fabrie
Computation of Incompressible Flows and Direct Simulation of
Transition to Turbulence … 320

W.J. Decker, J. Dorning
Incomplete LU Decomposition in Nodal Integral Methods for Laminar
Flow … 325

R. Friedrich, C. Wagner
On Turbulent Flow in a Sudden Pipe Expansion and Its Reverse
Transfer of Subgrid-Scale Energy … 330

K.S. Muralirajan, K. Ghia, U. Ghia, B. Brandes-Duncan
Analysis of 3-D Unsteady Separated Turbulent Flow Using a
Vorticity-Based Approach and Large-Eddy Simulation … 337

K. Ravi, K. Ramamurthi
Penalty Finite Element Method for Transient Free Convective Laminar
Flow … 343

M. Si Ameur, J.P. Chollet
Numerical Simulation of Three Dimensional Turbulent Underexpanded
Jets … 348

J. Steelant, E. Dick
Conditioned Navier-Stokes Equations Combined with the K-E Model
for By-pass Transitional Flows … 353

V. Theofilis, D. Dijkstra, P.J. Zandbergen
Numerical Aspects of Flow Stability in the Veldman Boundary Layers   358

S. Venkateswaran
Efficient Implementation of Turbulence Modeling for CFD   363

M. Winkelsträter, H. Laschefski, N.K. Mitra
Transition to Chaos in Impinging Three-Dimensional Axial and Radial Jets   368

G. Zhou, L. Davidson, E. Olsson
Transonic Inviscid/Turbulent Airfoil Flow Simulations Using a Pressure Based Method with High-Order Schemes   372

f) Miscellaneous Techniques

M. Arora, P.L. Roe
On Oscillations Due to Shock Capturing in Unsteady Flows   379

H. Dang-Vu, C. Delcarte
An Example of Hopf Bifurcation and Strange Attractor in Chebyshev Spectral Approximations to Burgers Equation   385

J. Fontaine, Loc Ta Phuoc
Simulation of the Unsteady Separated Flow Around a Finite Rectangular Plate Using a Free-Divergence Constraint Velocity-Vorticity Formulation   391

Z. Gao, F.G. Zhuang
Time–Space Scale Effects in Computing Numerically Flowfields and a New Approach to Flow Numerical Simulation   396

Y. Li, M. Rudman
An Evaluation of a Multi-dimensional FCT Algorithm with Some Higher-Order Upwind Schemes   402

F. Nasuti, M. Onofri
Numerical Study of Unsteady Compressible Flows by a Shock Fitting Technique   407

J.M. Pallis, J.-J. Chattot
Implementation of a Nonequilibrium Flow Solver on a Massively Parallel Computer   413

B.B. Paulsen, M.D. Holst, T.J. Condra, J. Rusaas
Lagrangian Prediction of Particulate Two-Phase Flows at High Particle Loadings   418

Jong-Youb Sa
A Numerical Method for Incompressible Viscous Flow Using Pseudocompressibility Approach on Regular Grids   423

S.J. Sherwin, G.E. Karniadakis
Tetrahedral Spectral Elements for CFD — 429

4. Applications

L. Bai, N.K. Mitra, M. Fiebig
Computation of Unsteady 3D Turbulent Flow and Torque Transmission in Fluid Couplings — 435

V.V. Bogolepov, I.I. Lipatov, L.A. Sokolov
Flow Investigation Near the Point of Surface Temperature and Catalytic Properties Discontinuity — 441

P.Y.P. Chen, M. Behnia, B.E. Milton
Computation of Air and Fuel Droplet Flows in S.I. Engine Manifolds — 447

K. Engel, F. Eulitz, M. Faden, S. Pokorny
Validation of TVD Schemes for Unsteady Turbomachinery Flow — 454

J. Fořt, M. Huněk, K. Kozel, J. Lain, M. Šejna, M. Vavřincová
Numerical Simulation of Steady and Unsteady Flows Through Plane Cascades — 461

E. Guilmineau, P. Queutey
Computation of Unsteady Viscous Flows past Oscillating Airfoils Using the CPI Method — 466

C. Hartmann, K.G. Roesner
Investigation of 3-D Shock Focusing Effects on a Massively Parallel Computer — 472

J. Mathew, A.J. Basu
The Mechanism of Entrainment in Circular Jets — 476

G. Mejak
Numerical Computation of Critical Flow over a Weir — 481

Y. Mochimaru
Numerical Simulation of a Natural Convection in a Cavity of an Ellipsoid of Revolution, Using a Spectral Finite Difference Method — 485

V.J. Modi, S.R. Munshi, G. Bandyopadhyay, T. Yokomizo
Multielement Systems with Moving Surface Boundary-Layer Control: Analysis and Validation — 490

N. O'Shea, C.A.J. Fletcher
Prediction of Turbulent Delta Wing Vortex Flows Using an RNG k-ε Model — 496

K. Oshima, H. Chen
Evaporative Flow in Heat Pipes — An Object Oriented System Analysis — 502

B. Ramakrishnananda, M. Damodaran
Numerical Study of Viscous Compressible Flow Inside Planar and
Axisymmetric Nozzles … 508

N. Satofuka, T. Nishitani
Numerical Simulation of an Air Flow past a Moving Body with/
without Deformation … 513

S.K. Saxena, K. Ravi
Computation of Three Dimensional Supersonic and Hypersonic Blunt
Body Flows Using High Resolution TVD Schemes Based on Roe's
Approximate Riemann Solver … 520

S. Sekar, M. Nagarathinam, R. Krishnamurthy, P.S. Kulkarni,
S.M. Deshpande
3-D KFVS Euler Code "BHEEMA" as Aerodynamic Design and
Analysis Tool for Complex Configurations … 525

W.L. Sickles, M.J. Rist, C.H. Morgret, S.L. Keeling,
K.N. Parthasarathy
Separation Analysis of the Pegasus XL from an L-1011 Aircraft … 530

K.P. Singh, K. Murali Krishna, S. Saha, S.K. Mukharjea
Application of an Euler Code on a Modern Combat Aircraft
Configuration … 535

D.J. Song, K.D. Kwon, S.Y. Won
Study on Hypersonic Finite-Rate Chemically Reacting Flows Using
Upwind Method … 540

K.R. Sreenivas, P.K. Dey, Jaywant H. Arakeri, J. Srinivasan
Study of the Erosion of Stably Stratified Medium Heated from Below … 551

G.R. Srinivasan
Advances in Euler and Navier-Stokes Methods for Helicopter
Applications … 556

C. Tenaud, Ta Phuoc Loc
Numerical Simulation of Unsteady Compressible Viscous Flow
NACA 0012 Airfoil-Vortex Interaction … 562

S. Tokarcik-Polsky, J.-L. Cambier
A Numerical Study of Unsteady Flow Phenomena in the Driven and
Nozzle Sections of Shock Tunnels … 568

Hsin-Hua Tsuei, C.L. Merkle
The Dynamics of Reacting Shear Layers Adjacent to a Wall … 574

Wang Wensheng, Zhang Fengxian, Xu Yanji, Chen Naixing
Validation of Two 3-D Numerical Computation Codes for the Flows in
an Annular Cascade of High-Turning-Angle Turbine Blades … 579

5. Author Index … 586

Aerodynamic Applications of Newton-Krylov-Schwarz Solvers

David E. Keyes

Institute for Computer Applications in
Science and Engg., NASA
Langley Research Center, Hampton, VA 23681-0001 USA
and
Department of Computer Science, Old Dominion
University, Newfolk, VA 23529-0162 USA

Abstract : Parallel implicit solution methods are increasingly important in aerodynamics, since reliable low-residual solutions at elevated CFL number are prerequisite to such large-scale applications of aerodynamic analysis codes as aeroelasticity and optimization. In this chapter, a class of nonlinear implicit methods and a class of linear implicit methods are defined and illustrated. Their composition forms a class of methods with strong potential for parallel implicit solution of aerodynamics problems. Newton-Krylov methods are suited for nonlinear problems in which it is unreasonable to compute or store a true Jacobian, given a strong enough preconditioner for the inner linear system that needs to be solved for each Newton correction. In turn, Krylov-Schwarz iterative methods are suited for the parallel implicit solution of multidimensional systems of linearized boundary value problems. Schwarz-type domain decomposition preconditioning provides good data locality for parallel implementations over a range of granularities. These methods are reviewed separately, illustrated with CFD applications, and composed in a class of methods named Newton-Krylov-Schwarz.

1 Introduction and Basic Methods

Two trends in the industrial practice of computational fluid dynamics demand algorithmic attention: a need for more rapid and reliable low-residual solutions to elevated CFL number problem, and a desire to take advantage to high-latency distributed computing environments, such as the often already existing infrastructure of high-performance workstations on a local area network. These factor suggest a niche for Newton-Krylov-Schwarz (NKS) domain decomposition methods - a niche that we are exploring experimentally with a variety of fluid dynamic models, from full potential to Navier-Stokes, on both structured and unstructured grids, with several collaborators [5, 10, 27].

A discussion of NKS methods assumes familiarity with each of the three basic constituent methods, reviewed in this section, and of the pairwise compositions of

Newton-Krylov and Newton-Schwarz, which are defined and illustrated in Section 2. Section 3 describes the full Newton-Krylov-Schwarz method.

1.1 Newton Methods

Newton methods for solution of the discretized governing equations

$$f(u) = 0, \qquad (1)$$

given an initial guess u^0, iterate for $l = 0, 1, \ldots$, on

$$u^{l+1} = u^l + \lambda^l \delta u^l, \qquad (2)$$

until the norm of the nonlinear residual, $\| f(u^l) \|$, is sufficiently small, where δu^l approximately solves the linear Newton correction equation

$$J(u^l)\delta u^l = -f(u^l) \qquad (3)$$

based on jacobian matrix $J \equiv \frac{\partial f}{\partial u}$. In these equations, u^l and f are vectors of dimension n and $J(u^l)$ is an $n \times n$ matrix. The Jacobian is generally *sparse*, reflecting the local stencil operations that define the components of f, *nonsymmetric*, reflecting the presence of large first-order advective terms in f and the multicomponent nature of u and f, and *ill-conditioned*, reflecting the second-order derivative terms in f and the very small grid cell dimensions used to resolve rapidly varying features of u. The parameter λ^l is selected by some line search or trust region algorithm [12]. A full Newton method, in which (3) is solved exactly, enjoys quadratic convergence asymptotically, effectively doubling the number of significant digits in the result in each step. Newton's method is important, in part, because a Newton-based analysis code allows convenient solution for first-order sensitivity data in design optimization.

A full Newton method usually cannot be sued far from the solution, because the two terms of the Taylor series for $f(u)$ upon which (3) is based do not provide reliable information there about the direction of the root. Through an example for which Newton converges unassisted is furnished later in Section 3, a continuation procedure or a mesh sequencing procedure may be required to provide an initial guess form which the Newton iteration (2) will converge. As a side benefit, these procedures may improve the conditioning of the Newton correction equation (3), which is important when it is solved by linear subiterations. Pseudo-transient continuation is often employed, whereby (1) is replaced with

$$D\frac{\partial u}{\partial t} + f(u) = 0 \qquad (4)$$

where D is a diagonal matrix. Implicit time leads to a modification of (3):

$$D\frac{\delta u^l}{\delta t} + f(u^l + \delta u^l) = 0$$

hence, linearised,

$$\left[\frac{D}{\delta t} + \frac{\partial f}{\partial u}\right]\delta u^l \approx -f(u^l). \qquad (5)$$

With $D/\delta t$ as a parameter, the linear conditioning of the Newton correction equation can be improved, simultaneously with the nonlinear robustness.

Newton's method has some disadvantages, besides lack of a global convergence theory: (1) the Jacobian is often expensive to form and store, (2) some popular discretizations (e.g., using limiters) are nondifferentiable, and most popular preconditioner family for structured and unstructured two-dimensional and three-dimensional grids, ILU(ℓ, ϵ), is difficult to be parallelized efficiently. We address the first two of these disadvantages in the context of Newton-Krylov methods, and the last in the context of NKS.

1.2 Krylov Methods

Krylov methods may be thought of as either Galerkin or minimal residual methods (or both, when applied to self-adjoint problems) for finding best approximations to the solutions of linear systems $Ax = b (A, n \times n)$, in subspaces. The subspaces are Krylov spaces, expended incrementally according to convergence requirements, generated from an initial vector by means of matrix-vector multiplication. The initial vector is usually the residual of the initial iterate, $r_0 = b - Ax_0$, so the first step of most Krylov methods is in the direction of steepest descent. Because Jacobians of aerodynamic problems are invariably nonsymmetric, we have concentrated experimental attention on the three most popular nonsymmetric Krylov solvers GMRES [30], BiCGCTAB [33], and TFQMR [18], and we limit our brief review to GMRES. To solve $Ax = b$, let $x = V_y$ (V, $n \times m$, $m \ll n$), and minimize the residual for increasing $m = 1, 2, \ldots$. The vectors $\{v_1, v_2, \ldots, v_m\}$ are the result of a Gram-Schmidt process on the Krylov space $\{r_0, Ar_0, \ldots, A^{m-1}r_0\}$. The coefficients y_k satisfy a small least squares problem. The algorithm requires only matrix-vector products with A, vector dot products (for Gram-Schmidt), and the solution of a small overdetermined system.

In exact arithmetic GMRES has a nonincreasing residual, and cannot break down short of delivering the exact solution. However, carrying GMRES to an exact solution for a large sparse Jacobian system is generally infeasible, since it is equivalent to performing a dense QR factorization. In practice, a memory limit is set for the size of the Krylov space m, and GMRES is restarted if convergence is not attained after m steps. In restarted mode, GMRES is not guaranteed to converge, and it is easy to construct pathological examples in which it will not. However, convergence is guaranteed if the symmetric part of A (that is, $(A + A^T)/2)$) is positive definite, and this requirement is often met in upwind-biased discretizations of convection-diffusion equations.

GMRES in exact precision requires a number of iterations that is at most the number of distinct eigenvalues of A, and when the eigenvalues are well clustered, it essentially devotes at most one step to each cluster. This leads to the convergence enhancement strategy known as preconditioning, namely, pre- or post-multiplication of A by operators that approximate A^{-1}.

GMRES and other Krylov solvers are often proclaimed to be "Parameterfree," in the sense that, unlike other iterative solvers, they require no *a priori* knowledge of the spectrum of the preconditioned operator and are less dependent on the quality of the initial iterate. In fact, however, the convergence tolerance, restart frequency, and preconditioner quality are difficult parameters that GMRES brings into an overall NK or NKS method. The accuracy with which the matrix-vector products are approximated is an additional parameter that cannot be taken for granted.

Finally, with a view towards optimization or design applications, we mention that it is difficult to amortize the work done in one GMRES cycle over multiple sequential right-hand sides except in special circumstances. Exceptions include cases in which the memory to store a full orthogonalization is available and cases in which successive right-hand sides are related in the sense that the initial residuals of later systems have only small projections into the orthogonal complement or earlier Krylov spaces. Cases of the first type arise in practice if the dimension of the system being solved by GMRES is small compared to the dimension of an overall model [15], or if the original system is already dense [28]. Cases of the second type may arise in transient problems that are smoothly resolved in time or in a perturbative series of problems [17, 28].

1.3 Schwarz Methods

Schwarz methods are divide-and-conquer methods for solving the PDE BVP $\mathcal{L}u = f$ in Ω through solving a sequence of problems $\mathcal{L}'_k u'_k = f'_k$ in subdomains Ω_k covering Ω, and iteratively combining the partial solutions u'_k to form u. They constitute an important class of iterative domain decomposition methods for PDEs, but also have a purely algebraic interpretation that can be exploited more generally. The two main advantages of divide-and-conquer applied to domains is that the subdomain problems are individually smaller than the original, and may also have simpler structure. The disadvantage to be overcome is that they are generally slowly convergent. A simple argument for domain decomposition methods is as follows: Given a problem of size N and a solver with arithmetic complexity $c.N^\alpha$, partition the problem into P subproblems of size N/P. The complexity of applying the solver once to the entire set of subproblems is $P.c.(N/P)^\alpha$. Even in serial, therefore, $I = P^{\alpha-1}$ iterations can be afforded to coordinate the solutions of the subproblems, while breaking even. If $\alpha = 1$, there is no "headroom" for extra iterations in serial on the basis of arithmetic complexity alone. However, there may still be headroom even if $\alpha = 1$ and: (1) the $\alpha = 1$ method involves too much communication to parallelize efficiently, (2) a hierarchical data structure is "natural" for modeling or implementation reasons, or (3) memory limitations, cache thrashing, or I/O costs demand decomposition anyway.

Schwarz methods are a special case of "iterative subspace correction methods" [35]. A general iterative correction algorithm for $Au = f$ proceeds as follows:

1. Compute the residual: $r^l = f - Au^l$

2. Solve $Ae = r^l$ approximately for an estimate of the error: $\tilde{e} = B^{-1}r^l$

3. Update $u^{l+1} = u^l + \tilde{e}$

An iterative subspace correction method accomplishes the middle step above by:

2(a) Decomposing the space of the solution $u : \mathcal{U} = \sum_k \mathcal{U}_k$

2(b) Finding the restriction of A to each $\mathcal{U}_k$: $A_k = R_k A R_k^T$, for some restriction operators R_k and extension operators R_k^T

2(c) Forming B^{-1} from the A_k^{-1}

Special cases include Jacobi and Gauss-Seidel, in which the "subspaces" are individual unit vectors or disjoint blocks thereof, multigrid, in which the subspace are hierarchically nested, and Schwarz-style domain decomposition, which combines subdomain block Jacobi or Gauss-Seidel methods (with possibly overlapping blocks), possibly with solutions on one or more coarse grids.

In Schwarz-style domain decomposition, the Jacobi-like method, in which all subdomains may be computed simultaneously is known as "additive," whereas the Gauss-Seidel-like method is known as "multiplicative." The origin of these terms can be seen in the error propagation operators for the two methods. Let $B_k^{-1} \equiv R_k^T A_k^{-1} R_k$. For the case of just two subspaces, the error in the multiplicative method satisfies $e^{l+1} = (I - B_2^{-1}A)(I - B_1^{-1}A)e^l$, whereas the error in the additive method satisfies $e^{l+1} = (I - B_2^{-1}A - B_1^{-1}A)e^l$. Comparing these expressions with the standard global iteration matrix $(I - B^{-1}A)$, we see that the multiplicative method has $B^{-1} = B_2^{-1} + B_1^{-1} - B_2^{-1}AB_1^{-1}$, whereas the additive method drops the triple product term. The spectral radius of $(I - B^{-1}A)$ is not guaranteed to be less than unity in general, so without acceleration, Schwarz methods may not converge. Schwarz preconditioners are more useful for clustering eigenvalues, in the context of being accelerated by Krylov methods, as shown in Table 1.

1.4 Implicit versus Explicit?

In concluding this introduction, we comment on the obviously parallelizable alternative of fully explicit methods, which depart from (4) by evaluating $f(u)$ only at the current time:

$$D \frac{\delta u^l}{\delta t} = -f(u^l).$$

The problem of evaluating the right-hand side efficiently in parallel is nothing more than the problem of load-balanced partitioning - trivial on structured grids, and by now well understood on unstructured grids and reduced to practice [16]. Having solved the partitioning problem, a time step in an explicit method is a set of local stencil operations comparable in communication cost to a single matrix-vector multiplication in an implicit method. In an implicit method, there may be many matrix-vector multiplications per time step, which prompts the question of how many inner iterations as implicit method may afford before the total number of accesses of the Jacobian matrix is comparable to the number required to march out the problem subject to explicit stability limitations. Just as an implicit method may be preconditioned by the choice of D and by other preconditioners that act on the sum $\frac{D}{\delta t} + \frac{\partial f}{\partial u}$ to cluster the spectrum, so an explicit method may be preconditioned (in the sense of [32]) by a matrix D, which is typically block diagonal, to locally equilibrate the wave speeds. Our preferred distinction between implicit and explicit methods does not lie in the way they access $f(u)$ or in their amenability to preconditioning, but in the fact that implicit methods make use of some global information, in addition to the stencil operations. This global information may come in the form of an inner product (for computing a step length in a Krylov method), or in the form of a global coarse grid solve (as a component of a multilevel preconditioner), or in the form of a pivot broadcast (as in a matrix factorization). Global information may be more or less important, depending upon $D/\delta t$ (see Figure 1).

Since there is sometimes no nonlinearly robust alternative to a small timestep at the outset of the steady solution of highly nonlinear problem, linear methods making use of varying amounts of global information may be employed along the route as Δt advances from 0 through ∞. This issue was explored theoretically for the heat equation and experimentally for a two-dimensional laminar reacting flow in [14], where it was concluded that a polyalgorithmic linear solver - simple iterations like point-Gauss-Seidel at the outset and accelerated methods like preconditioned GMRES in the endgame - will often lead to the best overall complexity in a difficult steady nonlinear problem.

An interesting way in which to visualize the parallel challenge of implicit methods is through the Green's function of the linearized operator (at any given nonlinear iteration stage), implicitly semi-discretized in time. For the constant-coefficient parabolic problem in one dimension,

$$\frac{\partial u}{\partial t} - \frac{\partial^2 u}{\partial x^2} = 0, \quad \text{with} \quad \frac{\partial u}{\partial t} \approx \frac{u(x,t) - u(x, t-\Delta t)}{\Delta t},$$

there results the backwards Helmholtz operator $-u'' + k^2 u = f$, with $k^2 \equiv 1/\Delta t$. The Green's function for the backwards Helmholtz operator on a finite segment, evaluated for a source at the midpoint for a variety of k ranging from small to large time steps, is shown in Figure 1. Recalling that the magnitude of the function shown is a measure of the relative influence of the data at the midpoint on the solution everywhere, or, reciprocally, of the relative influence on the solution at the midpoint of the data everywhere else, we see that the small step limit is trivially parallelizable: as $k \longrightarrow \infty$ only data in an infinitesimal neighborhood of the point of evaluation is necessary for accurate computation of the solution, just as with an explicit method.

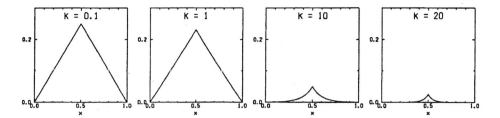

Figure 1. A series of Green's functions for a modified Helmholtz operator arising from implicit time discretisation of parabolic problem for several $k^2 \equiv 1/\Delta t$.

On the other hand, as $k \longrightarrow 0$ in the large time step limit, data from all other portions of the domain must be communicated to the evaluation point to enter in with some non-neglible weight. The only possible way to avoid all-to-all communication among all of the points of the discretization in this limit is to fix a desired precision in the current linearized update and to take advantage of the smoothness and the rate of delay of the Green's function to coarse-grain the communication at a distance. This is the role of the coarse grid in Sections 2.2 and 3.

2 Combination Methods

2.1 Newton-Krylov Methods

Newton-Krylov methods are born of using a Krylov method to solve (3). Two synergisms occur immediately. One is the freedom to solve for the Newton correction only approximately while controlling the degradation of the asymptotic rate of convergence. So-called "inexact Newton methods" were put on a practical theoretical footing in [11]. Another is freedom from explicit computation of the elements of the Jacobian matrix, through directional differencing. Early references to so-called "matrix-free" methods from the ODE literature may be found in [3, 8]. important introductions into the computational aerodynamics community occurred in [22, 34] and a general PDE treatment was given in [4].

Instead of a full Newton correction, inexact Newton methods work only to satisfy

$$\| J(u^l)\delta u^l + f(u^l) \|_2 \leq \eta \| f(u^l) \|_2,$$

for some sequence $\eta \geq 0$. It may be shown [11] that, for iterates in the domain of convergence of Newton's method,

1. $\eta < 1$ is sufficient to guarantee convergence,

2. $\eta \longrightarrow 0$ guarantees superliner convergence, and

3. $\eta = \mathcal{O}(\| f(u^l) \|_2)$ gives quadratic convergence.

Iterative linear solvers should work only as hard as outer loop progress warrants for more reasons than economy. Inexact Newton correction is a form of nonlinear damping at stages far from Newton convergence.

Matrix-free approximation to the matrix-vector multiplication required by a Krylov method is given by

$$J(u^l)v \approx \frac{1}{h}\left[f(u^l + hv) - f(u^l)\right] \tag{5}$$

to first-order in the differencing parameter h, or to second order by

$$J(u^l)v \approx \frac{1}{2h}\left[f(u^l + hv) - f(u^l - hv)\right], \tag{6}$$

etc. The selection of a suitable h balances two opposing disadvantages and is thus far trivial. As with any numerical differentiation, step/length must remain comfortably larger than the machine precision to avoid catastrophic loss of significant digits in the result. On the other hand, h should be small for low truncation error in (5,6). Part of the difficulty of choosing a single h with which to compute the entire directional derivative, rather than adapting h to probe a single column of the Jacobian, is that h must balance both u and v. Between

$$h = \sqrt{\epsilon_{mach}} \cdot (\| u^n \|_2 + 1) \quad \text{and} \quad h = \sqrt{\epsilon_{mach}} \cdot \frac{|(u^n, v)|}{\| v \|_2^2}$$

we found the second technique preferable. For GMRES, we are guaranteed that $\| v \|_2 = 1$ for all Krylov vectors because of the Gram-Schmidt orthonormalization, which makes their distinction moot. For this reason, GMRES stands alone among other Krylov methods in robustness in the matrix-free context in [26]. Diagonal scaling are addressed in [4], and are especially needed when problem is not well "non-dimensionalized" initially.

Preconditioning may easily be incorporated into matrix-free Newton-Krylov methods assuming that the action of a preconditioner $\tilde{J}_u^{-1}$, an approximation to the inverse of the Jacobian of $f(u)$ at u, is available from any source in subroutine call form. We consider first the left-preconditioned system

$$\mathcal{F}(u) \equiv \tilde{J}_u^{-1} f(u) = 0,$$

and iterate (2) with δu^l an approximate solution of the linear system

$$\mathcal{J}(u^l)\delta u^l = -\mathcal{F}(u^l), \quad \text{where} \quad \mathcal{J} \approx \frac{\partial \mathcal{F}}{\partial u}.$$

The action of $\mathcal{J}$ on a vector v can be approximated by

$$\mathcal{J}(u^l)v \approx \frac{1}{h}\left[\mathcal{F}(u^l + hv) - \mathcal{F}(u^l)\right]$$

Right-preconditioning is also possible in solving $f(u) = 0$, via matrix-free Newton iteration. To approximate the action of $J_u \tilde{J}_u^{-1}$ on a vector w, form

$$\frac{1}{h}\left[f((u^l + h\tilde{J}_u^{-1} w)) - f(u^l)\right].$$

After the solution δ_u^{-1} of the right-preconditioned system is found by the Krylov process, the preconditioning is unwound with $\tilde{J}_u \delta u^l = \delta_u^{-1}$. Note that right-preconditioning will, in general, destroy the normalization of the probing vector, v.

One of the major synergisms of the Newton-Krylov formulation, its matrix-free formulation, is lost if the Jacobian has to be formed explicitly for preconditioning purposes. As mentioned in the introduction, this may be difficult, or computationally complex, or storage-intensive; or any combination of the three for certain high-order monotype upwind discretizations in aerodynamics. We therefore consider basing preconditioner $\tilde{J}$ on a more convenient discretization than is required for the nonlinear residual f itself. Obvious possibilities include the Jacobian of a lower-order discretization of the same continuous problem, a finite-differenced Jacobian lagged for expensive-to-calculate terms of lesser magnitude, and the Jacobian of a related discretization that permits analytical evaluation of elements.

Such preconditioners often exist already : "implicit" CFD codes that operate on the principle of defect correction, including the commonly employed Steger-Warming-type splitting [31]. Such codes employ a left-hand side matrix (not a true Jacobian) in whose construction computational short-cuts are employed, and which may be stabilized by first-order upwinding that would not be acceptable in the discretization of the residual itself. We denote this generic distinction in (7) below by subscripting the residual "high" and the left-hand side matrix "low":

$$J_{low}\, \delta u = -f_{high} \tag{7}$$

Often, J_{low} is based on a low-accuracy residual:

$$J_{low} = \frac{D}{\delta t} + \frac{\partial f_{low}}{\partial u}.$$

The inconsistency of the left- and right-hand sides prevents the use of large timesteps. using the built-in capability to solve systems with J_{low} as a preconditioner, we advocate instead [5]

$$(J_{low})^{-1} J_{high}\, \delta u = -(J_{low})^{-1} f_{high}, \tag{8}$$

in which the action of J_{high} on a vector is obtained through directional differencing, as in (5). Note that the operators on both sides of (8) are based on consistent high-order discretizations; hence timesteps can be advanced to arbitrarily large values, recovering a true Newton method in the limit.

Because any inadequacy of preconditioner J_{low} appears only in the linear solution process, and not in the converged solution, J_{low} may be lagged, saving the expense of frequent Jacobian updates. The frequency of re-evaluation of the preconditioner is one of two parameters in a Newton-Krylov method, besides the parameters in the Newton and Krylov methods themselves, the other new one being the differencing parameter h.

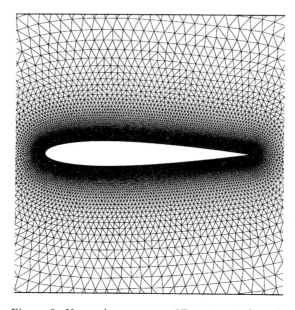

Figure 2: Near-wing zoom on 2D unstructed mesh used in the Navier-Stokes tests of the Newton-Krylov method.

We illustrate the use of (8) in some recent tests of a two-dimensional Navier-Stokes NACA0012 airfoil problem on an unstructured grid, obtained with Kyle Anderson. The problem is laminar ($Re = 5.000$) and subsonic ($Ma = 0.5$), with an angle of attack of $3°$. An unstructured triangular grid of approximately 35K vertices is used, as depicted in Figure 2. On the given scale, print media do not permit resolution of the triangles in the boundary layer, whose aspect ratios can be as high as 10^2 approximately. The discretization is vertex-based with a first-order Roe scheme as the "low order" discretization, and a second-order Roe scheme as the "high order" discretization with which the residual is evaluated. A standard Galerkin treatment is used for the viscous terms. The purpose of the tests is to compare two baseline methods based on (7) with a Newton-Krylov method based on (8). All of the methods employ a residual-adaptive CFL number, in which progress in reducing the residual norm in iteration l leads to an increase in the local CFL number in iteration $l+1$, up to a modest CFL bound in the case of the inconsistently discretized methods and without bound in (8). The first baseline method is a single-grid code based on defect correction and hybrid point GS/Jacobi as a relaxation process [1]. The second is a multigrid code that uses four non-nested coarse grids in FAS V-cycle with Jacobi smoothing on descent. The matrix-free Newton-Krylov method uses ILU(0) as the preconditioner and GMRES as the accelerator, with a small Krylov space of 25 vectors and up to 2 cycles.

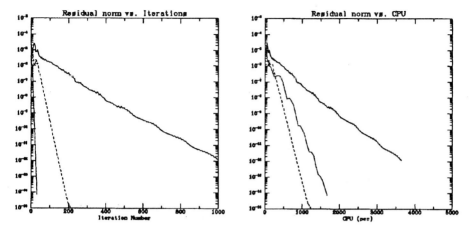

Figure 3. Norm of steady-state residual vs. iterations(a,left) & vs. CPU time(b,right) (Cray YMP) for a single grid Gauss-Seidel scheme(solid), a 4-V-cycle multigrid scheme using Gauss-Seidel scheme as a smoother(dashed), & a Newton-Krylov scheme using ILU(0) as a preconditioner(dotted).

Figure 3(a), which plots residual norm versus "outer iterations," shows that the multigrid and NK methods can achieve close to machine-zero residual reductions in very many fewer iterations than the single code. NK actually appears significantly superior until it is recalled that each "iteration" for NK involves a set of Krylov subiterations. Plots of residual norm versus sequential execution time on a Cray are shown in Figure 3(b). Here it is seen that NK trails multigrid by a factor of about 1.5 for this problem. This penalty in sequential contexts can be tolerated when it is realized that the NK method has the advantage of doing all of its computation without user generation of coarse unstructured grids and that it lends itself to easily parallelizable preconditioners. See [5,27] for further practical discussion of NK methods.

2.2 Krylov-Schwarz Methods

Krylov-Schwarz Methods are born of using Schwarz splitting as a preconditioner for Krylov method. We define

$$B^{-1} \equiv \sum_k R_k^T (\tilde{A}_k)^{-1} R_k, \qquad (9)$$

where $(\tilde{A}_k)^{-1}$ is a convenient approximation to $A_k = R_k A R_k^T$, and apply a Krylov method to

$$B^{-1} A u = B^{-1} f.$$

Because Schwarz is the least familiar component of NKS, we provide trivial illustrations before giving some results from CFD. Let A correspond to the Laplacian on a 1D interval containing four interior points (omitted elements are zero):

$$A = \begin{pmatrix} 2 & -1 & & \\ -1 & 2 & -1 & \\ & -1 & 2 & -1 \\ & & -1 & 2 \end{pmatrix}$$

Let R_1 gather the unknowns in the left subdomain, and R_2 gather the unknowns in the right:

$$R_1 = \begin{pmatrix} 1 & & & \\ & 1 & & \end{pmatrix}; \quad R_2 = \begin{pmatrix} & & 1 & \\ & & & 1 \end{pmatrix}.$$

Then, $A_1 \equiv R_1 A R_1^T$ and $A_2 \equiv R_2 A R_2^T$ are the same:

$$\begin{pmatrix} 1 & & & \\ & 1 & & \end{pmatrix} \begin{pmatrix} 2 & -1 & & \\ -1 & 2 & -1 & \\ & -1 & 2 & -1 \\ & & -1 & 2 \end{pmatrix} \begin{pmatrix} 1 & \\ & 1 \\ & \\ & \end{pmatrix} =$$

$$\begin{pmatrix} & & 1 & \\ & & & 1 \end{pmatrix} \begin{pmatrix} 2 & -1 & & \\ -1 & 2 & -1 & \\ & -1 & 2 & -1 \\ & & -1 & 2 \end{pmatrix} \begin{pmatrix} & \\ & \\ 1 & \\ & 1 \end{pmatrix} = \begin{pmatrix} 2 & -1 \\ -1 & 2 \end{pmatrix}$$

The inverses if A_1 and A_2 are:

$$A_1^{-1} = A_2^{-1} = \begin{pmatrix} \frac{2}{3} & \frac{1}{3} \\ \frac{1}{3} & \frac{2}{3} \end{pmatrix}.$$

The overall Schwarz preconditioner, with zero overlap, is:

$$B^{-1} = R_1^T A_1^{-1} R_1 + R_2^T A_2^{-1} R_2 = \begin{pmatrix} \frac{2}{3} & \frac{1}{3} & & \\ \frac{1}{3} & \frac{2}{3} & & \\ & & \frac{2}{3} & \frac{1}{3} \\ & & \frac{1}{3} & \frac{2}{3} \end{pmatrix}.$$

Whereas $k(A)$, the condition number of A, is approximately 9.47, $k(B^{-1}A)$ is approximately 5.45 Only local subproblems need to be solved to gain this improvement.

It is informative to write out $B^{-1}A$ for this example:

$$B^{-1}A = \begin{pmatrix} 1 & & -\frac{1}{3} & \\ & 1 & -\frac{2}{3} & \\ -\frac{2}{3} & & 1 & \\ -\frac{1}{3} & & & 1 \end{pmatrix}.$$

Note that $B^{-1}A$ is just a rank-2 perturbation of the identity. Conjugate gradients on this system will converge in just two steps, even if each subinterval is extended indefinitely. This result may also be generalized to higher dimensional regions, leading to bounds on the number of steps required for exact solution (in exact arithmetic), based on the number of non-identity columns in $B^{-1}A$, which is the number of stencils "cut" by the decomposition into subdomains.

Now consider operators R_1 and R_2 that gather overlapping left and right subintervals:

$$\begin{pmatrix} 1 & & & \\ & 1 & & \\ & & 1 & \end{pmatrix} ; \quad R_2 = \begin{pmatrix} & 1 & & \\ & & 1 & \\ & & & 1 \end{pmatrix}$$

Similar to the preceding, we have for A_1 and A_2:

$$\begin{pmatrix} 1 & & \\ & 1 & \\ & & 1 \end{pmatrix} \begin{pmatrix} 2 & -1 & & \\ -1 & 2 & -1 & \\ & -1 & 2 & -1 \\ & & -1 & 2 \end{pmatrix} \begin{pmatrix} 1 & & \\ & 1 & \\ & & 1 \end{pmatrix} =$$

$$\begin{pmatrix} 1 & & \\ & 1 & \\ & & 1 \end{pmatrix} \begin{pmatrix} 2 & -1 & & \\ -1 & 2 & -1 & \\ & -1 & 2 & -1 \\ & & -1 & 2 \end{pmatrix} \begin{pmatrix} 1 & & \\ & 1 & \\ & & 1 \end{pmatrix} = \begin{pmatrix} 2 & -1 & \\ -1 & 2 & -1 \\ & -1 & 2 \end{pmatrix},$$

and

$$A_1^{-1} = A_2^{-1} = \begin{pmatrix} \frac{3}{4} & \frac{1}{2} & \frac{1}{4} \\ \frac{1}{2} & 1 & \frac{1}{2} \\ \frac{1}{4} & \frac{1}{2} & \frac{3}{4} \end{pmatrix}.$$

whereby

$$B^{-1} = R_1^T A_1^{-1} R_1 + R_2^T A_2^{-1} R_2 = \begin{pmatrix} \frac{3}{4} & \frac{1}{2} & \frac{1}{4} & \\ \frac{1}{2} & \frac{7}{4} & 1 & \frac{1}{4} \\ \frac{1}{4} & 1 & \frac{7}{4} & \frac{1}{2} \\ & \frac{1}{4} & \frac{1}{2} & \frac{3}{4} \end{pmatrix}.$$

The condition number of this overlapping $k(B^{-1}A)$ improves further to 3.82 at the extra cost of larger local subproblems. For self-adjoint problems, the greater the overlap, the better the condition number and the fewer the number of iterations, but the greater the cost per step because of redundant work in the overlap region.

If a *sufficiently fine* global coarse space is included in the sum of terms (9) then even for nonsymmetric and/or indefinite systems, $k(B^{-1}A)$ is independent of both the mesh spacing and the number of subdomains. The coarse space need not be related to the decomposition into subdomains, as is demonstrated in Section 3.

More generally, let a selfadjoint elliptic problem with smoothly varying coefficients be discretized with spacing $\mathcal{O}(h)$ and divided into subdomains of diameter $\mathcal{O}(H)$. Then A has condition number $\mathcal{O}(h^{-2})$ [2], a non-overlapped Schwarz preconditioned problem has condition number $k(B^{-1}A) = \mathcal{O}(h^{-1}H^{-1})$ [9], and a generously overlapped Schwarz preconditioned problem has condition number $\mathcal{O}(H^{-2})$ [13]. Note that condensing out the subdomains converts one power of h to H, replacing the fine mesh scale of the problem with the subdomain scale. Overlapping the subdomains converts another power of h to H, dramatically improving the condition number. Since $1/H$ is roughly the number of subdomains in each coordinate direction when a unit-aspect ratio domain is partitioned into unit-aspect ratio subdomains, we see that in two dimensions the condition number scales like the number of subdomains. Addition of a coarse grid problem to generously overlapped problem, with approximately one degree of freedom per subdomain, brings the condition number to $\mathcal{O}(1)$ [13]. All of the cited condition number bounds hold in both two and three dimensions.

A major motivation for the investigation of additive Schwarz methods is their perfect parallelism: the problem in all subspaces, including that of the global coarse can be solved concurrently. The only impediments to parallelism are then the setting up of the local right-hand sides, the assembly of the subdomain solutions into a global vector, and the inner product and stencil operations associated with advancing the Krylov method. Given an order estimate for the number of iterations (from the square-roots of the condition numbers above, according to classical Krylov theory for the symmetric problem, see e.g. [21]), one can construct a cost function for the total amount of arithmetic and the total amount of data exchange between subdomain-oriented processes required to bring a Krylov-Schwarz method to convergence for the decomposition of given granularity. This complexity information can be combined with a model of parallel computer architecture to estimate the total running time (computation plus communication) of a parallel implementation of a Krylov-Schwarz method, as a function of the size of the problem, n, and the number of subdomains/processors, p. By taking the partial derivative of this cost function with respect to the number of processors and setting it to zero, one can derive an optimal order processor granularity, P_{opt}, and hence a optimal order running time T_{opt}.

Such a parallel computational complexity exercise was undertaken in [9] for (among other architectures) a 3D mesh of processors and a hypercube. The results for a self-adjoint elliptic problem in three-dimensions, solved by nonoverlapping Schwarzs preconditioned conjugate gradients are collected below (see [9] for derivation). Shown in the table is the time per iteration for a three-dimensional scalar elliptic problem posed on a rectilinear paralleliped domain with n smoothly distributed points on a side, decomposed by $p^{1/3} \times p^{1/3} \times p^{1/3}$ cutting planes into p cubes subdomains. The number of iterations is determined by the condition number of the block PCG preconditioned system, which is independent of architecture; in this case, it is $\mathcal{O}(n^{1/2}p^{1/6}\log n)$. An exact bandsolve is used for each subdomain block of the preconditioner on the hypercube; nested dissection is used on the 3D mesh, since the highest-order term is involved. We note that the hypercube architecture is scalable in the sense that the optimal number of processors is proportional to the memory (n^3) required to store the problem. However, the overall running time is not constant, but grows as n for this particular non-overlapped coarse-grid-free preconditioner.

arch	leading terms in $T(n,p)$ per iteration	P_{opt}	$T_{opt}(n)$
mesh	$\left(\frac{n^3}{p}\log n + \frac{n^3}{p} + p^{1/3} + \frac{n^2}{p^{2/3}}\right)$	$n^{9/4}$	$n^{13/8}\log^2 n$
hyper	$\left(\frac{n^5}{p^{5/3}} + \frac{n^3}{p} + \log p + \frac{n^2}{p^{2/3}}\right)$	n^3	$n\log^2 n$

Table 1: Parallel complexity of block PCG for a 3D elliptic problem on two different parallel architectures: a three-dimensional mesh of processors and a binary hypercube. n: number of points in each dimension; p: total number of processors.

Krylov-Schwarz methods have been reported on extensively, together with other types of domain decomposition methods, such as Krylov-Schur (or "iterative substructuring") since 1987 in the proceedings of the nearly annual International Conference on Domain Decomposition Methods for Partial Differential Equations [6, 7, 19, 20, 23, 25, 29], and also in [24].

Here we summarize some results of [5] for a two-dimensional body-fitted structured grid code from the United Technologies Research Center, obtained jointly with Driss Tidriri. The Euler code was run on a NACA0012 case study until the nonlinear residual norm had been reduced a couple of magnitude. Then a single typical Jacobian/right-hand side system at a local CFL of 100 was written to disk for some linear Krylov-Schwarz experiments. The right-hand side was evaluated using a Roe scheme and the Jacobian using a Van Leer scheme.

Approximately 300 tests were conducted on this same linear system of approximately 16K unknowns, with different Krylov accelerators and different Schwarz preconditioners. Two accelerators and different Schwarz preconditioners. Two accelerators, GMRES and BiCGSTAB, are employed. The partitioning of the Schwarz preconditioner includes strip-in-x (the streamwise direction), strip-in-y (the normal direction) and boxes. The number of partitions created by the various cutting schemes, representing the potential degree of parallelism in the preconditioner, varies from 1 to 64. Overlaps of 0, 1, and 2 subintervals are tested. The subdomain solutions ($\tilde{A}_k^{-1}$) are either exact, using LU decomposition, or inexact, using incomplete LU with zero fill. Finally, the full range of concurrency from multiplicative, through multi-coloring, through full additive is tried. We did not investigate the use of a coarse grid for the Euler equations.

Observations for this typical CFD Jacobian system include the following salient points:

- No Schwarz method beats global ILU(0) as a serial preconditioner, but no zero-overlap Schwarz method ever lags in executing time by more than a factor of two. (This is similar to Boeing's experience with preconditioning the Jacobians of the TRANAIR code [36].)

- Box decompositions are superior to strip decompositions in either direction, for the same number of subdomains, in both iterations and time.

- The effect of parallel granularity on convergence depends on overlap, exactness, and multiplicativity. The fastest (but still sublinear) iteration count growth as the number of subdomains is increased occurs for low overlap/exact/additive preconditioners. The slowest (best) growth rate occurs for high overlap / inexact / multiplicative preconditioners.

- Increasing overlap nearly always enhances convergence rate, but the effect saturated, and the cost per iteration escalates quickly, making zero or one cell of overlap the practical optimum.

- Using inexact (ILU(0)) versus exact subdomain solutions always worsens convergence rate, but always improves execution time, for the same granularity.
- GMRES always loses to BiCGSTAB in terms of number of iterations, but never by as much as a factor of two (which is the ratio of matrix-vector multiplications per iteration by which BiCGSTAB exceeds the cost of GMRES); and GMRES usually wins in terms of serial execution time, but rarely by more than 10%.

3 Newton-Krylov-Schwarz Method

Newton-Krylov-Schwarz is a straightforward combination of Newton-Krylov and Krylov-Schwarz. The $\tilde{J}$ of (8) is formed from a parallelizable Schwarxian approximation:

$$\mathcal{F}(u^l) = \left(R_0 J_{0,u^l}^{-1} R_0^T + \sum_{k=1}^{K} R_k J_{k,u^l}^{-1} R_k^T \right) f(u^l), \quad \text{where} \quad J_{k,u} = \left\{ \frac{\partial f_i(u)}{\partial u_j} \right\}$$

is the Jacobian of $f(u)$ of i and j in subdomain k ($k=0$ corresponds to the coarse grid space). As long as the subdomain solutions with $J_{k,u}$ are performed exactly, there are no new parameters introduced by the composition. However, a variety of new preconditioners are suggested by combining one of the approximate global preconditioners described under Newton-Krylov with the domain decomposition strategy of Krylov-Schwarz.

Coarse Grid	0 × 0	4 × 5	8 × 9	12 × 13	16 × 17	20 × 21
Analytical	117	35	28	27	24	21
Matrix-free	183	41	28	27	25	23

Table 2. Average number of GMRES(20) steps per Newton step for full potential Newton-Krylov-Schwarz solver with varying coarse grid size.

Here we summarize some results of [5] for a two-dimensional model Cartesian full potential grid code obtained jointly with Xiao-Chuan Cai. The domain is an upper half plane for which the bottom boundary is a line of symmetry containing a transpiration profile for one-half of a NACA0012 airfoil. Bilinear rectangular Galerkin finite elements are used for the discretization. The fine grid contains approximately 16K points. The NKS solver uses GMRES (with a restart size of 20) and two-cell overlap additive Schwarz with exact solvers on each subdomain. The number of subdomains of the fine grid is held fixed at 64 in an 8 x 8 array. The number of points in the coarse grid is varied from zero to approximately 400. The main objective of the tests is to ascertain the magnitude of the penalty for doing matrix-free Jacobian updates and to quantify the benefits of a coarse grid for a very elliptic case (with a Mach number of 0.1). From a uniform flow initial iterate, Newton's method converges in just three iterations in each case in the table, which shows the average number of inner iterations over the three Newton steps.

We observe that the penalty for using the matrix-free formulation for Jacobian-vector product evaluation in GMRES is minor (rarely more than 10%). More significantly, we observe that the introduction of a coarse grid can dramatically reduce the number of iterations required for convergence of the Krylov method, but that the effect saturates with approximately one coarse grid point per subdomain. Beyond one point per subdomain case here (the 8 × 9), iteration count reductions are small and coarse grid work per iteration increases.

4 Conclusions and Outstandings Issues

We have shown that a variety of CFD applications are (or have inner) nonlinear elliptically-dominated problems amenable to solution by Newton-Krylov-Schwarz algorithms characterized by: locally concentrated data dependencies and zero or small overlaps between the preconditioner blocks. The addition of a coarse grid in the Schwarz preconditioner may be effective. Parametric tuning is important to performance, but robust choices are not difficult. Nonsymmetry, nonlinearity, and multicomponent structure hinder theoretical development, but experimental development continues at industrial, NASA, and DOE laboratories.

Our Krylov-Schwarz solvers have been ported to a tightly-coupled distributed memory machine (the Intel Paragon) and an Ethernetwork of Sun SPARC stations, with worthwhile efficiencies in granularities of up to 16 subdomains [10] and more work of this nature is underway.

Acknowledgment

For both perspective and proof of concept, the author is indebted to several people: Dr. W. Kyle Anderson of NASA for trying equation (8) in his already option-laden code and for freshly implementing level-scheduling in the global ILU preconditioner, Mr. Eric Nielson of Varginia Tech for a summer of fast learning leading to research frontier contributions, Dr. Dana Knoll of the Idaho National Energy Laboratory for trying Newton-Krylov on function problems, and Professor Xiao-Chuan Cai of UC-Boulder, Dr. Bill Gropp of Argonne National Laboratory, and Drs. Driss Tidriri and V. Venkatakrishnan of ICASE for their constant stimulation.

This research was supported in part by the National Science Foundation under grant ECS-8957475, and in part by the National Aeronautics and Space Administration while the author was in residence at the Institute for Computer Applications in Science and Engineering (ICASE), NASA Langley Research Center, Hampton, VA 23681-0001.

The invitation of Professors R. Narasimha and S.M. Dehpande and Dr. S.S. Desai to present this work at the graciously organized and meticulously executed Fourteenth International Conference on Numerical Methods in Fluid Dynamics is greatly appreciated.

References

1. W.K. Anderson and D.L. Bonhaus (1994): *An Implicit upwind Algorithm for Computing Turbulent Flows on Unstructured Grids*, Computers & Fluids 23 1-21.

2. O. Axelsson and V.A. Barker (1983): *Finite Element Solution of Boundary Value Problems: Theory and Computation*, Academic Press, New York.

3. P.N. Brown and A. Hindmarsh (1987): *Matrix-free Methods for Stiff Systems of ODEs*, SIAM J. Numer. Anal. 24 610-638.

4. P.N. Brown and Y. Saad (1990): *Hybrid Krylov Methods for Nonlinear Systems of Equations*, SIAM J. Sci. Stat. Comp. 11 450-481.

5. X.- C. Cai, W.D. Gropp, D.E. Keyes and M.D. Tidriri (1994): *Newton-Krylov-Schwarz in CFD*, in "Proceedings of the International Workshop on Numerical Methods for the Navier-Stokes Equations" (F. Hebeker & R. Rannacher, eds.), Notes in Numerical Fluid Mechanics, Vieweg Verlag, Braunschweig, pp. 17-30.

6. T.F. Chan, R. Glowinski, J. Periaux, and O.B. Widlund, eds. (1989): *Proc. of the Second Intl. Symp. on Domain Decomposition Methods for Partial Differential Equations*, SIAM, Philadelphia.

7. T.F. Chan, R. Glowinski, J. Periaux, and O.B. Widlund, eds. (1990): *Proc. of the Third Intl. Symp. on Domain Decomposition Methods for Partial Differential Equations*, SIAM, Philadelphia.

8. T.F. Chan and K.R. Jackson (1986): *The Use of Iterative Linear Equation Solvers in Codes for Large Stiff System of IVPs for ODEs*, SIAM J. Sci. Stat. Comp. 7 378-417.

9. M. Y.-M. Chang and M.H. Schultz (1994): Bounds on Block Diagonal Preconditioning, Parallel Algs. and Applics. 1 141-164.

10. J.G. Chefter, C.K. Chu and D.E. Keyes (1995): *Domain Decomposition for the Shallow Water Equations*, in "Proceedings of the Seventh International Conference on Domain Decomposition Methods" (D.E. Keyes & J. Xu, eds.), AMS, Providence (to appear).

11. R. Dembo, S. Eisenstat and T. Steihaug (1982): *Inexact Newton Methods*, SIAM J. Numer. Anal. 19 400-408.

12. J.E. Dennia, Jr. and R.B. Schnabel (1983): *Numerical Methods for Unconstrained Optimization & Nonlinear Equations*, Prentice-Hall, Englewood Cliffs, NJ.

13. M. Dryja and O.B. Widlund (1987): *An Additive Variant of the Alternatively Method for the Case of Many Subregions*, TR 339, Courant Institute, NYU.

14 A. Ern, V. Giovangigli, D.E. Keyes and M.D. Smooke (1994): *Towards Polyalgorithmic Linear System Solvers for Nonlinear Elliptic Problems*, SIAM J. Sci. Comp. 15 681-703.

15 C. Farhat, L. Crivelli and F.-X. Roux, *extending Substructure-based Iterative Solvers to Multiple Load and Repeated Analyses*, Comput. Meths. Appl. Mech. Engrg. *37 195-209.

16 C. Farhat, S. Lanteri and H.D. Simon (1994): *TOP/DOMDEC: A Software Tool for Mesh Partitioning and Parallel Processing*, J. Comput. Sys. Engrg. (to appear).

17 P.F. Fischer (1993): *Projection Techniques for Iterative Solution of $Ax = b$ with Successive Right-hand sides*, Technical Report 93-90, ICASE, NASA Langley Res. Ctr.

18 R.W. Freund and N.M. Nachtigal (1991) QMR: A Quali-minimal Residual Methods for Non-Hermitian Linear Systems, *Numer. Math.* 60 315-339.

19 R. Glowinski, G.H. Golub, G.A. Meurant, and J. Periaux, eds. (1988): *Proc. of the First Intl. Symp. on Domain Decomposition Methods for Partial Differential Equations*, SIAM, Philadelphia.

20 R. Glowinski, Yu. A. Kuznetsov, G.A. Meurant, and O.B. Widlund, eds. (1991): *Proc. of the Fourth Intl Symp. on Domain Decomposition Methods for Partial Differential Equations*, SIAM. Phiadelphia.

21 G.H. Golub and C.F. van Loan (1989): *Matrix Computations*, Johns Hopkins, Baltimore.

22 Z. Johann, T.J.R. Hughes and F. Shakib (1991): *A Globally Convergent Matrix-Free Algorithm for Implicit Time-Marching Schemes Arising in Finite Element Analysis in Fluids*, Comp. Meths. Appl. Mech. Engrg. 87 281-304.

23 D.E. Keyes, T.F. Chan, G.A. Meurant, J.S. Scroggs, and R.G. Voigt, eds. (1992): *Proc. of the Fifth Intl. Symp. on Domain Decomposition Methods for Partial Differential Equations*, SIAM, Philadelphia.

24 D.E. Keyes, Y. Saad and D.G. Truhlar eds. (1995): *Domain-based Parallelism and Problem Decomposition Methods in Science and Engineering*, SIAM, Philadelphia (to appear).

25 D.E. Keyes and J. Xu, eds. (1995): *Proc. of the Seventh Intl. Symp. on Domain Decomposition Methods in Science and Engineering*, AMS, Providence (to appear).

26 D.A. Knoll and P.R. McHugh (1993): *Inexact Newton's Method Solutions to the Incompressible Navier-Stokes and Energy Equations Using Standard and Matrix-Free Implementations*, AIAA Paper 93-3332.

27 E.J. Nielsen, R.W. Walters, W.K. Anderson and D.E. Keyes (1995): *Application of Newton-Krylov Methodology to a Three-Dimensional Unstructured Euler Code*, in "Proceedings of the 12th AIAA Computational Fluid Dynamics Conference" (to appear).

28 K.G. Prasad, D.E. Keyes and J.H. Kane (1994): *GMRES for Sequentially Multiple Right-hand Sides*, SIAM J. Sci. Comp. (submitted).

29 A. Quarteroni, J. Periaux, Yu. A. Kuznetsov, and O.B. Widlund, eds. (1994): *Proc. of the Sixth Intl. Symp. on Domain Decomposition Methods in Science and Engineering*, AMS, Providence.

30 Y. Saad and M.H. Schultz (1986): *GMRES: A Generalized Minimal Residual Algorithm for Solving Nonsymmetric Linear Systems*, SIAM J. Sci. Stat. Comp. 7 865-869.

31 J.L. Steger and R.F. Warming (1981): *Flux Vector Splitting of the Inviscid Gasdynamic Equations with Applications of Finite-Difference Methods*, J. Comp. Phys. 40 263-293.

32 E. Turkel, A. Fiterman and B. van Leer (1993): *Preconditioning and the Limit to the Incompressible Flow Equations*, Technical Report 93-42, ICASE, NASA Langley Res. Ctr.

33 H.A. Van der Vorst (1992): *Bi-CGSTAB: A More Smoothly Converging Variant of CG-S for the Solution of Nonsymmetric Linear Systems*, SIAM J. Sci. Stat. Comp. 13 631-644.

34 L.B. Wigton, N.J. Yu and D.P. Young (1985): *GMRES Acceleration of Computational Fluid Dynamics Codes*, AIAA Paper 85-1494.

35 J. Xu (1992): *Iterative Methods by Space Decomposition and Subspace Correction*, SIAM Review 34 581-613.

36 D.P. Young, C.C. Ashcraft, R.G. Melvin, M.B. Bieterman, W.P. Huffman, T.F. Johnson, C.L. Hilmes, and J.E. Bussoletti (1993): *Ordering and Incomplete Factorization Issues for Matrices Arising from the TRANAIR CFD Code*, TR BCSTECH-93-025, Boeing Computer Services, Seattle.

Vortex Methods for Three-Dimensional Separated Flows

[1]A. Leonard, [1]P. Koumoutsakos and [2]G. Winckelmans

[1]Graduate Aeronautical Laboratories
California Institute of Technology, USA
[2]Mechanical Engineering Department
University of Sherbrooke
Sherbrooke, Quebec, Canada

Abstract : Traditionally, vortex methods have been used to model unsteady, high Reynolds number incompressible flow by representing the fluctuating vorticity field with a few tens to a few thousand Langrangian elements of vorticity. Now, with the advent of fast vortex algorithms, bringing the operating count per timestep down to O(N) from O(N^2) for N computational elements, and recent developments for the accurate treatment of viscous effects, one can use vortex methods for high resolution simulations of the Navier-Stokes equations. Their classical advantages still hold – (1) computational elements are needed only where the vorticity is nonzero (2) the flow domain is grid free (3) rigorous treatment of the boundary conditions at infinity is a natural byproduct and, (4) physical insights gained by dealing directly with the vorticity field–so that vortex methods have become an interesting alternative to finite difference and spectral methods for unsteady separated flows.

1 Basic Equations

To describe vortex methods we start with the three-dimensional vorticity equation for the vorticity field ω in a constant density, incompressible flow,

$$\frac{\partial \omega}{\partial t} + (\mathbf{u} \cdot \nabla)\omega = (\omega \cdot \nabla)\mathbf{u} + \nu \nabla^2 \omega \ . \tag{1}$$

where ν is the kinematic viscosity. Combining the incompressibility condition for the velocity field $\mathbf{u}$, $\nabla \cdot \mathbf{u} = 0$, and the definition of vorticity, $\nabla \times \mathbf{u} = \omega$, we find that

$$\nabla^2 \mathbf{u} = - \nabla \times \omega \tag{2}$$

We will consider vortex methods in the context of bluff-body flows and so we are interested in solutions to (1) and (2) corresponding to the no-slip at the surface of a rigid body moving with velocity $\mathbf{U}_b$,

$$\mathbf{u}\bigg|_{\text{surface}} = \mathbf{U}_b \bigg|_{\text{surface}} \tag{3}$$

and free stream conditions at infinity

$$\mathbf{u} \to \mathbf{U}_\infty(t) \quad \text{as} \quad |\mathbf{x}| \to \infty \tag{4}$$

To simplify the description of the method we restrict the discussion to non-rotating bodies. For the treatment of rotating bodies see Koumoutsakos (1993). The solution to (2) satisfying (3) and (4) is given in terms of the infinite medium Green's function

$$\mathbf{u}(\mathbf{x}) = -\frac{1}{4\pi} \int \frac{(\mathbf{x} - \mathbf{x}') \times \omega(\mathbf{x}')d\mathbf{x}'}{|\mathbf{x} - \mathbf{x}'|^3} + \mathbf{U}_\infty(t) \tag{5}$$

In our numerical approach described below it is important to recognize that the no-slip boundary condition is maintained by a continual flux of vorticity from the body surface into the fluid. Mathematically this flux is such that ω is nonzero only in the fluid (i.e., external to the body) and that (5) gives the result that

$$\mathbf{u} \cdot \mathbf{t} \Big|_{\text{surface}} = \mathbf{U}_b \cdot \mathbf{t} \Big|_{\text{surface}} \tag{6}$$

where $\mathbf{t}$ is any tangent vector at the body surface. That (5) and (6) imply (3) may be shown as follows. Let ψ, the stream function for the imaginary fluid within the body, be given by

$$\psi = \frac{\mathbf{U}_b \times \mathbf{x}}{2} + \psi' \tag{7}$$

Thus within the body B

$$\mathbf{u} = \nabla \times \psi = \mathbf{U}_b + \nabla \times \psi' \tag{8}$$

and

$$\nabla^2 \psi = \nabla^2 \psi' = -\omega = 0 \tag{9}$$

because all vorticity is external to B. Now we use Green's identity

$$\int_B [(\nabla u) \cdot (\nabla v) + u \nabla^2 v]\, d\mathbf{x} + \int_{\partial B} u \frac{\partial v}{\partial n}\, d\sigma = 0 \tag{10}$$

with $u = v = \psi'_i$. Here n is in the outward normal direction. From (6) and (8)

$$(\nabla \times \psi') \cdot \mathbf{t} = 0 \tag{11}$$

at every point on the surface ∂B for any tangent vector $\mathbf{t}$. Thus, $\frac{\partial \psi'}{\partial n} = 0$ so that (10) reduces to

$$\int_B |\nabla \psi'_i|^2 d\mathbf{x} = 0 \quad (i = 1, 2, 3) \tag{12}$$

Hence $\psi' = \underline{\text{const}}$ and

$$\mathbf{u} = \mathbf{U}_b \text{ in B}. \tag{13}$$

2 Vortex Particles

In the present method the vorticity field is represented by a dense collection of N moving, vector-valued computational elements or particles.

$$\omega(\mathbf{x},t) = \Sigma_k \alpha_k(t) \zeta_\sigma(\mathbf{x} - \mathbf{x}_k(t)) \tag{14}$$

where the particle vector amplitudes, α_k, have units of circulation times length and each particle has a radially symmetric spatial distribution defined by

$$\zeta_\sigma(\mathbf{x}) = \frac{1}{\sigma^3} \zeta\left(\frac{|\mathbf{x}|}{\sigma}\right) \tag{15}$$

where σ is the effective core radius of the particle and ζ has unit volume integral. See Leonard (1985) and Winckelmans and Leonard (1993) for further discussion including choices for the function ζ. For example, the distribution for the gaussian particle is given by

$$\zeta(z) = \frac{\exp(-z^2)}{\pi^{3/2}} \tag{16}$$

To satisfy the inviscid components of motion of the fluid each particle moves with the local velocity

$$\frac{d\mathbf{x}_k}{dt} = \mathbf{u}(\mathbf{x}_k(t), t) \tag{17}$$

and its vector α_k is stretched and rotated according to

$$\frac{d\alpha_k}{dt} = (\alpha_k \cdot \nabla)\mathbf{u}\Big|_{\mathbf{x}_k} \tag{18}$$

where the velocity field $\mathbf{u}$ is obtained by substituting (14) into (15) to obtain

$$\mathbf{u}(\mathbf{x}) = -\frac{1}{4\pi} \frac{\Sigma_k [\mathbf{x} - \mathbf{x}_k(t)] \times \alpha_k q\left(\frac{|\mathbf{x} - \mathbf{x}_k(t)|}{\sigma}\right)}{|\mathbf{x} - \mathbf{x}_k(t)|^3} + \mathbf{U}_\infty(t) \tag{19}$$

The function q is given by

$$q(y) = 4\pi \int_0^y \zeta(r) r^2 dr \tag{20}$$

For a variety of simple choices of the distribution function ζ (e.g., gaussian) the truncation error is $O(\sigma^2)$ if the typical interparticle distance, h, satisfies $h < \sigma$.

Higher order schemes are possible. See Leonard (1980) and Beale and Majda (1982) for more details.

Although equation (18) for the inviscid evolution of the $\mathbf{x}_k$ seems most natural, other possibilities, based on the fact that

$$(\omega \cdot \nabla)\mathbf{u} = (\omega \cdot \nabla)^T \mathbf{u} \tag{21}$$

also yield convergent, accurate schemes as discussed by Winckelmans and Leonard (1993).

3 Viscous Diffusion

For viscous diffusion we need to update the vorticity field following the equation

$$\frac{\partial \omega}{\partial t} = \nu \nabla^2 \omega \tag{22}$$

with the boundary condition implied by (3). The random walk method, as discussed by Chorin (1973), has been applied widely for a number of years but has the disadvantage of being slowly convergent. More recently the technique of particle strength exchange (see Degond and Mas-Gallic (1989)) has been proposed. This method has good convergence properties but requires an occasional remeshing of the particles as discussed below. In this algorithm we approximate ∇^2 as an integral operator and discretize the integration over the particles. Thus, we use

$$\nabla^2 \omega \approx \frac{1}{\sigma^2} \int G_\sigma(|\mathbf{x} - \mathbf{y}|)[\omega(\mathbf{y}) - \omega(\mathbf{x})] d\mathbf{y} \tag{23}$$

where, for example, with gaussian particle distributions,

$$G_\sigma(z) = 4\zeta_\sigma(z) = \frac{4}{\pi^{3/2}\sigma^3}\exp(-z^2/\sigma^2) \tag{24}$$

Consider first the application to an infinite domain. The approximation (23) is good to second order in σ as we can see by taking the Fourier transform of both sides,

$$-k^2\hat{\omega} \approx \frac{4}{\sigma^2}[\exp(-k^2\sigma^2/4) - 1]\hat{\omega} = -k^2\hat{\omega}(1 + O(k^2\sigma^2)) \tag{25}$$

Using (23) and the particle representation we find that (22) can be written

$$\Sigma_i \frac{d\alpha_i}{dt} \zeta_\sigma |\mathbf{x} - \mathbf{x_i}| = \frac{\nu}{\sigma^2} \int G_\sigma(|\mathbf{x} - \mathbf{y}|)$$

$$\times \left[\Sigma_j \alpha_j \zeta_\sigma(|\mathbf{y} - \mathbf{x_j}|) - \Sigma_i \alpha_i \zeta_\sigma(|\mathbf{x} - \mathbf{x_i}|)\right] d\mathbf{y} \tag{26}$$

Note now that by approximating the integral on the right hand side of (26) over the particles and using (24) we obtain the evolution equation for the problem (22) in an infinite domain,

$$\frac{d\alpha_i}{dt} = \frac{\nu v}{\sigma^2} \Sigma_i G_\sigma(|\mathbf{x_i} - \mathbf{x_j}|)(\alpha_j - \alpha_i) \tag{27}$$

where v is the particle volume.

The above algorithm conserves total vorticity exactly, i.e., $\Sigma_i^N d\alpha_i/dt = 0$. Furthermore it is found that if (27) is integrated in time with Euler timestepping and

$$\Delta t = \frac{\sigma^2}{\nu} \tag{28}$$

then the results are superaccurate (Pepin, 1990) (Leonard and Koumoutsakos, 1993). This phenomenon may be explained by noting that the exact solution to (22) in an infinite domain at time $t + \Delta t$ is given by

$$\omega(\mathbf{x}, t + \Delta t) = \int \frac{\exp(-|\mathbf{x}-\mathbf{y}|^2/4\nu \Delta t)}{(4\pi\nu \Delta t)^{3/2}} \omega(\mathbf{y}, t) d\mathbf{y} \tag{29}$$

For wall-bounded flows we enforce the no-slip condition in the form (6). In addition, however, for a substep of the diffusion step we use (27), that is we do not enforce $\partial \omega / \partial n = 0$ at the wall during this substep (Koumoutsakos et al, 1994). This leads to an additional spatially varying flux of vorticity from the surface. This effect along with the pressure gradient produces a wall slip ($\mathbf{U}_{\text{slip}}$) at time $t + \Delta t$. We compute this slip as an average over computational panels on the wall directly from the Green's function integral for the velocity potential, with a form analogous to (5), using the fast algorithm. Thus the total flux to be emitted into the flow for the other substep of the diffusion process is given by

$$\nu \frac{\partial \omega}{\partial n} = \frac{\mathbf{n} \times \mathbf{U}_{\text{slip}}}{\Delta t} \tag{30}$$

This flux is then distributed to the particles by discretizing the Green's integral for the inhomogeneous Neumann problem for the diffusion equation. See Koumoutsakos et al (1994) and Winckelmans et al (1995).

4 Remeshing

In order for the numerical simulation to be accurate, the particles must, to a certain degree, be uniformly distributed. This is required for accuracy in the convection step as well as the diffusion step. On the other hand the local strain rate following a particle may generate a substantial contraction or crowding of particles in one or two directions accompanied by an expansion in the other directions, similar to the situation at a hyperbolic stagnation point in steady flow. When remeshing is deemed necessary we overlay a regular grid (the new particles) over the old particles, keeping the average particle density constant, and interpolate the old vorticity onto the new particle locations. We use a 27-point scheme to distribute the old vorticity to the new mesh. Specifically, away from a wall, the ith old vortex with circulation Γ_i and location (x_i, y_i, z_i) contributes $\Delta \Gamma_{j,i}$ circulation to new mesh point $(\tilde{x}_j, \tilde{y}_j, \tilde{z}_j)$ according to

$$\Delta \Gamma_{j,i} = \Gamma_i \Lambda \left(\frac{\tilde{x}_j - x_i}{h} \right) \Lambda \left(\frac{\tilde{y}_j - y_i}{h} \right) \Lambda \left(\frac{\tilde{z}_j - z_i}{h} \right) \tag{31}$$

where the interpolation kernel Λ is given by

$$\Lambda(u) = \begin{cases} 1 - u^2, & 0 \leq |u| < 1/2; \\ (1-u)(2-u)/2, & 1/2 \leq |u| < 3/2; \\ 0, & \text{otherwise.} \end{cases} \tag{32}$$

If the old particle is less then a distance h from a wall the interpolation is modified to maintain the same conservation properties. See Koumoutsakos (1993) and Hockney and Eastwood (1981) for further discussion. This scheme conserves the circulation, linear and angular momentum.

5 Fast Vortex Methods

The straightforward method of computing the right hand sides of (17) and (18), using (19) for every particle, requires $O(N^2)$ operations for N vortex elements. This precludes high resolution studies of bluff body flows with more than say 50,000 elements.

Recently fast methods (see e.g., Barnes and Hut (1986), Greengard and Rohkin (1987) and Salmon et al (1994)) have been developed that have operation counts of $O(N \log N)$ or $O(N)$ depending on the details of the algorithm. The basic idea of these methods is to decompose the element population spatially into clusters of particles and build a hierarchy of clusters ("tree") - smaller neighboring clusters combine to form a cluster of the next size up in the hierarchy and so on. Fig. 1 shows an example of particle clustering in two dimensions.

The contribution of a cluster of particles to the velocity of a given vortex can then be computed to desired accuracy if the particle is sufficiently far from the cluster in proportion to the size of the cluster and a sufficiently large number of terms in the expansion is taken. This is the essence of the particle/box (PB) method, requiring $O(N \log N)$ operations. One then tries to minimize the work required by maximizing the size of the cluster used while keeping the number of terms in the expansion within a reasonable limit and maintaining a certain degree of accuracy.

The box-box (BB) scheme goes one step further as it accounts for box-box interactions as well. These interactions are in the form of shifting the expansions of a certain cluster with the desired accuracy to the center of another cluster. Then those expansions are used to determine the velocities of the particles in the second cluster. This has as an effect the minimization of the tree traversals for the individual particles requiring only $O(N)$ operations.

6 Applications

To illustrate the use of the viscous vortex method in two dimension we show in Figure 2 the vorticity distribution in accelerating flow past a flat plate at 90° angle-of-attack. (P. Koumoutsakos, private communication 1994). For similar applications to cylinder flow see Koumoutsakos and Leonard (1994). A three-dimensional example is shown in Figure 3. Shown is the deformation of an initially spherical vortex sheet corresponding to potential flow past a sphere.

References

1. Barnes, J. E., and Hut, P., 1986. "A hierarchical O (N log N) force calculations algorithm", Nature, **324**, pp. 446-449.

2. Beale, J. T. and Majda, A., 1982. "Vortex methods II: High order accuracy in two and three dimensions", Math. Comp. **39**, pp. 29-52.

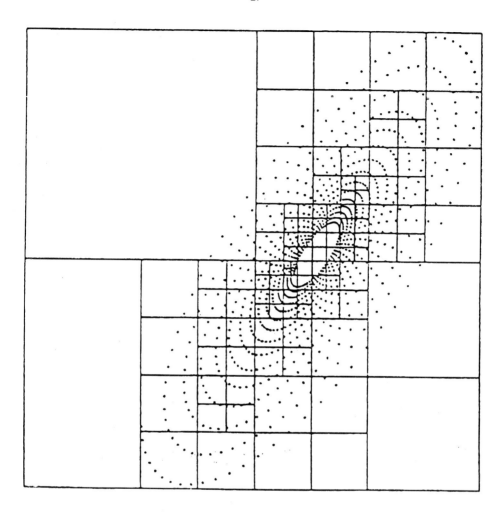

Fig. 1. Example of particle clustering for an elliptical spiral distribution of 1000 particles.

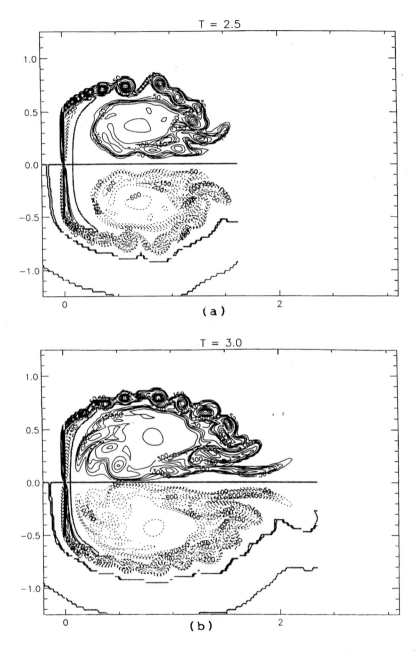

Fig. 2. Vorticity contours for accelerating freestream flow past a flat plate at 90° incidence. Viscous vortex method. Reynolds number based on acceleration, $a = \sqrt{aL^3/\nu} = 1,296$. (a) Non-dimensional time $= t\sqrt{a/L} = 2.5$; Number of vortex particles $\cong$ 210,000 (b) Non-dimensional time $= 3.0$; Number of vortex particles $\cong$ 290,000. (P. Koumoutsakos, 1994, private communication.)

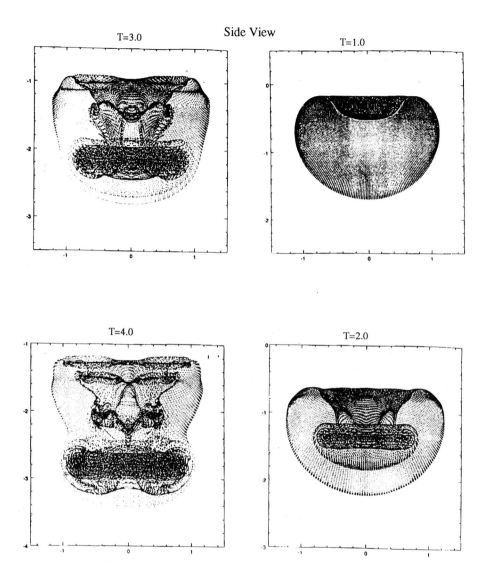

Fig. 3. Evolution of a spherical sheet of vorticity corresponding to potential flow past a sphere. Three-dimensional, inviscid vortex particle method with 81920 particles (Salmon et al, 1994).

3. Chorin, A., 1973. "Numerical study of slightly viscous flow", J. Fluid Mech. **57**, pp. 380-392.

4. Degond, P., and Mas-Gallic, S., 1989. "The weighted particle method for convection-diffusion equations, Part I: the case of isotropic viscosity, Part II: the anisotropic case", Math. Comp. **53**, pp. 485-526.

5. Greengard, L. and Rohklin, V., 1987. "A fast algorithm for particle simulations", J. Comput. Phys. **73**, pp. 325-348.

6. Hockney, R. W., and Eastwood, J. W., 1981. "Computer simulations using particles", McGraw-Hill New York.

7. Koumoutsakos, P. and Leonard, A., 1994. "High resolution simulations of the flow around an impulsively started cylinder using vortex methods", J. Fluid Mech. (to appear).

8. Koumoutsakos, P., Leonard, A., and Pepin, F., 1994. "Boundary conditions for viscous vortex methods", J. Comput. Phys., **113**, pp. 52-61.

9. Koumoutsakos, P. 1993. "Large scale direct numerical simulations using vortex methods". Ph.D. thesis, Caltech.

10. Leonard, A., 1985. "Computing three-dimensional incompressible flows with vortex elements", Ann. Rev. Fluid Mech., **17**, pp. 523-529.

11. Leonard, A., 1980. "Vortex methods for flow simulation", J.Comput. Phys. **37**, pp. 289-335.

12. Leonard, A. and Koumoutsakos, P., 1993. "High resolution vortex simulation of bluff body flows", J. Wind Eng. and Indust. Aero., **46** and **47**, pp. 315-325.

13. Pepin, F., 1990. "Simulation of the flow past an impulsively started cylinder using a discrete vortex method", Ph.D. thesis, Caltech.

14. Salmon, J. K., Warren, M. S., and Winckelmans, G. S., 1994. "Fast parallel tree codes for gravitational and fluid dynamical N-body problems", Int. J. Supercomputer Applications, **8**, pp. 129-142.

15. Winckelmans, G. S., Salmon, J. K., Warren, M. S., and Leonard, A., 1995. "The fast solution of three-dimensional fluid dynamical N-body problems using parallel tree codes: vortex element method and boundary element method," Seventh SIAM Conf. on Parallel Processing for Scientific Comp., Feb. 1995, San Francisco.

16. Winckelmans, G. S. and Leonard, A., 1993. "Contributions to vortex particle methods for the computation of three-dimensional incompressible unsteady flows", J. Comput. Phys., **109**, pp. 247-273.

On Wall Laws in CFD

Bijan Mohammadi[1], Olivier Pironneau[2]

[1]INRIA, 78153 Le Chesnay, France
[2]University Paris 6 and INRIA

Abstract : Wall laws are one possibility to remove zones with sharp gradients that lie near solid walls for turubulent flows. We shall review some known methods to deal with the problem in CFD, then show that wall laws work numerically even when boundary layers separate. In an attempt to measure the validity of wall laws we shall review their use in other contexts such as electromagnetism and show that it is really a domain decomposition method. This will allow us to conjecture that wall laws can withstand boundaries with small curvature and that it is a first order method.

1 Introduction

Turbulence is not the only difficulty in CFD. Boundary layers also require too many points in simulations. Wall Laws are an attempt at removing from the computational domain the laminar and turbulent boundary layers.

Wall laws are relations between a function and its normal derivative; they appear in many places such as

- Turbulent boundary layers: the log law.

- To take into account surface roughness.

- To model the interface between ocean and atmosphere in meteorology.

- To take into account electromagnetically absorbant paint for stealth airplanes.

In electromagnetism the wall laws, known as Leontovitch's conditions, can be proved to be correct. The question then is: can we extend the result and apply the method to turbulent wall laws?

In Fluid Mechanics wall laws are used for turbulent flows near walls where the no-slip boundary condition $u = 0$ is replaced by

$$u.n = 0, \quad \frac{u.s}{\sqrt{\nu|\frac{\partial u}{\partial n}|}} - \frac{1}{\chi}\log\left(\delta\sqrt{\frac{1}{\nu}|\frac{\partial u}{\partial n}|}\right) + 5.5 = 0,$$

It is used with the $k - \varepsilon$ model:

$$\mu_T = c_\mu \rho \frac{k^2}{\varepsilon} \quad D_t = \partial_t + u\nabla$$
$$E = \frac{1}{2}|\nabla u + \nabla u^T|^2 - \frac{2}{3}|\nabla \cdot u|^2$$
$$D_t k - \frac{\sigma_k}{\rho}\nabla \cdot (\mu_T \nabla k) + k(\frac{\varepsilon}{k} + \frac{2}{3}\nabla \cdot u) = c_\mu \frac{k^2}{\varepsilon} E$$
$$D_t \varepsilon - \frac{\sigma_\varepsilon}{\rho}\nabla \cdot (\mu_T \nabla \varepsilon) + \varepsilon(c_2 \frac{\varepsilon}{k} + \frac{2c_1}{3c_\mu}\nabla \cdot u) = c_1 k E$$

This model contains the log law as a special solution but is not valid very near walls; 3 ways out:

- Low Reynolds number corrections where all constants are primed and $(\nu = \mu/\rho)$:

$$c'_\mu = f_\mu c_\mu \quad c'_1 = f_1 c_1 \quad c'_2 = f_2 c_2$$

$$f_\mu = \left(1 - e^{-0.017y\nu^{-1}\sqrt{k}}\right)\left(1 + 20.5\frac{\nu\varepsilon}{k^2}\right)$$

$$f_1 = 1 + \left(\frac{0.05}{f_\mu}\right)^3 \quad f_2 = 1 - e^{\nu\varepsilon k^{-2}}$$

- Use only the k-equation with $\varepsilon = \frac{k^{3/2}}{l_\varepsilon}$ and l_ε a decaying function of y

- Use the wall law as boundary condition together with

$$k = u^{*2} c_\mu^{-1/2}, \quad \varepsilon = \frac{u^{*3}}{\chi\delta} \quad \text{or} \quad \frac{\partial k}{\partial n} = 0 \quad \varepsilon = c_\mu^{-3/4}\frac{k^{3/2}}{\chi\delta}.$$

Low Reynolds corrections are difficult to integrate because they also produce sharp gradients near boundaries: too many discretization points are needed in the boundary layers.

The best way seems to be to use instead a one equation model near walls: the equation for k is unchanged. For ε there are two zones: Near walls where

$$\varepsilon = \frac{k^{3/2}}{y(1 - e^{-\alpha y})} \quad \nu_T = c_\mu^{1/4}\chi\sqrt{k}\, y(1 - e^{-\beta y})$$

and far from walls where the usual $k - \varepsilon$ is used.

It is worth pointing out that this is not difficult to implement numerically with a semi implicit scheme (see Mohammadi-Pironneau[1994]).

2 On the Log Laws

2.1 Implementation

It seems at first sight that it is necessary to know n and s to implement the log law. But this is not so.

Indeed consider the mass conservation equation

$$\partial_t \rho + \nabla \cdot (\rho u) = 0, \text{ in } \Omega \quad u \cdot n|_{\partial\Omega} = 0$$

Then it is equivalent to

$$\forall r(x): \quad \int_\Omega (\partial_t \rho) r - \int_\Omega \rho u \nabla r = 0$$

Similarly in the discrete case, with finite volumes, cell centered in ρ: if for all cells σ

$$\int_\sigma \partial_t \rho + \int_{\partial\sigma - \partial\Omega} \rho u \cdot n = 0,$$

then this contains a numerical approximation of

$$\int_{\partial\sigma \cap \partial\Omega} u \cdot n = 0.$$

Now for the other part of the law, written as $\partial_n u \cdot s = f(u \cdot s) u \cdot s$, to integrate it in a semi implicit scheme in time in the viscous step which usually involve the solution of

$$\frac{1}{\delta t} \rho u - \cdot (G(\rho, u, E) u) = g$$

we apply Green's formula and use

$$\int_\Omega \left[\frac{\rho}{\delta t} u \cdot v + G \nabla u : \nabla v \right] - \int_{\partial\Omega} f(|u|) u \cdot v = \int_\Omega g v$$

2.2 Questions

The log law is an experimental evidence for flows over flat plates. A change of scale

$$u^* = \sqrt{\nu \frac{\partial U \cdot s}{\partial n}}, \quad u^+ = u \,/\, u^* \quad y^+ = y \,/\, \frac{\nu}{u^*}$$

shows a viscous sublayer ($0 \leq y^+ \leq 5$):

$$u^+ = y^+$$

and a logarithmic layer ($10 \leq y^+ \leq 100$):

$$u^+ = \frac{1}{\chi} \log y^+ + \beta, \quad \chi = 0.41 \quad \beta = 5.5$$

The following questions arise:
- What if the radius of curvature of the surface is large ?
- What if the boundary layer separates ?
- What is the order of the method?

The problem is somewhat similar to the Euler-boundary layer matching in laminar flows. There the technique nearest to a wall law is Polhausen's matching scheme (see Cousteix[1990]). The velocity in the boundary layer is approximated as :

$$u = U_e(A_0 + A_1 y + A_2 y^2 + A_3 y^3 + A_4 y^4)$$

and A_i are chosen to satisfy the boundary conditions at $y = 0$ and $y = \delta$ and the Navier-Stokes equations at $y = 0$.

To test these methods we have performed the following numerical simulations:
- 1. Laminar flow over a flat plate at Re=1000.
- 2. Turbulent flow at $Re = 10^6$ over a flat plate
- 3. " " hitting a vertical wall
- 4. Flow over a transonic hump, $M_\infty = 0.635$, $Re = 1.35 \times 10^7$

The conclusion (see figures) is that wall laws work even when there is separation, that low Reynolds corrections are not necessary but there is a need for more investigations and a need for justifications.

3 Other Wall Laws

3.1 Rough Boundaries

Carrau[1991] studied the flow over a rough surface. A wing profile Γ has very fine periodic irregularities. He considered 2 regions, separated by an interface §: B_δ below § , near the wing profile and Ω_δ above Σ. B_δ is thin and periodic so it is reasonable to assume that the flow also is $s-$ periodic below Σ.

Let Y be the cell of periodicity.

If $U_\Sigma \equiv u_1|_\Sigma$ is known, then the flow below Σ is determined by

$$u\nabla u + \nabla p - \nu\Delta u = 0 \quad \nabla.u = 0 \text{ in } Y,$$

$$u|_\Gamma = 0 \quad u_1|_\Sigma = U_{1\Sigma} \quad u_2|_\Sigma = 0,$$

u, p x_1-periodic. Let $g(U_\Sigma)$ be $\partial_n u_1|_\Sigma$; it is some nonlinear function of U_Σ.

Above Σ, u is also solution of the Navier-Stokes equations. Matching u and ∇u on Σ gives:

$$\partial_n u_1 = g(u_1), \quad u_2 = 0 \text{ on } \Sigma.$$

This is also a wall law.

3.2 Ocean-atmosphere interface

In climatology, at the interface between sea and air, the following is used (Lions[1994], the [ ] stands for the jump across the interface):

$$u.n = 0 \quad [\partial_n u.s] = f([u.s])$$

That such a relation exists can be understood by the same domain decomposition but this time with 3 domains:

- A periodic cell with Navier-Stokes equations following a representative train of water waves; the density is ρ_s below the free boundary and ρ_a above;

- a domain above with the Navier Stokes equations for the atmosphere and
- a domain below for the sea with the Navier Stokes equations also with constant density.

The periodic cell gives $\partial_n u.s$ on the upper and lower interface as a function of the fluid velocities on these interfaces. But due to translation invariance, these have to be functions of the difference of velocities above and below.

3.3 Electromagnetic coating for stealth airplanes

Electromagnetic absorbing paints made of homogeneous material, require too many grid points for numerical simulations. In the case of *T.E. Polarization* plane incident monochromatic electromagnetic wave of complex amplitude u, Leontowitch's *effective condition* is

$$u + \delta \frac{\mu_1}{\mu_\infty} \frac{\partial u}{\partial n} = 0$$

where, Γ_δ being the upper surface of the paint layer,

$$\nabla.(\frac{1}{\mu}\nabla u) + \epsilon u = 0 \quad [\frac{1}{\mu}\frac{\partial u}{\partial n}]_{\Gamma_\delta} = 0$$

$$u|_\Gamma = 0 \quad \lim_{r \to \infty} r\left(\frac{\partial u}{\partial n} - i\omega\sqrt{\mu\epsilon}u\right) = g$$

The proof is simple; it assumes u linear in the paint and does the matching by hand.

Remark: The method works also for accoustic dampers and break waters.

For non-homogeneous paints containing metallic impurities Γ_b the derivation of such an effective condition is much harder (Achdou [1989]).

Consider a domain with a thin periodic layer of thickness δ attached to the wall Γ and

$$\nabla.\frac{1}{\mu}\nabla u^\delta + au^\delta = 0$$

$u = 0$ on the wall $u \to u_\infty$ at infinity

An asymptotic expansion

$$u^\delta(x) = u^0(x) + \delta u^1(x, \frac{x}{\delta}) + \delta^2 u^2(x, \frac{x}{\delta})$$

with

$$\nabla.\frac{1}{\mu_\infty}\nabla u^0 + a_\infty u^0 = g \text{ in } \Omega$$
$$u^0|_\Gamma = 0 \quad u^0 \to 0 \text{ at } \infty$$

gives

$$\left[\frac{1}{\mu}\frac{\partial u^0}{\partial n}\right]_{\Gamma_\delta} \neq 0, u^0|_{\Gamma_b} \cong -x_2\frac{\partial u^0}{\partial n}|_\Gamma \neq 0$$

So correct it with u^1 by

$$\nabla \cdot \frac{1}{\mu}\nabla u^1 + au^1 = -\left(\nabla \cdot \frac{1}{\mu_1}\nabla u^0 + au^0\right)I_{\Omega_\delta}$$

$$\left[\frac{1}{\mu}\frac{\partial u^1}{\partial n}\right]_{\Gamma_\delta} = -\frac{1}{\delta}\left[\frac{1}{\mu}\right]\frac{\partial u^0}{\partial n} \quad u^1|_\Gamma = 0 \quad u^1|_{\Gamma_b} = \frac{x_2}{\delta}\frac{\partial u^0}{\partial n}|_\Gamma$$

Next use the fact that $u^1(x, x/\delta)$ has a fast scale $y = x/\delta$ to define it on a semi-infinite vertical slab Y

$$\nabla_y \cdot \frac{1}{\mu}\nabla_y u^1 = 0 \quad \left[\frac{1}{\mu}\frac{\partial u^1}{\partial n_y}\right]_{\Gamma_\delta} = -\left[\frac{1}{\mu}\right]\frac{\partial u^0}{\partial n} \quad u^1|_\Gamma = 0 \quad u^1|_{\Gamma_b} = y_2\frac{\partial u^0}{\partial n}|_\Gamma$$

So u^1 is of the form

$$u^1 = \chi\frac{\partial u^0}{\partial n} \quad \text{Hence } u_h = u^0 + \delta u^1 = \delta(\chi - 1)\frac{\partial u^0}{\partial n} \cong \delta(\chi - 1)\frac{\partial u_h}{\partial n}$$

So finally

$$u_h + \delta(\chi - 1)\frac{\partial u_h}{\partial n} = 0, \quad \|u_h - u^\delta\| \leq c\delta$$

where

$$\nabla_y \cdot \frac{1}{\mu}\nabla_y \chi = 0 \quad \left[\frac{1}{\mu}\frac{\partial \chi}{\partial n_y}\right]_{\Gamma_\delta} = -\left[\frac{1}{\mu}\right]$$

$$\chi|_\Gamma = 0 \quad \chi|_{\Gamma_b} = y_2 \quad \chi \to \text{ a constant at } \infty$$

Problem: Such analysis is non-trivial, hard to generalize to curved boundaries and hard to sell to engineers.

Remark: $\chi \to$ a constant at infinity may be replaced by $\partial_n \chi \to 0$?, or equivalently $\partial_n(\chi - y_2) \to 1$. Notice that $(\chi - y_2)|_\Gamma = 0$.

4 Homogenization by Patching

4.1 Electromagnetism

Following Carrau-Le Tallec, Call u_∞ and u_1 the solution u restricted to Ω_∞ and Ω_δ. Obviously

away from the paint: $\quad \nabla(\frac{1}{\mu_\infty}\nabla u_\infty) + a_\infty u_\infty = 0, \quad u_\infty - u_{inc} \to 0$

in the paint and close to it: $\quad \nabla(\frac{1}{\mu}\nabla u_1) + a_1 u_1 = 0$
$\quad u_1|_\Gamma = 0 \quad u_1|_{\Gamma_b} = 0$

and at the interface:
$$u_1 = u_\infty \qquad \frac{1}{\mu_\infty}\frac{\partial u_1}{\partial n} = \frac{1}{\mu_\infty}\frac{\partial u_\infty}{\partial n} \quad \text{on } \Gamma_\delta$$

But u_1 is defined in a narrow band with periodicity. So if u_∞ is smooth then u_1 is also periodic and satisfies:
$$\nabla(\tfrac{1}{\mu}\nabla u_1) + a_1 u_1 = 0 \quad u_1|_\Gamma = 0 \quad u_1|_{\Gamma_b} = 0 \quad u_1|_{\Gamma_\delta} = u_\infty$$

Therefore $u_1 = \chi u_\infty$ with
$$\frac{1}{\mu_\infty}\frac{\partial u_1}{\partial n} = u_\infty \left(\frac{1}{\mu_\infty}\frac{\partial \chi}{\partial n}\right) = \frac{1}{\mu_\infty}\frac{\partial u_\infty}{\partial n}$$

which is the wall law.

Theorem
Let R be the minimal radius of curvature of the boundary, then
$$\|u^\delta - u^\delta\|_1 \leq C\left(\delta^{3/2} + \frac{\delta}{R_{min}} + \delta\|\partial_s\chi\|_{0,\Gamma_\delta}\right)$$

Proof: see Achdou-Pironneau [2].

4.2 Application to turbulent flows

For a $k - \varepsilon$ model with low Reynolds corrections let us use the same approach: a domain decompostion with
- Domain 1: the boundary layer
- Domain 2: above the boundary layer

and match u and the normal stress tensor $\sigma \cdot n$ at the interface §.

In domain 1, assume periodicity, i.e. horizontal translation invariance. Then the equations reduce to
$$-\partial_y(\nu_T \partial_y u_1) + \partial_x p = 0$$
$$\nu_T = c_\mu \frac{k^2}{\varepsilon} \quad E = |\partial_y u|^2$$
$$-\partial_y(\nu_T \partial_y k) + k\frac{\varepsilon}{k} = c_\mu \frac{k^2}{\varepsilon} E$$
$$-\sigma_\varepsilon \partial_y(\nu_T \partial_y \varepsilon) + \varepsilon c_2 \frac{\varepsilon}{k} = c_1 k E$$

The solution is again a log law but because of the pressure term :
$$u^+ = \frac{1}{\chi}\log y^+ + \pm y^+ \frac{\nu}{\chi u^{*2}}\frac{\partial p}{\partial s}$$

We tested the effect of this pressure correction term on the transonic hump problem and found it to be small.

5 Conclusion

While surveying numerical methods for turbulent boundary layers we found that a simple log law and a removal of the boundary layer is adequate even if the boundary layer does not exist in some places or separates. Though the low Reynolds number corrections are necessary in theory, they may be forgotten in computations because the coupling is well taken care of by the log law.

We also made an attempt to show that such wall laws belong to a more general domain decomposition technique, particularly well studied in the context of radar cross section effects of thin absorbing paints. This may lead us to believe that the method is first order with respect to boundary layer thickness and half order with respect to curvature of the surface. But these conclusions rely on the ellipticity of the equations and for turbulence models it defenitely requires more investigations.

References

[1] **Achdou Y.** Effect of metallized coating on the reflection of an electromagnetic wave. INRIA report 1136.

[2] **Achdou Y., O. Pironneau** Analysis of wall laws. UPMC report R94018.

[3] **Carrau R.** Thèse de doctorat, Université de Bordeaux I.

[4] **Cousteix J.** Couches limites laminaires et turbulentes. (I) & (II). Cepadues editions.

[5] **Lions J.L.** Cours du Collège de France.

[6] **Mohammadi B., O. Pironneau** Analysis of the $k-\varepsilon$ turbulence model. Wiley.

Status of CFD in India

T. S. Prahlad

Aeronautical Development Agency, P.O. Box No. 1718
Bangalore–560 017, India

Abstract : This paper gives an overview of the status of CFD activity that is being carried out in a number of organisations in India. It discusses the drive for CFD in India and its relevance, the general pattern of growth for CFD in the country, the present CFD scenario, some applications and directions for the future. The emphasis is mainly on aerospace related CFD, eventhough other applications are briefly mentioned. The paper indicates how the problem of computational capability is being tackled through the development of parallel computing platforms.

1 Introduction

Computational Fluid Dynamics (CFD) has been making rapid strides in India over the last fifteen years. The reasons have been essentially twofold. Firstly, there has been a strong tradition of research in Fluid Mechanics in India for over forty years which was pioneered by the Indian Institute of Science at Bangalore and which has now spread to many other organisations. CFD is seen by these organizations as an emerging powerful tool of basic and applied research in Fluid Dynamics. Secondly, over the last ten to twelve years, the country has embarked on a number of aerospace programmes in the areas of military and civilian aircraft, advanced helicopters, satellite launch vehicles, missiles, unmanned aircraft, high performance aircraft engines etc. These programmes have given a real boost to the development of CFD in the country, especially from an application angle.

The present paper discusses briefly this drive for CFD activity in India, its relevance, the general pattern of growth of CFD in the country, present CFD scenario, some applications and directions for the future. The emphasis is mainly on aerospace related CFD where there has been significant progress over the last few years.

2 Overview of CFD Capability in India

Because of the heavy concentration of aerospace organizations in Bangalore, a whole range of CFD activities are in progress in a number of centres in this city which include the Indian Institute of Science (IISc), National Aerospace Laboratories (NAL), Aeronautical Development Agency (ADA), Hindustan Aeronautics Limited (HAL), Aeronautical Development Establishment (ADE) and Gas Turbine Research Establishment (GTRE). Launch vehicle related CFD activities are in progress at Vikram

Sarabhai Space Centre (VSSC) at Trivandrum and missile related CFD at Defence Research and Development Laboratory (DRDL) at Hyderabad. The five Indian Institutes of Technology (IIT) at Madras, Bombay, Kanpur, Kharagpur and Delhi are involved in basic and applied CFD.

Fig. 1 indicates the growth of computing power in India for CFD work over the last ten years. The several parallel platforms shown in the figure are all built in the country, pioneered by the FLOSOLVER (FS) development at NAL, specifically tailored for CFD applications. The 32 node machine (BPPS) built by Bhabha Atomic Research Centre (BARC), and based on Intel 80860 processor, gives around 120 MFLOPS speed and has been used extensively for several aerospace CFD applications. A SPARC-10 based 32 node machine (PACE) is under development at Advanced Numerical Research & Analysis Group (ANURAG) under a project from ADA to give a computing power of 700-800 MFLOPS. The Centre for the Development of Advanced Computing (CDAC) has been building a series of parallel processor computing platforms (PARAM). While powerful serial workstations are very useful, especially at the development stages of the code, the future for India for advanced CFD seems to be clear - Build Parallel and Think Parallel !

Fig. 2 attempts to give a broad summary of the present CFD capability in India with X-axis representing the geometrical complexity and the Y-axis the complexity of flow physics. The applications associated with the hierarchy of CFD codes are indicated in this figure. The Light Combat Aircraft (LCA) has been a major driver in the development of this capability. To illustrate this further, the role played by CFD in the aerodynamic design evolution of LCA and its analysis is illustrated in Fig. 3. CFD has played similar roles in the country's advanced satellite launch vehicle, missile and light transport aircraft programmes also.

Eventhough there are a number of CFD codes available in the market today with varying claims, a deliberate decision has been taken in India that the indigenous development of codes is the route for us to follow, as it is only then that we will have a firm grip on this emerging technology. We can tailor and modify the codes to suit our own applications and really know where we can use them and where we cannot.

The current work going on in India in the development and application of the hierarchy of CFD codes from panel methods to Reynolds averaged Navier Stokes solutions - are briefly covered in the remaining part of the paper. Only a few examples are indicated for the sake of brevity. Basic work under progress in algorithm development, direct numerical simulation etc., is also mentioned.

3 Panel and Full Potential Codes

A number of panel codes that address the full aerospace vehicle configuration have been developed in the country and have been employed usefully in the preliminary design phases of these vehicles. The panel code applications for a fighter aircraft configuration like LCA includes preliminary wing camber and twist optimisation with leading edge slats, effect of drop tanks and other stores, control surface deflection

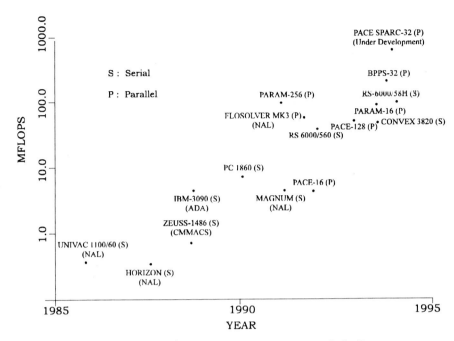

Fig. 1 Growth in speed of computers in India

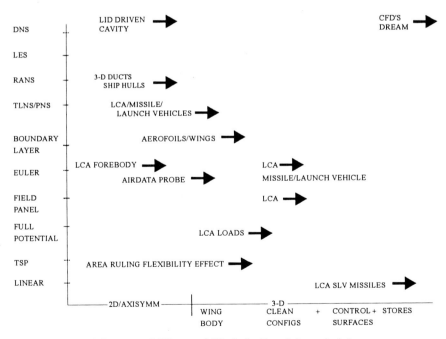

Fig. 2 Present CFD capability in India - A broad picture

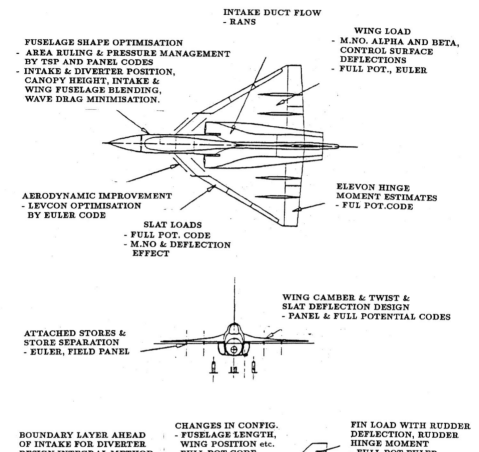

Fig. 3. Major CFD application for LCA

effects, estimation of ground effect etc. Panel codes have also been used at NAL to optimise the wing-body junction of their trainer aircraft 'Hansa' and for studying the aerodynamic characteristics of their light transport aircraft 'Saras'. VSSC has used panel codes for aerodynamic load distribution estimates for their satellite launch vehicles with strap-on boosters and DRDL for their missiles.

A finite element full potential code (FETRAN) has been the workhorse to tackle many aerodynamic design problems of a complex configuration like LCA upto about 10° incidence. Some of the typical applications have been load distributions over the LCA wing and fin at transonic Mach numbers for structural design, estimation of forces and moments acting on leading edge slats at different deflections, elevon hinge moment calculations with both wing incidence and control surface deflection effects etc. Fig. 4 shows the lift and pitching moment comparison for a low aspect ratio fighter aircraft wing. While the lift comparison is good even upto higher angles of incidence, the pitching moment comparison is good only at lower incidences; as can be expected, the full potential formulation cannot capture the pitch up tendencies. However, a code like this has been very useful to estimate the incremental effects due to small design changes. For example, the incremental change in stability margin due to marginal configuration changes for a fighter aircraft, involving increase in fuselage length and moving the wing aft, is shown in Fig. 5 in comparison with experiments. A three dimensional integral method boundary layer code has been coupled with this potential code and is used for the design of a laminar supercritical wing. A field integral method has also been developed between IIT, Kanpur and ADA which uses a paneled surface and a rectangular field grid and employs a finite volume scheme for field potentials. This can handle complex configurations and is being used for stores separation analysis of LCA. Another interesting application of a full potential code at NAL is in estimating the effect of the open area ratio of their trisonic wind tunnel with slotted test section for transonic tests.

4 Euler Solvers

Euler solvers are being increasingly used in India in both design and analysis modes. Significant work is going on in almost all the CFD centres in the country with particular stress on applications.

The major Euler Code applications in Indian aerospace programmes include the following: (a) LCA - aerodynamic loads with effect of Mach number, incidence, sideslip, control surface deflections etc., calibration of fuselage mounted air data sensors, aerodynamic improvement studies through leading edge devices, effect of stores (b) Launch Vehicles - effect of strap-on boosters, secondary injection thrust vector control (c) Missiles - supersonic and hypersonic flow simulations, intake flow simulation for ramjet applications, mixing of supersonic streams. (d) Engines - axial flow compressors, turbine cascades.

An Euler Code that is extensively used in the LCA context is the Aircraft Multi-block Euler Solver (AMES) developed jointly between IIT, Kharagpur and ADA. This

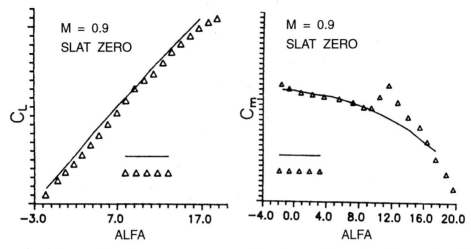

Fig. 4 Lift and Pitching moment comparision for a delta wing fignter aircraft (FETRAN CODE)

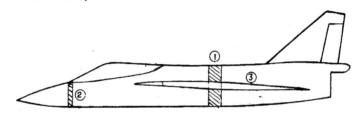

STABILITY MARGIN AS A PERCENTAGE OF MAC : CINFIGURATION - 2		
M	FETRAN	EXPERIMENT (CALSPAN)
0.7	- 0.4 %	- 0.34 %
0.9	- 1.85 %	- 1.64 %
	FEBRUARY - 1990	SEPTEMBER - 1990

Fig. 5 Incremental changes in stability margins due to configuration changes (from FETRAN)

is an explicit finite volume code and has features like multi-block structured grid, local time stepping, implicit residual averaging etc. It has been extensively used for estimation of aerodynamic pressure distribution and loads, control surface deflection effects, effect of stores, preflight calibration of air data sensors etc. Typical application of AMES for elevon deflection along with the associated grids and the resultant changes in lift, pitching moment and drag due to the deflection are shown in Fig. 6. VSSC has done fairly extensive work on the application of Euler Codes for their launch vehicle problems. Some of the applications include aerodynamics of strap-on boosters with overlap grids, two phase flow in rocket nozzles with secondary injection for thrust vector control, impingement of rocket jet over a deflector during lift off, intake flow studies of air breathing rockets etc. Results from the jet impingement calculations over a deflector are shown in Fig. 7 in comparison with experimental results. NAL has developed a finite volume, cell-centred multiblock Euler Code and applied the same for aircraft and launch vehicle studies. Typical results from the application of this Euler Code to the Augmented Satellite Launch Vehicle of VSSC is shown in Fig. 8.

IISc at Bangalore has been developing a number of new algorithms for Euler Solvers and, in association with design organisations like DRDL and ADA, has been applying the same for flow analysis of practical configurations. These include Kinetic Flux Vector Splitting (KFVS) scheme, Peculiar Velocity based Upwind (PVU) method, Boltzmann - Taylor - Galerkin Finite Element (BTG - FEM) method, Acoustic Flux Vector Splitting (AFVS) scheme, Genuinely Multidimensional Upwind Boltzmann (GMUB) scheme etc.

Based on KFVS, a 3D Euler Code called BHEEMA (Boltzmann Hypersonic Euler Equation Solver for Missile Aerodynamics) has been developed between DRDL & IISc and applied to a variety of missile related problems in the Mach number range of 0.1 to 14.

5 Navier - Stokes Codes

With the availability of increasing computing capability, Reynolds Averaged Navier-Stokes (RANS) calculations have been taken up at many places. While educational institutions are mainly concentrating on 2D calculations, 3D application oriented computations are being done at ADA & NAL with full RANS as well as Thin-Layer Navier-Stokes (TLNS) and Parabolized Navier-Stokes (PNS) equations.

The application of an implicit finite difference RANS code (with Baldwin-Lomax turbulence model) at ADA to a natural laminar flow supercritical aerofoil, along with comparison with experimental results is shown in Fig. 9. Results from another compressible implicit thin layer Navier-Stokes solver at ADA for variation of aerodynamic characteristics of a fighter aircraft upto about $16°$ incidence are shown in Fig. 10 and compared with wind tunnel measurements. An explicit finite volume compressible RANS code has been developed between ADA, IISc and IIT, Bombay to analyse the flow in the bifurcated LCA duct using a multiblock algebraic grid. The parallel ma-

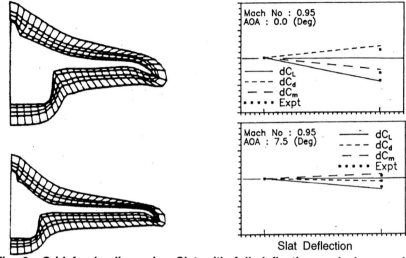

Fig. 6 Grid for leading edge Slat with full deflection and changes in aerodynamic coefficients due to Slat deflection

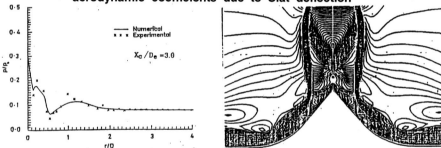

Fig. 7 Jet impingement on a deflector during launch vehicle take-off

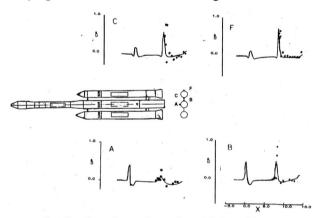

Fig. 8 Pressure distribution for a launch vehicle with strap-on booster $M\infty = 2.09$, $\alpha = 4°$

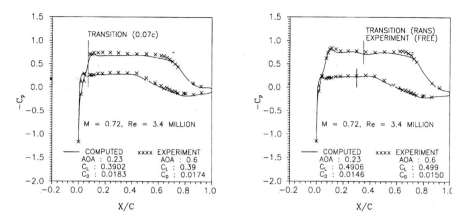

Fig. 9 Analysis of Natural Laminar Flow (NLF) Supercritical Aerofoil with RANS Solver

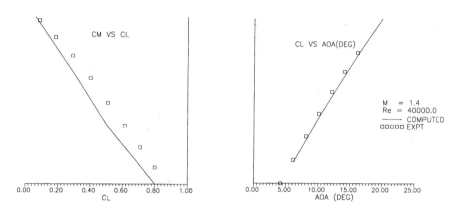

Fig. 10 Thin-Layer N-S Solution for a Fighter Aircraft

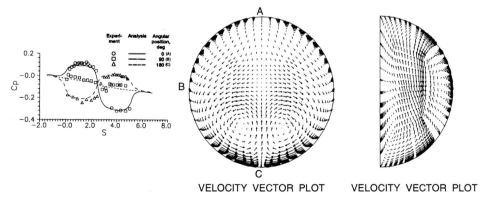

Fig. 11 RANS Solution for Curved Duct

chines of BARC have been extensively used for fine tuning and validating the code for an S-shaped curved duct and for generating LCA intake duct results. Typical results showing experimental comparisons and capture of secondary flows in ducts (under certain upstream flow assumptions) are shown in Fig. 11.

There is a vigorous effort at NAL on the development and application of Navier-Stokes codes. This includes a cell-vertex finite volume RANS code applied to delta wings, a finite volume incompressible N-S code with pressure velocity formulation applied to a variety of problems like flow in ducts, submarine flow problems, flow in an autoclave etc., explicit Runge-Kutta time stepping Navier-Stokes solver with application to separated flow computations over aerofoils, compressible thin-layer cell-centred finite volume scheme applied to hypersonic flow over satellite launch vehicle and missile forebodies, parabolized Navier-Stokes code used for equilibrium and non-equilibrium calculations over axisymmetric bodies etc. Results from two applications are shown in Fig. 12. Euler and Navier-Stokes 2D turbine cascade calculations are also done at NAL.

VSSC has been working on the application of N-S solvers for their satellite launch vehicle problems. Some of the problems addressed are stage separation problems involving multiple flow interactions with a fluid-in-cell approach and an assessment of sound pressure level and unsteady effects on payload shrouds. An interesting recent application of a 2D/axisymmetric N-S code is the computation of pressure difference with time in a payload shroud separation test conducted in ambient air, so that the necessary correction to the test could be applied to assess clean separation under vacuum conditions. The computed delta pressure variation (between the ambient and inside the shroud) with time is shown in Fig. 13.

Typical Navier-Stokes solver capability developed in IITs include unsteady Navier-Stokes computations for 2D bodies moved impulsively from rest with and without rotation, wake calculations behind bodies placed in ducts, 3D flow in heat exchanger ducts with vortex generators, 2D transonic turbine cascade flows, compressible interacting flows between coolant jets and hot cross flow, curved duct computations, flow through 2D slots etc. One such calculation of unsteady flow past a cylinder being done at IIT, Kanpur, is shown in Fig. 14.

6 Vortex Dynamics

Interesting work on vortex dynamics has been going on at NAL through discrete vortex simulation where discrete vortex blobs are released from the surface at short time intervals, satisfying the no slip condition on the surface and simulating the viscous diffusion using a random walk method. A major task undertaken by NAL for UK Civil Aviation Authority is the prediction of the characteristics of aircraft trailing vortices, so as to decide the safe separation distance between a leader and a follower aircraft. Illustrative calculations are represented in Fig. 15. This approach is being extended for the study of helicopter wake vortices also.

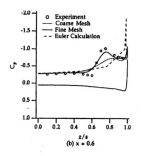
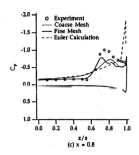
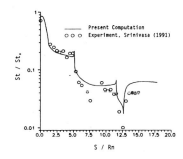

a) Flow over a Delta wing

b) Flow past the bulbous heat shield of a satellite launch vehicle

Fig. 12. Application of Navier-Stokes Codes at NAL

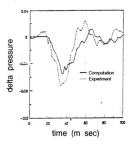
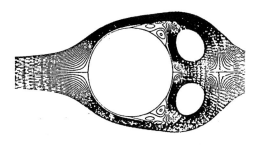

Fig.13 Application of Navier-Stokes Code to payload shroud separation problem at VSSC

Fig. 14 Unsteady Navier-Stokes calculation of flow past cylinder

- LIGHT TRASPORT AIRCRAFT
- PASQUILL STABILITY CLASS C
- NONUNIFORM CROSSWIND (2.5 m/s)
- TURBULENT DECAY
- MERGER EVENTS
- WAKE BANKING
- DISPERSED WAKE

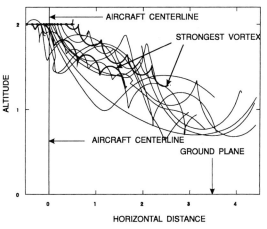

Fig. 15 Vortex tranjectory for a transport aircraft under landing configuration under crosswind

7 Direct Numerical Simulation (DNS)

The increased computational capability on serial and parallel platforms presently available in the country have prompted research groups at NAL and IISc to take up computational work on the direct numerical simulation of turbulent flows. Unsteady transitional and turbulent flows have been simulated at Reynolds numbers upto 10000 for two and three–dimensional driven cavities where the upper wall moves parallel to itself. Both uniform and clustered grids have been used. Some typical results are shown in Fig. 16. Direct numerical simulation of shear layers is also done at NAL using spectral methods.

8 Some Interesting Non-aerospace Problems

While many of the recent major applications of CFD in India have been in the area of aerospace, some interesting non-aerospace problems are also being addressed at various research centres. Work is under progress at the Centre for Mathematical Modelling And Computer Simulation (CMMACS) on a variety of interesting problems like ocean modelling, thermosolutal convection, tropical cyclone early warning calculations etc. Some of the tropical vortex modelling results are shown in Fig. 17 for a mean thermodynamic state representative of the Bay of Bengal region for premonsoon, monsoon and post monsoon months. The x–axis shows the size of the incipient system and the y–axis the pressure difference in the vortex between centre and far away, time taken to reach the maximum ΔP, maximum low level tangential wind V_{max} and time taken to attain V_{max}.

An interesting application of the parallel computer FLOSOLVER MK3 which NAL has taken up for the Department of Science and Technology is parallelisation of the global circulation model T–80 which is presently running on a CRAY machine. As part of this work, new and very effective graphic tools have been added to the FLOSOLVER for good post processing and visualisation.

9 Grid Generation

A surface modeller capability has been developed between the National Centre for Software Technology (NCST) at Bombay and ADA using a fifth order B-spline method. There is reasonable capability in the country in generating multiblock 3D structured grid. Work has been initiated at NCST on 3D unstructured grid generation for complex configurations. However, there is need for much further work to be done in India in the area of grid generation.

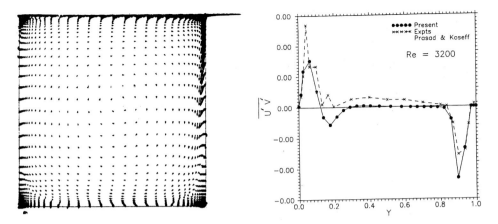

Fig. 16 DNS results for a driven cavity
(a) velocity vector plot (IISc) at Re No. = 10000
(b) shear stress distribution at Re No. = 3200 (NAL)

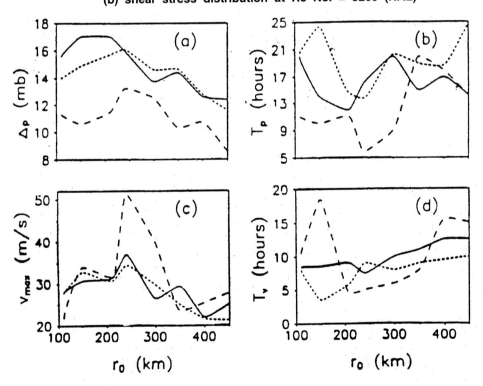

Fig. 17 Dependence of intensification of tropical vortex over Bay of Bengal on the size of incipient vortex:

———— Pre-monsoon; ---------- during monsoon ;
— — — Post monsoon

10 Future Perspective

Some of the future application CFD tasks that need further thrust in India are indicated below:

- Modelling of complex geometry and 3D structured and unstructured grid generation.

- Capability to handle through RANS a full aircraft configuration with stores and at high incidence.

- Realising the full potential of CFD for engine and other internal flow applications.

- Unsteady flow phenomena like transonic buffet, supersonic intake buzz and pulsation, vortex burst etc.

- Further push to DNS for complex geometries and at higher Reynolds numbers; modelling of subgrid scales for Large Eddy Simulation (LES).

- Multidisciplinary Computational Dynamics using CFD approaches:

    Computational Electromagnetics

    Computational aeroelasticity and aeroservoelasticity.

    Computational propulsion and chemistry

- CFD for helicopter, naval, industrial, atmospheric and educational applications.

- Multi-Gigaflop computing platforms for solving the complex real life CFD problems.

All these provide tremendous challenges to the CFD community in the country and work has been initiated in several directions. Multidisciplinary computations in the area computational electromagnetics has been taken up between ADA and IISc. Computational combustion and propulsion calculations are underway at IISc and several other organisations like VSSC, DRDL, GTRE and NAL. Powerful parallel computing platforms are under development at several places using different architectural approaches.

There is an urgent need for increasing the number of CFD workers in the country, with application of CFD for complex real life problems as the main thrust. The Aeronautical Research and Development Board (AR&DB) is planning to create Centres of Excellence in CFD with reasonably good funding for establishment of computing infrastructure and for coordinating the national effort, especially in CFD applications.

11 Concluding Remarks

This paper attempts to give a broad sweep of the Indian CFD scene and to indicate where the country stands today in this emerging technology. In spite of the known limitations and practical difficulties of CFD like modelling of transition and turbulence, grid generation and grid sensitivity for complex configurations, ever increasing demands on computing power, etc., a high potential is seen in India for using CFD intelligently in practical design applications. The day CFD can totally replace wind tunnel testing is still very far off (if at all); however, CFD even now can very usefully complement wind tunnel testing to achieve a better and a more intelligent design. It is this potential that needs to be harnessed fully and this is the direction in which the Indian CFD community is broadly moving. The current aerospace programmes of the country have been a major driving force in this attempt. The future certainly looks bright and promising for the Indian CFD.

Acknowledgements

The author would like to express his immense gratitude to all the groups in ADA, NAL, IISc, HAL, VSSC, DRDL, IITs and CMMACS who have given their invaluable help in the preparation of the material for the invited talk on 'Status of CFD in India' at the 14th ICNMFD . It was not possible to cover all the CFD work going on in India in the invited talk; it has been much less so in this paper. Also, for convenience and for avoiding a long list of references, only organisations are mentioned and not the individual contributions. This in no way is meant to minimize the significance of individual contributions.

The Development of High Quality Unstructured Grids on Parallel Computers

N. P. Weatherill[1], D. L. Marcum[2], A. J. Gaither[2], O. Hassan[1], and M. J. Marchant[1]

[1] Department of Civil Engineering,
University of Wales, Swansea,
Singleton park, Swansea, SA2 8PP, UK
[2] NSF/Engineering Research Center,
Mississippi State University, PO Box 6176
Mississippi State, MS 39762, USA

Summary : The paper highlights some new work and trends related to unstructured grid generation and flow simulation. The generation of unstructured grids using iterative point insertion routines is discussed and it is shown that a general class of methods, of which one is the Delaunay triangulation, can be developed to give high quality grids. A framework is presented for the easy implementation of these methods into a parallel computer environment which significantly improves computational performance. Finally, the issue of the generation of grids suitable for viscous flows is discussed and results presented for high Reynolds number flows.

1 Introduction

Unstructured grid technology is a promising approach offering geometric flexibility for handling complex geometry and physics. The approach has been successfully demonstrated for use in the simulation of inviscid flows. However, to fully exploit the potential of this technology and increase its applicability, further improvements in efficiency, robustness and accuracy of algorithms are needed. In this paper, several aspects of unstructured grid technology applied to aerospace flow simulation will be discussed.

2 Grid Generation by Iterative Point Insertion

Unstructured grid generation procedures for computational fluid dynamics are typically based upon either the advancing front[1,2] or Delaunay[3,8]. The advancing front approach offers the advantage of high quality point placement and integrity of the boundary. The Delaunay approach offers the advantages of computational efficiency and a sound mathematical basis. A thorough discussion of these methods is given by Mavriplis[9].

Recently[4,10,11], our work on unstructured grid generation has concentrated upon automatic point insertion strategies for the Delaunay triangulation. However, further work[12] has indicated that a general framework can be established which utilises insertion procedures followed by connection strategies, of which the Delaunay triangulation is an option. A general framework for the generation of unstructured grids which has emerged is:

1. Generate a boundary surface grid for the given configuration.

2. Obtain a valid triangulation of the boundary points and recover all boundary surfaces.

3. Assign a point distribution function to each initial boundary point.

4. Initially make all elements active.

5. Turn off each active element which satisfies the point distribution function.

6. Create a new point for each active element.

7. For each new point interpolate the point distribution function.

8. Add the new point into the triangulation.

9. Repeat the point insertion strategy until no elements are active.

In step 3, the computation of the point distribution function, which represents the typical length scale at each node, is performed as

$$dp_i = \frac{1}{N_p} \sum_{j=1}^{N_p} \{(x_i - x_j)^2\}^{\frac{1}{2}} \qquad (1)$$

where N_P is the number of points surrounding the point i.

In a manner similar to Rebay[13], elements, in step 5, are turned off during the grid generation process when they satisfy the point distribution function. Typically this is so when the length of an edge of an element is less than 1.5 times the average distribution function of the edge points.

New field points are created by generating a candidate for each active element in the current grid. Several possibilities are available. A technique previously used is to insert a point at the centroid of the active element. Another technique is to place a point at the circumcentre of the active element[7]. An alternative method[12] is that, for an active element, a point is placed along a line which is normal to an edge of the active element. This is very similar to the method used in the advancing front technique[1,2]. The advantage here is that the point is placed into an existing triangulation and thus search routines can be more efficient.

In step 7, the point distribution function at any point in the domain can be determined in a number of ways. The default is to linearly interpolate the point distribution from the boundary nodes for which equation(1) is applied. However, for greater flexibility it is possible to use the concept of the background mesh[14] or the use of sources[4,10,11,15]. The background mesh consists of a mesh which covers the domain and at whose nodes a given point distribution function is specified. At any given position in the domain the spacing can be derived from interpolation from the background mesh. A source is defined as a position with an attached point spacing and decay function. Sources can be defined pointwise or as lines, curves, planes or volumes.

In step 8, a new point created can be connected into an existing triangulation by using either the Delaunay triangulation algorithm of through local reconnection procedures. The method which utilises the Delaunay triangulation has been extensively discussed[3-8, 10-12]. Local reconnection schemes can also be used to both connect the point and to optimize the connectivities. A scheme which has been explored is based upon the edge swapping algorithm of Lawson[16]. In this approach, a point is inserted into the element which contains it and the grid is repetitively reconnected or swapped to locally satisfy a desired quality criterion. The process is repeated until no new reconnections are possible and terminates in a finite number of iterations. A limited subset of points are locally evaluated using Lawson's result[17] that there are no more than two possible connectivities of $n+2$ points in n dimensions. There are several quality criteria which can be used. For example, a min-max (minimize the maximum angle) criterion [18] has desirable characteristics for high-aspect ratio grids in that it maximizes the minimum element edge weight for a Laplacian as shown by Barth[19].

A detailed discussion[12] can be found on aspects of point insertion and reconnection in both 2 and 3 dimensions. It is clear from this work that high quality grids can be obtained by utilising iterative point insertion and different connectivity strategies. An example of a grid generated using point placement based upon advancing in a given direction from selected edges in the triangulation is shown in Figure 1a. This is compared with point placement based upon the positioning of points at centroids of active elements in Figure 1b. A comparison of element angle distributions for the two grids is also presented.

One of the significant advantages of the iterative point insertion procedures, in addition to the quality of the grids, is the speed of generation. The key to this is that when a point is derived the elements which will be modified in the insertion are either directly known without a search or can be located with a very short search procedure. Hence the grid generation procedures can be very fast. Table 1 gives typical times for grid generation.

Computer	CPU time in minutes
CRAY YMP	4.96
IBM Risc 6000 550	12.28
SGI PI	28.84
SGI Indigo XS 24/4000	14.75
SUN Sparc II	26.98

Table 1. Times for grid generation of 1 million tetrahedral elements

3 Parallel Unstructured Grid Generation

Although the times for the generation of grids, as shown in Table 1, are fast compared with most other unstructured grid generation times, it is likely in the near future that very large grids will be required. In our own work in computational electromagnetics grids exceeding 5 million elements are presently being used. There is, therefore, a continuing requirement to improve computational efficiency.

It is now generally acknowledged that to solve large scale computational field simulation problems it will be necessary to perform computations on parallel computer platforms. The solution of so-called 'Grand Challenge' problems, involving millions of elements within the grid, is the motivation behind the considerable efforts now being expended in parallel processing.

Many years ago, it was necessary to focus effort on modifications of algorithms to fully capitalise on vector computer architecture. Today, computational engineers must pay due attention to ensuring maximum benefit from parallel computer platform. With the advent of message passing library routines, such as PVM[20] and MP[21] the parallelization of codes is now practical. To this end, some initial work has begun on the parallelizing of unstructured grid generators and flow algorithms. The work presented here uses an 'easy parallelization' strategy, which although simple to implement yields significant improvements in computational performance.

4 Strategy for Parallel Unstructured Grid Generation

Techniques for the construction of structured grids for realistic applications are based upon the multiblock approach, where the domain is subdivided into regions which are topologically equivalent to squares or cuboids. Grids on the boundaries of these blocks are often generated prior to the construction of the field grid within the blocks. After generation, the boundary and field grid is globally smoothed to remove any discontinuities within the grid. In fact, the generation of boundary grids often constrains the grid and leads to grids of a higher quality than would be produced if all interior boundaries were free to evolve. The grid lines effectively act as grid control boundaries.

Unstructured grid algorithms do not usually use interior control boundaries since the generation mechanisms are very general and do not require the additional effort required in defining 'artificial' or 'false' boundaries. However, the strategy outlined for the generation of block structured grids is more amenable to 'easy parallelization' than the unstructured approaches. our strategy for the 'easy parallelization' of unstructured grid generations follows the 'false' boundary approach of the structured grid generators.

The strategy for building grids in parallel using the Delaunay or other point insertion methods previously outlined is straight forward.

5 Parallelization Strategy

1. Perform boundary curve discretization.

2. Use the Delaunay grid generator to connect the boundary points.

3. Perform the domain decomposition.

4. Gererate unstructured grids within each domain in parallel.

5. Assemble all the domain grids to form the global grid.

6. Perform a local smoothing on the interface boundaries.

Each gridded domain uses the sequential grid generation approach previously mentioned. This is significant in that a new grid generator does not need to be implemented, just a framework that enables the parallel decomposition and communication between the host and children processors. The major challenge is the requirement for a good domain decomposition routine which can ensure good grid quality and good load balancing.

6 Domain Decomposition and Load Balancing

The Delaunay method can be used to decompose a domain into a coarse triangulation using only boundary points. This initial grid then provides all of the information needed to establish domain boundaries.

Currently the decomposition is done by computing the area of the entire domain, then computing a fraction of the area depending upon the number of processors available for the parallel computation and estimates of the grid point density. The triangles from the initial grid are then summed to obtain a fraction of the total area. Additional points on 'false' boundaries can then be created consistent with point and the line sources or a background mesh. In this way, the complete boundary enclosing each subdomain is constructed and is then in a form appropriate for the construction of subdomain grids. The algorithm for domain decomposition is as follows:

1. Using the coarse triangulation of the boundary points compute the area of the domain.

2. Determine the area of each subdomain.

3. a. Identify an element not already assigned to a subdomain - create a new subdomain.
   b. Determine a direct neighbour element. Add the area of the element to the area of the subdomain.
       If the area for the subdomain equals or exceeds the processor area then;
           Is the number of subdomains equal to the required number?
               If yes then go to step 4.

If no then start a new subdomain by identifying an element which has not been attached to a subdomain and return to step 3b.

4. Check to ensure that all elements have been attached to a subdomain.
   If all have been included then go to step 5.
   If an element has been found which has not been identified then attach the element to a neighbour subdomain.

5. Domain decomposition complete.

The appropriate creation of boundary points on inter-subdomain boundaries is a key step in ensuring that the resulting global grid is of high quality. The following procedure is used in 2 dimensions.

1. Find the common edges between subdomains

2. Compute the points distribution function at the end points

3. Identify the required point spacing along the edge, taking into account information from the background grid and source data.

4. Generate the new grid points on the boundary edges.

The key to even load balancing is how to determine the effective 'size' of each subdomain to be gridded on the distributed computers. Ideally, the effective work associated with gridding each of the domains should be equal. Several approaches to the computation of the appropriate 'sizes' of the domains have been studied.

Equal Area Criterion
In this approach the size of each domain, A_i, is given by,

$$A_i = \frac{\text{Total Area}}{\text{No. of domains}}.$$

This approach is only likely to give good load balancing for a uniform mesh.

Equal Point Criterion
In this approach the size of each domain, A_i, is given by

$$A_i = \frac{\text{Estimate of total points}}{\text{No. of domains}}.$$

where the estimate of the number of points within a triangle is given by

$$\frac{4A}{\pi d_{p_{\text{average}}}}$$

Where A is the area of the triangle and $d_{p_{\text{average}}}$ average value of the required point spacing for the triangle. The estimate for the total points within the domain is the sum of the estimates for all the initial triangles. This is an approximate estimate for the number of points.

Figure 2 shows a grid generated on 16 processors around a multicomponent airfoil using the parallel strategy. Shown are the individual domain grids and the final global grid after inter-domain boundaries have been smoothed. An example of the performance achieved using this very simple strategy using PVM on a network of workstations connected with an ethernet link is shown in Table 2.

No. of processors	Factor
1	1
2	1.57
4	5.9
6	6.8
8	6.4

Table 2. Grid of 82,800 triangles

The ideas described above are presently being extended for applications in 3 dimensions.

7 Grids for Viscous Flows

Regular unit aspect ratio elements are optimal for inviscid flow simulations. Such elements are also probably optimal for viscous flow simulation, but the approach is deemed not to be valid, particularly for applications in 3 dimensions, since, in order to resolve boundary and shear layer with elements of unit aspect ratio, the number of grid points for a complex configuration would be impractically high. For turbulent flow over a 2-dimensional aerofoil, where the hight of the elements on the boundary is, for example, of the order of 10^{-4} of the chord, and it is estimated that about 20 'layers' of elements are required within the viscous dominated region, then approximately 450,000 points will be needed. In computational fluid dynamics such a grid, for a relatively simple configuration, is generally deemed to be excessive. If such an approach is applied for typical wing then grids of the order of 1,000 million grid points would be required - clearly well beyond computational capabilities either now or within the forseable future. It is such estimates of grid sizes which have motivated, if not forced, researchers to investigate techniques with inbuilt adaptation. i.e. methods which produce stretched elements with very high aspect ratios in boundary and shear layers. unfortunately, the advancing front and point insertion algorithms like the Delaunay triangulation are not well suited for the generation of very high aspect ratio elements either in 2 or 3 dimensions. Hence, in most attempts at the generation of unstructured viscous grids an 'a priori' adaptation strategy has been adopted and stretched elements generated which assume a particular dominant flow direction. Using this strategy, which, it should be noted, is also the philosophy adopted for the creation of structured viscous grids, it would appear possible to generate grids of practical size on which viscous flows can be computed using present day computational resources.

However, even with this assumption about the characteristic properties required of the grid, the simulation of viscous laminar and turbulent transonic flows is still a major computational challenge.

Given that high aspect ratio elements are required, the performance and accuracy of flow solvers on such elements is important. In sensitive regions of flow in, for example, a boundary layer there is some concern that irregularities in the grid give rise to inaccuracies in the flow simulation. For example, Figure 3a shows a variety of grid structures close to a solid wall. All the grids are formed from connecting, in different ways, the points from the structured quadrilateral grid. It is desirable that a flow algorithm can perform equally well on all these grids. If this were the case then this would provide useful information on the types of grids which should be used close to solid walls and for discretizing boundary layers.

The governing equations, taken to be the Reynolds-averaged Navier-Stokes equations for time-dependent two and three-dimensional compressible flows for a simple system in thermodynamic equilibrium in the absence of body forces, are discretized in space using a Galerkin weighted-residual approximation[22]. The discretization is vertex based and the solution vector is stored and solved for at the element vertices or nodes. Time discretization is obtained using an explicit Lax-Wendroff scheme. Boundary conditions are implemented using a method of characteristics. Artificial dissipation is used in regions containing severe flow gradients. For this solver an adaptive second-order dissipation model is used.

Figure 3b shows boundary layer profiles and skin friction coefficient for the compressible flow over a NACA0012 airfoil computed on the 3 grids shown in Figure 3a. It is clear that the differences between the results is very small. Similar results have been obtained for other transonic and supersonic viscous flows.

With some confidence that flow solvers can successfully handle features exhibited in the grids shown in Figure 3a, suitable strategies can be defined for the generation of grids for viscous flows.

Several methods which are capable of generating directly a mesh suitable for the analysis of viscous flows have been studied[23]. All these approaches are based upon a common philosophy, which involves the direct contribution of high aspect ratio elements within the region close to solid walls. The generic algorithm for this strategy is outlined below.

1. Place nodes along the solid walls, according to a distribution of mesh spacing specified via a mesh control function.

2. Propagate the boundary points into the field according to a prespecified criterion.

3. Construct a triangulation of these points, terminating the propagation of the points when unit aspect ratio elements are achieved.

4. Determine the outer loop of the propagated points.

5. Complete the triangulation of the remaining domain using an unstructured grid approach.

6. Create the final grid by combining the propagated triangulation with the regular unstructured grid.

Within this framework, a number of approaches have investigated as to how best to propagate the boundary points[23].

One such approach advances points along normal vectors from the solid boundaries.

1. Specify a distance for the distance of the first points from the boundary.

2. Form the normal to the boundary at each boundary point.

3. Moving in a given direction along the boundary, advance the boundary points along the normal vector, creating points and elements.

4. Before a point is created, ensure that overlap with the advancing layers does not occur. If overlap occurs, or, the point is within a prespecified distance of a layer, then the propagating normal is terminated.

5. When an element is created with an aspect ratio greater than or equal to unity, propagation along the normal is also terminated.

6. Return to step 3, until either the required number of layers has been created or all elements have reached an aspect ratio of unity.

A variation of this method can be implemented whereby the boundary normals are smoothed. In such cases, the normal can be smoothed using a Laplacian smoothing operator of the form.

$$v_p^{t+1} = (1-\omega) v_p^t + \frac{\omega}{N_p} \sum_{n=1}^{N_p} (v_n^t) \qquad (2)$$

where v_p is the boundary vector for point p, v_n is the nth neighbouring vector of p (in two dimensions $Np = 2$), ω is a relaxation parameter and t is an iteration level. The smoothing operator has the effect of removing large discontinuities in the normal vector directions between neighbouring nodes. To achieve orthogonality close to the solid boundary the relaxation parameter ω is set to a value given by

$$\omega = \min\left(1, \frac{n-1}{\alpha N}\right) \qquad (3)$$

where N is the final number of layers and α controls the rate of relaxation of the orthogonality constraint. If α is set to less than unity the orthogonality constraint will be fully relaxed before the final layer is reached and the converse applies if α is greater than unity. Typically α is set at 0.9.

In regions of smooth flow above walls it would seem to be a reasonable basis for grid elements to be highly stretched. however, there are regions around a geometry where the assumption of a dominant flow direction with which elements can be aligned is not always valid. For example, in the region of junctions between different surfaces it is not apparent, a priori, in which direction elements should be stretched. Furthermore, the construction of grid lines are approximately normal to surfaces is also problematic in such regions. In view of the uncertainty of the basic assumption of element construction it would appear that it may, in the longer term, be necessary to relax the condition of grid element stretching in such regions and resort to elements with an aspect ratio close to unity. The approach outlined can be modified to provide the flexibility to explore this approach in that trailing edges can be automatically identified and then the point spacing on the body reduced to give unit aspect ratio elements. An example of this approach is discussed later.

To demonstrate the flexibility and suitability of the viscous grid generation procedure proposed, the turbulent flow for a 2 airfoil high-lift configuration has been performed for the freestream flow of Mach number 0.25 and 1.0° of incidence at a Reynolds number of 5 million. Initially a grid was generated which did not attempt to resolve the wakes formed from the trailing edges of the airfoils. The grid, which is shown in Figure 4a, was generated using the advancing boundary normal approach and contains 25,400 elements and 12,750 points, of which 447 points define the boundary and 159 and 165 are on the main and component airfoils, respectively. The height of the initial spacing close to the wall was $.5 \times 10^{-5}$, with a grid point expansion ratio of 1.5. The outer boundary was 25 main airfoil chords away. The flow simulation was performed using the Lax-Wendroff solver with the Baldwin-Bath one-equation turbulence model[23]. Figure 4b shows the flowfield contours of Mach number.

Following this computation, the flexibility of the use of sources was applied to the generation of a grid with an adequate resolution of the wake. In practice, the unstructured grid generation is implemented within a graphics user interface which allows the user to set input parameters using push button menus and also to visualise, where appropriate, input data. Figure 5a shows the configuration with curves, defined interactively by the user, indicating the likely position of the wakes. Then line sources are defined by selecting points on the screen. In this case the formulation of the source requires a position and 2 specified radii. The grid density is constant within the inner radius and varies linearly from the inner circle to the boundary of the outer circle. Figure 5b shows what appears of the workstation. The effect of the line sources will affect the region of the grid shown by shading in Figure 5c. For each airfoil 5 line sources were positioned. The new grid contained 40,591 elements and 20,524 points and is shown in Figure 5d. the effect of the line sources is clearly visible. The results from the computation performed on this grid are shown in Figure 5e and the pressure coefficient on the surface is presented in Figure 5f. It is clear by comparing the flowfield contours the beneficial effect of introducing the line sources to increase the grid resolution in the wake region.

An example of the construction of a grid for a 3-dimensional configuration and the simulation of laminar viscous flow at a freestream Mach number of 0.5 and an angle of attack of 3 degrees with a Reynolds number of 1 million based upon chord length is shown in Figure 6. Ten layers of elements have been generated in the vicinity of the solid walls. 16,414 triangles defined the surface grid. An initial spacing of 0.0005 of the chord was used for the height of the elements close to the wall. The total number of elements generated within the boundary layers was 277,500 with 46,230 points. To provide adequate resolution of the wake planar sources were included in the generation of the inviscid part of the grid. The final grid contained 910,622 elements and 159,634 points. Details of cuts through the mesh parallel to the symmetry plane are shown in Figure 6a. It is evident that the generation of the layers automatically stops at the appropriate height. The grid and computed contours on the surface of the wing are shown in Figure 6d. Sections through the grid at a variety of spanwise stations are also shown together with the contours of pressure. The distributions of the pressure coefficient at 2 sections on the wing are given in Figure 6c.

Summary

This paper has highlighted some issues related to grid generation and viscous flow analysis. A general framework for iterative point insertion schemes is emerging and high quality grids, generated very efficiently can be obtained. These grid generators can be implemented within a parallel computer environment to achieve even faster computational speeds. Viscous flow simulation on unstructured grids requires new modifications to standard grid generation procedures. However, general procedures, applicable in both 2 and 3 dimensions, which inherit all the flexibility of the more traditional unstructured grid generators can be constructed.

References

1. J. Peraire, J. Peiro, L. Formaggia, K. Mrogan, and O.C. Zienkiewicz, 'Finite Element Euler Computations in three Dimensions', Int. J. Num. Meths. Eng. vol. 26, p2135-2159, 1988.

2. R. Lohner and P. Parikn, 'Three-Dimensional Grid Generation by the Advancing Front Method', Int. J. Num. Meth. Fluids, No. 8, ppl135-1149, 1988.

3. N.P. Weatherill and O. Hassan, 'Efficient Three-dimensional Delaunay triangulation with automatic point creation and imposed boundary constraints', Int. J. Num. Meths in Eng., Vol. 37, p2005-2039, 1994.

4. A. Jameson, T.J. Baker and N.P. Weatherill, 'Calculation of Inviscid Transonic Flow over a Complete Aircraft', AIAA Paper 86-0103, 1986.

5. T.J. Baker, 'Three-Dimensional Mesh Generation by Triangulation of Arbitrary Point Sets', Proceedings of the AIAA 8th CFD Conference, Hawaii, June 1987.

6. D.G. Holmes and S.H. Lamson, 'Adaptive triangular Meshes for Compressible Flow Solution', Numerical Grid Generation in Computational Fluid Dynamics, Pineridge Press 1986.

7. P.L. George, F. Hecht and E. Saltel, 'Fully automatic mesh generator for 3D domains of any shape', Impact of Computing in Science and Engineering, Vol. 2, No. 3, pp.187-218, 1990.

8. D.J. Mavriplis, 'An advancing front Delaunay triangulation algorithm designed for robustness', AIAA 31st Aerospace Sciences Meeting and Exhibit, AIAA Paper 93-0671, January 1993.

9. N.P. Weatherill, O. Hassan and D.L. Marcum, 'Calculation of Steady Compressible flowfields with the finite element method', AIAA 31st Aerospace Sciences Meeting and Exhibit, AIAA Paper 93-0341, 1993.

10. N.P. Weatherill, O. Hassan, M.J. Marchant and D.L. Marcum, 'Adaptive inviscid flow solutions for aerospace geometries on efficiently generated unstructured tetrahedral meshes', 11th AIAA Computational Fluid Dynamics Conf., AIAA Paper 93-3390, 1993.

11. D.L. Marcum and N.P. Weatherill, 'Unstructured grid generation using iterative point insertion and local reconnection', 12th AIAA Applied Aerodynamics Conf., AIAA Paper 94-1926, 1994.

12. S. Rebay, 'Efficient unstructured mesh generation by means of Delaunay triangulation and Bowyer-Watson algorithm', J. Comput. Phys., 106 (1993), No. 1, p125-138.

13. J. Peraire, M. Vahdati, K. Morgan and O.C. Zienkiewicz, 'Adaptive remeshing for compressible flow computations', J. Comp. Phys. 72,449-466, 1987.

14. S. Pirzadeh, 'Structured background grids for generation of unstructured grids by the advancing front', AIAA J. 31.2. 1993.

15. C.L. Lawson, 'Software for C_1 surface interpolation', Mathematical Software III. (Ed. J. R. Rice), Academic Press, New York, 1977.

16. C.L. Lawson, 'Properties of n-dimensional triangulation', CAGD, Vol. 3, April 1986, p231-246.

17. R.E. Barnhill and F.F. Little, 'Three- and four- dimensional surfaces', Rocky Mt. J. Math., 14, 77-102, 1994.

18. T.J. Barth, N.L. Wiltberger and A.S. Gandhi, 'Three-dimensional unstructured grid generation via incremental insertion and local optimization'. In Software Systems for Surface Modeling and Grid Generation, Langely Research Center, hampton, VA, april 1992, NASA Conf. Publication 3143.

19. A. Beguelin, G.A. Geist, W. Jiang, R. Manchek, K. Moore and V. Sunderam, 'The PVM project', Technical Report, Oak Ridge National Laboratory, Feb. 1993.

20. W. Gropp. E. Lusk, A. Skjellum, 'Using MPI: Portable parallel programming with the message passing interface'. Pub. MIT Press, Cambridge, Massachusetts, 1994.

21. D.L. Marcum and N.P. Weatherill, 'Finite Element Calculations of Inviscid and Viscous Flowfields about Launch Vehicle configurations'. Proceedings of the VIII international Conf. on Finite Elements in Fluids, Barcelona, September 1993.

22. O. Hassan, J. Probert, N.P. Weatherill, D.L. Marcum, M.J. Marchant and K. Morgan, 'The numerical simulation of viscous transonic flow using unstructured grids', AIAA Paper -94-2346, June, 1994.

23. B. Baldwin and T. Barth, 'A one-equation turbulence transport model for high Reynolds wall-bounded flows', NASA TM 102847 (1990).

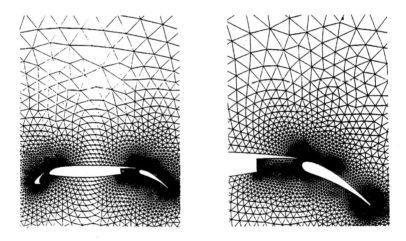

Figure 1a Multi-component airfoil grid with 19,301 elements generated by advancing point insertion.

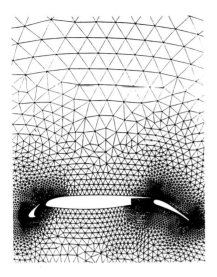

Figure 1b. Multi-component airfoil grid with 19,109 elements generated by centroid point insertion.

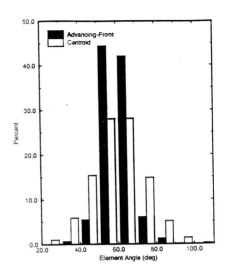

Figure 1c. Element angle distributions for the multi-component grids.

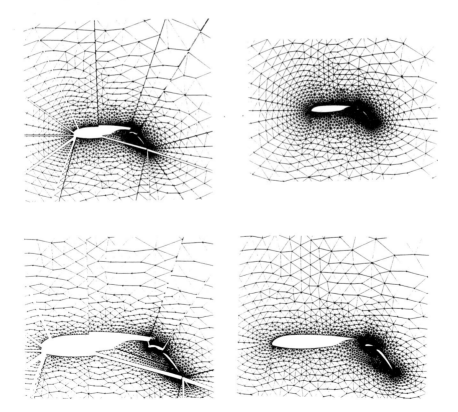

Figure 2. Multi-component airfoil grid generated using 16 processors. Shown are the domain grids and the global grid.

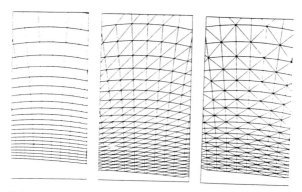

Figure 3a Boundary layer grids showing the different connectivities of points.

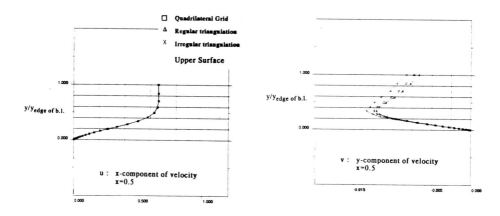

Figure 3b Boundary layer profiles for a NACA0012 airfoil at Mach number 0.5, zero incidence and Reynolds number of 5000.

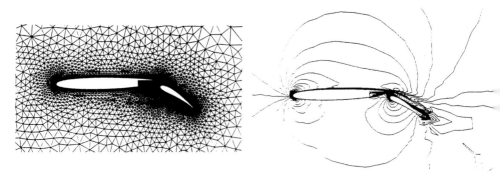

Figure 4a. Viscous grid   Figure 4b. Contours of Mach number.

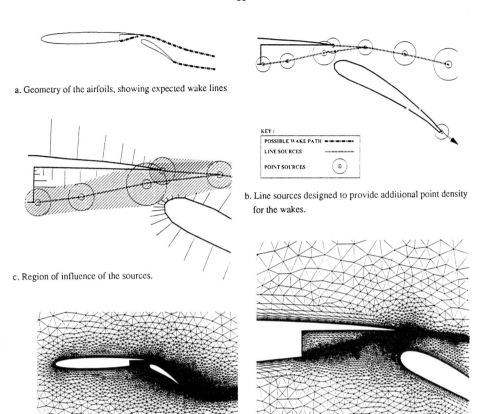

a. Geometry of the airfoils, showing expected wake lines

b. Line sources designed to provide additional point density for the wakes.

c. Region of influence of the sources.

d. The generated grid showing the effects of the line sources.

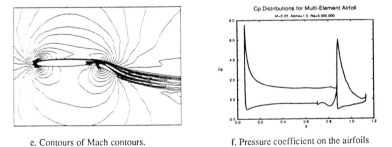

e. Contours of Mach contours.

f. Pressure coefficient on the airfoils

Figure 5

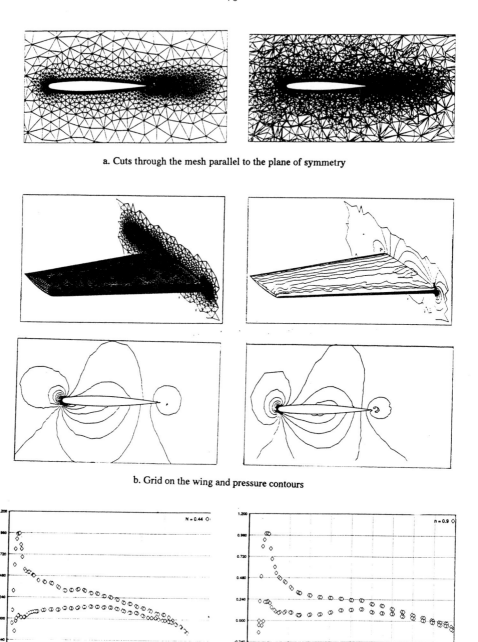

a. Cuts through the mesh parallel to the plane of symmetry

b. Grid on the wing and pressure contours

c. Pressure coefficient at 44% and 90% of the chord

Figure 6 ONERA M6 wing. Mach number 0.5, Angle of attack 3°, Reynolds No. 10^6

Computational Methods for Aerodynamic Design

Antony Jameson

Department of M.A.E, Princeton University
Princeton N.J. 08544, USA

1 Introduction : The Design as a Control Problem

The ultimate objective of the aerodynamic design is to optimize the geometric shape of a configuration taking into account the trade-offs between aerodynamic performance, structure weight, and the requirement for internal volume to contain fuel and payload. The subtlety and complexity of fluid flow is such that it is unlikely that repeated trials in an interactive analysis and design procedure can lead to a truly optimum design. Progress toward automatic design has been restricted by the extreme computing costs that might be incurred from brute force numerical optimization. However, useful design methods have been devised for various simplified cases, such as two-dimensional airfoils in viscous flows and wings in inviscid flows [11]. The computational costs for these methods result directly from the vast number of flow solutions that are required to obtain a converged design.

Alternatively, it has been recognized that the designer generally has an idea of the kind of pressure distribution that will lead to the desired performance. Thus, it is useful to consider the inverse problem of calculating the shape that will lead to a given pressure distribution. The method is advantageous, since only one flow solution is required to obtain the desired design. Unfortunately, a physical realizable shape may not necessarily exist, unless the pressure distribution satisfies certain constraints. Thus the problem must be very carefully formulated.

The problem of designing a two-dimensional profile to attain a desired pressure distribution was first studied by Lighthill, who solved it for the case of incompressible flow with a conformal mapping of the profile to a unit circle [7]. The speed over the profile is

$$q = \frac{1}{h}|\nabla \phi|,$$

Where ϕ is the potential which is known for incompressible flow and h is the modulus of the mapping function. The surface of h can be obtained by setting $q = q_d$, where q_d is the desired speed, and since the mapping function is analytic, it is uniquely determined by the value of h on the boundary. A solution exists for a given speed q_∞ at infinity only if

$$\frac{1}{2\phi}\oint q d\theta = q_\infty,$$

and there are additional constraints on q if the profile is required to be closed.

The differently that the objective may be unattainable can be circumvented by regarding the design problem as a control problem in which the control is the shape of the boundary. A variety of alternative formulations of the design problem can be then be treated systematically within the framework of the mathematical theory for control of systems governed by partial differential equations [8]. This approach of optimal aerodynamic design was introduced by Jameson [4,5], who examined the design problem for compressible flow with shock waves, and devised adjoint equations to determine the gradient for potential flow and also flows governed by the Euler equations. More recently Ta'asan, Kuruvila, and Salas implemented a one shot approach in which the constraint represented by the flow equations is only required to be satisfied by the final converged solution [14]. Pironneau has studied the use of control theory for optimum shape design of systems governed by elliptic equations [9], while adjoint methods have also been used by Baysal and Eleshaky [1].

Suppose that the control is defined by a function $\mathcal{F}(\xi)$ of some independent variable ξ or in the discrete case a vector with components $\mathcal{F}_i$. Also suppose that the desired objective is measured by a cost function I. This may, for example, measure the deviation from a desired surface pressure distribution, but it can also represent other measures of performance such as lift and drag. Thus the design problem is recast into a numerical optimization procedure. This has the advantage that if objective, say, of a target pressure distribution, is unattainable, it is still possible to find a minimum of cost function. Now a variation $\delta\mathcal{F}$ in the control produces a variation δI in the cost. Following control theory, δI can be expressed to first order as an inner product.

$$\delta I = (\mathcal{G}, \delta\mathcal{F}),$$

where the gradient $\mathcal{G}$ is independent of the particular variation $\delta\mathcal{F}$, and can be determined by solving an adjoint equation. For a discrete system of equations

$$(\mathcal{G}, \delta\mathcal{F}) \equiv \sum \mathcal{G}_i \delta \mathcal{F}_i$$

and for an infinite dimensional system

$$(\mathcal{G}, \delta\mathcal{F}) \equiv \int \mathcal{G}(\xi)\delta\mathcal{F} d\xi,$$

In either case, if one makes a shape change

$$\delta\mathcal{F} = -\lambda\mathcal{G}, \tag{1}$$

where λ is sufficiently small and positive, then

$$\delta I = -\lambda(\mathcal{G}, \mathcal{G}) < 0$$

assuring a reduction in I.

For flow about an airfoil or wing, the aerodynamic properties which define the cost function are functions of the flow-field variables (ω) and the physical location of the boundary, which may be represented by the function $\mathcal{F}$, say. Then

$$I = I(\omega, \mathcal{F}),$$

and a change in $\mathcal{F}$ results in a change

$$\delta I = \frac{\partial I^T}{\partial \omega}\delta\omega + \frac{\partial I^T}{\partial \mathcal{F}}\delta\mathcal{F}, \tag{2}$$

in the cost function. Brute force methods evaluate the gradient by making a small change in each design variable separately, and then recalculating both the grid and flow-field variables. This requires a number of additional flow calculations equal to the number of design variables. Using control theory, the governing equations of the flowfield are introduced as a constraint in such a way that the final expression for the gradient does not require revaluation of the flow field. In order to achieve this $\delta\omega$ must be eliminated from (2). The governing equation R expresses the dependence of ω and $\mathcal{F}$ within the flowfield domain D,

$$R(\omega, \mathcal{F}) = 0,$$

Thus $\delta\omega$ is determined from the equation

$$\delta R = \left[\frac{\partial R}{\partial \omega}\right]\delta\omega + \left[\frac{\partial R}{\partial \mathcal{F}}\right]\delta\mathcal{F} = 0 \tag{3}$$

Next, introducing a Lagrange Multiplier ψ, we have

$$\begin{aligned}\delta I &= \frac{\partial I^T}{\partial \omega}\delta\omega + \frac{\partial I^T}{\partial \mathcal{F}}\delta\mathcal{F} - \psi^T\left(\left[\frac{\partial R}{\partial \omega}\right]\delta\omega + \left[\frac{\partial R}{\partial \mathcal{F}}\right]\delta\mathcal{F}\right) \\ &= \left\{\frac{\partial I^T}{\partial \omega} - \psi^T\left[\frac{\partial R}{\partial \omega}\right]\right\}\delta\omega + \left\{\frac{\partial I^T}{\partial \mathcal{F}} - \psi^T\left[\frac{\partial R}{\partial \mathcal{F}}\right]\right\}\delta\mathcal{F}\end{aligned}$$

Choosing ψ to satisfy the adjoint equation

$$\left[\frac{\partial R}{\partial \omega}\right]^T \psi = \frac{\partial I}{\partial \omega} \tag{4}$$

the first term is eliminated, and we find that

$$\delta I = \mathcal{G}\delta\mathcal{F} \tag{5}$$

where

$$\mathcal{G} = \frac{\partial I^T}{\partial \mathcal{F}} - \psi^T\left[\frac{\partial R}{\partial \mathcal{F}}\right].$$

The advantage is that (5) is independent of $\delta\omega$, with the result that the gradient of I with respect to an arbitrary number of design variables can be determined without the need for additional flow-field evaluations. The main cost is in solving the adjoint equation (4). In general, the adjoint problem is about as complex as a flow solution. If the number of design variables is large, the cost differential between one adjoint solution and the large number of flowfield evaluations required to determine the gradient by brute force becomes compelling. Instead of introducing a Lagrange multiplier, ψ, one can solve (3) for δw as

$$\delta w = - \left[\frac{\partial R}{\partial w}\right]^{-1} \left[\frac{\partial R}{\partial \mathcal{F}}\right] \delta \mathcal{F},$$

and insert the result in (2). This is the implicit gradient approach, which is essentially equivalent to the control theory approach, as has been pointed out by Shubin and Frank [12,13]. In any event there is advantage in determining the gradient $\mathcal{G}$ by the solution of the adjoint equation.

After making such a modification, the gradient can be recalculated and the process repeated to follow a path of steepest descent (1) until a minimum is reached. In order to avoid violating constraints, such as a minimum acceptable wing thickness, the gradient may be projected into the allowable subspace within which the constraints are satisfied. In this way one can devise procedures which must necessarily converge at least to a local minimum, and which can be accelerated by the use of more sophisticated descent methods such as conjugate gradient or quasi-Newton algorithms. There is the possibility of more than one local minimum, but in any case the method will lead to an improvement over the original design. Furthermore, unlike the traditional inverse algorithms, any measure of performance can be used as the cost function.

In order to illustrate the application of control theory to aerodynamic design problems the next section presents the method for three-dimensional wing design using the inviscid Euler equations as the mathematical model for compressible flow.

2 Three Dimensional Design Using the Euler Equations

It proves convenient to denote the Cartesian coordinates and velocity components by $x1$, $x2$, $x3$ and $u1$, $u2$, $u3$, and to use the convention that summation over $i = 1$ to 3 is implied by a repeated index i. The three-dimensional Euler equations may be written as

$$\frac{\partial w}{\partial t} + \frac{\partial f_i}{\partial x_i} = 0 \text{ in } D, \tag{6}$$

where

$$w = \begin{Bmatrix} \rho \\ \rho u_1 \\ \rho u_2 \\ \rho u_3 \\ \rho E \end{Bmatrix}, \quad f_i = \begin{Bmatrix} \rho u_i \\ \rho u_i u_1 + p\delta_{i1} \\ \rho u_i u_2 + p\delta_{i2} \\ \rho u_i u_3 + p\delta_{i3} \\ \rho u_i H \end{Bmatrix} \tag{7}$$

and δ_{ij} is the Kronecker delta function. Also,

$$p = (\gamma - 1)\rho \left\{ E - \frac{1}{2}(u_i^2) \right\}, \tag{8}$$

and

$$\rho H = \rho E + p \tag{9}$$

where γ is the ratio of the specific heats. Consider a transformation to coordinates ξ_1, ξ_2, ξ_3 where

$$K_{ij} = \left[\frac{\partial x_i}{\partial \xi_i}\right], J = \det(K), K_{ij}^{-1} = \left[\frac{\partial \xi_i}{\partial x_j}\right].$$

Introduce contravariant velocity components as

$$\left\{\begin{array}{c} U_1 \\ U_2 \\ U_3 \end{array}\right\} = K^{-1}\left\{\begin{array}{c} u_1 \\ u_2 \\ u_3 \end{array}\right\}$$

The Euler equations can be written as

$$\frac{\partial W}{\partial t} + \frac{\partial F_i}{\partial \xi_i} = 0 \text{ in } D, \tag{10}$$

with

$$W = J\left\{\begin{array}{c} \rho \\ \rho u_1 \\ \rho u_2 \\ \rho u_3 \\ \rho E \end{array}\right\}, \quad F_i = J\left\{\begin{array}{c} \rho U_i \\ \rho U_i u_1 + \frac{\partial \xi_i}{\partial x_1}p \\ \rho U_i u_2 + \frac{\partial \xi_i}{\partial x_2}p \\ \rho U_i u_3 + \frac{\partial \xi_i}{\partial x_3}p \\ \rho U_i H \end{array}\right\} \tag{11}$$

Assume now that the new computational coordinate system conforms to the wing in such a way that the wing surface B_W is represented by $\xi_2 = 0$. Then the flow is determined as the steady solution of equation (10) subject to the flow tangency condition

$$U_2 = 0 \text{ on } B_W. \tag{12}$$

At the far field boundary B_F, conditions are specified for incoming waves, as in the two-dimensional case, while outgoing waves are determined by the solution.

Suppose now that it is desired to control the surface pressure by varying the wing shape. It is convenient to retain a fixed computational domain. Variations in the shape then result in corresponding variations in the mapping derivatives defined by H. Introduce the cost function

$$I = \frac{1}{2}\int\int_{B_W}(p - p_d)^2 d\xi_1\, d\xi_3,$$

where p_d is the desired pressure. The design problem is now treated as a control problem where the control function is the wing shape, which is to be chosen to minimize I subject to the constraints defined by the flow equations (10-11). A variation in the shape will cause a variation δp in the pressure and consequently the a variation in the cost function

$$\delta I = \int\int_{B_W}(p - p_d)\delta p\, d\xi_1\, d\xi_3. \tag{13}$$

Since p depends on w through the equation of state (8-9), the variation δp can be determined from the variation δw. Define the Jacobian matrices

$$A_i = \frac{\partial f_i}{\partial w}, \quad C_i = JK_{ij}^{-1}A_j. \tag{14}$$

Then the equation for $\delta\omega$ in the steady state becomes

$$\frac{\partial}{\partial \xi_i}(\delta F_i) = 0, \tag{15}$$

where

$$\delta F_i = C_i \delta\omega + \delta\left(J\frac{\partial \xi_i}{\partial x_j}\right) f_j.$$

Now, multiplying by a vector co-state variable ψ and integrating over the domain

$$\int_{D_j} \psi^T \left(\frac{\partial \delta F_i}{\partial \xi_i}\right) d\xi_j = 0,$$

and if ψ is differentiable this may be integrated by parts to give

$$\int_{D_j} \left(\frac{\partial \psi^T}{\partial \xi_i}\delta F_i\right) d\xi_j = \int_B (n_i \psi^T \delta F_i) d\xi_B,$$

where n_i are components of a unit vector normal to the boundary. Thus the variation in the cost function may now be written

$$\delta I = \int\int_{B_W} (p - p_d)\delta_p d\xi_1 d\xi_3 - \int_{D_j} \left(\frac{\partial \psi^T}{\partial \xi_i}\delta F_i\right) d\xi_j \\ + \int_B (n_i \psi^T \delta F_i) d\xi_B, \tag{16}$$

On the wing surface B_W, $n_1 = n_3 = 0$ and it follows from equation (12) that

$$\delta F_2 = J \left\{\begin{array}{c} 0 \\ \frac{\partial \xi_2}{\partial x_1}\delta p \\ \frac{\partial \xi_2}{\partial x_2}\delta p \\ \frac{\partial \xi_2}{\partial x_3}\delta p \\ 0 \end{array}\right\} + p \left\{\begin{array}{c} 0 \\ \delta\left(J\frac{\partial \xi_2}{\partial x_1}\right) \\ \delta\left(J\frac{\partial \xi_2}{\partial x_2}\right) \\ \delta\left(J\frac{\partial \xi_2}{\partial x_3}\right) \\ 0 \end{array}\right\} \tag{17}$$

Suppose now that ψ is the steady state solution of the adjoint equation

$$\frac{\partial \psi}{\partial t} - C_i^T \frac{\partial \psi}{\partial \xi_i} = 0 \text{ in } D. \tag{18}$$

At the outer boundary incoming characteristics for ψ correspond to outgoing characteristics for $\delta\omega$. Consequently, as in the two-dimensional case, one can choose boundary conditions for ψ such that

$$n_i \psi^T C_i \delta\omega = 0.$$

Then if the coordinate transformation is such that $\delta(JK^{-1})$ is negligible in the far field, the only remaining boundary term is

$$-\int\int_{B_W} \psi^T \delta F_2 \, d\xi_1 \, d\xi_3.$$

Thus by letting ψ satisfy the boundary conditions,

$$J\left(\psi_2 \frac{\partial \xi_2}{\partial x_1} + \psi_3 \frac{\partial \xi_2}{\partial x_2} + \psi_4 \frac{\partial \xi_2}{\partial x_3}\right) = (p - p_d) \text{ on } B_W, \quad (19)$$

We find finally that

$$\begin{aligned}\delta I &= \int_D \frac{\partial \psi^T}{\partial \xi_i} \delta\left(J \frac{\partial \xi_i}{\partial x_j}\right) f_j d\xi D \\ &- \int\int_{B_W} \left\{\psi_2 \frac{\partial \xi_2}{\partial x_1} + \psi_3 \frac{\partial \xi_2}{\partial x_2} + \psi_4 \frac{\partial \xi_2}{\partial x_3}\right\} = p \, d\xi_1 \, d\xi_3.\end{aligned} \quad (20)$$

Fig. 1.a. x, y-Plane. **Fig. 1.b.** ξ, η-Plane.

A convenient way to treat a wing is to introduce sheared parabolic coordinates as shown in figure 1 through the transformation

$$x = x_0(\zeta) + \tfrac{1}{2} a(\zeta) \left\{\xi^2 - (\eta + S(\xi,\zeta))^2\right\}$$

$$y = y_0(\zeta) + a(\zeta)\xi(\eta + S(\xi,\zeta))$$

$$z = \zeta.$$

Here $x = x_1, y = x_2, z = x_3$ are Cartesian coordinates, and ψ and $\eta + S$ correspond to parabolic coordinates generated by the mapping

$$x + iy = x_0 + iy_0 + \frac{1}{2} a(\zeta) \left\{\xi + i(\eta + S)\right\}^2$$

at a fixed span station ζ. $x_0(\zeta)$ and $y_0(\zeta)$ are the coordinates of a singular line which is swept to lie just inside the leading edge of a swept wing, while $a(\zeta)$ is a scale factor to allow for spanwise chord variations. The surface $\eta = 0$ is a shallow bump corresponding to the wing surface, with a height $S(\xi,\zeta)$ determined by the equation

$$\xi + iS = \sqrt{2(x_{B_W} + iy_{B_W})},$$

where $x_{Bw}(z)$ and $y_{Bw}(z)$ are coordinates of points lying on the wing surface. We now treat $S(\xi,\zeta)$ as the control.

In this case the transformation matrix $\dfrac{\partial x_i}{\partial \xi_j}$ becomes

$$K = \begin{bmatrix} a(\xi - (\eta + S)S_\xi) & -a(\eta + S) & \mathcal{A} - a(\eta + S)S_\zeta \\ a(\eta + S + \xi S_\xi) & a\xi & \mathcal{B} + a\xi S_\zeta \\ 0 & 0 & 1 \end{bmatrix}$$

$$= \begin{bmatrix} x_\xi & x_\eta & \mathcal{A} + x_\eta S_\zeta \\ y_\xi & y_\eta & \mathcal{B} + y_\eta S_\zeta \\ 0 & 0 & 1 \end{bmatrix}, \text{ where}$$

$$\mathcal{A} = a_\zeta \frac{x - x_0}{a} + x_{0_\zeta}, \qquad \mathcal{B} = a_\zeta \frac{y - y_0}{a} + y_{0_\zeta}.$$

Now,
$$J = x_\xi y_\eta - x_\eta y_\xi = \xi^2 + (\eta + S)^2$$

and

$$JK^{-1} = \begin{bmatrix} y_\eta & -x_\eta & x_\eta \mathcal{B} - y_\eta \mathcal{A} \\ -y_\xi & x_\xi & y_\xi \mathcal{A} - x_\xi \mathcal{B} - JS_\zeta \\ 0 & 0 & J \end{bmatrix}.$$

Then under a modification δS

$$\begin{aligned} \delta x_\xi &= -a(\delta S S_\xi + (\eta + S)\delta S_\xi) \\ \delta x_\eta &= -a\delta S \\ \delta y_\xi &= a(\delta S + \xi \delta S_\xi) \\ \delta y_\eta &= 0. \end{aligned}$$

Thus
$$\delta J = 2a^2(\eta + S)\delta S$$

and

$$\delta J K^{-1} = \begin{bmatrix} 0 & a\delta S & -a\mathcal{B}\delta S \\ -\delta y_\xi & \delta x_\xi & \mathcal{D} \\ 0 & 0 & \delta J \end{bmatrix}$$

where

$$\mathcal{D} = \delta y_\xi \mathcal{A} - \delta x_\xi \mathcal{B} - a\zeta \frac{J}{a} \delta S - \delta J S_\zeta - J \delta S_z.$$

Inserting these formulas in equation (20) we find that the volume integral in δI is

$$\iiint \frac{\partial \psi^T}{\partial \xi} \delta S f_2 d\xi d\eta d\zeta$$

$$- \iiint \frac{\partial \psi^T}{\partial \eta} \{-\delta y_\xi f_1 + \delta x_\xi f_2 + \mathcal{D} f_3\} d\xi d\eta d\zeta$$

$$+ \iiint \frac{\partial \psi^T}{\partial \zeta} \delta J f_3 d\xi d\eta d\zeta$$

where S and δS are independent of η. Therefore, integrating over η, the variation in the cost function can be reduced to a surface of the form

$$\delta I = \int\int_{B_W} (P(\xi,\zeta)\delta S - Q(\xi,\zeta)\delta S_\xi - R(\xi,\zeta)\delta S_\zeta)d\xi\, d\zeta$$

Here
$$\begin{aligned}
P &= a(\psi_2 + S_\xi\psi_3 + C\psi_4)p \\
&- \int \frac{\partial \psi^T}{\partial \xi}\{\xi f_1 + (\eta+S)f_2 + (\xi\mathcal{A}+(\eta+S)\mathcal{B}f_3\}\,d\eta \\
&- \int \frac{\partial \psi^T}{\partial \eta}f_1 + S\xi f_2 + Cf_3)d\eta \\
&- \int \frac{\partial \psi^T}{\partial \zeta}J\,d\eta
\end{aligned}$$

$$\begin{aligned}
Q &= a(\xi\psi_2 + (\eta+S)\psi_3)p \\
&+ \int \frac{\partial \psi^T}{\partial \eta}\{\xi f_1 + (\eta+S)f_2 + (\xi\mathcal{A}+(\eta+S)\mathcal{B}f_3\}\,d\eta
\end{aligned}$$

$$\begin{aligned}
R &= H\psi_4 p \\
&+ \int \frac{\partial f_3}{\partial \eta}J\psi_4 d\eta,
\end{aligned}$$

Where
$$C = 2a(\eta+S)S_\zeta - \mathcal{A} - \mathcal{B}S_\xi + \frac{J}{a}.$$

Also shape change will be confined to a boundary region of the $\xi - \zeta$ plane, so we can integrate by parts to obtain

$$\delta I = \int\int_{B_W} \left(P + \frac{\partial Q}{\partial \xi} + \frac{\partial R}{\partial \zeta}\right)\delta S d\xi d\zeta.$$

Thus to reduce I we can choose

$$\delta S = -\lambda\left(P + \frac{\partial Q}{\partial \xi} + \frac{\partial R}{\partial \zeta}\right),$$

where λ is sufficiently small and non-negative.

In order to impose a thickness constraint we can define a baseline surface $S_0(\xi,\zeta)$ below which $S(\xi,\zeta)$ is not allowed to fall. Now if we take $\lambda = \lambda(\xi,\zeta)$ as a non-negative function such that

$$S(\xi,\zeta) + \delta S(\xi,\zeta) \geq S_0(\xi,\zeta).$$

Then the constraint is satisfied, while

$$\delta I = -\int\int_{B_W} \lambda \left(P + \frac{\partial Q}{\partial \xi} + \frac{\partial R}{\partial \zeta}\right)^2 d\xi d\zeta \leq 0.$$

3 Implementation for Swept Wings

Since three dimensional calculations require substantial computational resources, it is extremely important for the practical implementation of the method to use fast solution algorithms for the flow and the adjoint equations. In this case the author's FLO87 computer program has been used as the basis of the design method. FLO87 solves the three dimensional Euler equations with a cell-centered finite volume scheme, and uses residual averaging and multigrid acceleration to obtain very rapid steady state solutions, usually in 25 to 50 multigrid cycles [2.3]. Upwind biasing is used to produce nonoscillatory solutions, and assure the clean capture of shock waves. This is introduced through the addition of carefully controlled numerical diffusion terms, with a magnitude of order Δx^3 in smooth parts of the flow. The adjoint equations are treated in the same way as the flow equations. The fluxes are first estimated by central differences, and then modified by downwind biasing through numerical diffusive terms which are supplied by the same subroutines that were used for the flow equations.

The method has been tested for the optimization of a swept wing. The wing planform was fixed while the sections were free to be changed arbitrarily by the design method, with a restriction on the minimum thickness. The wing has a unit-semi-span, with 38 degrees leading edge sweep. It has a modified trapezoidal planform, with straight taper from a root chord of 0.38, and a curved trailing edge in the inboard region blending into straight taper outboard of the 30 percent span station to a chord of 0.10, with an aspect ratio of 9.0. The initial wing sections were based on a section specifically designed by the author's two dimensional design method [4] to give shock free flow at Mach 0.78 with a lift coefficient of 0.6. The pressure distribution is displayed in figure 2. This section, which has a thickness to chord ratio of 9.5 percent, was used at the tip. Similar sections with an increased thickness were used inboard. The variation of thickness was non-linear with a more rapid increase near root, where the thickness to chord ratio of the basic section was multiplied by a factor of 1.47. The inboard sections were rotated upwards to give the initial wing 3.0 degrees twist from root to tip.

The two dimensional pressure distribution of the wing section at its design point was introduced as a target pressure distribution uniformly across the span. The target is presumably not realizable, but serves to favor the establishment of relatively benign pressure distribution. The total inviscid drag coefficient, due to the combination of vortex and shock wave drag, was also included in the cost function. Calculations were performed with the lift coefficient forced to approach a fixed value by adjusting the angle of attack every fifth iteration of the flow solution. It was found that the computational costs can be reduced by using only 15 multigrid cycles in each flow solution, and in each adjoint solution. Although this is not enough for full convergence, it proves sufficient to provide a shape modification which leads to an improvement. Figures 3 and 4 show the result of a calculation at Mach number of 0.85, with the lift coefficient forced to approach a value of 0.5. This calculation was performed on a mesh with 192 intervals in the ζ direction wrapping around the wing, 32 intervals in the normal η direction and 48 intervals in the spanwise ζ direction, giving a total of 294912 cells. The wing was specified by 33 sections, each with 128 points, giving a total of 4224 design variables. The plot show the initial wing geometry and pressure distribution, and the modified geometry and pressure distribution after 10 design cycles. The total inviscid drag was reduced from 0.0209 to 0.0119. The initial design exhibits a very strong shock wave in the inboard region. It can be seen that this is completely eliminated, leaving a very weak shock wave in the inboard region. The drag reduction is mainly accomplished in the first four design cycles but the pressure distribution continues to be adjusted to become more like the target pressure distribution.

To verify the solution, the final geometry, after 10 design cycles, was analyzed with another method using the computer program FLO67. This program uses a cell-vertex formulation, and has recently been modified to incorporate a local extremum diminishing algorithm with a very low level of numerical diffusion [6]. When run to full convergence it was found that the redesigned wing has a drag coefficient of 0.0096 at Mach 0.85 at a lift coefficient of 0.5, with a corresponding lift to drag ratio of 52. The result for $\alpha = 0.0°$ and $C_L = 0.505$ is illustrated in Figure 5: this seems to be the nearest to a shock free condition. A calculation at Mach 0.50 shows a drag coefficient of 0.0089 for a lift coefficient of 0.5. Since in this case the flow is entirely subsonic, this provides an estimation of the vortex drag for this planform and lift distribution, which is just what one obtains from the standard formula for induced drag, $C_D = C_L^2/\epsilon \phi AR$, with an aspect ratio $AR = 9$, and an efficiency factor $\epsilon = 0.97$. Thus the design method has reduced the shock wave drag coefficient to about 0.0007 at a lift coefficient of 0.5. For a representative transport aircraft the parasite drag coefficient of the wing due to skin friction is about 0.0045. Also the fuselage drag coefficient is about 0.0050, the nacelle drag coefficient is about 0.0015, the empennage drag coefficient is about 0.0020, and excrescence drag coefficient is about 0.0010. This would give a total drag coefficient $C_D = 0.0236$ for a lift coefficient of 0.5, corresponding to a lift to drag ratio $L/D = 21$. This would be a substantial improvement over the values obtained by currently flying transport aircraft.

4 Conclusion

In the period since this approach to optimal shape design was first proposed by the author [4], the method has been verified by numerical implementation for both potential flow and flows modeled by the Euler equations. It has been demonstrated that it can be successfully used with a finite volume formulation to perform calculations with arbitrary numerically generated grids [10]. The first results which have been obtained for swept wings with the three dimensional Euler equations suggest that the method has now matured to the point where it can be a very useful tool for the design of new airplanes. Even in the case of three dimensional flows, the computational requirements are so moderate that the calculations can be performed with workstations such as the IBM RISC 6000 series. A design cycle on a $192 \times 32 \times 48$ mesh takes about $1\frac{1}{2}$ hours on an IBM model 530 workstation, allowing overnight completion of a design calculation for a swept wing.

Acknowledgments :

The author is grateful to James Reuther for his assistance in assembling the text with TEX. This research has benefited greatly from the generous support of the AFOSR under grant number AFOSR-91-0391, ARPA under grant number N00014-92-J-1976, USRA through RIACS, and IBM. The warm hospitality of the Aeronautics and Astronautics Department of Stanford University, and NASA Ames Research Center, provided a very favorable environment for the pursuit of this research while the author was on leave from Princeton University.

References

1. O. Baysal and M.E. Eleshaky. Aerodynamic design optimization using sensitivity analysis and computational fluid dynamics. *AIAA paper 91-0471*, 29th Aerospace Sciences Meeting, Reno, Nevada, January 1991.

2. A. Jameson. Solution of the Euler equations by a multigrid method. *Applied Mathematics and Computations*, 13:327-356, 1983.

3. A. Jameson. Multigrid algorithms for compressible flow calculations. In W. Hackbusch and U. Trottenberg, editor, *Lecture Notes in Mathematics, Vol. 1228*, pages 166-201. Proceedings of the 2nd European Conference on Multigrid Methods, Cologne, 1985, Springer-Verlag, 1986.

4. A. Jameson. Aerodynamic design via control theory. *Journal of Scientific Computing*, 3:233-260, 1988.

5. A. Jameson. Automatic design of transonic airfoils to reduce the shock induced pressure drag. In *Proceedings of the 31st Israel Annual Conference on Aviation and Aeronautics, Tel Aviv*, pages 5-17, February 1990.

6. A. Jameson. Artificial diffusion, upwind biasing, limiters and their effect on accuracy and multigrid convergence in transonic and hypersonic flows. *AIAA paper 93-3359*, AIAA 11th Computational Fluid Dynamic Conference, Orlando, Florida, July 1993.

7. M.J. Lighthill. A new method of two-dimensional aerodynamic design. R & M 1111, Aeronautical Research Council, 1945

8. J.L. Lions. *Optimal Control of systems Governed by Partial Difference Equations*. Springer-Verlag, New York, 1971. translated by S.K. Mitter.

9. J. O. Pironneau. *Optimal Shape Design for Elliptic Systems.* Springer-Verlag, New York, 1984.

10. J. Reuther and A. Jameson. Control theory based airfoil design for potential flow and finite volume discretization. *AIAA paper 91-499*, 32th Aerospace Sciences Meeting and Exibit, reno, Nevada, January 1994.

11. J. Reuther, C.P. van Dam, and R. Hicks. Subsonic and transonic low-Reynolds-number airfoils with reduced pitching moments. *Journal of Aircraft*, 29:297-298, 1992.

12. G.R. Shubin. Obtaining cheap optimization gradients from computational aerodynamics codes. *Internal paper AMS-TR-164*, Boeing Computer Services, June 1991.

13. G.R. Shubin and P.D. Frank. A comparison of the implicit gradient approach and the variational approach to aerodynamic design optimization. *internal paper AMS-TR-164*, Boeing Computer Services, April 1991.

14. S. Ta'asan, G. Kuruvila, and M.D. Salas. Aerodynamic design and optimization in one shot. *AIAA paper 91-005*, 30th Aerospace Meeting and Exibit, Reno, Nevada, January 1992.

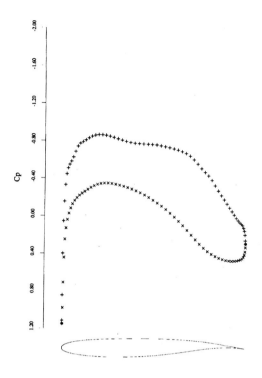

Fig. 2. Initial Wing Section and Target Pressure Distribution

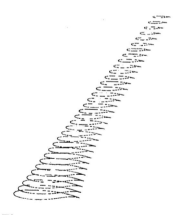

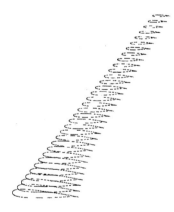

Fig. 3.a. Lifting Design Case,
$M = 0.85$, Fixed Lift Mode.
Initial Wing
$C_l = 0.5000$, $C_d = 0.0209$, $\alpha = -1.349°$

Fig. 3.b. Lifting Design Case,
$M = 0.85$, Fixed Lift Mode.
10 Design Iterations
$C_l = 0.5000$, $C_d = 0.0119$, $\alpha = 0.033°$

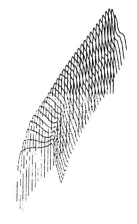

UPPER SURFACE PRESSURE

UPPER SURFACE PRESSURE

Fig. 4.a. Lifting Design Case,
$M = 0.85$, Fixed Lift Mode.
$C_L = 0.5000, C_D = 0.0209, \alpha = -1.349°$
Drag Reduction

Fig. 4.b. Lifting Design Case,
$M = 0.85$, Fixed Lift Mode.
$C_L = 0.5000, C_D = 0.0119, \alpha = 0.033°$
Drag Reduction

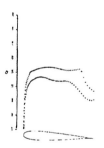

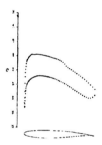

Fig. 5.a. FLO67 check on redesigned wing,
$M = 0.85, C_L = 0.5051, C_D = 0.0099$
$\alpha = 0.0°$
Span station $z = 0.00$

Fig. 5.b. FLO67 check on redesigned wing.
$M = 0.85, C_L = 0.5051, C_D = 0.0099$
$\alpha = 0.0°$
Span station $z = 0.312$

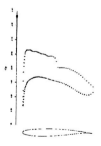

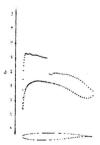

Fig. 5.c. FLO67 check on redesigned wing.
$M = 0.85, C_L = 0.5051, C_D = 0.0099$
$\alpha = 0.0°$
Span station $z = 0.625$

Fig. 5.d. FLO67 check on redesigned wing.
$M = 0.85, C_L = 0.5051, C_D = 0.0099$
$\alpha = 0.0°$
Span station $z = 0.937$

A Kinetic Flux-splitting Scheme for the MHD Equations

Jean-Pierre Croisille[1], Rabia Khanfir[2], Gérard Chanteur[2]

[1]Laboratoire d'Analyse Numérique d'Orsay, Bât. 425, Université Paris Sud, 91405 Orsay Cedex, France

[2]CETP/CNRS/UVSQ, 10-12 av. de l'Europe, 78140 Vélizy Cedex, France

1 Introduction

The numerical approximation of the conservative system of the MHD has received since a few years an increasing interest, especially for the study of cosmic plasmas. Since this system appears as hyperbolic with many features of the Euler equations, it is natural to try numerical methods which have been applied with success to these equations. Several works have been recently devoted to the construction of Approximate Riemann Solvers for the MHD [1],[2]. The aim is here to define a simple flux-splitting formula with in mind the well-known strategy of the kinetic-scheme for the Euler equations.

Recall that this procedure requires two ingredients, an equilibrium repartition function $\bar{f}(w,\xi)$ and a kinetic equation. The basic time-scheme reads [3],[4]

1. **Relaxation step.**

    If $w^n(x)$ is given at time t^n, construct $\bar{f}(w^n(x),\xi)$.

2. **Transport step.**

    Evolve $f(x,\xi,t)$ by the free Boltzmann equation $\partial_t f + \xi \cdot \nabla_x f = 0$, and define the average value $w^{n+1}(x) = \int_\xi f(x,\xi,t^{n+1})K(\xi)$ where $K(\xi) = (1,\xi,|\xi|^2/2)^t$.

If $w^n(x)$ is constant in each cell M_j, we get the classical first order finite-volume scheme, by averaging onto M_j, $t \in [0,\Delta t]$, and $\xi \in \mathbb{R}$.

$$w_j^{n+1} = w_j^n - (\Delta t/\Delta x)(F_{nu,j+1/2}^n - F_{nu,j-1/2}^n) \qquad (1)$$

The numerical flux formula is given in flux-splitting form as

$$F_{nu}(w_1,w_2) = \underbrace{\int_{\xi>0} \xi \bar{f}(w_1,\xi)K(\xi)}_{F^+(w_1)} + \underbrace{\int_{\xi<0} \xi \bar{f}(w_2,\xi)K(\xi)}_{F^-(w_2)} \qquad (2)$$

2 Flux-splitting for the MHD equations

The MHD equations in conservative form are

$$(MHD) \begin{cases} \partial_t \rho + \nabla \cdot \rho u = 0, & \text{mass} \\ \partial_t \rho u + \nabla \cdot (\rho u \otimes u + \bar{p}I - B \otimes B) = 0 & \text{momentum} \\ \partial_t \rho \bar{e} + \nabla \cdot ((\rho \bar{e} + \bar{p})u - (B \cdot u)B) = 0 & \text{total energy} \\ \partial_t B + \nabla \cdot (B \otimes u - u \otimes B) = 0 & \text{Faraday} \end{cases}$$

$\bar{e} = \varepsilon + \frac{1}{2}|u|^2 + |B|^2/2\rho$ is the total energy, ε the internal energy, B the magnetic field, $\bar{p} = p(\rho, \varepsilon) + |B|^2/2$ is the sum of the fluid and magnetic pressure. Moreover, the involutive constraint $\nabla \cdot B = 0$ has to be added to (MHD). In the sequel we denote the mass, momentum and energy equations by (E) and the Faraday equation by (F). The vector of the conservative variables is $w = (w_E, B)$ and the total flux is $F(w)$.

We have not found a simple way to write globally (MHD) as the kinetic average of one single kinetic equation. So we write a flux-splitting formula in two steps.

2.1 Flux-splitting for (E)

We start with the following property of the well-known half-Maxwellian flux-splitting formula for the Euler equations introduced in [6],[3] in the case of a polytropic gas. If $p = \rho T$ is any perfect gas pressure law with caloric law $T = f(\varepsilon)$ where f is an increasing function, define the thermal velocity by $C = (2T)^{1/2}$ and, for any unitary vector n, pose $z_n = -u.n/C$. Then a remarkable form of $F^+(w, n)$ in (2) is

$$F^+(w, n) = (u^+.n)w + (0, p^+, \frac{1}{2}p^+(u.n) + \frac{1}{2}p(u^+.n))^t \qquad (3)$$

where $u^+ = A(z_n)u + CB(z_n)n$ and $p^+ = A(z_n)p$. The functions A and B are $A(z) = erfc(z)/2$ and $B(z) = exp(-z^2)/2\pi^{\frac{1}{2}}$. With this form, the half-Maxwellian flux-splitting formula appears very convenient and general. It clearly shows the upwind effect resulting of the kinetic scheme: it is encompassed at the macroscopic level in the upwind velocity u^+ and pressure p^+. Note that the thermal speed is the sound speed only if $p = (\gamma - 1)\rho\varepsilon$ with $\gamma = 2$.

Concerning the part (E) of (MHD), we generalize (3) in the following way. Noting $\bar{T} = \bar{p}/\rho$, $\bar{C} = (2\bar{T})^{1/2}$ and $z_n = -u.n/\bar{C}$ we define

$$F_E^+(w, n) = (u^+.n)w_E + (0, P^+.n, \frac{1}{2}u.P^+.n + \frac{1}{2}u^+.P.n)^t \qquad (4)$$

where $P = \bar{p}I - B \otimes B$ and $P^+ = A(z_n)P$. Note that the magnetic field appears in (4) as a frozen variable.

Finally, we emphazise that (4) is only a formal extension of (3). It has no longer a kinetic interpretation.

2.2 Flux-splitting for (F)

In order to introduce a kinetic interpretation of the advection of the magnetic field, we introduce the following kinetic equation

$$\frac{\partial f_B}{\partial t} + \nabla . (f_B \otimes \xi - \xi \otimes f_B) = 0 \tag{5}$$

where $f_B(x, \xi, t) \in \mathbb{R}^3$ is a vectorial velocities distribution function. Although this equation has to our knowledge no physical meaning, $f_B(x, \xi)$ can be seen as a density of magnetic field convected by the speed ξ at the place x. The solution of (5) with initial condition $f^0(x, \xi)$ is

$$f_B(x, \xi, t) = f_B^0(x, \xi) + \int_0^t \nabla \wedge (\xi \wedge f_B^0(x - \xi s, \xi)) ds \tag{6}$$

Averaging on ξ and x, we get the following finite volume scheme for B

$$B_\omega^{n+1} = B_\omega^n - \frac{\Delta t}{a(\omega)} \sum_k l(A_k)(F_B^+(w_\omega^n, n_k) + F_B^-(w_k^n, n_k)) \tag{7}$$

with ω a cell, $l(A_k)$ the length of the edge A_k, n_k the outgoing normal to A_k, and the positive part of the flux is (see [5] for details)

$$F^+(w, n) = -n \wedge (u^+ \wedge B) \tag{8}$$

Finally the global basic scheme stands in form (1) where the positive part of the flux is

$$F^+(w, n) = (u^+ . n)w + (0, P^+ . n, \frac{1}{2} u . P^+ . n + \frac{1}{2} u^+ . P . n, -u^+(B . n)) \tag{9}$$

3 Numerical results

The preceding scheme is extended to second order in a classical way using a second-order space interpolation and a second-order Runge-Kutta scheme in time, [5]. The interpolation of the gradients is done by a local least-square procedure [7], and the limitation of the slopes is given by a min-mod limitation carried out on each edge of the cell.

3.1 1-D test case

The test case of Brio and Wu is an MHD coplanar Riemann problem with initial conditions: $\rho_l = 1., u_l = 0., p_l = 1., B_{x,l} = 0.75, B_{y,l} = 1.$ for the left state and $\rho_r = 0.125, u_r = 0, p_r = 0.1, B_{x,r} = 0.75, B_{y,r} = -1.$ for the right state. The adiabatic index is $\gamma = 2$. There are 800 cells of length $\Delta x = 1$. The Courant number is 0.4 and the results are displayed at time $T = 80$. We observe on the density the following waves propagating to the left: a fast rarefaction wave, a compound wave propagating to the left and made up by an intermediate shock attached to a slow rarefaction, [1]. The intermediate shock is characterized by changing of sign of B_y. We observe next a contact discontinuity, a slow shock wave and finally a fast rarefaction wave propagating to the right.

3.2 2-D test cases

We show some two-dimensional wave propagation tests on a square domain with biperiodic boundary conditions. We test the ability of the code to accurately propagate the three MHD modes on a triangular unstructured grid.(3700 meshes, CFL=0.4). The two frames of Fig. 2 display the density and the magnetic field of a finite amplitude plane fast wave. The uniform background magnetic field makes a 45° angle with the wave vector. The projections of the profiles are shown at $T = 0$ and after two theoretical periods. The profile is made of a cloud of points. The coordinates of each point are the abscissa of the projection of the center of the cell on the direction of propagation and the value of the physical field. Notice first the accuracy of the phase of the computed wave and the nonlinear steepening of the profile. Fig.3 displays the magnetic field of a small amplitude Alfven wave after 2 theoretical periods. Note that there is no nonlinear steepening according to the theory. The history of B_y is displayed on the right panel of Fig.3 and shows an exponential damping which allows to estimate the value of a numerical diffusion coefficient (4.10^{-3} in this case).

4 Conclusion

The aim of this work is to provide a very simple scheme with hopefully the good robustness properties of the finite-volume kinetic schemes of the Euler equations. We have emphazised on wave propagation test problems in order to precisely estimate the dissipation of the scheme. Other problems involving Alfvenic exact solutions are currently tested. It would be interesting to construct higher accurate versions of this scheme preserving its stability properties.

References

[1] M. Brio, C.C. Wu, *J. Comp. Phys.*, 75, 400-422, 1988.

[2] A.L. Zachary, P. Colella, *J. Comp. Phys.* ,99, 341-347, 1992.

[3] S.M. Deshpande, AIAA - 86 - 0275.

[4] B. Perthame, *SIAM J. Num. Anal.*, 29, 1, 1-19, 1992.

[5] J.P. Croisille, R. Khanfir, G. Chanteur, *J. of Scient. Comp.*, to appear.

[6] D.I. Pullin, *Jour. of Comp. Phys.*, 34, 231-244, 1980.

[7] R. Khanfir, *Thesis*, to appear.

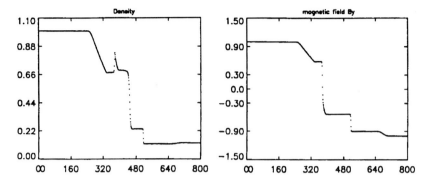

Fig1. Brio & Wu test case at time T = 80.

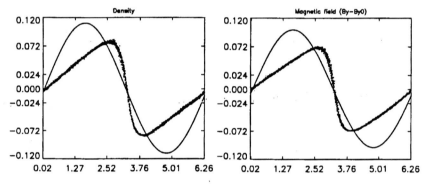

Fig2. MHD Fast mode : solution at t=0 and after two periods

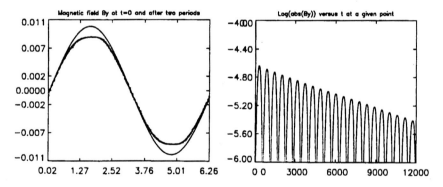

Fig3. MHD Alfven mode

A Boltzmann Taylor Galerkin FEM for Compressible Euler Equations

Sanjay S. Deshpande

C.F.D. Laboratory, Department of Aerospace Engineering,
Indian Institute of Science, Bangalore, India

Abstract : A new Taylor Galerkin approach for compressible Euler equations based on the Kinetic Theory of Gases has been presented. The new Boltzmann Taylor Galerkin (BTG) FEM has been shown to be as accurate and as fast as the two step Taylor Galerkin and less dissipative than other Boltzmann schemes for continuum gas dynamics.

1 Introduction

Recent years have seen a remarkable progress in the solution of convection dominated problems by the finite element method. The Taylor Galerkin FEM of Donea[1] and especially its two step version of Swansea group[2] and the Streamline Upwind Petrov Galerkin (SUPG) method of Hughes [3] have been applied to a variety of problems ranging from simple convection equation to the compressible Navier-Stokes equations of fluid dynamics. Yet there is no agreement about the optimal finite element formulation for hyperbolic system of conservation laws which admit discontinuous solutions such as shocks and contact discontinuities. In this paper, a new line of research for developing novel finite element method for compressible Euler equations is put forward. The method exploits the rich connection between the Boltzmann equation of Kinetic Theory of Gases and the equations of continuum gas dynamics. This new method which we call as the Boltzmann Taylor Galerkin FEM is based on the well known fact that the Euler equations are moments of the Boltzmann equation when the velocity distribution is a Maxwellian.

2 Basic Theory of Boltzmann Schemes

Let us illustrate the basic idea with reference to one dimensional unsteady Euler equations

$$\frac{\partial U}{\partial t} + \frac{\partial G}{\partial x} = 0 \tag{1}$$

The vector of conserved variables U and flux vector G are given by $U = [\rho, \rho u, \rho e]^T$ and $G = [\rho u, p + \rho u^2, (\rho e + p) u]^T$, where $\rho =$ mass density, $u =$ fluid velocity, $p =$ pressure, $e =$ total specific energy given by $e = \frac{p}{\rho(\gamma-1)} + \frac{1}{2}u^2$. Equation (1) can be obtained as the Ψ moment of the 1-D Boltzmann equation

$$\frac{\partial F}{\partial t} + v\frac{\partial F}{\partial x} = 0 \tag{2}$$

where F is the Maxwellian velocity distribution given by,

$$F = \frac{\rho}{I_0}\sqrt{\frac{\beta}{\pi}} exp\left[-\beta(v-u)^2 - \frac{I}{I_0}\right] \tag{3}$$

here $\beta = \frac{1}{2RT}$, R = gas constant per unit mass, v=molecular velocity, I = internal energy variable corresponding to non-translational degrees of freedom and I_0 is defined as $I_0 = \frac{3-\gamma}{4(\gamma-1)\beta}$. The moment function vector ψ is defined by $\psi = \left[1, v, I + \frac{v^2}{2}\right]^T$ and is precisely the vector of fundamental collisional invariants. The Euler equations (1) can then be cast in the compact form

$$\langle\psi, \frac{\partial F}{\partial t} + v\frac{\partial F}{\partial x}\rangle = 0 \tag{4}$$

where the scalar product $\langle\psi, f\rangle$ is defined by

$$\langle\psi, f\rangle = \int_0^\infty dI \int_{-\infty}^\infty dv\,\psi f(v) \tag{5}$$

We thus see that Euler equations (which are a system of nonlinear hyperbolic system of conservation laws) are the moments of Boltzmann equation without the collision term which is a linear and scalar hyperbolic equation.

3 The Boltzmann Taylor Galerkin Scheme

We now proceed to develop the BTG scheme scheme. The various steps in the formulation are as follows.
Step 1: We start with the Lax-Wendroff step for the Boltzmann eqn.(2).

$$F^{n+1} = F^n + \Delta t F_t^n + \frac{\Delta t^2}{2} F_{tt}^n \tag{6}$$

$$F^{n+1} = F^n - \Delta t (vF)_x + \frac{\Delta t^2}{2} \left(v^2 F\right)_{xx} \tag{7}$$

Step 2: Now we apply the standard Galerkin FEM to equation (7) to obtain

$$\int_0^L \left(F^{n+1} - F^n\right) W dx = -\Delta t \int_0^L W(vF)_x\,dx + \frac{\Delta t^2}{2}\int_0^L W\left(v^2 F\right)_{xx} dx \tag{8}$$

where W is the weighting function, L is the length of the domain. Integrating by parts the last term in eqn.(8), we obtain

$$\int_0^L W\Delta F^{n+1} dx = -\Delta t \int_0^L W(vF)_x\,dx - \frac{\Delta t^2}{2}\int_0^L \frac{\partial W}{\partial x}\frac{\partial\left(v^2 F\right)}{\partial x} dx + R_b \tag{9}$$

Here $\Delta F^{n+1} = F^{n+1} - F^n$ and R_b is the boundary term. Now we put $W = N_l$ and approximate F by a piecewise linear basis function, $F = \sum_m F_m N_m$ to get

$$M \Delta F^{n+1} = \Delta t \, (vF) \, K_1 + \frac{\Delta t^2}{2} \left(v^2 F\right) K_2 + R_b \qquad (10)$$

where the standard mass matrix M, the convection stiffness matrix K_1 and the diffusion stiffness matrix K_2 are given by $M = \sum_e \int N_l N_m dx$, $K_1 = \sum_e \int N_l \frac{\partial N_m}{\partial x} dx$ and, $K_2 = \sum_e \int \frac{\partial N_l}{\partial x} \frac{\partial N_m}{\partial x}$

It may be noted that for piecewise linear finite element on a uniform mesh of size h, the scheme given by eqn (10) reduces to the fully discrete equations

$$\left(1 + \frac{1}{6}\delta^2\right) \Delta F^{n+1} = -\lambda \Delta_0 \, (vF)_j^n + \frac{\lambda^2}{2} \delta^2 \left(v^2 F\right)_j^n \qquad (11)$$

where $\lambda = \frac{\Delta t}{\Delta x}$ and Δ_0 and δ^2 are central difference operators.

Step 3: We apply the moment method strategy so as to get the BTG scheme for Euler equations. *i.e.* we take the inner product of the scheme given by eqn (10) with the moment function vector ψ to get

$$M \langle \Delta F^{n+1}, \psi \rangle = \Delta t K_1 \langle vF, \psi \rangle + \frac{\Delta t^2}{2} K_2 \langle v^2 F, \psi \rangle + \langle R_b, \psi \rangle \qquad (12)$$

Or alternately
$$M \Delta U^{n+1} = R \qquad (14)$$

where R = load vector appearing on the right hand side of eqn (14). It may be noted that $\langle vF, \psi \rangle = G$ and

$$\langle v^2 F, \psi \rangle = D = \left[p + \rho u^2, 3pu + \rho u^3, \frac{5\gamma - 3}{2(\gamma - 1)} pu^2 + \frac{\gamma}{\gamma - 1} \frac{p^2}{\rho} + \frac{\rho u^4}{2}\right]^T \qquad (15)$$

It may further be noted that by taking moments of the scheme given by eqn(11), we get the final form of BTG finite element scheme as

$$\left(1 + \frac{1}{6}\delta^2\right) \Delta U^{n+1} = -\lambda \Delta_0 \, (G)_j^n + \frac{\lambda^2}{2} \delta^2 (D)_j^n \qquad (16)$$

This completes the formulation of the BTG scheme in one dimension. The extension to two dimensional flows is straightforward. It is interesting to note that the Lax-Wendroff step applied directly to the Euler equations involves computation of the flux jacobian matrices whereas the BTG scheme is free from this problem. The BTG scheme is a central difference type schemes and requires the addition of artificial viscosity or a TVD formulation for capturing strong shocks. For a detail discussion of the BTG method, the reader is referred to [7].

4 Results

The BTG scheme presented in the previous section was applied to the shock tube problem due to Sod[4].The sharp resolution of the shock and the contact surface can be seen in figure 1 . The expansion fan is captured accurately. One hundred elements were used in the computation.Fig.2 shows a comparison of the BTG scheme with the Kinetic flux vector splitting (KFVS) scheme[5]. The BTG-FEM is less dissipative than the KFVS method. Fig.3 shows the results obtained by the BTG scheme with addition of artificial viscosity based on the second derivative of pressure for shock reflection problem. The results obtained by using a characteristic TVD approach suggested by Selmin[6] are shown in Fig.4.The later approach produces a sharp resolution of shocks.A grid of 61 by 21 was used in the computations with piecewise bilinear quadrilateral elements.

5 Conclusions

A novel finite element method for inviscid compressible flows within the Kinetic framework has been presented. It has been shown that the Taylor Galerkin approach applied to the Boltzmann equation is capable of producing computational schemes of good overall accuracy for the numerical solution of inviscid compressible flows. The BTG scheme is less dissipative than the other Boltzmann schemes such as KFVS method and is as fast as the two step Taylor Galerkin scheme of Swansea. A new methodology for developing novel FEM based on the Boltzmann equation of Kinetic Theory of Gases has opened new vistas for further research.

Acknowledgement

I would like to express my sincere thanks to Prof. S. M. Deshpande for his valuable guidance during this work.

References

[1] J. Donea: Int J Num Meth Engg, **20** 101 (1984)

[2] J. Peraire, K. Morgan, J. Peiro, O.C. Zienkiewicz: AIAA paper 87-0558 (1987)

[3] T.J.R. Hughes: Int J Num Meth Fluids, **7** 1261 (1987)

[4] G.A. Sod: J Comp Phys, **27** 1 (1978)

[5] S.M. Deshpande: AIAA paper 86-0275 (1986)

[6] V. Selmin: INRIA report 655 (1987)

[7] S.S. Deshpande, S.M. Deshpande: Fluid Mechanics Lab. Report 94 FM 01,Department of Aerospace Engg., Indian Institute of Science, Bangalore (1994)

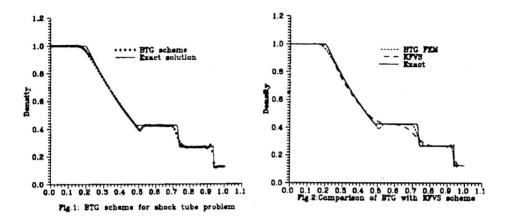

Fig.1: BTG scheme for shock tube problem

Fig.2: Comparison of BTG with KFVS scheme

Fig.3: BTG scheme with artificial viscosity

Fig.4: BTG scheme with characteristic TVD

A new second order positivity preserving kinetic schemes for the compressible Euler equations

Jean-Luc Estivalezes[1], Philippe Villedieu[2]

[1]ONERA-CERT, 2 Av Edouard Belin, 31055 Toulouse cedex, France
[2]MIP, Unité mixte CNRS, Université de Toulouse III
118 Route de Narbonne, 31062 Toulouse cedex, France

Abstract : We present a new second order kinetic flux-splitting schemes for the compressible Euler equations and we prove that this scheme is positivity preserving (i.e ρ and T remain ≥ 0). Our first order kinetic scheme is based on the Maxwellian equilibrium function and was initially proposed by Pullin. Our higher order extension can be seen as a variant of the so called corrected anti-diffusive flux approach. The necessity of a limitation on the antidiffusive correction appears naturally in order to satisfy the constraint of positivity.

1 Introduction

In this paper, we present a new theoretical way to construct a second order Kinetic Flux Vector Splitting scheme for the Euler equations. Different approaches have already been proposed in Deshpande, Perthame, Prendergast.The interesting point here is that density and pressure can be proved to remain non-negative under a CFL-like condition. Higher order extensions following the same approach can be easily performed with the same properties.

Here, for the sake of simplicity, we will only give a kinetic interpretation of our scheme. The mathematical tools which enable to prove the non-negativity of density and pressure are given in Estivalezes and Villedieu.

The paper is organized as follow. First we recall the main features of the first order scheme, then we show how to build a truly second order KFVS scheme and we give the main theoretical results. Lastly numerical results for various 1D test cases (Sod test, Blast Waves interactions, Vacuum apparition) are shown.

2 First order scheme

For the sake of simplicity, we only consider the case of the one dimensional Euler equations for a gas with one degree of freedom ($\gamma = 3$). It is well known that the Euler equations are the **K**-moment of the Boltzmann equation when the distribution fonction is Maxwellian, **K** being the collision vector of components $(1, \xi, \xi^2/2)$.

The first order scheme is constructed in two step like in Mazet.

- First the free kinetic transport equation is solved for each ξ by the explicit Courant scheme :

$$f_i^{n+1} = f_i^n - \lambda\ max(\xi,0)(f_i^n - f_{i-1}^n) - \lambda\ min(\xi,0)(f_{i+1}^n - f_i^n) \qquad (1)$$

where $\lambda = \Delta t/\Delta x$, f_i^n is the Maxwellian distribution associated to the macroscopic state of the gas at the time t^n, and ξ is the microscopic velocity.

$$f_i^n(\xi) = \frac{\rho_i^n}{\sqrt{2\pi r T_i^n}} \exp \frac{-(\xi - U_i^n)^2}{2 r T_i^n} \qquad (2)$$

Where ρ_i^n, $\rho_i^n U_i^n$ and T_i^n are respectively cells values of the density, the macroscopic velocity and the temperature.

- The second step consists in taking the **K**-moment of (1). We get the following Kinetic Flux Splitting scheme for the Euler equations:

$$\mathbf{w}_i^{n+1} = \mathbf{w}_i^n - \lambda(\mathbf{g}_{i+1/2} - \mathbf{g}_{i-1/2}) \qquad (3)$$

where :

$$\begin{array}{ll} \mathbf{w}_i^n = (\rho_i^n, \rho_i^n U_i^n, E_i^n) & \mathbf{g}_{i+1/2} = \mathbf{F}^+(\mathbf{w}_i^n) + \mathbf{F}^-(\mathbf{w}_{i+1}^n) \\ \text{and} & \mathbf{F}^+(\mathbf{w}_i^n) = \int_0^\infty \xi \mathbf{K}(\xi) f_i^n(\xi) d\xi \end{array}$$

The entropy consistency and the non-negativity of this scheme can be proved under a CFL like condition (Villedieu and Mazet).

3 Second order scheme

The main idea of our approach is to use a second order scheme at the kinetic level. Our extension can be derived in three steps :

- First (1) is replaced by a second order scheme :

$$f_i^{n+1} = f_i^n - \lambda\ max(\xi,0)(f_{i-1/2,R}^{n+1/2} - f_{i-1/2,L}^{n+1/2}) - \lambda\ min(\xi,0)(f_{i+1/2,R}^{n+1/2} - f_{i+1/2,L}^{n+1/2}) \qquad (4)$$

where $f_{i+1/2,R}^{n+1/2}$ and $f_{i+1/2,L}^{n+1/2}$ are respectively right and left second order estimates of f at the node $i + 1/2$. For exemple, we have:

$$f_{i+1/2,L}^{n+1/2} = f_i^n + \Delta f_i^n \text{ with } \Delta f_i^n = p_i^n \Delta x/2 + q_i^n \Delta t/2 \qquad (5)$$

p_i^n and q_i^n being respectively second order estimates of $(\partial f/\partial x)_i^n$ and $(\partial f/\partial t)_i^n$
Note that (2) is equivalent to:

$$f_i^n = C \exp -\frac{1}{r}\left[s_i^n + \frac{(\xi - U_i^n)^2}{2T_i^n}\right] \qquad (6)$$

Where $s_i^n = -r \log \rho_i^n + 1/2r \log T_i^n$ is the specific entropy. So by (6) we get the following expressions for p_i^n and q_i^n :

$$p_i^n = f_i^n(-\frac{p_s}{r} + \frac{p_U}{rT_i^n}(\xi - U_i^n) + \frac{p_T}{2rT_i^{n2}}(\xi - U_i^n)^2) \qquad (7)$$

$$q_i^n = f_i^n(-\frac{q_s}{r} + \frac{q_U}{rT_i^n}(\xi - U_i^n) + \frac{q_T}{2rT_i^{n\,2}}(\xi - U_i^n)^2) \tag{8}$$

where $p_s, p_U, p_T, q_s, q_U, q_T$ are respectively first order estimates of $(\partial s/\partial x)_i^n$, $(\partial U/\partial x)_i^n$, $(\partial T/\partial x)_i^n$, $(\partial s/\partial t)_i^n$, $(\partial U/\partial t)_i^n$, $(\partial T/\partial t)_i^n$. Furthermore q_s, q_U, q_T are replaced by their expression in function of ρ_i^n, $\rho_i^n U_i^n$, T_i^n, p_s, p_U and p_T, by using the non conservative form of the Euler equations.

- Since f can be seen as a probability density, it is quite natural to ensure its non-negativity at the nodes of the mesh. To do so, the following constraint is imposed on Δf_i^n :

$$-f_i^n \leq \Delta f_i^n \leq f_i^n \tag{9}$$

Since, by (5), (6), (7) Δf_i^n is a second order polynomial of $(\xi - U_i^n)$, it is necessary to impose $q_i^n = 0$ and $p_i^n = 0$ for large values of $|\xi - U_i^n|$. Let ξ_{max} be the maximum value of $|\xi - U_i^n|$ such that the constraint (9) is satisfied. For the values of $|\xi - U_i^n| \geq \xi_{max}$ one takes $\Delta f_i^n = 0$ and the kinetic equation for f is only solved by the first order scheme (1)

- Lastly, by taking the **K**-moment of (4), we get the following scheme :

$$\mathbf{w}_i^{n+1} = \mathbf{w}_i^n - \lambda(\mathbf{g}_{i+1/2} - \mathbf{g}_{i-1/2}) \tag{10}$$

where: $\mathbf{g}_{i+1/2} = \mathbf{F}^+(\mathbf{w}_i^n) - \mathbf{F}^-(\mathbf{w}_{i+1}^n) + \Delta\mathbf{F}^+(\mathbf{w}_i^n) - \Delta\mathbf{F}^-(\mathbf{w}_{i+1}^n)$ and:

$$\Delta\mathbf{F}^+(\mathbf{w}_i^n) = \int_{|\xi - U_i^n| \leq \xi_{max}} max(\xi, 0) \mathbf{K}(\xi) \Delta f_i^n(\xi) d\xi \tag{11}$$

This scheme can be proved to be second order accurate in space and time despite of the limitation on Δf_i^n. Moreover, thanks to the constraint (9) and the stability properties of the first order scheme, density and can be proved (Estivalezes and Villedieu) to remain non-negative under a CFL like condition (exactly corresponding to the half stability condition of the first order KFVS scheme). The extension of this approach to the multidimensional Euler equations for perfect gases with several degrees of freedom is described in Estivalezes and Villedieu.

4 Numerical results

In order to validate our approach, several one dimensional test cases have been run and compared to analytical solutions. Here, 200 points were used for the first two cases and 400 for the last one. All test cases have been run with the same numerical code without any adjustable parameters. The first one is the classical Sod test case. To prove the robustness of the scheme, Sjögreen test case has been performed (cf Einfeld and al). For this test, initial data are chosen so that the density in the theoretical solution evolves very close to zero. It is well known that classical approximate Riemann solvers produce non physical states with negative density or internal energy (cf Einfeld and al). We have not encountered this problem as it was theoretically predicted. The

DENSITY
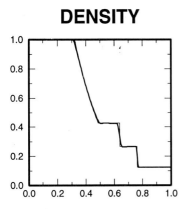

VELOCITY
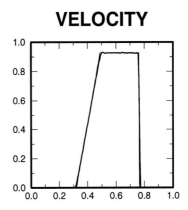

DENSITY
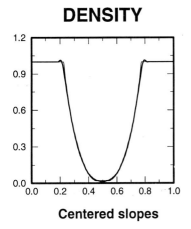

Centered slopes

DENSITY
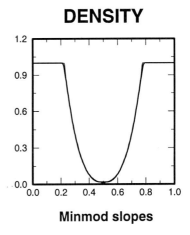

Minmod slopes

DENSITY
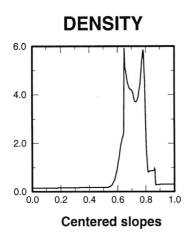

Centered slopes

DENSITY
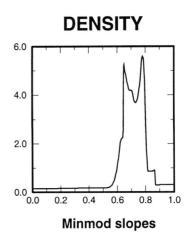

Minmod slopes

numerical results agree very well with the analytical solution. To evaluate p_s, p_U and p_T, we have used both slopes reconstructions: the centered one and the classical one based on the min-mod function. Note that numerical results using the centered slopes reconstruction are very good. The third test case is now a classical one after the work of Woodward and Collela. Our numerical results agree very well with those obtained with other numerical algorithms.

5 Conclusions

We have presented in this paper a new second order positivity preserving kinetic scheme for the compressible Euler equations. Classical 1D test cases have been run, wich have given very satisfactory results. Although only one dimensional results have been shown here, our scheme is multidimensional and the 2D extension is currently tested. Furthermore, higher order accurate schemes, which are also positivity-preserving under a CFL-like condition, can be derived (at least formally) with the same approach.

References

[1] Deshpande, S. (1986): "On the Maxwellian distribution, symmetric form and entropy conservation for the Euler equations", *NASA Langley report 2583*

[2] Einfeld, B., Munz, C., Roe,P. ,Sjögreen,B. (1991): "On Godunov type methods near low density ", *Journal of Computational Physics*, Vol 92, pp. 273-295

[3] Estivalezes, J.L., Villedieu,P. (1994): "High order positivity preserving kinetic schemes for the compressible Euler equations", *SIAM Journal of Numerical Analysis*, submitted

[4] Perthame, B. (1992): "Second order Boltzmann schemes for gas compressible Euler equations in one and two space dimensions", *SIAM Journal of Numerical Analysis*, Vol 27, pp. 1405-1421

[5] Prendergast, K. H., Xu, K. (1993): "Numerical hydrodynamics from gas kinetic theory", *Journal of Computational Physics*, Vol 109, pp. 53-66

[6] Pullin, D. (1980): "Direct simulations methods for compressible inviscid ideal gas-flows", *Journal of Computational Physics*, Vol 34, pp. 231-244

[7] Villedieu, P., Mazet, P. (1994): "Schémas cinétiques pour les équations d'Euler hors équilibre thermochimique", *La Recherche Aérospatiale*, accepted for publication

[8] Woodward, P., Collela, P. (1984): "The numerical simulation of two-dimensional fluid flow with strong shocks", *Journal of Computational Physics*, Vol 54, pp. 115-173

A Robust Least Squares Kinetic Upwind Scheme for Euler Equations

A K Ghosh[1] , S M Deshpande[2]

[1]ADA, Post Box 1718, Bangalore 560017, India
[2] Dept. of Aerospace Engg., IISc, Bangalore 560012, India

Abstract : A new second order accurate grid-free scheme called Least Squares Kinetic Upwind Method (LSKUM) for the solution of Euler Equations has been developed. This method works just on a distribution of points and only the local connectivity information at each node needs to be stored. Weighted least squares method has been used to obtain the discrete approximation of the spatial derivatives. Then upwinding is done by appropriately choosing the weights. This method is made second order accurate by a two step formula. In this kinetc upwind method, both Kinetic Flux Vector Splitting (KFVS) and Pecular Velocity based Upwinding (PVU) are used to obtain results for a wide range of two dimensional flow problems. These results demonstrate amply the capability and the robustness of the present method.

1 Introduction

Numerical solution of Euler equations for flow past complex shapes on unstructured meshes are becoming increasingly popular in recent times. It is also expected that the Euler solver on unstructured mesh along with grid adaptation can capture many high angle of attack flow phenomena of practical interest. Here, one new grid free Euler solver is presented which works on just arbitrary distribution of points. The greatest advantage in this method is that only a local connectivity information at each node is required to be stored and it is not necessary to tesselate these arbitrary distribution of points to form unstructured mesh. The novelty in this method lies in the way upwinding has been done within the frame work of least squares for the arbitrary distribution of points and thus making the present method totally different from the existing finite volume and finite element methods. Also, this node based kinetic upwind method is made second order accurate using a two step formula.

2 Basic Theory

The basis of the present method lies in successfully exploiting the fact that the Euler equations are the moments of Boltzmann equation of kinetic theory of gases when the distribution function is a Maxwellian [1-6]. The 2-D Boltzmann equation in the

Euler limit is given by

$$\frac{\partial F}{\partial t} + v_1 \frac{\partial F}{\partial x} + v_2 \frac{\partial F}{\partial y} = 0 \qquad (1)$$

Now, at the heart of the present method is the weighted least squares discrete approximation to the derivatives F_{x_0} and F_{y_0} of the velocity distribution function F which is a Maxwellian and v_1 and v_2 are the molecular velocities along x and y directions respectively[3]. Referring to Fig. 1 and denoting the coordinates of the various nodes P_i by (x_i, y_i), $i = 1, 2, ..., n$, the weighted least squares method gives the discrete approximation [2,4]

$$F_{x_0} = \frac{\|w\Delta y\|^2 (w\Delta F, \Delta x) - (w\Delta x, \Delta y)(w\Delta F, \Delta y)}{\|w\Delta x\|^2 \|w\Delta y\|^2 - (w\Delta x, \Delta y)^2} \qquad (2)$$

and

$$F_{y_0} = \frac{\|w\Delta x\|^2 (w\Delta F, \Delta y) - (w\Delta x, \Delta y)(w\Delta F, \Delta x)}{\|w\Delta x\|^2 \|w\Delta y\|^2 - (w\Delta x, \Delta y)^2} \qquad (3)$$

where

$$\|w\Delta x\|^2 = \sum_{i=1}^{n} w_i \Delta x_i^2, \quad \|w\Delta y\|^2 = \sum_{i=1}^{n} w_i \Delta y_i^2, \quad (w\Delta F, \Delta x) = \sum_{i=1}^{n} w_i \Delta F_i \Delta x_i,$$

$$(w\Delta F, \Delta y) = \sum_{i=1}^{n} w_i \Delta F_i \Delta y_i, \quad (w\Delta x, \Delta y) = \sum_{i=1}^{n} w_i \Delta x_i \Delta y_i$$

$\Delta x_i = x_i - x_0, \Delta y_i = y_i - y_0, \Delta F = F_i - F_0$ The approximations (2) are only first order accurate. In order to get second order accurate formula, we use two step method which is given below

a. **First Step**: We use formula (2) to get first order accurate scheme.

b. **Second Step**: We again use the formula (2) but replace ΔF by $\Delta \widehat{F}$ where $\Delta \widehat{F}$ is given by

$$\Delta \widehat{F}_i = \Delta F_i - \frac{1}{2}\Delta x_i \Delta F_{x_i} - \frac{1}{2}\Delta y_i \Delta F_{y_i} \qquad (4)$$

Here, $\Delta F_{x_i} = F_{x_i} - F_{y_0}$, $\Delta F_{y_i} = F_{y_i} - F_{y_i}$.

After obtaining the discrete approximation of the spatial derivatives, we now define the local connectivity information at each node. Let n_0 be the set of all nodes which are first neighbours of the node P_0. A first neighbour of P_0 is a node which is connected to P_0 by one straight line segment. Referring to Fig. 1, we note that the nodal points $P_1, P_2, ..., P_n$ are all first neighbours of P_0. We can if necessary augment the set n_0 such that it contains the first and second neighbours of P_0. A second neighbour of P_0 is a node connected to P_0 by two straight line segments. In the grid-free version of this method, it is not necessary to obtain neighbours by insisting upon connection through one straight segment. We may just define the connection set as consisting of points P_i falling in the r- neighbourhood, that is, $d(P_i, P_0) \leq r$. Such a connection set can be determined by a suitable search algorithm. The LSKUM therefore does not require interconnection of points by straight segment.

It is now possible to construct an upwind scheme for the Boltzmann equation (1) by using eqns.(2) and (3) to obtain a first order method [2,3,4,5] and eqns.(2), (3) and (4) for a second order accurate method [6,7]. Now, to get an upwind discrete approximation of F_{x_0}, we take the stencil to the left of point P_0 for $v_1 < 0$ and to the right of P_0 for $v_1 > 0$ (see Fig. 1). Similarly, to get an upwind discrete approximation of F_{y_0}, we take upper and lower stencil of P_0 for $v_2 < 0$ and $v_2 > 0$ respectively. We call this approach as X-Y splitting. It is important to note that we can locally rotate the coordinate axis at each node as the formulae (2) and (3) are valid for any arbitrary frame. It is possible to reduce the dissipation in this method by locally rotating the frame so that x- axis coincides with the streamline and y- axis normal to it. We then do upwinding along the rotated x- axis and take all the neighbouring nodes ie n_0 in the least squares expression for the rotated y- derivative (this is similar to central differencing on a regular mesh). It is also possible to control the artificial dissipation by choosing the weights in eqns.(2) and (3) as a function of distance. Once the equation (1) is upwind differenced and the local connectivity information at each node is defined, the corresponding Least Squares Kinetic Upwind Method for Euler equations is obtained by moment method strategy with KFVS [2,3,4] or PVU splittings [6,8]. Also, monotonicity of the higher order method is accomplished by enforcing that the field variables at a node are always a convex combination of the neighbouring data. Apart from X-Y splitting, it is possible to split the kinetic fluxes quadrantwise. In fact, the solid wall boundary condition is imposed using quadrantwise splitting and is based on specular reflection model as in the Kinetic Characteristic Boundary Condition (KCBC)[2,3].

3 Results and Discussions

LSKUM has been tested for a wide range of two dimensional flow problems. However, limitation of space has made us to present only a few results. Fig. 2 shows a portion of the unstructured mesh used for the airfoil calculation. Results for flow past NACA 0012 at diferent Mach No. and α are shown in Fig. 3-8. This includes subcritical flow past NACA 0012 at M=0.63 and $\alpha = 2^0$, transonic flow past NACA 0012 at M=0.8, $\alpha = 1.25^0$ and M=0.85, $\alpha = 1^0$ and supersonic case of M=1.2, $\alpha = 0^0$. These results compare very well with those given in AGARD AR 211. Fig. 9 shows the Mach contours and Fig. 10 shows the Mach No. along the centerline for flow past a half cylinder at M=4.6. In this calculation, it has been found that the stagnation pressure losses are very small. Apart from these calculations, two important studies are done on LSKUM. These are : (1) Effect of increase in stencil size on the convergence rate, (2) Effect of weight function on the convergence rate of the solution and also its effect on the accuracy of the results. In our calculation weight is taken to be proportional to $\frac{1}{s^p}$ where s is the distance and p=0, 2, 4, 6. These results are shown in Fig. 9 and Fig. 10. These results show the exellent potential of the present method. But one of the most important feature of this method is that it is not very sensitive to the point distribution as the least squares method enables to overcome any problem caused by mesh distortion (i.e. lack of regularity in the point distribution) by simply enhancing the stencil size.

References

[1] S.M. Deshpande: AIAA Paper 86-0275 (1986)

[2] S.M. Deshpande, J.C. Mandal, A.K. Ghosh: in Proc. CAARC Specialists' Meeting at NAL, India (1988)

[3] J.C. Mandal: Ph.D Thesis, Dept. of Aerospace Engg., IISc, Bangalore, India (1989)

[4] S.M. Deshpande, J.C. Mandal, A.K. Ghosh: Fluid Mechanics Report 89 FM 4, Dept. of Aerospace Engg., IISc, Bangalore, India (1989)

[5] S.M. Deshpande, J.C. Mandal, A.K. Ghosh: in Rarefied Gas Dynamics, ed. by A.E. Beylich, pp.305-312, (1991)

[6] S. M. Deshpande : in Sadhana, Vol. 18, Parts 3 and 4, pp. 405-430, (1993)

[7] A.K. Ghosh: Ph.D Thesis (in preparation), Dept. of Aerospace Engg., IISc, Bangalore, India(1994)

[8] S.V. Raghurama Rao: Ph.D Thesis (in preparation), Dept. of Aerospace Engg., IISc, Bangalore, India

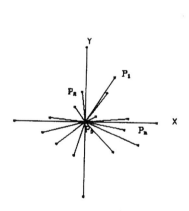

Fig. 1 CONNECTIVITY

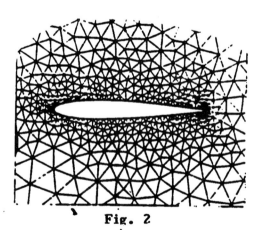

Fig. 2
A Portion of Unstructured Mesh

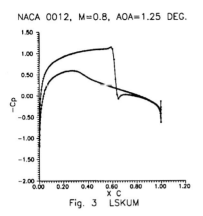

Fig. 3 LSKUM

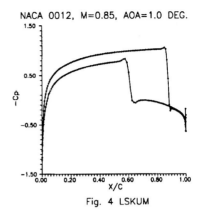

Fig. 4 LSKUM

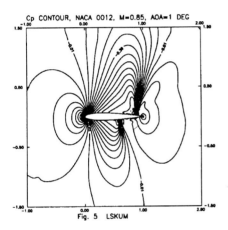

Fig. 5 LSKUM

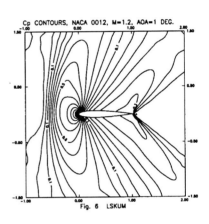

Fig. 6 LSKUM

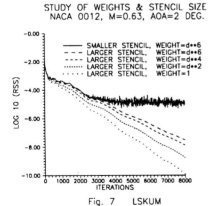

Fig. 7 LSKUM

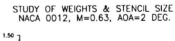

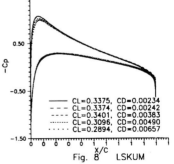

Fig. 8 LSKUM

Gas-Kinetic Finite Volume Methods

K. Xu, L. Martinelli, A. Jameson

Department of M.A.E, Princeton University,
Princeton N.J. 08544, USA

1 Introduction

Gas-kinetic schemes developed from the BGK model have been successfully applied to 1-D and 2-D flows [1,2]. One of the advantages of the gas-kinetic approach over more conventional methods is realized when one considers the multidimensional Euler and Navier-Stokes equations. In fact, by requiring only a single scalar gas distribution function f, the gas-kinetic approach greatly simplifies the calculation of mass, momentum, energy densities and their fluxes.

In this paper we construct a novel 3-D method for the system of conservation laws of fluid flow. A Lax-Wendroff type scheme is developed from gas-kinetics and applied to solve time dependent problems. In the limit of infinite collision time ($\tau = \infty$) our formulation reduces precisely to the kinetic representation of flux vector splitting for the Euler equations. Moreover, we introduce a new time-independent flux formulation which, when applied in conjunction with Jameson's multigrid time stepping scheme, yields efficient and accurate solutions of steady compressible flows.

2 Three Dimensional Finite Volume Gas-Kinetic Schemes

In the finite volume method, the discretization is accomplished by dividing the flow into a large number of small subdomains, and applying the conservation laws in the integral form

$$\frac{d}{dt}\int_\Omega U\,dV + \int_{\partial\Omega} \mathbf{F}\cdot d\mathbf{S} = 0$$

to each subdomain Ω with boundary $\partial\Omega$. In this equation U is the macroscopic state vector, defined as $U = (\rho, P_x, P_y, P_z, \epsilon)^T$, where ρ, $\vec{P}$, and ϵ are the mass, momentum, and energy densities, and $\mathbf{F}$ is the flux vector. In a gas-kinetic finite volume scheme the flux vectors across cell boundaries are constructed by computing the gas distribution function f. In three dimensions, we assume the BGK model as the governing equation for the distribution function f

$$f_t + uf_x + vf_y + wf_z = (g-f)/\tau \tag{1}$$

Here f is a function of space (x,y,z), time t, particle velocity (u,v,w) and internal variable ξ with K degrees of freedom(i.e. $K = 2$ for $\gamma = 1.4$ gases). The relations between mass ρ, momentum $\vec{P}$ and energy ϵ densities with the distribution function f are

$$(\rho, P_x, P_y, P_z, \epsilon)^T = \int \psi_\alpha f d\Xi, \qquad \alpha = 1, 2, ...5 \qquad (2)$$

where ψ_α is the vector of moments $\psi_\alpha = \left(1, u, v, w, \frac{1}{2}(u^2 + v^2 + w^2 + \zeta^2)\right)^T$ and $d\Xi = dudvdwd\xi$ is the volume element in the phase space. In the BGK model the equilibrium state is described by a Maxwellian distribution

$$g = Ae^{-\lambda((u-U)^2 + (v-V)^2 + (w-W)^2 + \xi^2)} \qquad (3)$$

where U, V and W are macroscopic gas velocities. For an equilibrium gas flow with $f = g$, the Euler equations in three dimensional space can be obtained by taking the moments of ψ_α in Eq.(1). On the other hand, to the first order of τ, the Navier-Stokes equations, with a dynamic viscosity coefficient of $\nu = \tau p$ (where p is the pressure), can be derived from the Chapman-Enskog expansion. Since mass, momentum and energy are conservative quantities in the process of gas evolution, f and g have to satisfy the conservation constraint

$$\int (g - f)\psi_\alpha d\Xi = 0, \qquad \alpha = 1, 2, ...5 \qquad (4)$$

at any point in space and time.

The general solution for f in Eq.(1) in three dimensions at the position of (x, y, z) and t is

$$\begin{aligned} f(x, y, z, t, u, v, w, \xi) &= \frac{1}{\tau} \int_0^t g(x', y', z', t, u, v, w, \xi) e^{-(t-t')/\tau} dt' \\ &+ e^{-t/\tau} f_0(x - ut, y - vt, z - wt) \end{aligned} \qquad (5)$$

where $x' = x - u(t - t'), y' = y - v(t - t'), z' = z - w(t - t')$ are the trajectory of a particle motion, and f_0 is the initial nonequilibrium gas distribution function f at the beginning of each time step $(t = 0)$. Two unknowns g and f_0 must be determined in the above equation to obtain the solution f, so as to evaluate the fluxes across cell boundaries. We will consider the evaluation of fluxes across a boundary separating two cells in the x direction and, to simplify the notation, the point for evaluating fluxes at the cell boundary will be assumed at $(x = 0, y = 0, z = 0)$.

Generally, f_0 and g can be expanded around the cell boundary as

$$f_0 = \begin{cases} g^l(1 + a^l x + b^l y + c^l z), & x < 0 \\ g^r(1 + a^r x + b^r y + c^r z), & x > 0 \end{cases} \qquad (6)$$

and

$$g = g_0(1 + \bar{a}x + \bar{b}y + \bar{c}z + \bar{A}t) \qquad (7)$$

where g^l, g^r and g_0 are local Maxwellian distribution functions. The dependence of $a^l, b^l, ..., \bar{A}$ on the particle velocity can be obtained from the Taylor expansion of a Maxwellian such as

$$\bar{A} = \bar{A}_1 + \bar{A}_2 u + \bar{A}_3 v + \bar{A}_4 w + \bar{A}_5(u^2 + v^2 + w^2 + \xi^2) \qquad (8)$$

In order to get all the parameters in Eq.(6) and Eq.(7) at time $t = 0$, we need first interpolate the initial macroscopic variables. The interpolation is carried out by using a Symmetric LImited Positive (SLIP) technique originally developed by Jameson from Local Extremum Diminishing (LED) considerations [3]. All the coefficients in f_0 can be obtained directly. Then, g_0 in Eq.(7) at $(x = 0, y = 0, z = 0, t = 0)$ can be evaluated automatically by taking the limit of Eq. (5) as $t \to 0$ and substituting into Eq.(4) to obtain

$$\int g_0 \psi_\alpha d\Xi = \int_{u>0} \int g^l \psi_\alpha d\Xi + \int_{u<0} \int g^r \psi_\alpha d\Xi, \qquad \alpha = 1, 2, ...5 \qquad (9)$$

The other terms of $\bar{a}, \bar{b}$ and $\bar{c}$ in Eq.(8) at $t = 0$ can be computed from the new mass, momentum and energy interpolations which are continuous across the cell boundary in all three directions. Now, the only unknown term left in Eq.(8) is $\bar{A}$. This can be evaluated as follows by substituting Eq.(6) and Eq.(7) into Eq.(5), we get

$$\begin{aligned} f(0,0,0,t,u,v,w) &= (1 - e^{-t/\tau})g_0 + (\tau(-1 + e^{-t/\tau}) + te^{-t/\tau})(u\bar{a} + v\bar{b} + w\bar{c})g_0 \\ &+ \tau(t/\tau - 1 + e^{-t/\tau})\bar{A}g_0 + e^{-t/\tau}f_0(-ut, -vt, -wt) \end{aligned} \qquad (10)$$

with

$$f_0(-ut, -vt, -wt) = \begin{cases} g^l(1 - a^l ut - b^l vt - c^l wt), & u > 0 \\ g^r(1 - a^r ut - b^r vt - c^r wt), & u < 0 \end{cases} \qquad (11)$$

Both f (Eq.(10)) and g (Eq.(7)) contain $\bar{A}$. After applying the condition (4) at $(x = 0, y = 0, z = 0)$ and integrating it over the whole time step T, such as

$$\int_0^T \int (g - f)\psi_\alpha d\Xi dt = 0 \qquad (12)$$

five moment equations of $\bar{A}$ can obtained, from which the five constants in $\bar{A}$ of Eq.(8) can be uniquely determined. Finally, the time-dependent numerical fluxes at a cell boundary can be computed. These fluxes satisfy the consistency condition $\mathcal{F}(U, U) = \mathcal{F}(U)$ for a homogeneous uniform flow, where $\mathcal{F}(U)$ is identical to the corresponding Euler fluxes in the 3-D case.

Eq.(10) gives explicitly the time-dependent gas distribution function f at the cell boundary. Several limiting cases can be obtained, in particular: (i) in the hydrodynamic limit of $\tau \ll T$ and in a smooth region one obtains schemes which are identical to these obtained from the one-step Lax-Wendroff scheme for the Navier-Stokes equations, (ii) in the limit of $\tau = \infty$ corresponding to the collisionless Boltzmann equation, our formulation reduces precisely to the kinetic representation of flux vector splitting [4] for the Euler equations, (iii) for steady state calculations, the relaxation process can be simplified by ignoring all high-order spatial and temporal terms in the expansion of f and g to yield

$$f = g_0 + e^{-t/\tau}(f_0 - g_0) = g_0 + \epsilon(f_0 - g_0)$$

The first term g_0 represents the Euler fluxes, while the second term $\epsilon(f_0 - g_0)$ gives rise to a diffusion term which should be large near discontinuities to help fix a nonequilibrium state. Thus, $e^{-t/\tau}$ can be regarded as a diffusive parameter which should be adapted to the local flow.

It can be proved [5] that the entropy of g_0 is always larger than that of f_0, which guarantees that the local physical system will approach the state with larger entropy. This, for example, prevents the formation of unphysical rarefaction shocks

3 Unsteady Flow Calculations

A forward-facing step test is carried out on a uniform mesh of 240 × 80 cells and the numerical results are presented in Fig.(1.a) and Fig.(1.b) for the density and entropy distribution. Here, the collision time used is equivalent to a Reynolds number of $Re \simeq 50,000$ for the upstream gas flow, taking the wind tunnel height as the characteristic length scale. A slip boundary condition is imposed in order to avoid using finer meshes close to the boundary. There is no special treatment around the corner, and we never found any expansion shocks emerging from the corner.

4 Steady Flow Calculations with Multigrid Acceleration

The time independent gas-kinetic discretization scheme formulated in a previous section has been implemented in the full approximation multigrid time stepping scheme of Jameson [6]. Steady state transonic flow calculations for several airfoils have been performed to validate the simplified relaxation scheme. In these calculations, the selective parameter ϵ is determined by a switching function calculated from local pressure gradients. Using subscripts i and j to label the mesh cells, the switching function for fluxes in the i direction is

$$\epsilon = 1 - e^{-\alpha \max(P_{i+1,j}, P_{i,j})}$$

where α is a constant, and $P_{i,j}$ is an appropriate pressure switch. The computational domain is an O-mesh with 160 cells in the circumferential direction and 32 cells in the radial direction. This is a fine enough mesh to produce accurate answers with standard high resolution difference schemes. Fig. (2.a) shows the computed pressure distribution on a RAE2822 airfoil at $M_\infty = .75$ and 3.0 degrees angle of attack: A sharp two point shock is obtained. Fig. (2.b) shows the computed pressure distribution on a supercritical Korn airfoil at the design condition: A shock-free solution is obtained.

5 Conclusion

Blending the collisional gas-kinetic BGK model into the fluxes of a finite volume discretization of the conservation laws offers a promising new approach to the development of numerical hydrodynamics codes. The new scheme contains dissipation naturally through the kinetic flux vector splitting in the initial nonequilibrium term f_0. At the same time, the high resolution and multidimensionality of the traditional central difference schemes are recovered from the equilibrium state g. For steady state calculations, the coupling of a simple gas-kinetic relaxation scheme with a multigrid strategy provides fast convergence together with the favorable shock capturing properties of the gas-kinetic scheme.

Acknowledgment :

The research in this paper is supported by Grant URI/AFOSR F49620-93-1-0427.

References

[1] K.H. Prendergast and K. Xu , "Numerical Hydrodynamics from Gas-Kinetic Theory", *J. of Comput. Phys.* **109**, 53, 1993.

[2] K. Xu , "Numerical Hydrodynamics from Gas-Kinetic Theory", Ph.D. thesis, Columbia University, 1993.

[3] A. Jameson, "Artificial Diffusion, Upwind Biasing, Limiters and their Effect on Accuracy and Multigrid Convergence in Transonic and Hypersonic Flows", AIAA paper 93-3359, 1993.

[4] J.C. Mandal andd S.M. Deshpande, "Kinetic Flux Vector Splitting for Euler Equations", Computers and Fluids, Vol.23, No.2, P447, 1994.

[5] K. Xu , L. Martinelli and A. Jameson, " Gas-Kinetic Finite Volume Methods, Flux-Vector Splitting and Artificial Diffusion", submitted to JCP, July, 1994.

[6] A. Jameson, "Transonic Flow Calculations", MAE Report # 1651, Princeton University, March 1984.

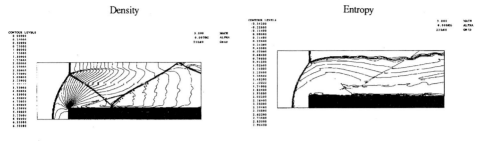

Fig. 1. Density Distribution for Mach 3 Wind Tunnel Test on Mesh 240 × 80

Fig. 1.b Entropy Distribution for Mach 3 Wind Tunnel Test on Mesh 240 × 80

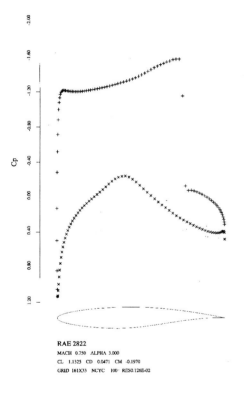

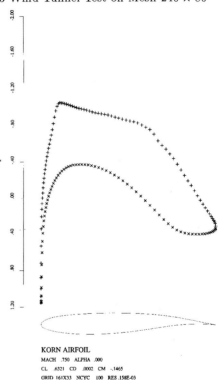

Fig. 2.a Airfoil Rae2822 with $M = 0.75$, $\alpha = 3.0$

Fig. 2.b Korn Airfoil with $M = 0.75$, $\alpha = 0.0$

Peculiar Velocity based Upwind Method for Inviscid Compressible Flows

S.V.Raghurama Rao

Joint Advanced Technology Programme, Indian Institute of Science,
Bangalore - 560 012, India

Abstract : A new upwind Boltzmann scheme termed *Peculiar Velocity based Upwind (PVU) method* is developed for solving Euler equations. Upwinding is done based on *peculiar velocity* of Kinetic Theory of Gases. This new method is more efficient and physically more meaningful than its predecessor, the Kinetic Flux Vector Splitting (KFVS) method. The PVU method is applied to some standard test problems. The results demonstrate the soundness of the new idea.

The motivation for this work came from the idea of improving the Kinetic Flux Vector Splitting (KFVS) method [1] which was demonstrated to be very robust by its application to a wide variety of 2-D and 3-D problems using structured and unstructured meshes[2,3,4,5]. In the KFVS method upwinding is done based on whether molecular velocity is greater or less than zero. Here, a new upwind Boltzmann scheme, in which upwinding is done based on *peculiar velocity*, which is the relative velocity of the molecule with respect to the fluid, is presented. The upwind method for Euler equations is obtained by taking moments of the upwind scheme applied to the Boltzmann equation of Kinetic Theory of Gases.

1 Peculiar Velocity based Upwind (PVU) method

In the KFVS method, a rest frame of reference is implicitly assumed as upwinding is done based on molecular velocity (v) greater or less than zero. Using the *peculiar velocity* ($c = v - u$), where u is the fluid velocity, as the basis for upwinding, the rest frame of reference can be avoided. Besides, replacing v by $c + u$ separates the unidirectional and multidirectional terms since u is a deterministic variable and c is a random variable with zero mean and a variance of $\frac{1}{2\beta}$ $\left(c \sim \left(0, \frac{1}{2\beta} \right) \right)$, where $\beta = \frac{1}{2RT}$.

The basis of the Boltzmann schemes is the fact that Euler equations can be derived as moments of the Boltzmann equation using a Maxwellian distribution F. Consider 1-D Euler equations given by

$$\frac{\partial U}{\partial t} + \frac{\partial G}{\partial x} = 0 \qquad (1)$$

where $U = [\rho, \rho u, \rho E]^T$, $G = [\rho u, p + \rho u^2, pu + \rho u E]^T$ and $E = \frac{p}{\rho(\gamma-1)} + \frac{u^2}{2}$. The 1-D Boltzmann equation is given by

$$\frac{\partial f}{\partial t} + \frac{\partial (vf)}{\partial x} = J(f, f) \qquad (2)$$

F is defined by

$$F = \frac{\rho}{I_0}\sqrt{\frac{\beta}{\pi}}\left[-\beta(v-u)^2 - \frac{I}{I_0}\right] \quad (3)$$

where ρ is density, I is internal energy variable corresponding to nontranslational degrees of freedom, $\beta = \frac{1}{2RT}$, $I_0 = \frac{3-\gamma}{4\beta(\gamma-1)}$, R is gas constant and T is temperature. For the Maxwellian distribution F, the collision term $J(f,f)$ becomes zero. The process of taking moments can be mathematically expressed as

$$\int_0^\infty dI \int_{-\infty}^\infty dv\ \Psi \left[\frac{\partial F}{\partial t} + \frac{\partial (vF)}{\partial x}\right] = 0 \quad (4)$$

where Ψ, called the moment functions vector, is defined by $\Psi = \left[1, v, I + \frac{v^2}{2}\right]^T$. Completing the integrations in the equation (4) yields the equation (1). Using F and peculiar velocity ($c = v - u$), the equation (2) becomes

$$\frac{\partial F}{\partial t} + \frac{\partial (cF)}{\partial x} + \frac{\partial (uF)}{\partial x} = 0 \quad (5)$$

Before deriving the PVU method, it is interesting to note that by taking moments of the equation (5) we obtain the Euler equations

$$\frac{\partial U}{\partial t} + \frac{\partial G^a}{\partial x} + \frac{\partial G^t}{\partial x} = 0 \quad (6)$$

When the variable is changed from v to c, the flux vector G is naturally split into two parts G^a and G^t, where $G^a = [0, p, pu]^T$ and $G^t = [\rho u, \rho u^2, \rho u E]^T$. G^a, termed as the *acoustic flux*, is the pressure part and G^t, termed as the *transport flux*, is the convective part. G^t is unidirectional while G^a is bidirectional (multidirectional in 2-D and 3-D). The physical meanings of G^a and G^t based on the eigenvalues, physics of the flow and the derivations are available in [6].

In the equation (5), splitting c as $\frac{c+|c|}{2} + \frac{c-|c|}{2}$ and u as $\frac{u+|u|}{2} + \frac{u-|u|}{2}$, forward differencing negative c and u parts, backward differencing positive c and u parts, and taking moments, we get the PVU method as

$$\frac{\partial U}{\partial t} + \frac{G_j^{t+} - G_{j-1}^{t+}}{\Delta x} + \frac{G_{j+1}^{t-} - G_j^{t-}}{\Delta x} + \frac{G_j^{a+} - G_{j-1}^{a+}}{\Delta x} + \frac{G_{j+1}^{a-} - G_j^{a-}}{\Delta x} = 0 \quad (7)$$

The split fluxes are given by

$$G^{t\pm} = \frac{u \pm |u|}{2} U \ \& \ G^{a\pm} = \left[\pm\frac{\rho}{2\sqrt{\pi\beta}}, \frac{p}{2} \pm \frac{\rho u}{2\sqrt{\pi\beta}}, \frac{pu}{2} \pm \frac{1}{2\sqrt{\pi\beta}}\left(\frac{p}{2} + \rho E\right)\right]^T \quad (8)$$

The split fluxes are functions of macroscopic variables only. The Kinetic Theory is used only at the conceptual level. The PVU method is found to be about 3 times faster (for 2-D problems) than its predecessor, the Kinetic Flux Vector Splitting (KFVS) method [1] as the error functions and exponential terms present in the KFVS method are absent in the PVU method. The PVU method is also physically more meaningful than the KFVS method as no rest frame of reference is assumed and the unidirectional and multidirectional terms are separated in the PVU method. Even though only grid-aligned upwinding is used in this work, a genuinely multidimensional upwind scheme using peculiar velocity based upwinding can be developed following the approach given in [7]. The PVU method is applied to shock reflection problem for supersonic and hypersonic Mach numbers and for NACA0012 airfoil for supersonic, transonic and subsonic flows using unstructured meshes and mesh adaptation. The results demonstrate the capability of this new method in capturing the flow features for a wide range of Mach numbers, thus establishing the soundness of the idea.

Acknowledgments

I thank Prof. S.M.Deshpande, Dept. of Aerospace Engg., Indian Institute of Science, under whose inspiring guidance this work was done as a part of my doctoral programme. I also thank J.S.Mathur, N.A.L., Bangalore for providing me his unstructured mesh code using the KFVS scheme.

References

[1] J.C.Mandal and S.M.Deshpande, Comp. and Fluids, vol.23, no.2, pp 447-478 (1994).

[2] J.S.Mathur and N.P.Weatherill, Int. J. Num. Methods in Fluids, 15, 59-82 (1992).

[3] N.P.Weatherill, J.S.Mathur and M.J.Marchant, Int. J. Num. Methods in Engg., 37, 623-643 (1994).

[4] S.M.Deshpande, S.Sekhar, S.Nagarathinam, M.Krishnamurthy, R.Sinha and P.S. Kulkarni, Lecture Notes in Physics, 414, 105-109, Springer-Verlag (1992).

[5] S.Sekhar, M.Nagarathinam, R.Krishnamurthy, P.S.Kulkarni & S.M.Deshpande, 14th ICNMFD, Bangalore, India (1994).

[6] S.V.Raghurama Rao, Ph.D.thesis, Dept. of Mech. Engg., Indian Insitute of Science, Bangalore, India (1994).

[7] S.V.Raghurama Rao and S.M.Deshpande, 91 FM 6, Fluid Mechanics report, Dept. of Aerospace Engg., Indian Institute of Science, Bangalore (1991).

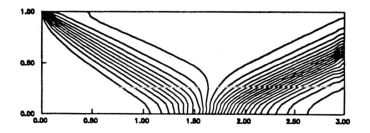

Fig. 1 Shock reflection with first order PVU method
Grid : 60x20, M = 2.9, $\theta = 29^0$

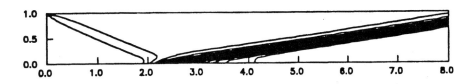

Fig. 2 Shock reflection with first order PVU method
Grid : 80x20, M = 20.0, $\theta = 20^0$

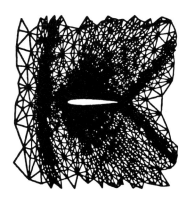

Fig.3 Adapted mesh for supersonic flow

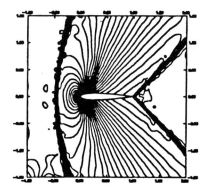

Fig.4 Supersonic flow around NACA 0012 airfoil
M = 1.2, $\alpha = 0^0$

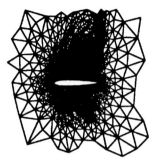

Fig.5 Adapted mesh for transonic flow

Fig.6 Transonic flow around NACA 0012 airfoil
$M = 0.85, \alpha = 1^0$

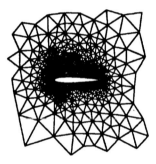

Fig.7 Adapted mesh for subsonic flow

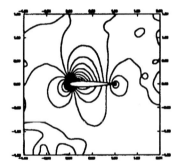

Fig.8 Subsonic flow around NACA 0012 airfoil
$M = 0.63, \alpha = 2^0$

Discrete Boltzmann Equation for Solving Rarefied Gas Problems

François Rogier - Jacques Schneider

Direction des Moyens d'Informatique,
ONERA, Châtillon, FRANCE

1 Introduction

The kinetic approach involves determining a particle density at a given time t at position $\vec{x}$, having a velocity $\vec{v}$, by solving an integro-differential (Boltzmann) equation in which seven variables are used to determine the unknown.

We propose a new method for solving Boltzmann's equation based on discrete Boltzmann equations [1],[2]. These equations have most of the physical properties of the continuous Boltzmann equation and therefore are a priori suitable to approximate its solutions (though the link between the two has not been proved).

Our approach is quite similar but is derived directly from the continuous Boltzmann equation. We describe a method for solving the inhomogeneous equation. The numerical scheme that we obtain is close to a discrete model.

2 The general method

We consider the Boltzmann equation:

$$\frac{\partial f}{\partial t} + \vec{v}.\nabla_{\vec{x}} f = Q(f,f), f(0,\vec{x},\vec{v}) = f_0(\vec{x},\vec{v}),$$

$$Q(f,f)(\vec{v_i}) = \int_{S_2}\int_{\Re^3} (f(\vec{v_a})f(\vec{v_b}) - f(\vec{v_i})f(\vec{v_j}))q(|\vec{v_i}-\vec{v_j}|,\vec{\omega})\,d\vec{v_j}d\vec{\omega}$$

$$\vec{v_a} = \vec{v_i} + ((\vec{v_i}-\vec{v_j}).\vec{\omega})\vec{\omega}, \quad \vec{v_b} = \vec{v_j} + ((\vec{v_j}-\vec{v_i}).\vec{\omega})\vec{\omega},$$

where $f(t,\vec{x},\vec{v})$ is the density function at time t, location $\vec{x}$ and velocity $\vec{v}$. S_1 is the unit sphere in $\mathbf{R}^2$ and $q(\vec{v_i}-\vec{v_j},\vec{\omega})$ is the cross differential section.

As Monte-Carlo methods, our scheme consists [4] of splitting the equation in the transport phase and the collision phase.

3 The collision phase

The aim of this section is to present a numerical scheme for integrating the collision phase.

A discretization of the Boltzmann equation yields to the following approximation.

Details of the technics can be found in [4].
Let V be a regular grid: $V = \{(ih, jh, kh); (i,j) \in \Omega\}$, Ω is a subset of $\Re^3$ the space of velocities.

$$\begin{cases} Q(\omega,\omega)(\vec{v_i}) \approx \sum_{j,a,b=1}^{(n+1)^2} \Gamma_{ab}^{ij}(\omega_a\omega_b - \omega_i\omega_j) \\ \omega_i(0) = \omega_i^0 \quad i = 1, ..., (n+1)^2 \end{cases} \quad (1)$$

where $(n+1)^2$ is the number of lattice points belonging to Ω. Here, the function f has been replaced by its approximation $\omega_i(t)$ at time t and velocity $\vec{v_i}$.

The coefficients Γ_{ab}^{ij} depend on $\tilde{q}$ and the parameters of the discretization, the number of coefficients nonequal to 0 is finite.

We now want to show some important results concerning the solution of (1).

Proposition 1 Let $(\omega_i(t))_{i=1,...,(n+1)^2}$ be the solution of (1), then the following properties hold: 1. $\sum_{i=1}^{(n+1)^2} \omega_i(t)\varphi(\vec{v_i}) = \sum_{i=1}^{(n+1)^2} \omega_i^0 \varphi(\vec{v_i})$ for $\varphi(\vec{v}) = 1, \vec{v}$ or $|\vec{v}|^2$.

2. Let $H(t) = \sum_{i=1}^{(n+1)^2} \omega_i(t) \log \omega_i(t)$, then $H(t)$ is a nonincreasing function of the time.

Proof:
One can find a sketch of the proof in [2]. It is interesting to notice that one has a link between the Boltzmann equation and a discrete velocity model ([4]). On the other hand, Theorems 3 show that all the properties of the Boltzmann equation are also valid for this model: in particular the space of invariants is reduced to the one expected.

A first order scheme in time is enough to solve the system 1, provided that the whole scheme (transport and collision) is of the first order. Thus, an explicit Euler scheme has been used:

$$\omega_i^{n+1} = \omega_i^n + \Delta t \sum_{j,a,b=1}^{(n+1)^2} \Gamma_{ab}^{ij}(\omega_a^n\omega_b^n - \omega_i^n\omega_j^n) \quad \forall i \in (1, ..., (n+1)^2) \quad (2)$$

Now, the stability of the scheme is given by the condition:

$$\Delta t \leq \Delta t^\star = \frac{1}{\max_{i,j} \sum \Gamma_{ab}^{ij} M} \quad \text{with } M = \sum_i \omega_i^0,$$

$$\text{then } \sum_i |\omega_i^n| = \sum_i |\omega_i^0|$$

3.1 The randomly alternative

As it can be easily remark, the cost computational of the collision phas eis about $O(n^2)$. In order to reduce the cost of the collision phase we propose to modify the previous deterministic algorithm in the following way. Before to do that, we have rewritten the scheme (2) in the following way :

$$\{\omega_i^{n+1} = \omega_i^n, i \in (1,...,(n+1)^2)\}$$

For $p = 1, N_{coll}$ compute the collision kernel Q_p :

$$Q_p = \Delta t \Gamma_p(\omega_a^n \omega_b^n - \omega_i^n \omega_j^n)$$

and increment simultaneously the values of

$$\omega_i^{n+1} = \omega_i^{n+1} + Q_p \qquad (3)$$
$$\omega_j^{n+1} = \omega_j^{n+1} + Q_p \qquad (4)$$
$$\omega_a^{n+1} = \omega_a^{n+1} - Q_p \qquad (5)$$
$$\omega_b^{n+1} = \omega_b^{n+1} - Q_p \qquad (6)$$

where N_{coll} is the number of collisions and (i,j,a,b) is the pre and post collisional velocities (depending on p).

The general idea consists in randomly choosing a fixed number N_{rand} of collisions in the possible set of collisions and then modifying the above loop by :

Choose randomly a subset $(p_1, p_2, ..., p_{N_{rand}})$ of collisions in the set of N_{coll} collisions. Then, compute the renormalized coefficients Γ as :

$$\tilde{\Gamma}_p = \Gamma_p \times \frac{\sum_{r=1}^{N_{coll}} \Gamma_r}{\sum_{r=1}^{N_{rand}} \Gamma_{p_r}}$$

For $p_r, r = 1, N_{rand}$ compute the collision kernel $\tilde{Q}_{p_r}$:

$$\tilde{Q}_{p_r} = \Delta t \tilde{\Gamma}_{p_r}(\omega_a^n \omega_b^n - \omega_i^n \omega_j^n)$$

and replace Q_p by $\tilde{Q}_{p_r}$ in (3),(4),(5),(6). The renormalization of the coefficients Γ_p is crucial for preserving the homogeneity of the collision kernel with respect the number of points. The form of computation of the collisions allows to conserve the momentum and the energy. At this moment, there is no proof of convergence of this algorithm, but numerical results have shown a good agreement with the deterministic solution.

3.2 Approximation of the transport phase

In this section, our purpose is the way for solving the transport equation:

$$\frac{\partial f}{\partial t} + \vec{v}.\nabla_x f = 0$$
$$f(0, \vec{x}, \vec{v}) = f^0(\vec{x}, \vec{v})$$

After discretisation, the discrete set of velocities is given by the grid V, and this problem is replaced by the system:

$$\frac{\partial f_k}{\partial t} + \vec{v}_k.\nabla_x f_k = 0 \qquad (7)$$
$$f_k(0, \vec{x}) = f_k^0(\vec{x}, \vec{v}), k = 1, (n+1)^2 \qquad (8)$$

In order to solve the previous system we use an two or first order explicit finite volume method. This scheme is conditionally stable (CFL number ≤ 1).

4 Numerical results

4.1 The homogeneous case

As we are concerned with resolution of homogeneous Boltzmann equation, we are going to compare results obtained with our numerical scheme to an analytical solution which has been given by A.V. Bobylev, M. Krook and T.T. Wu [5] in the case of Maxwellian molecules with isotropic scattering. we define the quadratic error by :

$$E(t) = 2\sqrt{\frac{\sum_{a=1}^{p^3}(\omega_a(t) - f(t, \vec{v_a}))^2}{\sum_{a=1}^{p^3}(\omega_a(t) + f(t, \vec{v_a}))^2}} \qquad (9)$$

where ω_a and f are respectively the solutions of the discrete problem and the Boltzmann equation.

Let us remark that it is not necessary to compare the first order moments because the algorithm conserves them. Error on mass, momentum and energy are connected with the representation of the solution on the grid at the initial time. The computational grid is a regular $10*10*10$ grid. The total number of coefficients Γ_p is 20499. We have computed on the figure 1 the relative residual $E(t)$. The evolution of $E(t)$ is plotted along the time for differents choice of $N_{rand} = \{10, 100, 1000, 10000, 20499\}$. The case of $N_{rand} = 20499$ correspond to the deterministic method and the others cases for the random version.

It can be seen that the relative residual decrease as N_{rand} increase. In other hand, the relative residual are quasi the same when $N_{rand} \geq 1000$.

4.2 The inhomogeneous case

Let us look at possible applications of our method : a compression ramp, a plane rarefied supersonic jet into the vacuum and a supersonic flow around an ellipse. For both cases, the macroscopic variables at a point x and time $n\Delta t$ are given by:

$$\text{Density}: n(x) = \sum_{i,k} \omega_{ik}^n \phi_i(x),$$
$$\text{Gas velocity}: \vec{u}(x) = \frac{1}{n(x)} \sum_{i,k} \omega_{ik}^n \phi_i(x) \vec{v_k},$$
$$\text{Temperature}: T(x) = \frac{1}{n(x)} \sum_{i,k} \omega_{ik}^n \phi_i(x) |\vec{v_k}|^2 - |\vec{u}(x)|^2,$$
$$\text{Heat flux}: \vec{q}(x) = \frac{1}{2} \sum_{i,k} \omega_{ik}^n \phi_i(x) |\vec{v_k} - \vec{u}(x)|^2 (\vec{v_k} - \vec{u}(x))$$

References

[1] Cabannes H. (1980) The Discrete Boltzmann Equation (Theory and Application) Lectures Notes, University of California, Berkeley.

[2] Gatignol R. (1975) Théorie Cinétique des gaz à répartition discrète des vitesses, Lecture Notes in Physics, 36, Springer Verlag.

[3] Cercignani C. (1975) Theory and Application of the Boltzmann Equation, Scottish Academic Press.

[4] Rogier F., Schneider, A direct method for solving the Boltzmann equation, J. Transport Theory and Statistical physics, vol 23, 1994

[5] Krook M., Wu T.T. (1977) Exact Solutions of the Boltzmann Equation, The Physics of Fluids, Vol. 20, No. 10, Pt. 1, October 1977.

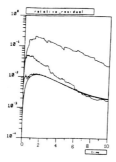

Figure 1 : Relative residual for different choice of N_r and

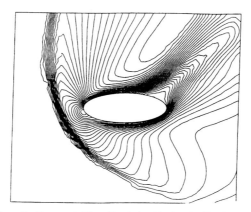

Figure 2 : Hypersonic Flow (density) : $Mach = 4.$, $Knudsen = 0.036$

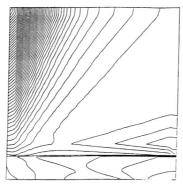

Figure 3 : Jet into vaccum (density): $Mach = 2.$, $Knudsen = 1$.

Spurious Entropy Generation as a Mesh Quality Indicator[1]

J. G. Andrews and K. W. Morton

Oxford University Computing Laboratory,
Wolfson Building, Parks Road,
Oxford. OX1 3QD, U.K.

Abstract : We present an *a posteriori* error sensor for inviscid, compressible flow. The error sensor is based on an analysis of the phenomenon of spurious entropy generation. We apply the estimator to the case of a Cell Vertex discretisation on quadrilaterals. On the basis of the resulting estimates the mesh points are moved using a hybrid equidistribution method known as the *LPE* method. The modified meshes are shown to be greatly improved, in the sense that pressure loss and spurious entropy generation are reduced. For subcritical flows, drag is also reduced.

1 Introduction

The two central issues in adaptive mesh generation are, firstly, the development of a sensor which indicates regions of relatively poor solution accuracy and, secondly, the development of an efficient procedure which uses the sensor to concentrate mesh nodes in such regions. In this paper, we develop a sensor and use it in conjunction with an existing mesh movement method.

The most frequently used sensors are those which use the principle of "solution activity" as the criterion for mesh adaptation. Such sensors usually involve the modulus of the first or second derivative of a flow variable such as Mach number. The approach has proved its worth over a long period of time and can be partially justified theoretically, see for example Mackenzie *et al.* (1992). An alternative approach, which is the one investigated here, attempts to indicate the local accuracy of a solution using some measure of the error in the vicinity. This method has the advantage of being directly related to the actual error inherent in the solution.

The particular class of problems to which we direct our attention are those of two dimensional, inviscid compressible flow calculations around aerofoils or turbine blades. In such situations the four conservation laws for mass, momentum and energy imply an extra conservation law for an entropy variable in regions which exclude shocks. We choose this entropy variable as the basis for development of a mesh adaptation sensor. A link between values of the sensor and actual nodal errors in entropy is established

---

[1] The authors would like to thank Dr. P.I. Crumpton for invaluable computing assistance. J. G. Andrews gratefully acknowledges the financial support of Rolls-Royce plc, D.R.A. Farnborough and the E.P.S.R.C

and the formula for the sensor is analysed qualitatively to indicate desirable mesh features.

2 Outline of the Analysis

2.1 Local Error

For an inviscid flow, away from shocks and moving with velocity **u**,

$$\mathbf{u} \cdot \nabla S = 0 \tag{1}$$

where S is some measure of the entropy. We shall choose S to be given by P/ρ^γ where P is the static pressure, ρ is the density and γ is the ratio of specific heats. Equation (1) follows directly from combining the conservation laws to obtain the following identity:

$$a_1 \times (\text{Mass Conservation}) +$$
$$a_2 \times (x \text{ Momentum Conservation}) +$$
$$a_3 \times (y \text{ Momentum Conservation}) +$$
$$a_4 \times (\text{Energy Conservation}) \equiv \mathbf{u} \cdot \nabla S. \tag{2}$$

Here, the a_i are functions depending on the density, the momenta and the enthalpy. Obtaining the right-hand side from the left-hand side involves repeated application of the Leibnitz rule for differentiation, i.e.,

$$\frac{\partial(\phi\psi)}{\partial z} = \phi \frac{\partial \psi}{\partial z} + \psi \frac{\partial \phi}{\partial z}. \tag{3}$$

A discrete approximation to (1), on a cell α will give the *algebraic* equation

$$\overline{\mathbf{u}_h^\alpha} \cdot \nabla_h^\alpha S_h = E^\alpha, \tag{4}$$

where the subscript h on S and $\mathbf{u}$ indicates numerical approximation, $\overline{\mathbf{u}_h}^\alpha$ is some discrete average of $\mathbf{u}_h$ on α and $\nabla_h^\alpha S_h$ is some discrete average of ∇S_h on α. The local error term E^α may be estimated in terms of the original discrete conservation laws by replacing the left-hand side of (2) with a discrete equivalent to obtain:

$$A_1^\alpha \times (\text{Discrete Mass Conservation}) +$$
$$A_2^\alpha \times (\text{Discrete } x \text{ Momentum Conservation}) +$$
$$A_3^\alpha \times (\text{Discrete } y \text{ Momentum Conservation}) +$$
$$A_4^\alpha \times (\text{Discrete Energy Conservation}) \equiv \overline{\mathbf{u}_h^\alpha} \cdot \nabla_h^\alpha S_h - E^\alpha, \tag{5}$$

where the A_i^α are discrete equivalents to the a_i, and E^α is defined uniquely in (5) by making specific choices for the A_i^α and the operators $\overline{(\cdot)}$, $D_x^\alpha(\cdot)$ and $D_y^\alpha(\cdot)$. These operators correspond to an average over the cell and the x, y components of the gradient ∇_h^α respectively.

2.2 Global Error

For sub-critical external flows, there is a close relationship between the local error E and the global error $e = S_h - S$ since all the streamlines originate in the far field and hence share a common entropy value. Consequently, the truncation error for the entropy is zero throughout the domain:

$$\overline{\mathbf{u}_h^\alpha} \cdot \nabla_h^\alpha S = 0. \tag{6}$$

Note that (6) differs from (4) as it is the exact solution S which has been substituted. Subtraction of (4) from (6) leads to the important equation

$$\overline{\mathbf{u}_h^\alpha} \cdot \nabla_h^\alpha e = E^\alpha. \tag{7}$$

which is an *algebraic* expression relating the local error to the global error. Consequently, insight into the behaviour of E^α gives corresponding insight into the behaviour of the global error e. In particular, for situations in which the entropy is known on the domain boundary, a global entropy error of zero may be ensured by setting $E^\alpha = 0$ in all cells.

2.3 Truncation Error for Mass, Momentum and Energy

We may regard E^α as an estimate of a linear combination of the truncation errors in the four conservation laws; this follows from substituting the true solution into the relation (5). However, it is a much more compact expression than conventional truncation error expressions; and it appears to contain the important effects of geometry and flow parameters on the accuracy of the computational method.

3 Application to the Cell Vertex Method

Numbering the nodes locally in an anti-clockwise fashion, the Cell Vertex finite volume residual $\mathbf{R}_\alpha$, on a two dimensional quadrilateral cell α for inviscid flows, may be expressed as

$$\mathbf{R}_\alpha = \frac{1}{2}[(\mathbf{F_1} - \mathbf{F_3})\delta y_{24} + (\mathbf{F_2} - \mathbf{F_4})\delta y_{31} - (\mathbf{G_1} - \mathbf{G_3})\delta x_{24} - (\mathbf{G_2} - \mathbf{G_4})\delta x_{31}], \tag{8}$$

where the $\mathbf{F_i}$, $\mathbf{G_i}$ are vectors representing the nodal values of the mass, momentum and energy fluxes; δx_{ij}, δy_{ij} represent the change in x and y from vertex i to vertex j. (For further details, see Crumpton *et. al.* (1993).) The aim of the Cell Vertex method is to set the right hand side of (8) to zero and, for the purposes of this analysis, we shall assume that this has been accomplished.

In order to apply (5) to this method we need to substitute the different components of (8) into (5) and apply some discrete form of the Leibnitz rule to obtain the expression on the right. We have shown that, if $D_x^\alpha \phi$, $D_y^\alpha \phi$ and $\overline{\phi^\alpha}$ are linear combinations of the four vertex values of ϕ, it is impossible to choose these discrete operators so that an exact discrete Leibnitz rule exists for arbitrary vertex values ϕ_i, ψ_i - which can be done in one dimension. Instead we choose definitions of these operators which exactly equal their continuous counterparts for a bilinear function at one point in a cell. (Such a policy arises naturally from noting that our Cell Vertex method may be reformulated as a Petrov-Galerkin finite element method with a trial space consisting of isoparametric bilinear functions; this point is then the origin of the local coordinate system.) This results in the following definitions:

$$D_x^\alpha \phi = \frac{(\phi_1 - \phi_3)\delta y_{24} + (\phi_2 - \phi_4)\delta y_{31}}{2V_\alpha}, \quad D_y^\alpha \phi = \frac{(\phi_1 - \phi_3)\delta x_{24} + (\phi_2 - \phi_4)\delta x_{31}}{-2V_\alpha} \quad (9)$$

$$\overline{\phi} = (\phi_1 + \phi_2 + \phi_3 + \phi_4)/4. \quad (10)$$

There is then an error in each application of the discrete Leibnitz rule and the accumulation of these in deriving an expression for $\overline{\mathbf{u}_h^\alpha} \cdot \nabla_h^\alpha S_h$ gives the error E^α in (4).

3.1 Analysis of E^α for the Cell Vertex Method

The expression obtained for the quantity E^α, by assuming that the discrete conservation laws are exactly satisfied over each cell, is dominated by terms of the general forms

$$\frac{\overline{q_h^\alpha}}{V_\alpha} \delta \mathbf{u}_h^e \wedge \delta \mathbf{x}_\pm \quad \text{and} \quad \frac{\overline{q_h^\alpha}}{V_\alpha} \overline{\mathbf{u}_h^\alpha} \wedge \delta \mathbf{x}_\pm, \quad (11)$$

where $\overline{q_h^\alpha}$ is the cell average speed, $\delta \mathbf{u}_h^e$ is the change in $\mathbf{u}_h$ along an edge of a cell, $\delta \mathbf{x}_+$ is the sum of the diagonal vectors and $\delta \mathbf{x}_-$ is the difference of the diagonal vectors of α. On cells which are close in shape to parallelograms, we find that, as expected from Süli (1992) and Morton and Paisley (1989), $E^\alpha = O(h)^2$, where h characterises the mesh spacing. The desirablity of an orthogonal mesh is also encapsulated, in some sense, in these general forms.

4 Results and Conclusions

The sensor E^α has been combined with the LPE mesh movement technique due to Catherall (Catherall (1993)). The resulting meshes have displayed significantly lowered levels of pressure loss (and spurious entropy generation.) (See figures.) Furthermore, for typical subcritical cases the drag has been reduced by up to half. It appears that the techniques outlined in this paper are of great use in obtaining definitive Euler solutions to simple external flows. Their extension to more complicated situations is, however, unclear.

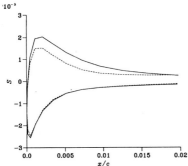

Fig. 1. Predicted (−) versus actual (...) error on body surface.

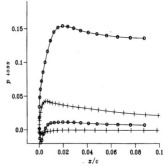

Fig. 2. Pressure loss; Original mesh (o), Adapted mesh (+).

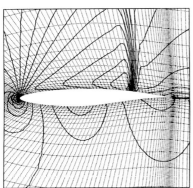

Fig. 3. Original mesh with contours of Mach number.

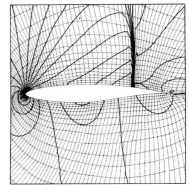

Fig. 4. Adapted mesh with contours of Mach number.

References

[1] Catherall, D. (1993): "Solution Adaptivity with Structured Grids", Numerical Methods for Fluid Dynamics IV, Ed. Baines M.J. and Morton K.W., pp 19-37, Oxford University Press 1993

[2] Crumpton, P.I.; Mackenzie, J.A. and Morton, K.W. (1993): " Cell Vertex Algorithms for the Compressible Navier-Stokes Equations", Journal of Computational Physics, 109:1-15,1993.

[3] Mackenzie, J.A.; Mayers, D.F. and Mayfield, A.J. (1992):" Error Estimates and Mesh Adaption for a Cell Vertex Finite Volume Scheme", Oxford University Computing Laboratory Report No. 92/10

[4] Morton, K.W. and Paisley, M.F. (1989): "A Finite Volume Scheme with Shock Fitting for the Steady Euler Equations", Journal of Computational Physics, 80:168-203,1989

[5] Süli, E. (1992): "The Accuracy of Cell Vertex Finite Volume Methods on Quadrilateral Meshes", Mathematics of Computation, 59(200):359-382,1992

Dynamic Mesh Adaption for Tetrahedral Grids

Rupak Biswas[1], Roger C. Strawn[2]

[1]RIACS, Mail Stop T27A-1
NASA Ames Research Center, Moffett Field, CA 94035, U.S.A.
[2]US Army AFDD, ATCOM, Mail Stop 258-1
NASA Ames Research Center, Moffett Field, CA 94035, U.S.A.

1 Introduction

Two types of solution-adaptive strategies are commonly used with unstructured-grid methods. The first is grid regeneration, where the mesh is recreated in regions where a change in resolution in desired. Although computationally intensive, the resulting grids are usually well-formed with smooth transitions between regions of coarse and fine mesh spacing. The second strategy involves local refinement and/or coarsening of the existing grid. Grid points are individually added to regions where an error indicator is high, and removed from regions where the indicator is low. The advantage is that relatively few mesh points need to be deleted or added at each coarsening/refinement step. This makes it attractive for unsteady problems where the solution usually changes rapidly.

The numerical scheme in this paper aims to solve problems in helicopter aerodynamics and acoustics. Because of the ultimate importance of flowfield unsteadiness for these problems, local mesh refinement and coarsening has been chosen as the solution-adaptive strategy. This paper examines such a mesh adaption scheme, paying particular attention to the quality of the resulting grids.

2 Adaptive Scheme

Biswas and Strawn [2] describe the basic mesh adaption strategy used in this paper. The code, called 3D_TAG, has its data structure based on edges that connect the vertices of a tetrahedral mesh. This edge data structure makes the mesh adaption compatible with Barth's Euler solver [1] and facilitates efficient refinement and coarsening.

Only three subdivision types are allowed for each tetrahedral element. The standard 1:8 isotropic subdivision is implemented by adding a new vertex at the mid-point of each of the six edges. The 1:4 and 1:2 subdivisions are used in two ways. First, they can result because the edges of a parent tetrahedron were targeted anisotropically. Second, they are used as buffers between the refined elements and the surrounding unrefined grid. These buffer elements are required to form a valid connectivity for the new mesh.

Each tetrahedral element is defined in terms of its six edges. Mesh refinement is performed by first assigning a bit value to each edge that is targeted for subdivision (R). The edge markings for each element can then be combined to form a binary pattern as shown in Fig. 1. Once the edge marking is completed, each element is subdivided based on this binary pattern.

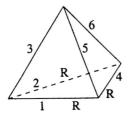

6	5	4	3	2	1	Edge number
0	0	1	0	1	1	Pattern = 11

Fig. 1. Sample edge-marking pattern for element subdivision

Mesh coarsening is also performed using the binary patterns. If a child element has any of its edges marked for coarsening, this element and its siblings are removed and their parent element is reinstated. The edge-marking patterns for these reinstated parents are altered to reflect that some edges have been coarsened. The parents are then subdivided based on the new patterns. As a result, the coarsening and refinement procedures share much of the same logic.

In [2], the data structure was implemented in C as a series of dynamically-allocated linked lists. This facilitated the addition and deletion of mesh points, but the linked lists made it very difficult to pass information directly to the Fortran flow solver. In order to reduce the communication overhead, the linked lists have been replaced with arrays and a *garbage collection* algorithm is used to compact free space when mesh points are removed. In addition to these changes, an option to control the quality of the subdivided meshes has also been developed.

3 Mesh Quality

One of the problems with anisotropic mesh refinement is that repeated subdivision can lead to poor mesh quality. Poor mesh quality is defined as a grid deficiency that leads to inaccurate flowfield solutions. Poor meshes can have disparate element sizes, large face angles, and high vertex degrees. The degree of a vertex is defined as the number of edges that are incident on it. Note that an element with a high aspect ratio does not necessarily have poor quality.

Edge swapping is commonly used to improve the quality of a mesh. It is not a panacea however, and its effects can sometimes propagate with undesirable results. This is particularly true for grids with high aspect ratio elements.

A relatively simple strategy is used to control the mesh quality for the calculations in this paper. First, it is assumed that the initial mesh has acceptable element quality and yields smooth flowfield solutions. This means that children of elements that are isotropically subdivided (1:8) will be similar to their parents. Thus, the quality problem is limited only to elements that are anisotropically refined (1:4 and 1:2).

There are no initial restrictions when an element is subdivided 1:4 or 1:2; however, these anisotropic elements cannot be further refined. To prevent this, any 1:4 or 1:2 element marked for refinement first reverts to an isotropic 1:8 element. Subsequent 1:8, 1:4, or 1:2 refinement proceeds as before. As a result, buffer elements with poor quality are never subdivided. This provides an upper bound on element face angles and controls the growth of the maximum vertex degree. This mesh quality algorithm is similar to that used by Rausch et al. [3].

Application of the mesh quality option is limited to isotropic subdivision. Any possibility for increased efficiency with the anisotropic strategy is lost if the mesh quality logic is invoked. However, for fluid dynamics problems on tetrahedral meshes, truly anisotropic subdivision is difficult to realize in practice. Implementation of our mesh quality algorithm does not significantly increase the time required for the refinement code.

4 Results

The effects of the mesh quality logic are tested for the mesh shown in Fig. 2. The initial mesh is $11 \times 11 \times 6$ with some stretching in the vertical direction. A geometric error indicator targets every element up to the second layer for uniform refinement. This subdivision strategy is a demanding test for grid quality since 1:4 and 1:2 buffer elements will be continuously marked for further refinement.

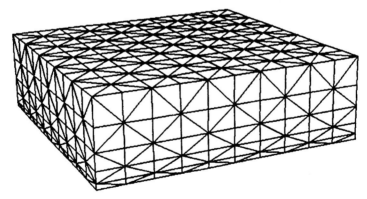

Fig. 2. Initial mesh for the first test case

Figure 3 shows two-dimensional sections of the resulting meshes after four refinement levels. The grid in Fig. 3a did not employ the mesh quality logic, while the grid in Fig. 3b did. The two meshes have significant differences that can be explained with the help of Table 1. The largest face angle for the grid without mesh quality continues to increase with each refinement. However, when the mesh quality logic is used, the largest face angle is bounded after two refinement levels. Similarly, the maximum vertex degree for the grid without mesh quality increases exponentially whereas the vertex degree with mesh quality increases only slightly with each refinement. High vertex degrees result in flow solver inaccuracies and poor efficiency on vector and parallel computers.

In addition to the above improvements, the mesh quality logic has had two other beneficial effects on the mesh. First, the mesh in Fig. 3a shows a large disparity in the sizes of adjacent elements. A much smoother transition, as shown in Fig. 3b, is obtained with the mesh quality logic. Finally, in addition to a lower overall maximum angle, the mesh in Fig. 3b has a smaller percentage of elements with large angles ($> 130°$) than its counterpart in Fig. 3a.

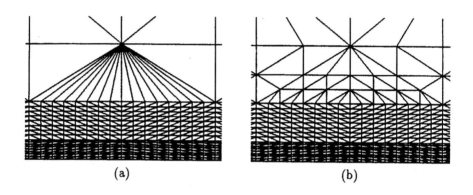

Fig. 3. Final mesh after four refinement levels (a) without and (b) with mesh quality

Table 1. Maximum face angle and vertex degree versus number of refinement levels

	Adaption strategy	Number of refinement levels				
		0	1	2	3	4
Max. face angle	Mesh quality OFF	90.00	129.81	141.34	145.56	147.38
	Mesh quality ON	90.00	129.81	139.68	139.68	139.68
Max. vertex degree	Mesh quality OFF	18	26	54	158	558
	Mesh quality ON	18	26	24	26	34

The 3D_TAG code has been recently used to model high-speed impulsive noise for a UH-1H helicopter rotor in hover [5]. A coarse mesh is initially generated. Additional mesh points are then dynamically added to capture the shock near the tip and the resulting acoustic signal as it propagates to the far field. In addition to mesh refinement, mesh coarsening is used to redistribute existing grid points in a more efficient manner. A total of three refinement levels and two coarsening steps are performed. The rotary-wing Euler solver [4] was used in all calculations. The mesh quality logic is used for this case, and Fig. 4 shows the final adapted grid and solution. An error indicator based on acoustic pressure is used to target regions for adaption. Computed results show excellent agreement with experimental microphone data for far-field acoustic pressures. The final mesh refinement step added approximately 70,000 nodes and required 29 CPU seconds on a Cray C-90. The final grid contains approximately 140,000 nodes and 783,000 tetrahedra.

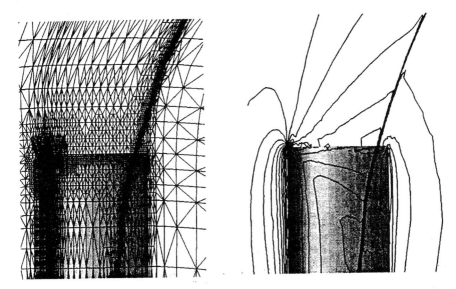

Fig. 4. Final mesh and computed pressure contours for a tip Mach number of 0.95

The helicopter acoustics problem in Fig. 4 was also run without mesh quality in the 3D_TAG code. The mesh shows a slightly higher maximum vertex degree and largest face angle, but these differences have no significant effects on either the flow solver efficiency or computed solution. Clearly, the mesh quality logic has a much smaller effect here than for the earlier test case. This is probably because the first test case presented a worst-case mesh refinement scenario that does not usually occur in practical problems. Typical refinement strategies for steady problems target smaller geometrical regions at each refinement step. This means that the anisotropically-refined elements are rarely targeted for further refinement. As long as this is the case, the mesh quality logic is not required. This is not true, however, for mesh refinement in unsteady flows.

5 Summary

Dynamic mesh adaption in three dimensions provides a powerful framework for computing steady and unsteady problems that require localized mesh refinement. This paper describes such a scheme and presents results for a practical problem in helicopter acoustics. A grid quality algorithm has been developed to ensure that the mesh does not deteriorate after many levels of refinement. Although it is not required for all steady-state problems, this mesh quality logic should be important for unsteady problems with several refinement levels.

References

[1] T.J. Barth: AIAA-91-1548 (1991)

[2] R. Biswas, R.C. Strawn: Appl. Numer. Math. **13** 437 (1994)

[3] R. Rausch, J. Batina, H. Yang: AIAA-93-0670 (1993)

[4] R.C. Strawn, T.J. Barth: AHS J. **38** 61 (1993)

[5] R.C. Strawn, M. Garceau, R. Biswas: AIAA-93-4359 (1993)

Extensive analysis and cross-comparisons of past and present numerical formulations for flow problems on unstructured meshes

Frédéric Chalot, Claudine Kasbarian, Marie-Pierre Leclercq,
Michel Mallet, Michel Ravachol, Bruno Stoufflet

Dassault Aviation
78, quai Marcel Dassault-92214 Saint-Cloud (France)

1 Introduction

Over the past years, intensive research has been devoted to the development of numerical discretization for the solution of the Euler and Navier Stokes equations on unstructured meshes. This raises a number of difficulties including (i) the multidimensional character of the problem, (ii) the fact that these equations form a system of coupled equations. While highly accurate schemes can be derived for scalar multidimensional problems or one dimensional systems, the extension to coupled multidimensional systems is still a challenging problem. The aim of this paper is to analyse the different schemes proposed and their link and to compare their respective accuracy on model problems. A large class of the methods described will be interpreted as Lagrange Galerkin formulations. Moreover this unified vision of different approaches may promote the combination of several formulations to derive new approximations.

A first class of methods is constructed from a Finite Volume interpretation of a Galerkin form of conservation equations combined with a MUSCL extension [1] and have been applied to the simulation of 3D compressible flows [6].

A second approach is constructed by stabilizing a Galerkin formulation with the addition of a least squares term of the residual. These methods are called SUPG or Galerkin least squares (GLS) and have been widely studied in [3], [4], [5].

Recently, very accurate schemes for multidimensional scalar problems have been proposed by Roe et al. [8]. They are based on distributive formulations (see van Leer [9] for a review). A natural extension to systems is still lacking although the use of various wave decomposions is possible and leads to a family of schemes which are presently actively investigated.

Recent studies have been performed where a convective/pressure splitting is used. With this approach, the convective term reduces to a set of scalar convection equations and accurate multidimensional schemes could be used. The main contributions in this direction are the upwind flux splitting scheme named AUSM developed by Liou [10] and the related CUSP formulation recently proposed by Jameson [11]. One should also mention an earlier formulation [12] combining an SUPG approximation for convective terms with a centered approximation for the pressure term.

2 Lagrange-Galerkin formulation

The following model convective/diffusive equation (or system of equations as Navier-Stokes model) is considered:

$$R(U) = U_t + \nabla . F(U) - \nabla . (K\nabla U) = 0 \tag{1}$$

Let $\mathcal{T}$ be a triangulation of the computational domain $\Omega \subset \mathbb{R}^2$ with boundary Γ of unit normal $\vec{\nu}_\Gamma$. We denote by T a current element, in case of a triangle by $\vec{n}_i^T$ the inward integrated normal opposite to node N_i, $K(i)$ the set of neighbouring nodes to node N_i and $supp(i)$ the support of the basis function ϕ_i associated to node N_i. Let V_h^d be a set of piecewise polynomial functions from $\mathbb{R}^2$ with values in $\mathbb{R}^d$ that are continuous. Further, the basis of V_h^d is the set of functions ϕ_j satisfying the Lagrange interpolation conditions. Let us consider the following abstract family of schemes for the spatial approximation of the hyperbolic system (4):

$$\int_\Omega U_t \phi \, d\nu + \int_\Omega \phi \nabla . F(U) d\nu + \int_\Omega K \nabla U \nabla \phi d\nu = 0 \tag{2}$$

where $F(U)$ is considered as an element of V_h^d (group representation) and ϕ is a test function.

3 Galerkin Least-Squares

This formulation has been developed into a general approach for a wide class of problems. The basic idea can be understood by considering the steady scalar advection-diffusion model problem:

$$R(u) = \alpha . \nabla u - \nabla . K\nabla_u = 0. \tag{3}$$

where α and K are constant parameters. For simplicity we assume that u vanishes on the boundary. The Galerkin method is defined as:

Find $u^h \in V_h$ such that for all
$w^h \in V_h$, $B(w^h, u^h) = \int_\Omega (w^h \alpha . \nabla u^h + \nabla w^h . K \nabla u^h) \, d\nu = 0$

The Galerkin/least-squares method can be defined by the following variational equation:

$$B(w^h, u^h) + \sum_T \int_T R(w^h) \tau R(u^h) d\nu = 0$$

The additional term is the sum of integrals over interior elements. It adds stability to the Galerkin formulation without upsetting the consistency of the method. For the multidimensional case, the numerical diffusion is characterized by the diffusivity matrix $K^{num} = \alpha \tau \alpha^T$ where $\tau = \frac{h}{2} \frac{f(Pe)}{|a|}$ and $f(Pe) = \coth(Pe) - 1/Pe$ is a doubly asymptotic function of the element Peclet number ($Pe = |a| h/2 | K |$) going to zero when diffusion dominates and to one when advection dominates.

The Galerkin/least-squares method can be extended to symmetric linear advective systems. In the case of a system of n equations we can write the eigenvalues decomposition of $\tau = \sum_{i=1}^{n} T_i \tau_i T_i^T$. In the presence of physical diffusion, the matrix τ is modified; it becomes $\tau = \sum_{i=1}^{n} T_i f(Pe_i) \tau_i T_i^T$ where Pe_i is the Peclet number corresponding to the ith mode, $Pe_i = \frac{h}{2}\frac{T_i}{K_i}$. The doubly asymptotic behavior is present in each mode in the numerical diffusion. This ingredient of the method is critical in establishing the convergence results presented in [4] for linear systems of advection-diffusion equations. The formulation can be applied to the compressible Navier-Stokes equations which can be written in the form of a symmetric advective-diffusive system in terms of entropy variables.

Formally, the formulation on system (1) is given by:

$$\int^{\Omega} \phi R(U) dv + \sum_T \int_T R(U) \tau R(\phi) \, dv = 0 \qquad (4)$$

Remark 1: the GLS approach is combined with a DC operator that brings more stability to the method without upsetting the consistency.

Remarks 2: a complete convergence analysis can be performed using classical Finite Element analysis of the GLS method for linear multidimensional system.

4 Finite Volume Galerkin

Lagrange-Galerkin methods may be interpreted as finite-volume schemes in some extended sense; indeed the divergence operator can be written as (boundary terms excluded):

$$\int_{\Omega} \phi_i \nabla \cdot F(U) = \sum_{j \in K(i)} \Phi^{\text{centered}} (U_i, U_j, \vec{\eta}_{ij}) \qquad (5)$$

where $\vec{\eta}_{ij} = \int_{supp(i) \cap supp(j)} \left(\phi_i \vec{\nabla} \phi_j - \phi_j \vec{\nabla} \phi_i \right) dv$.

Upwinding can then be introduced by replacing the centered flux in considering a Riemann problem with U_i and U_j as left and right status and η_{ij} defining the interface normal. Extensions of previous schemes to second-order accuracy are performed through MUSCL Finite Element interpolations as performed in [1] with specific limiters involving the element Peclet number as in the previous section in the presence of diffusion [2]. An improved formulation with a modified definition of control cells has been proposed in [7].

5 Distributive formulations

In this section, we restrict to linear finite element on triangles (P_1). Distributive compact schemes introduced by Roe et al [8] have proved to give an appropriate frame to derive accurate monotone schemes for scalar conservation equation. If we come back to the Galerkin formulation given in equation (5) and formally apply the group representation to the advection problem (3).

$$\int_\Omega \phi_i \nabla \cdot \vec{f}(u) d\nu = \int_\Omega \phi_i \sum_j \nabla \phi_j \vec{f_j}(u) d\nu = \frac{\text{area}(T)}{3} \int_\Omega \nabla \cdot \vec{f}(u) d\nu \quad (6)$$

The mass-lumped variant can then be written in a compact residual distributive formulation

$$V_i \frac{(u_i^{n+1} - u_i^n)}{\Delta t} - \sum_T \frac{1}{3} \mathcal{R}(T) = 0 \quad (7)$$

where the residual $\mathcal{R}(T)$ is given by $\mathcal{R}(T) = - \int_{\partial T} \vec{f} \cdot \vec{n}$. Assuming now a continuous piecewise linear approximation of u (P_1 elements), the above formulation suggests to define a general class of distributive scheme of the following form:

$$V_i \frac{(u_i^{n+1} - u_i^n)}{\Delta t} - \sum_T \beta_{T,i} \, R(\vec{\lambda}, T, u) = 0 \quad \text{with} \quad \sum_{i=1}^{3} \beta_{T,i} = 1 \quad (8)$$

where the residual is integrated using the linear representation of u is

$$R(\vec{\lambda}, T, u) = \sum_{i=1}^{3} k_i \, u_i \text{ with and } \vec{\lambda} = \frac{1}{\text{area}(T)} \int_T \frac{D\vec{f}}{Du} d\nu \text{ and } k_i = \frac{1}{2} \vec{\lambda} \cdot \vec{n}_i^T.$$

This formulation has been advocated in [8] in order to derive upwind schemes based on advection speed $\vec{\lambda}$ where only downwind nodes receive a contribution. To have simultaneously positive and linearly-preserving approximations, schemes have to be non-linear as so-called NN scheme, PSI scheme.

6 Generalized Flux Vector Splitting

We have shown that a Lagrange-Galerkin approximation of a general hyperbolic system can be interpreted consistently in other formulations which lead to different integration schemes. Combination of above formulations applied to an adequate splitting of the Euler equations of gas dynamics are investigated.

In that direction, an attractive approach is to investigate schemes based on a separation of convective and pressure fluxes as mentioned in the introduction. The convection part can be discretized by either distributive or SUPG schemes which have truely multidimensional property whereas the pressure part can be treated by Finite Volume Galerkin techniques. An early study proves to be promising.

7 Numerical results

Different set of comparisons have and will be performed in order to evaluate the respective accuracy of the different classes presented below. In this paper, some representative comparisons are shown.

A first set of results concerning the scalar advection of a rotated spot is presented in Fig. 1. The computation has been realized on a triangular mesh. The solution obtained with respectively SUPG, SUPG + discontinuity capturing, Finite Volume Galerkin and distributive scheme PSI is plotted in Fig.1. Clearly, this kind of simple numerical diffusion, monotonically, ... ?.

A second result demonstrates the influence of the integration scheme of the k and ϵ equations of a two-equations turbulence model Fig.2 shows the distribution of turbulent viscosity in a boundary layer obtained respectively using a linear distributive scheme (N-scheme) and a non-linear one (second-order NN scheme).

A third result compares the inviscid flow around a RAE2822 airfoil ($M_\infty = .30$, $\alpha = 2°$) obtained by a Finite-Volume Galerkin scheme and by a generalized flux vector splitting scheme using a distributive integration for the convective term. In both solutions (Fig.3) spurious entropy is generated at the leading edge region while the splitting scheme gives a better level at the winward side.

References

1 Fezoui, L. and Stoufflet, B.- A class of implicit upwind schemes for Euler simulations with unstructured meshes, *Journal of Computational Physics*, 84(1), 174-206 (1989).

2 Rostand, P. and Stoufflet, B.- TVD Schemes to Compute Compressible Viscous Flows on Unstructured Meshes, Notes on Numerical Fluid Mechanics, **vol 24**, Vieweg, Braunchweig, 1989,p.510.

3 Johnson, C. - Streamline Diffusion Methods for problem in Fluid Mechanics, R. H. Gallager et al (eds), Finite Element in Fluids, Vol.VI, Wiley, London, pp. 251-261, 1986.

4 Hughes, T.J.R., Franca, L.P., and Mallet, M.- A New Finite Element Formulation For Computational Fluid Dynamics: VI convergence analysis of the generalized SUPG formulation for linear dependent multidimensional advective diffusive system, *Computer Methods in Applied Mechanics and Engineering*, vol. 63, pp. 97-112, 1987.

5 Hughes, T.J.R., Mallet, M. and Mizukami, A New Finite Element Formulation For Computational Fluid Dynamics: II beyond SUPG, *Computer Methods in Applied Mechanics and Engineering*, vol. 54,pp. 341-355, 1986.

6 Leclercq, M.P., Mantel, B., Periaux, J., Perrier, P. and Stoufflet, B.- On recent 3-D Euler computations around a complete aircraft using adaptive unstructured mesh refinements, Proceedings of Second World Congress on Computational Mechanics, Stuttgart (Germany), August 27-31, 1990.

7 Kasbarian, C., Leclercq, M.P., Ravachol, M. and Stoufflet, B.- Improvements of Upwind Formulations on Unstructured Meshes, in 4th International Conf. on Hyperbolic Problems, Taormina(Italy), 3-8 April 1992.

8 Paillere, H., Deconinck, H., Struijs, R., Roe, P.L., Mesaros, L. M., Muller, J.D., - Computations of inviscid Compressible Flows using Fluctuation-Splitting on Triangular Meshes, 11th AIAA Computational Fluid Dynamics Conference, Orlando (1993), AIAA Paper 93 - 9301.

9 Van Leer, B.- Progress in Multi-Dimensional Upwind Differencing, NASA Contractor Report 189708, ICASE Report No 92-43, September 1992.

10 Liou, M.S.- 3-D Hypersonic Euler Numerical Simulation around Space Vehicles using Adapted Finite Elements, 25th AIAA Aerospace Meeting, Reno (1987), AIAA Paper 86-0560.

11 Jameson, A.- 3-D Hypersonic Artificial, Upwind Biasing, Limiters and Their Effect on Accuracy and Multigrid Convergence in Transonic and Hypersonic Flows, 11th AIAA Computational Fluid Dynamics Conference, Orlando (1993), AIAA Paper 93-3359.

12 Ianelli, G.S. and Baker, A.J.- An Intrinsically N-Dimensional Generalized Flux Vector Splitting Implicit Finite Element Euler Algorithm, 29th AIAA Aerospace Meeting, Reno (1991), AIAA Paper 91-0123.

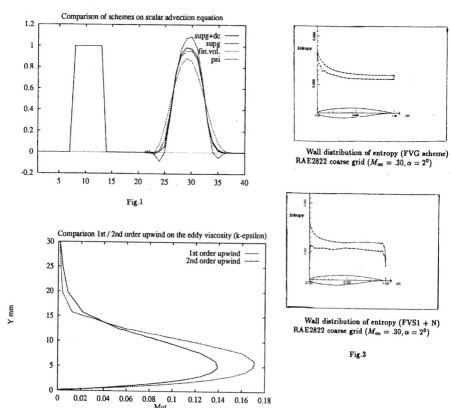

Fig.1

Wall distribution of entropy (FVG scheme)
RAE2822 coarse grid ($M_\infty = .30, \alpha = 2^0$)

Fig.2

Wall distribution of entropy (FVS1 + N)
RAE2822 coarse grid ($M_\infty = .30, \alpha = 2^0$)

Fig.3

A Time-Accurate Multigrid Algorithm for Euler Equations

A. Chatterjee and G.R. Shevare

Aerospace Engineering Deptt., Indian Institute of Technology,
Bombay, India.

Abstract : A time-accurate multigrid scheme is sought to be developed which allows one to track the evolution of hyperbolic conservation laws for substantial amount of time on coarser levels while retaining a fine-grid accuracy.

1 Introduction

The multigrid approach used in this formulation is of the Full Approximate Scheme (FAS) type [1,2], such schemes use the relative truncation error between the coarse and fine levels as a correction to the coarse-grid equations. This correction depends on the non-smooth components of the solution. When more than one time step is sought to be spent on coarser levels, the current and not too successful strategy would be to freeze this correction [3]. The ability of an algorithm to stay on the coarser levels for longer periods of time would depend on a large extent on the correction term being a reasonable mechanism for the transfer of fine-grid information. In this formulation we attempt to vary the correction term on the coarse level without recourse to the fine level, such that the new value of the correction term approximates the value that would be obtained from actual fine-grid information.

2 Scalar Linear Case

Consider the scalar linear convective equation

$$u_t + cu_x = 0. \qquad (1)$$

We consider two levels of discretization, both regularly spaced. A linear two-level scheme is used to solve Eqn. (1). We assume that the mesh spacing and the time step doubles on the coarse level. Let τ be the correction to the coarse-grid equations. It corrects our algorithm in space and time on the coarse grid, and can be represented as a weighted average of fine-grid values on the coarse grid. The weights will be a function of the Courant number (assumed to be same on both levels). The correction will go to zero if the shift condition is being satisfied. For this family of difference schemes, it can be easily shown that if more than one time step is sought on the coarse grid, then solving

$$\tau_t + c\tau_x = 0 \tag{2}$$

to a fine-grid accuracy by the same difference formula will give the new distribution for the correction on the coarse grid. The correction τ is a quantity defined on the coarse grid and it is clearly not possible to a solve it according to a fine-grid accuracy on the coarse grid. The scheme proposed solves Eq. (2) with a coarse-grid accuracy on the coarse grid to give a new distribution for the correction. Thus in an actual multigrid computation, this correction would necessitate periodic updating by going back to the finest grid though after a fairly long time. Our composite scheme to be solved on the coarse grid is given by

$$u_t + cu_x + \tau = 0. \tag{3}$$

$$\tau_t + c\tau_x = 0. \tag{4}$$

This describes a hyperbolic system with one characteristic along which $dx/dt = c$, but with two independent characteristic forms. Numerically if the above system is looked upon as a sequence of Riemann problems to be solved, then the local structure of the solution between any two grid points $(i, i+1)$ can be approximated by solving a Riemann problem with the corresponding grid values defining the pair of states for the Riemann problem. One can easily see from the Rankine-Hugoniot relation that the solution to the above Riemann problem is a single jump moving with a speed c. Equation (4) can be thought of as simulating the effect of non-smooth components on the coarse grid, whereas Eq. (3) models the behaviour of smooth components affected by a correction from the non-smooth terms. An interesting possibility would be to use a lower order monotone scheme such as the first order upwind scheme to solve for Eq. (4) on the coarse grid, while using a higher order scheme such as a second order central difference scheme for solving Eq. (3).

3 Scalar Nonlinear Case

Consider the scalar nonlinear hyperbolic conservation law $u_t + f(u)_x = 0$. For conservative schemes, addition of a correction term τ to the coarse grid, results in the numerical flux function on the coarse grid being calculated using fine grid accurate (in space and time) values. For the nonlinear case, we move τ nonlinearly on the coarse grid; clearly any such movement disregards the effect of nonlinear distortion on the fine grid. For the present case we locally enforce the formulations from the linear case, and the composite scheme used on the coarse grid is

$$u_t + f(u)_x + \tau = 0. \tag{5}$$

$$\tau_t + \hat{a}\tau_x = 0. \tag{6}$$

Here $\hat{a}$ is a piecewise constant value, constant in each interval $(i, i+1)$, and expressed as $(f(u_{i+1})-f(u_i))/(u_{i+1}-u_i)$. $\hat{a}$ is the actual shock speed for the nonlinear Riemann problem which is the conservation law $u_t + f(u)_x = 0$ with piecewise constant initial data u_{i+1} and u_i. $\hat{a}$ can be used to linearize this nonlinear Riemann problem to a scalar convective problem $u_t + \dot{a} u_x = 0$, the Riemann solution to which consists of a jump $u_{i+1} - u_i$ propagating with speed $\hat{a}$ (an exact weak solution to the nonlinear problem that may violate entropy conditions). As in the linear case, the locally linear structure between any two grid points $(i, i+1)$ obtained by considering a sequence of Riemann problems for the above system of Eqs. (5) and (6) consists of a single jump propagating with speed $\hat{a}$.

4 System of Linear Equations

For a linear system of hyperbolic equations given by

$$\underline{u}_t + A \underline{u}_x = 0 \tag{7}$$

the idea would be to apply the analysis of the scalar linear case to each scalar equation separately, obtained by decoupling the above system. The decoupled system can be written as

$$\underline{v}_t + \Lambda \underline{v}_x = 0 \tag{8}$$

where $A = R \Lambda R^{-1}$, $\underline{v} = R^{-1} \underline{u}$, R is the matrix of right eigenvectors of A, and Λ the diagonal matrix consisting of the eigenvalues of A. Applying the scalar linear analysis to each equation of the decoupled system Eq. (8), the system to be solved on the coarse grid can be written as

$$\underline{v}_t + \Lambda \underline{v}_x + \underline{\tau} = 0 \tag{9}$$

$$\underline{\tau}_t + \Lambda \underline{\tau}_x = 0 \tag{10}$$

where $\underline{\tau}$ is the vector consisting of corrections for the scalar equations. We can multiply Eqs. (9) and (10) by R to get coupled systems on the coarse grid

$$\underline{u}_t + A \underline{u}_x + \hat{\underline{\tau}} = 0 \tag{11}$$

$$\hat{\underline{\tau}}_t + A \hat{\underline{\tau}}_x = 0 \tag{12}$$

where $\hat{\underline{\tau}} = R \underline{\tau}$. It turns out that there is no need to actually deal with decoupled systems. This is possible because it can be proved that if $\underline{\tau}$ is the vector of correction terms for the system Eq. (8), then $R \underline{\tau}$ is the corresponding vector for the coupled system Eq. (7). The above relationship is true provided the restriction operators used in calculating the correction term are of the averaging or summation type. Fortunately while dealing with conservation laws, these are the type of restriction operators used.

5 System of Nonlinear Equations

Consider a system of nonlinear hyperbolic conservation laws represented by,

$$\underline{u}_t + A(\underline{u})\underline{u}_x = 0 \tag{13}$$

While writing down Eq. (6), we had used a piecewise constant value $\hat{a}$. The generalization of $\hat{a}$ while dealing with a system of equations given by Eq. (13) is $\hat{A}$ the Roe-averaged matrix defined by Roe [4]. So the composite system to be solved on the coarse grid is written as

$$\underline{u}_t + A(\underline{u})\underline{u}_x + \underline{\tau} = 0 \tag{14}$$

$$\underline{\tau}_t + \hat{A}\underline{\tau}_x = 0 \tag{15}$$

where $\underline{\tau}$ is the correction for the system of equations given by Eq. (13).

6 Numerical Results

The time-accuracy of the composite scheme on the coarse grid is tested by solving the shock tube problem of Sod [5]. We observe the evolution of this problem on the coarse grid, where the problem is solved for a considerable amount of time. In our numerical experiment the problem is first solved for half the total time on a fine grid and then for the second half on a coarse grid having twice the grid size (and consequently about twice the time step) without resorting to the fine grid. Equation (13) on the fine grid and Eq. (14) on the coarse grid are solved by the second order explicit MacCormack scheme with a third order artificial viscosity of Lapidus [6]. Equation (15) is solved by a first order upwind scheme.

We compare results on the coarse grid after the total time elapsed for a scheme which has its correction term frozen to its initially calculated value with that of the composite scheme. Figures 1 and 2 show the pressure profiles obtained from the frozen and the composite schemes respectively. The pressure profile for the frozen scheme in Fig. 1 has an appreciable kink exactly at the place where the shock was when restriction to the coarse grid took place. This can be attributed to the frozen discretization error at that place which is left behind while the solution has progressed. Our composite scheme on the contrary in Fig. 2 exhibits a smooth pressure profile without a kink.

In an actual multigrid calculation, the disadvantage of extra computational effort required to get a new approximation for the correction term on the coarser levels will be compensated by the substantially longer period of time one can stay on the coarser levels without going back to the finer levels to renew the correction term.

7 Multidimensional cases

This concept has been extended to multidimensional cases in the form of the two-dimensional Euler equations. The correction equation is solved by operator splitting.

Special treatment is required at the boundaries where boundary conditions for the correction equation needs to be specified.

References

[1] Brandt, A. (1982), Lecture Notes in Math (Springer), **960**, p. 22

[2] Jameson, A. (1986), Lecture Notes in Math (Springer), **1228**, p. 166

[3] Jespersen, D.C. (1985), AIAA 85-1493-CP, p. 58

[4] Roe, P.L. (1981), J. Comp. Phys, **43**, p. 357

[5] Sod, G. (1978), J. Comp. Phys, **27**, p. 1

[6] Lapidus, A. (1967), J. Comp. Phys, **2**, p. 154

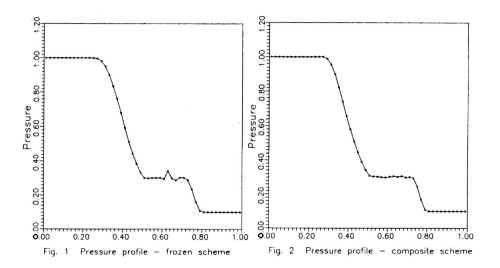

Fig. 1   Pressure profile – frozen scheme

Fig. 2   Pressure profile – composite scheme

Hypersonic Viscous Flow Computations with Solver and Grid Sensitivity Analysis

Michel Gazaix[1]

[1]Office National d'Etudes et de Recherches Aérospatiales
BP. 72, 92322 Châtillon Cedex, France

Abstract: A computational study of high speed viscous flows is described. The key features of this work are: the development of an efficient and robust implicit algorithm to solve the laminar/turbulent compressible Navier-Stokes equations, upwind discretization with carefully controlled numerical diffusion, the study of grid convergence, and the inclusion of real gas effects through a fully optimized numerical model of air in chemical equilibrium.

1 Introduction

Many technological projects require accurate numerical prediction of the hypersonic chemically reactive flow around a space vehicle flying in the earth atmosphere at high Mach number. In this context, we have undertaken a detailed computational study with the 3D multidomain block-structured finite volume code FLU3M, developed mainly by ONERA. We identify the upwind solvers and limiters which have to be used or rejected, depending upon their intrinsic diffusive properties. Both blunt body flows and complex ramp flows with separation are considered, with emphasis on the determination of peak heat flux and extension of the reversed flow region. A grid sensitivity study is then presented.

2 Numerical Algorithm

The computations have been performed with FLU3M, a three-dimensional, multi-domain, structured finite-volume code, in an upwind, cell-vertex, MUSCL formulation. The inviscid (Euler) fluxes can be computed with several solvers, of Flux Vector Splitting type (van Leer, Steger-Warming) or Flux Difference Splitting type (Godunov, Roe, Osher), and, recently, we have implemented the "hybrid" schemes introduced by Coquel and Liou [1]; these schemes combine the robustness of van Leer's scheme to capture strong stationary shocks, and the low diffusive properties of Osher's scheme, to accurately compute high gradient regions such as boundary layers. Real gas extensions of these upwind solvers have also been implemented. Viscous fluxes are discretized with a standard centered scheme. The implicit operator uses ADI factorization [2], with local time step.

3 Physical Model

High speed flows require accurate description of the chemical reactions occurring in the earth atmosphere. In many situations, chemical equilibrium hypothesis can be made. Computationally, it results in a system of non linear algebraic equations, which has to be solved for each grid point at each time step. A very large increase in numerical efficiency is obtained by using a numerical approximation of the thermodynamic functions (Mollier diagram) We have implemented an eleven species model of air in chemical equilibrium: $N_2, O_2, NO, N, O, N_2^+, O_2^+, N^+, O^+, NO^+, e^-$. By solving the resulting 5 × 5 non-linear system with Newton's method, one can construct a large thermodynamic data base. A very fast, computationally optimized, approximation is then built, using local 2D bicubic (spline or Hermite) interpolation, with variable knot spacing [3]. An example of a 2D viscous computation with chemical equilibrium assumption is given in Fig.1, showing friction and pressure coefficient computed for a Mach 16 flow over a cylinder (Re = 75 000) with adiabatic wall. Two grids (61 × 60 and 61 × 30) have been used. The computed values of the skin friction differ by less than 7%.

4 Control of Numerical Diffusion

4.1 Choice of Upwind Solver

The first test case is a 2D blunt body flow (M_∞=5.94, Re=400 000, T_∞=59.5 K, T_w=408.4 K). Fig.2 presents the heat flux and friction coefficient computed on the same grid with four different solvers: Roe with constant entropy correction (parameter ψ=0.3), Roe with entropic correction cancelled in the boundary layer, Coquel and van Leer. Whereas Coquel's and variable ψ Roe's solver give essentially identical results, constant ψ Roe's and van Leer's scheme are significantly overdiffusive and give poor estimates of the wall viscous coefficients. Let us remark that in this first example, the pressure coefficient was barely affected: the sensitivity to the solver was confined to viscous coefficients. However, it would be erroneous to extend this observation to all situations. Indeed, for more complex flows, and specially those with large reversed flow regions, even the pressure coefficient cannot be predicted faithfully with FVS or constant ψ Roe schemes. This is demonstrated by the second test case, which is a ramp flow (M_∞=14.1, α=24°, Re=104 000, T_∞=72.2 K, T_w=297 K), studied experimentally by Holden and Moselle. We have computed this flow with several upwind solvers. The limiter used here is van Albada. The plate leading edge region is computed separately, and provides the same upstream conditions for all the computations, thus avoiding any unwanted effects related to leading edge singularity. The minimum grid spacing ($dy = 2.10^{-5}$) is constant along the plate wall. The grid (199 × 99) is refined in the corner region. The size of the separated region, and consequently the location of the pressure plateau, is practically constant with the two variants of Osher and Coquel (hybrid) schemes. We obtain nearly the same result with the Roe scheme, provided that the entropic correction is canceled in the boundary layer. On the contrary, large variations are observed with van Leer's or Roe's scheme with ψ constant (Fig.3).

4.2 Choice of Limiter

One can observe on Fig.4 that, on the same grid (199 × 99), the size of the recirculation bubble predicted by *minmod* or *van Albada* limiter is significantly different. We have observed, however, that the effect of limiters is usually much smaller for attached flows. We have also observed that the convergence towards steady state is significantly faster with *minmod* limiter.

5 Study of Grid Convergence

Ideally, every computational result should be substantiated with a careful and detailed grid refinement study. However, the cost of a fully converged computation usually increases very rapidly with the number of grid points as well as with the inverse of the size of the smallest grid cell, making complete study very expensive. In this work, we wish to establish some criteria useful to choose a priori a rough estimation of the grid spacing necessary to compute viscous coefficients with a given accuracy. A parameter useful to describe the grid resolution is the wall cell Reynolds number, defined here as the ratio between the characteristic times of viscous and convective phenomena [4]:

$$\Delta T_V = \frac{\rho \, dy^2}{\mu} \quad , \quad \Delta T_E = \frac{dy}{V + a}$$

where V and a are characteristic convective and sonic speeds, respectively. At the wall, V vanishes, and in the case of an isothermal wall, a and μ are fully determined ($a_w = \sqrt{\gamma R T_w}$, $\mu_w = \mu(T_w)$), which leads to:

$$Re_c = \frac{\rho_w a_w dy_w}{\mu_w}$$

In hypersonic flows, an estimation *a priori* of the wall density ρ_w can be quite difficult, specially in separated flows where very large variations - typically two orders of magnitude - of the wall density can be found.

We focus on the blunt body 2D flow. The grid stretching is given by:

$$dy_{j+1} = dy_j.(1 + \epsilon \, sin(\pi(j-1)/(N-1)))$$

where N is the number of grid points on the normals to the wall, and ϵ is computed iteratively. Four grids have been used (40 × 25, 40 × 50, 40 × 100, 40 × 200), with the finest grid spacing at the wall (dy_w) varying from 10^{-5} to 8.10^{-5}; Re_c at the stagnation point varies between 2 on the finest grid, and 16 on the coarsest grid. Fig.5 shows that the discrepancies are higher for the heat coefficient than for the friction coefficient. The computed stagnation heat flux differs by 13 % on the two extreme grids (40 × 25 and 40 × 200).

For any quantity Q, an estimation of the numerical error can be obtained through Richardson's extrapolation [5]. Assuming a second order scheme, and using numerical results Q_N obtained with grid step size Δ and 2Δ, an estimation of the "exact" solution Q_E can be expressed as:

$$\begin{aligned} Q_E &= Q_N(\Delta) + \alpha\Delta^2 + \text{higher order terms} \\ Q_E &= Q_N(2\Delta) + \alpha(2\Delta)^2 + \text{higher order terms} \\ \Longrightarrow Q_E &= Q_N(\Delta) + \tfrac{1}{3}[Q_N(\Delta) - Q_N(2\Delta)] \end{aligned}$$

Fig.6 plots the estimated numerical error $Q_N - Q_E$ for the heat flux coefficient at the stagnation point as a function of the number of points. We observe that the error decrease is less than quadratic. To explain this behaviour, one can invoke limiter effects (MUSCL schemes are only first order in regions where limiters are active), or the possibility that higher order terms are not negligible (contrary to the assumption made to derive an "exact" numerical estimation).

6 Conclusion

In this work, detailed attention has been given to the control of numerical errors, arising from truncation error (grid sensitivity) and from the intrinsic dissipation of upwind schemes, in the context of high Reynolds number hypersonic flow computations. This work should help workers in CFD to choose the solver and limiter best suited to their needs, in terms of accuracy and numerical efficiency. An analysis of the grid requirements has been performed, and can be used as a guide to construct grids in other related computations.

References

[1] F. Coquel, M.S. Liou: AIAA 93-3302-CP (1993)

[2] D.Darracq, M. Gazaix: 19th Int. Symp. on Shock Waves, Marseille, France (1993)

[3] M. Gazaix: AIAA 93-0892 (1993)

[4] K. Khalfallah, G. Lacombe, A. Lerat, A. Raulot, Z.N. Wu: Proc. 1st European CFD Conf., 7-11 Sept. 1992, Brussels, ed. by Ch. Hirsch et al (Elsevier), p175

[5] F.G. Blottner: AIAA 89-1671 (1989)

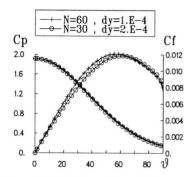

Fig. 1. Equilibrium Flow, $M_\infty = 16$

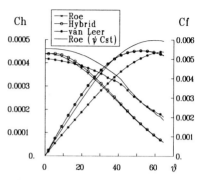

Fig. 2. Blunt body, $M_\infty = 5.94$
Influence of solvers

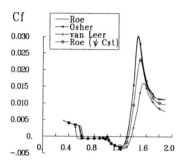

Fig. 3. Holden ramp.
Influence of upwind solvers

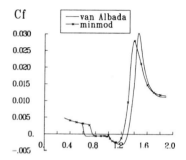

Fig. 4. Holden ramp.
Influence of limiters

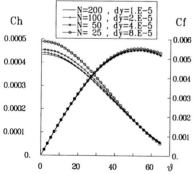

Fig. 5. Blunt body, $M_\infty = 5.94$
Grid sensitivity study

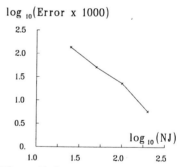

Fig. 6. Estimated numerical error
through Richardson's extrapolation

Speed-up of CFD codes using analytical FE calculations

J. P. Gregoire and G. Pot

D.E.R. Electricitè de France
Dèpartment MMN
1, avenue du Gènèral de Gaulle
92141 Clamart Cedex - France

Abstract : Today combining supercomputer performances with algorithmic optimizations large linear systems are solved in a very short time. Consequently reduction of FE calculations time becomes crucial. This paper presents how these calculations can be efficiently speed-up by replacing them by analytical ones and by fully vectorizing their assembling. Moreover the simplicity of these calculations enables on efficient parallelization on CRAY YMP.

1 Introduction

Today, combining supercomputer performances with algorithmic optimizations (full vectorization of conjugate gradient algorithm using jagged diagonal storage of matrices [2]), large linear systems are solved in a very short time. For example, using N3S code [1], a 3D flow simulation around a car (mesh composed of 70,000 tetrahedra with 100,000 velocity nodes) needs for each turbulent diffusion computation on a CRAY YMP:

- 30 seconds for one global matrix calculation,
- 57 second for the right hand side calculation (very complex because of turbulent $k - \epsilon$ model),
- 15 seconds for the STOKES problem solving (350 000 velocity-pressure unknowns).

Consequently reduction of FE calculations time becomes crucial for any production FE code. Usually, in these codes, several types of FE are implemented in a general way using coordinates transformation from the real FE to a reference FE and numerical integration performed by GAUSS formulae.

2 Restriction of FE types

Our approach is to restrict to triangle in 2D and tetrahedron in 3D although the same ideas can be extended to other kinds of FE. These FE, named simplicial elements, have the following simplifying properties:

- linear shape functions Ψ_I, used to discretize pressure, are identical to area or volume co-ordinates,
- their gradients $\overrightarrow{\nabla \varphi_I}$ are constant vectors on each FE,
- their formal intergation is done using ZIENKIEWICZ formula (triangle):

$$\int_T \Psi_1^{n_1} \Psi_2^{n_2} \Psi_3^{n_3} \, dx = \frac{n_1! \, n_2! \, n_3!}{(n_1 + n_2 + n_3 + 2)!} \, 2S$$

(analog formula for tetrahedron)

- quadratic shape functions Ψ_I in 2D and 3D, used to discretize velocity, are directly expressed in terms of linear ones,
- their gradients $\overrightarrow{\nabla \varphi_I}$ are expressed in terms of Ψ_I and $\overrightarrow{\nabla \Psi_I}$ on each FE. The shape functions of higher order are also expressed in the same way.

3 FE calculations by hand

We start these analytical FE calculations by hand. From the previous properties it follows directly that every elementary mass matrix M has the form (see at the end of the paper):

$$[M] = \frac{S}{N}[Z]$$

where S is the surface of the FE (volume in 3D), N a positive integer, Z a relative integer matrix; N and Z depend only on shape function degree and space dimension. The elementary Laplacian matrix has a similar form except that Z coefficients are now linear combinations of FE the parameters $A_k = \overrightarrow{\nabla \Psi_I} \cdot \overrightarrow{\nabla \Psi_J}$ (3 values in 2D and 6 values in 3D).

For the integration of the right hand side $f = \sum_f f_J \, \varphi_J$ on the FE, T, we reuse the elementary mass matrix as follows:

$$\int_T \varphi_I \, dx = \frac{S}{N} \sum_J Z_{IJ} \, f_J$$

The tests of these improvements were so promising (speed-up of a factor more than 100), that we went on more complex FE computations by developing an adequate formal preprocessor.

4 PREFN3S: a formal preprocessor for analytical FE calculations

Using adequate data structure to represent polynomials of the form:

$$\sum_{I,J,K} n_{IJK} \, \Psi_I^{n_I} \, \Psi_J^{n_J} \, \Psi_K^{n_K} \quad \text{(for 2D mass matrix)}$$

or $\quad \displaystyle\sum_{I,J,K,G} n_{IJKG} \, \Psi_I^{n_I} \, \Psi_J^{n_J} \, \Psi_K^{n_K} \, \overrightarrow{\nabla \Psi}_G \quad$ (for 2D Laplacian matrix)

we coded, in FORTRAN, the formal operations on polynomials previously executed by hand.

For example, in case of quadratic triangle and linear density ρ, the elementary mass matrix is :

$$\left[\int_T \rho \varphi_I \varphi_J \, dx \right]_{I,J} = \frac{S}{N} \left[\sum_{K=1}^{3} Z_{IJK} \, \rho_K \right]$$

and PREFN3S returns the value $N = 1260$ and the Z_{IJK} integers given in a FORTRAN common.

With this package we computed about 100 elementary matrices of different types including the 2D axisymmetrical case (dx is replaced by $2\pi R \, dx$ where R is a linear function).

5 Special right side calculations

In a turbulent thermal flow calculation, the equations governing the unknowns k and ϵ of the turbulence model contain terms of the following form [3].

$$\int_E^T \rho \, \epsilon \, \Phi_I \, dx$$

where T is a complex expression of quadratic shape function derivatives.

This integral is too complex to be analytically calculated for quadratic FE. If we restrict to P1-isoP2 sub-quadratic FE (same nodes as quadratic ones but the element is decomposed in sub-elements S-E on which shape function are linear) the expression T becomes constant by sub-element.

Consequently we can reuse the elementary linear mass matrix with linear coefficient ρ on each sub-element and the previous integral has the form:

$$\frac{S}{N} \sum_{S-E} T_{S-E} \sum_{J \in S-E} \left[\sum_{K=1}^{3} Z_{IJK} \rho_K \right] \epsilon_J$$

6 Full vectorization of assembling

The assembling of elementary matrices or vectors necessarily an indirect addressing which generates vector dependencies. The CRAY compiler generates a semi-vectorized code: a vectorized update followed by a scalar correction of vector dependencies.

In order to obtain full vectorization of assembling we manage dependencies through FE reordering so that these dependencies occur only between 2 different vector registers guaranteeing no vector dependency. While depending on the initial number of dependencies the local speed-up is always greater than 4.

7 Timings on CRAY YMP

1) **Computational speed**

   Combining analytical FE calculations with full vectorized assembling, as in N3S code, leads, on CRAY YMP, to a speed always greater than 120 Mflop/s for any kind of matrix computation and 150 Mflop/s for any kind of vector computation (peak is 330 Mflop/s).

2) **Computational speed-up**

   A new simulation of the 3D flow calculation around a car with the optimized N3S version leads to the following timings:

   0.22 seconds instead of 30 seconds for one global matrix calculation: speed-up of 136

   0.33 seconds instead of 57 seconds for the right hand side calculations: speed-up of 184

   same computational time for the STOKES problem solving.

Note that these high speed-ups result of the combination of I/O deletion, computational reduction and vectorization improvements and that the resulting benefit occurs at each time step.

8 Conclusions

By restricting to simplicial FE and by carrying analytical FE calculations, elementary matrices and right hand sides can be precomputed, either by hand or thanks to the formal preprocessor PREFN3S for more complex cases.

The use of these precalculations coupled with a full vectorized assembling of them leads to a local speed-up greater than 140 on a CRAY YMP.

The global speed-up is highly dependent on algorithmic choices. For N3S code, a 3D turbulent simulation, for one time step and 1,000 velocity nodes, needs 0.22 seconds instead of 0.70 seconds before, giving a global speed-up of 3.3.

Moreover, the simplicity directives and gives a local speed-up of 3.3 on 4 processor.

As FE calculations are explicit, FE information file becomes obsolete. So, memory requirement is reduced by a factor at about 100 which enables an in-core treatment even on a workstation.

Finally, the same approach can be applied to most FE codes with hopefully the same performance benefits.

References

1 Chabard, Metivet B., Pot G., Thomas B., An efficient finite element method for the computation of 3D turbulent incompressible flows

2 Gregoire J.P., Nitrosso B., Pot G., Conjugate gradient performances enhanced in the C.F.D. code N3S by using a better vectorization of matrix vector product, Proceedings of IMACs 91, Dublin, July 1991.

3 Moulin V., Caruso A., Daubert O., Pot G., Thomas B. Improvement of FE. algorithms implemented in CFD code N3S for turbulent or dilatable flows. Proceedings of 5th Int. Symp. on Refined Flow Modelling and Turbulence Measurements, Paris, September 1993.

EXAMPLES OF MASS MATRICES FOR TRIANGLE (ANALOG FOR TETRAHEDRON)

S/12

2	1	1
***	2	1
***	***	2

linear

S/60

$6\rho1+2\rho2+2\rho3$	$2\rho1+2\rho2+1\rho3$	$2\rho1+1\rho2+2\rho3$
***	$2\rho1+6\rho2+2\rho3$	$1\rho1+2\rho2+2\rho3$
***	***	$2\rho1+2\rho2+6\rho3$

linear with linear ρ

S/180

6	-1	-1		-4	
***	6	-1			-4
***	***	6	-4		
***	***	***	32	16	16
***	***	***	***	32	16
***	***	**	***	***	32

quadratic

EXAMPLES OF LAPLACIAN MATRICES FOR TRIANGLE (SURFACE S DISEAPPERS WHILE VOLUME IS PRESENT FOR TETRAHEDRON)

A2+A3	-A3	-A2
***	A1+A3	-A1
***	***	A1+A2

$(\rho 1+\rho 2+\rho 3)$

A2+A3	-A3	-A2
***	A1+A3	-A1
***	***	A1+A2

linear   linear with linear ρ

3A2+3A3	A3	A2	-4A3		-4A2
***	3A1+3A3	A1	-4A3	-4A1	
***	***	3A1+3A2		-4A1	-4A2
***	***	***	8s	-8A2	-8A1
***	***	***	***	8s	-8A3
***	***	***	***	***	8s

quadratic   (s=A1+A2+A3)

A Field Method for 3D Tetrahedral Mesh Generation and Adaption

D. Hänel, R. Vilsmeier

Institut für Verbrennung und Gasdynamik, Universität Duisburg,
Lotharstrasse 1, 47048 Duisburg, Germany

Abstract : A field method to generate 3D tetrahedral meshes is presented. After the initial triangulation between given boundaries a set of tools is employed to generate a suitable mesh for a first computation. The same algorithm is used for mesh adaptation as well. The adaptation can be seen as a continued mesh generation while additional information is available. This information is introduced via virtual stretching.

1 Introduction

Comparing methods in CFD employing structured or unstructured grids, each grid type shows its distinct advantages. Structured grids offer easy algorithmic structures, sequential access to the stored data and higher performance per grid node. The advantage of unstructured grids is the geometric flexibility, including the ease to adapt meshes. However the development of flexible and reliable mesh generation and adaptation algorithms is required.

Since the triangulation of given sets of points is unflexible, the attention is drawn towards methods, able to generate interior nodes by their own. Marching front algorithms generate elements and nodes upon existing boundaries. These methods are very common in use, for example [1].

Another approach to generate unstructured meshes are the field methods, for example [2]. These methods optimize an existing, closed triangulation. Although elliptic the complexity is low due to the locality of all performed operations. Our proposal here is such a field method. Since initial triangulations are required, the first topic is related to the generation of these. The second part is related to the optimization from the initial grid to a grid, suitable for a first flow computation. The approach allows to use the same algorithm for mesh adaptation as well.

2 Initial triangulation

Aim of the initial triangulation is to close the volume between given surface triangulations, being the boundaries of the domain.

The problem is solved employing a rising bubble Delaunay tetrahedra generator due to its property of respecting the boundary triangulations:

a) The boundary triangles of the domain are the initial front triangles. The boundary nodes are the set of points to be triangulated.

b) For each front triangle (3 nodes) an additional fourth node is sought, in order to obtain a tetrahedron, whose circumsphere does not contain other nodes in the triangulation direction. The chosen node is the one, which yields the tetrahedron with the rear-most centre of its circumsphere. At this item a restriction of the original Delaunay criterion has to be made: It is allowed to have nodes within the circumsphere of a tetrahedron, if there is a boundary of the domain between the tetrahedron and the node contained.

c) Some checks for the new tetrahedron produced are performed. Other front triangles may be intersected and some accuracy limits apply. In any trouble, the new tetrahedron is rejected.

d) The sides of the new tetrahedra are inserted in the list of front triangles. Front triangles touching each other are deleted from the list.

e) Continue at b) until the front list is empty.

Unfortunately in some cases it is impossible to close the triangulation in the way proposed. Beside the mentioned accuracy limits another reason is the existence of polyhedra with triangulated surfaces, that may not be filled by tetrahedra, unless crossing their surfaces. The simplest polyhedron with this property is an octahedron with six nodes. It is a distorted prism, whose "quadrilateral" sides are triangulated to obtain the minimal volume. If for any reasons it is impossible to close the triangulation an additional node is inserted at an optimized position, all tetrahedra, whose circumspheres contain the new node, are deleted, the list of front triangles is adjusted accordingly and the process continues.

No search reduction is yet employed, resulting in a complexity of $O(N_b^2)$. As the number of boundary nodes N_b is proportional to $N^{\frac{2}{3}}$ the complexity is $O(N^{\frac{4}{3}})$ based on the number of nodes of the later mesh. To reduce this complexity a space marching version is planed, analogous to the 2D method, [3].

3 Mesh optimization

After the initial triangulation the mesh is optimized employing a set of tools. Since this optimization problem is a mixed discrete analogue one, the tools solve the correspondent partitions of the optimization problem.

a)The local mesh density

is represented by the locally preferred length G_k of the edges. For boundary nodes this quantity is the length of the longest boundary edge adjacent to these nodes, multiplied with a statistic factor. For any other node the quantity G_k is computed solving a boundary value problem during the development of the mesh. As the boundary nodes and edges are known at the beginning of the generation process, the function G_k could be computed in advance. However there is no mesh yet to support the solution.

b) Point insertion:

Additional points are inserted in the centre of all tetrahedra if at least one of their edges is longer than the local quantity G_k and none of the neighbouring tetrahedra was already refined in the same insertion loop. The insertion is done in that sparse way to avoid a sudden point overflow.

The point insertion is also used to break up very flat boundary tetrahedra having boundary triangles as two of their faces. These tetrahedra may have been produced by the rising bubble algorithm and fit the modified Delaunay criterion, having a very large circumsphere, almost all outside the computational domain. Since their vertices are all boundary nodes, not shift-able by the smoothing procedure, these tetrahedra would else survive all other optimization steps.

After the insertion the new node is connected to the four nodes of the tetrahedron, in which it has been inserted, producing four new tetrahedra. The insertion of additional boundary nodes is only required at a later adaptation of the mesh.

c) Edge reconnection:

In the present version edge reconnection is used to fit Delaunay's criterion. The method is based on diagonal swapping. It corresponds in essential to that proposed in [4]. Local groups of tetrahedra, that together are build out of 5 nodes, are analyzed. These groups might consist of 2, 3 or 4 tetrahedra. The local group of 2 tetrahedra may swap to 3 ones by introducing an edge between the two previously not connected nodes. Opposite to this swap, a group of 3 tetrahedra may swap to 2 ones by deleting the common edge. The possible swaps are performed if the circumspheres of the 2 or 3 tetrahedra contain the fifth node of the local group. A truncation of the swapping sequence is possible, therefore achieving Delaunay's criterion is not save, however for the practical use this property does not impose any trouble.

As the discrete Delaunay criterion is not identical to the main optimization criterion given by the smoothing procedure, tests were made to use a discrete optimization for the swapping procedure. Unfortunately this was not successful as the swap sequences are often blocked at local minima.

The very flat boundary tetrahedra, broken up via insertion, would easily be reproduced by the reconnection procedure if the inserted node would have been shifted farther away from the boundaries. Therefore, the production of such tetrahedra is rejected if a limit of the smoothing criterion is exceeded.

d) Smoothing:

Smoothing is done to increase grid quality by recursively moving the points to optimal positions. The smoothing procedure is based on the circumsphere volumes of the tetrahedra and formulated as a minimization of a sensitive quantity T_S:

$$T_S = \sum_{i=1}^{ntet} \frac{V_c(i)}{(V_t(i))^w} \cdot G_k^{3 \cdot (w-1)}$$

where $V_c(i)$ is the circumsphere volume of a tetrahedron i, $V_t(i)$ is its own volume and $G_k(i)$ the average value for the quantity G_k at its vertices. $ntet$ is the number of tetrahedra and w is a weighting term.

The minimization is carried out by a point-wise Newton method, iterating derivatives of T_S to zero. In rare cases of ill condition a line search is performed.

This smoothing procedure consumes a remarkable amount of computational time, compared with other methods, e.g. laplacian smoothers. Its advantage is that too flat or invalid tetrahedra are avoided. The smoothing procedure may be used up to the convergence to a global minimum. However, for the applications here, only a limited amount of steps is performed.

The tools described above are used in recurrence up to moderate convergence of the whole process. It is useful to avoid spending to much work on smoothing at intermediate stages and to increase the number of smoothing steps towards the end of the generation process. The complexity of the algorithm is $O(N)$. However this low complexity can only be achieved, if smoothing procedure and relaxation of the G_k are used as local tools without global convergence.

The present version of the algorithm is able to generate around 1 million tetrahedra per CPU-hour on an HP735 workstation, where about 80 % of the CPU time is spent for the smoothing procedure. It is expected to enhance the performance by a factor of at least 5 in further versions without quality penalties.

4 Mesh adaptation

The aim of the generation procedure, as described above, is to generate a smooth homogeneous triangulation without any directionality. Only the density function was used to allow a smooth change of the element sizes.

Adaptivity is introduced via virtual stretching, that means, according to a previous flow solution and some adaption criteria a transformation from the physical to a wider virtual space is performed. The mesh optimization continues at this virtual space and in physical space the adaptive triangulation is obtained. The transformations are supported via symmetric 3x3 matrices, stored at the nodes of the previous mesh, being a background mesh for the adaption process. This concept enables anisotropic or isotropic refinements. A more detailed description, restricted to the analogous version in 2D can be found in [5].

In 3 dimensions however an additional difficulty is the adaption of the triangulated surfaces. At present, a subdivision algorithm for the boundary edges is used. Although no restrictions apply, the anisotropic adaption on curved boundaries can seriously "damage" the original geometry. A further problem is the production of tetrahedra with large interior angles, critical when evaluating second derivatives. These problems are subjects of ongoing investigations.

References

[1] Löhner R. and Parikh P.: Three- Dimensional Grid Generation by the Advancing Front Method. Int.J. Num.Meth. Fluids 8, pp 1135-1149, 1988.

[2] Weatherhill N. P., Hassan O., Marchant M. J., Marcum D. L.: Adaptive Inviscid Flow Simulations using Distributions of Sources. Num. Meth. in Laminar and Turbulent Flow, pp 1394-1407 Pineridge Press, 1993.

[3] Vilsmeier R., Hänel D.: Generation and Adaptation of 2-D Unstructured Meshes. Numerical Grid Generation in Computational Fluid Dynamics and Related Fields, pp. 55-66, North-Holland, Amsterdam, 1991.

[4] Joe B.: Construction of Three-Dimensional Delaunay Triangulations From Local Transformations. In CAGD, Vol. 8, pp 123-142, 1991.

[5] Vilsmeier R., Hänel D.: Adaptive Solutions for Compressible Flows on Unstructured, Strongly Anisotropic Grids. In: Ch. Hirsch et al.: Computational Fluid Dynamics 92, Volume 2, Elsevier Science Publishers, 1992.

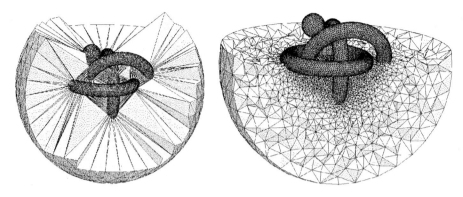

Test case consisting of three interlocked tori and a sphere placed inside a spherical domain. Broken partition of initial triangulation, rising bubble (top, left). Broken partition of final mesh 484'795 tet. (top, right).

Example for anisotropic adaptivity. Broken partition of a supersonic engine inlet after the first adaption, adaption criterion: gradient of density, 800'128 tet. (middle, right). Corresponding solution for inviscid flow from $-X$ direction, $Ma = 3.0$, central discretization, lines of constant pressure (bottom, right).

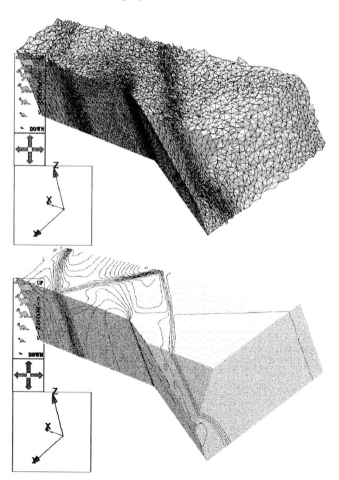

Multigrid Euler Solutions with Semi-Coarsening and Local Preconditioning

John F. Lynn and Bram van Leer

W. M. Keck Foundation Laboratory for Computational Fluid Dynamics,
The University of Michigan, Ann Arbor, Michigan, USA

1 Introduction

Explicit marching schemes are commonly used as multigrid relaxation schemes when solving the Euler and Navier-Stokes equations. These schemes must feature effective high-frequency damping under all flow conditions in order to be suited for use in multigrid marching. Multi-stage schemes provide the flexibility to achieve the desired smoothing properties. In (Lynn and Van Leer, 1993) we obtained optimal sequences of multi-stage time-step values for discretizations of the Euler or Navier-Stokes spatial operator. Though this method was a step forward from earlier optimization formulations, which based the design of the multi-stage schemes on either the scalar one-dimensional e.g. (Van Leer et al., 1989) or two-dimensional convection equation, the schemes obtained were not completely independent of flow conditions, such as Mach number and flow angle. With semi-coarsening (Mulder, 1989, 1991), the high-frequency domain over which the multi-stage scheme must be a good damper of errors is reduced to "hi-hi" combinations. This makes it possible to design optimal multi-stage schemes that are largely independent of flow conditions.

It is shown that such multi-stage schemes, in conjunction with the semi-coarsened multigrid algorithm and local preconditioning (cf. Van Leer et al., 1991), provide a fast and robust method for achieving steady-state solutions.

2 Semi-coarsening and local preconditioning

When multi-dimensional convection is aligned with one of the grid directions, single-grid relaxation schemes cannot damp high-frequency errors propagating in the normal direction that are coupled to low frequency errors in the convection direction. This is known as the grid-alignment problem. Semi-coarsening is a method meant to resolve this problem in a multigrid context. Mulder (Mulder, 1989, 1991) has developed an efficient solver for the steady 2-D Euler equations based on semi-coarsening.

Local preconditioning matrices attempt to remove the spread among characteristic speeds as much as possible. A recently derived matrix (Van Leer et al., 1991) achieves what can be shown to be the optimal condition number for the characteristic speeds, namely, $1/\sqrt{1 - \min(M^2, M^{-2})}$ where M is the local Mach number. The effect of local preconditioning on the discretizations of the spatial Euler operator is a strong concentration of the pattern of eigenvalues in the complex plane. This makes it possible to design multi-stage schemes that systematically damp most high-frequency waves admitted by a particular discrete operator (Lynn and Van Leer, 1993).

The matrix employed in this work is a modification by (Lee and Van Leer, 1993). It was realized that multigrid damping could be improved by modifying the matrix such that the highest-frequency modes of all the waves coincide.

3 Optimization procedure

The procedure for optimizing high-frequency damping aims at minimizing the maximum of the modulus of the scheme's amplification factor over a given set of high-frequency eigenvalues. The input parameters are the time-step values $\Delta t^{(k)}$, $k = 1,..,m$, of an m-stage algorithm. According to linear theory, one step with the full scheme multiplies each eigenvector of the operator $Res(U)$, with associated eigenvalue λ, by a factor of the form

$$P(z) = 1 + z + \sum_{k=2}^{m} c_k z^k, \qquad (3.1)$$

where $z = \lambda \Delta t$ generally is complex. The $m-1$ coefficients c_k relate to the time-step ratios $\alpha_k = \Delta t^{(k)}/\Delta t$; the actual time step Δt is the mth parameter.

The optimization procedure starts out by computing, for a fixed combination of M and ϕ ($\equiv$ flow angle), a discrete set of eigenvalues for wave-number pairs (β_x, β_y) in the high-frequency range, i.e.

$$|\beta_x| \in (\pi/2, \pi) \quad \text{and} \quad |\beta_y| \in (\pi/2, \pi). \qquad (3.2)$$

Assuming a set of starting values for the m-stage scheme, for instance Tai's values (Van Leer et al., 1989), the value of $|P(z)|$ is computed for all eigenvalues previously obtained, and its maximum is found. This is our functional $\sigma(\Delta t^{(1)}, .., \Delta t^{(m)}; M, \phi)$; it must be minimized by varying the m parameters.

It is not a priori clear that the $\Delta t^{(k)}$ will be insensitive to values of M and ϕ. If, as in (Lynn and Van Leer, 1993), we have to consider the whole high-frequency domain, i.e. $|\beta_x| \in (\pi/2, \pi)$ and/or $|\beta_y| \in (\pi/2, \pi)$, two problems arise: the alignment problem mentioned earlier and a singularity problem for $M \to 1$. Figure 1 gives evidence of both problems. There are lo-hi entropy/shear modes extending all the way into the origin, and hi-lo acoustic error modes at some distance to the origin that vanish as $\sqrt{1-M^2}$. These two effects go away if we restrict ourselves to the high-frequency range appropriate for semi-coarsening.

Any remaining flow-angle (ϕ) dependence may also be removed by appropriate definition of the Courant number ν. As in (Lynn and Van Leer, 1993), for the preconditioned Euler equations, with the characteristic speeds equal to or close to q, we define the Courant number as $\nu = q\Delta t/l(\Delta x, \Delta y, \phi)$, where l is a typical cell-width that depends on the flow direction. Figure 3 shows flow angle variations for a typical optimal scheme. The hi-hi frequency eigenvalues are largely unaffected by Mach number variations. This results in the ability to design schemes that are also Mach number independent. Figure 4 shows Mach number variations for a typical optimal scheme.

The optimal (in the L_∞ sense) m-stage scheme may hence be obtained as the solution to the following minmax problem:

$$\sigma_{opt} = \min_{(\vec{\alpha},\nu)} \left(\max_{|\beta_x|,|\beta_y| \in [\pi/2,\pi]} \|P(z(\beta_x,\beta_y,\nu),\vec{\alpha})\| \right). \quad (3.3)$$

This optimization problem is solved using the method of *simulated annealing* in conjunction with the *downhill simplex* algorithm of Nelder and Mead. Figure 2 shows an example of a scheme designed using this optimization procedure.

4 Multigrid solutions

Multi-stage schemes have been developed using the optimization procedure described above together with the high-frequency footprint of the modified Roe operator (cf. Van Leer *et al.*, 1991) for the Euler equations. Table 1 contains a family of these coefficients for use with first-order and second-order ($\kappa = 0$) upwind spatial discretizations.

Table 1. First and second-order multistage coefficients based on optimization over "hi-hi" frequency domain. σ_{opt} values are for $M = 0.1$ and $\phi = 0$.

	\multicolumn{9}{c}{Number of Stages}
	2	3	4	5	6	2	3	4	5	6
α_1	0.3333	0.1467	0.08125	0.05204	0.03441	0.5713	0.2239	0.1299	0.08699	0.05691
α_2	1	0.3979	0.2033	0.1240	0.08101	1	0.5653	0.2940	0.1892	0.1264
α_3		1	0.4226	0.2343	0.1476		1	0.5604	0.3263	0.2113
α_4			1	0.4381	0.2519			1	0.5558	0.3312
α_5				1	0.4486				1	0.5359
α_6					1					1
ν	1.000	1.525	2.105	2.682	3.042	0.6305	1.046	1.401	1.747	2.148
σ_{opt}	0.3333	0.1418	0.06328	0.03024	0.02723	0.6475	0.4279	0.2927	0.2047	0.1592
	\multicolumn{5}{c}{First order}	\multicolumn{5}{c}{Second order ($\kappa = 0$)}

A multigrid Euler code was written that implemented these coefficients along with semi-coarsening and local preconditioning. The problem of "flow past a semi-circular bump ($t/c = 0.042$) in a channel" was solved for $M_\infty = 0.35$, 0.85 and 1.4. Results of these runs are presented in Tables 2 and 3 in terms of equivalent work units. A work unit is the amount of work required to compute a single-stage update on the fine grid. These are the "best" single and multigrid results obtained from multiple runs using 2-6 (3-6 for second order) stages and 1-5 (1-4 for second order) grid levels. The second-order solutions made use of Van Albada's limiter and defect-correction multigrid cycles for the multigrid cases.

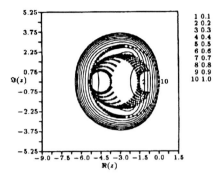

Fig. 1 Optimal scheme obtained by optimizing over entire high frequency domain minus wedge filter (cf. Lynn and Van Leer, 1993). $M = 0.9$, $\phi = 0$.

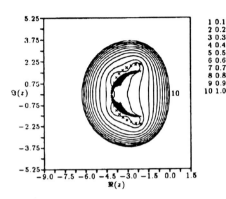

Fig. 2 Optimal scheme obtained by optimizing over "hi-hi" high frequency domain only. $M = 0.9$, $\phi = 0$. σ_{opt} is reduced to 0.06 from 0.41 in Figure 1.

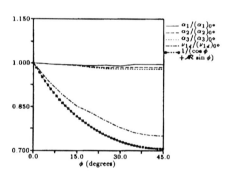

Fig. 3 Variation of multi-stage coefficients with flow angle (ϕ).

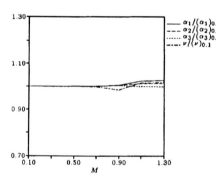

Fig. 4 Variation of multi-stage coefficients with Mach number.

Tables 2 and 3 indicate that local and matrix time-stepping require similar amounts of work in the single-grid subsonic and transonic runs. This is probably an artifact of the problem. The reflective wall-boundary conditions in the channel provide minimal attenuation of acoustic error-modes and this effect dominates single-grid convergence. This is not a problem for the supersonic cases, which have a largely convective nature. The speed-up of multigrid with local preconditioning over multigrid with local time-stepping is a lot more dramatic; the former is 3-4 times faster in all cases. Convergence in the subsonic and transonic cases is also an order of magnitude faster than in the corresponding single-grid cases. Local preconditioning performs admirably on a single-grid for the supersonic case, and there is not much improvement possible without modifying multigrid to better handle convection-dominated flow.

Table 2. Comparison of first-order convergence rates for flow past a semi-circular bump in a channel, 64x32 grid. Work units required to reduce the norm of the residual by five orders of magnitude. (Matrix TS ≡ local preconditioning).

	$M = 0.35$		$M = 0.85$		$M = 1.4$	
	Local TS	Matrix TS	Local TS	Matrix TS	Local TS	Matrix TS
Single	2940	2520	4140	4920	1550	355
Multigrid	1123	268	911	326	651	234

Table 3. Comparison of second-order ($\kappa = 0$) convergence rates for flow past a semi-circular bump in a channel, 64x32 grid. Work units required to reduce $\|TE\|_1$ (cf. Mulder, 1991) to 10^{-2} for $M = 0.35$ and $M = 1.4$ and to 5×10^{-2} for $M = 0.85$. Nested iteration with 5 defect-correction sweeps on each coarse grid level was used initially to improve robustness for multigrid solutions.

	$M = 0.35$		$M = 0.85$		$M = 1.4$	
	Local TS	Matrix TS	Local TS	Matrix TS	Local TS	Matrix TS
Single	3685	2380	5305	9600	1344	414
Multigrid	582	191	515	167	722	309

References

[1] Lee, D., van Leer, B. (1993): "Progress in Local Preconditioning of the Euler and Navier-Stokes Equations", 11th AIAA Computational Fluid Dynamics Conference (in Proceedings).

[2] Lynn, J. F., Van Leer, B. (1993): "Multi-Stage Schemes for the Euler and Navier-Stokes Equations with Optimal Smoothing", 11th AIAA Computational Fluid Dynamics Conference (in Proceedings).

[3] Mulder, W. (1989): "A New Multigrid Approach to Convection Problems", J. Comput. Phys., **93**.

[4] Mulder, W. (1991): "A High-Resolution Euler Solver Based on Multigrid, Semi-Coarsening and Defect Correction", J. Comput. Phys., **100**.

[5] Van Leer, B., Lee, W.-T., Roe, P. L. (1991): "Characteristic Time-Stepping or Local Preconditioning of the Euler Equations", 10th AIAA Computational Fluid Dynamics Conference (in Proceedings).

[6] Van Leer, B., Tai, C.-H., Powell, K. G. (1989): "Design of Optimally-Smoothing Multi-Stage Schemes for the Euler Equations", 9th AIAA Computational Fluid Dynamics Conference (in Proceedings).

CAGD Techniques in Structured Grid Generation and Adaption

Bharat K. Soni[1]
Hugh Thornburg[2]

NSF Engineering Research Center for
Computational Field Simulation
Mississippi State University
Mississippi State, MS 39762

Abstract : Applications of Computer Aided Geometric Design (CAGD) techniques for effficient and accurate boundary/surface/volume structured grid generation and adaption are presented. The Non-Uniform Rational B-Spline (NURBS) representation for parametric geometric entity definition is employed. The applications include redistribution, refinement, remapping, adaption and optimization of structured grids. Computational examples of practical interst are presented to demonstrate the success of these methodologies.

1 Introduction

In the last few years, numerical grid generation has evolved as an essential tool in obtaining numerical solutions of the partial differential equations of fluid mechanics. A multitude of techniques and computer codes [1-7] have been developed to support multiblock grid generation associated with complex configurations. Grid generation methodologies can be grouped into two main categories: direct methods, where algebraic interpolation techniques are utilized, and indirect methods, where a set of partial differential equations is solved. Both of these techniques are utilized either separately or in combination, to efficiently generate grids in the aforementioned codes.

The parametric based Non Uniform Rational B-Spline (NURBS) representation is widely utilized for geometrical entities in CAGD and CAD/CAM systems. The properties of NURBS are extremely attractive for engineering design applications and geometry definition. In fact, the NURBS representation is rapidly becoming the "defacto" standard for geometry description in many modern grid generation systems. Recently, the research concentration in algebraic grid generation has been placed on utilizing CAGD (Computer Aided Geometric Design) techniaques for efficient and/or accurate (boundary/surface/volume) grid generation.

The development of the algorithms for *grid refinement, redistribution, adaptation*, and *optimization* (in view of smoothness and orthogonality with precise desired

---
[1] Professor, Aerospace Engineering, Grid Generation Thrust Leader.
[2] Research Engineer.

distribution) developed by the author using CAGD techniques are discussed in this paper [5,8,11-13].

2 Basic Concepts

Structured generation involves the establishment of a one-to-one correspondence between the non-uniformly distributed physical space and the uniformly distributed computational space. Let $\underline{r} = (x_1(u,v,w),\, x_2\,(u,v,w)\, x_3\,(u,v,w))$ denote a parametric volume with Euclidean coordinates $(x_1,\, x_2,\, x_3)$ and parameter values (u,v,w). The grid associated with the computational space $\{(\xi^1,\xi^2,\xi^3, 1 \leq \xi^1 \leq r, 1 \leq \xi^2 \leq m, 1 \leq \xi^3 \leq n\}$ will be denoted by

$$r_{ijk} = ((x_1)_{ijk}, (x_2)_{ijk}, (x_3)_{ijk}),\ i = 1,2,\ldots,r;\ j = 1,2,\ldots,m;\ k = 1,2,\ldots,n.$$

Using a two part mapping there exists a one to one correspondence between the physical space and the distribution space, and between the distribution space and the computational space. These relations are demonstrated for a surface grid in Figure 1.

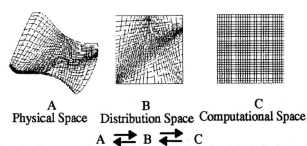

A   B   C
Physical Space   Distribution Space   Computational Space
A ⇌ B ⇌ C
Figure 1. Relationship between Physical Space, Distribution Mesh & Computational Space

The convex hull, local support, shape preserving forms, and variation diminishing properties of B-spline functions [9] contribute to the generation of a smooth well-distributed grid. Also, widely utilized geometrical entities can be exactly or approximately (with desired tolerance) converted to NURBS representation [13]. The derivatives $\underline{r}_s, \underline{r}_t, \underline{r}_q, \underline{r}_{ss}, \underline{r}_{st}, \underline{r}_{tt}, \ldots$ etc., etc. can be readily evaluated. The algorithms to evaluate a NURBS representation of a curve, surface or volume from a given sculptured data set, originally developed by Deboor [10] and then enhanced by Yu and Sone [11-13] is utilized in this work.

3 Redistribution & Remapping

Many practical applications in CFD require the redistribution of surfaces defined by an existing network of points. The distribution mesh associated with a NURBS surface is computed from the surface based on normalized arc length. The redistributed surface grid is obtained by mapping the desired distribution mesh to the NURBS surface. An example network of control points and the associated parametric space are presented in Figures 2a and 2b. A candidate desired distribution mesh is demonstrated in Figure 2D with the resulting redistributed surface grid in Figure 2c.

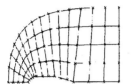 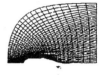

Figure 2a and 2b. Origianl Surface and Distribution Mesh.

Figure 2c and 2d. Modified Surface and Distribution Mesh.

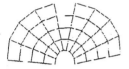 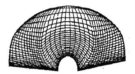

Figure 3a and 3b. Initial O-Type grid and Distribution Mesh..

Figure 3c and 3d. Resulting H-Type grid and Distribution Mesh.

However, for an $(nx \times ms)$ sculptured set of points this algorithm provides $(ns + 2) \times (ms + 2)$ control points. This would make evaluation of NURBS for redistribution extremely time consuming. To this end, a knot removal strategy developed by Lyche and Markey [14] is utilized. This technique provides an approximation of the surfaces (with desired tolerance) by removing redundant knots and is referred to as "approximation by data reduction". Several examples have been used to evaluate this technique. Typically the surfaces appear visually indistinguishable.

Remapping can also be used to change the topology of the grid, for example from a C grid to an O grid. This process is accomplished by creating a distribution space which, when evaluated, results in the desired grid. The creation of the distribution space for remapping is intuitive and highly application dependent. An example describing the process of remapping an O-type grid into an H-type grid is presented in Figures 3a-d. The remapped parametric surface is created by reestablishing the associated transformation between the physical space and the computational space. The generation of an H-type grid (Figure 3c) is accomplished by evaluating the parametric distribution presented in Figure 3d with the NURBS surface resulting from the sculptured O-type grid presented in Figure 3a and the parametric space of Figure 3b.

Remapping of a surface grid can also be used to blend interior object(s) into an existing grid. For example, consider the sculptured surface and the interior object presented in Figure 4. The desired distribution space and the resulting redistributed surface along with the interior object are presented in Figures 5a and 5b.

 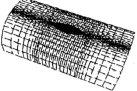

Figure 4. Intial Sculptured Surface with Interior Object.   Figure 5a. Desired Distribution Mesh.   Figure 5b. Redistributed Surface.

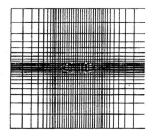 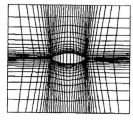 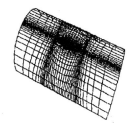

Figure 6. Desired distribution Space with associated parametric Values of Interior Object.   Figure 7. Reparameterized Distribution mesh.   Figure 8. Resulting Surface grid.

An interpolation - search algorithm based on the NURBS representation is used to evaluate the parametric surface associated with the interior object. This evaluation is demonstrated in Figure 6. The interpolation/search algorithm utilizes derivatives, and Taylor's expansion along with a Newton Raphson algorithm to inversely evaluate parameters. An automatic algorithm based on the weighted transfinite interpolation [5,12] in two dimensions is used to blend the parameters associated with interior object(s) into an overall distribution space. The resulting re-parameterized distribution space is demonstrated in Figure 7. The surface grid is then evaluated with respect to this new distribution space. The fidelity of the geometry assocaited with the interior object is kept precisely on the surface grid. The resulting remapped surface grid is presented in Figure 8.

4 Grid Optimization and Adaption

The transfinite interpolation (TFI) technique is a widely utilized algebraic grid generation method. The grid is developed by applying aBoolean sum of a selected one dimensional interpolation operator in each direction. The interpolation operators allow linear, quadratic, cubic Lagrange, cubic Hermite, quintic and Bezier polynomials, and piecewise polynomials-spline-B-spline-NURBS schemes to be utilized in grid generation.

Let $P_{\xi i}, i = 1, 2, 3$ be an interpolation operation in the *ith* direction.

The Boolean sum of the P_ξ^i results in the grid genrated by the TFI scheme. The redistributed surface grid is usually smooth and well distributed. To resolve orthogonality at the boundaries

$$r_{\xi i} \cdot r_{\xi i} = 0$$

must be satisfied. The approach taken here is to refine the parametric distrituion mesh to account for orthogonality on the boundaries. The following equations are solved for obtaining the respective slopes at the $\xi^3 = 1$ and $\xi^3 = \xi^3_{\max}$ boundaries:

$$r_{\xi 1} \cdot r_{\xi r} = 0 \qquad r_{\xi 2} \cdot r_{\xi 3} = 0$$

and $\mid (\mid r_{\xi 1} \times r_{\xi 2} \mid) r_{\xi 3} = r$ (desired cell volume off the boundary in ξ^3 direction).

These equations can be transformed to the parametric space by applying the chain rule. For example,

$$r_{\xi i} = (r_s S_{\xi i} + r_t t_{\xi i} + r_q q_{\xi i}) \cdot (r_s S_{\xi^2} + r_t t_{\xi^2}) = 0$$

Adaptive grid schemes have also been used in order to reduce the required number of grid points for adequate resolution. Both algebraic and elliptic equation based redistributional adaptive methods have been developed [15]. Figure 9a and 9b present an example of the algebraic technique on a rotating helicopter blade.

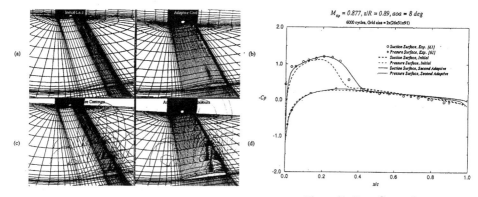

Figure 9a. Adaptive Solutions on a Helicopter Rotor.   Figure 9b. Data Comparisons.

The CAGD techniques described and applied to grid generation for CFD have been very useful. The procedures are much quicker than more traditional grid generation techniques, both in terms of labour and computer time. The NURBS representation speeds and generalizes geometry definition, while increasing the fidelity of the representation. In particular the NURBS representation has facilitated adaptive grid applications where surfaces have been continuously redistributed with normal spacing of 0.000001 at solid walls. These results were not obtainable without the high fidelity NURBS definition.

References

1. Thompson, J.F. and Gatlin, B., "Easy EAGLE: An Introduction too the EAGLE Grid Code", Mississippi State University, 1988.

2. Soni,B.K,"GENIE:Generation of Computational Geometry-Grids for Internal-External Flow Configurations", *Proceedings of the Numerical Grid Generation in Computational Fluid Mechanics' 88*, Miami, FL, 1988.

3. Steinbrenner, J. and Foitts, C., "GRIDGEN: User's Manual," Wright Research and Development Center and General Dynamics, Fort Worth Division, 1990.

4. Eiseman, Peter R. and Wang, Zhu, "Gridman: A Grid Manipulation System," *Proceedings of a workshop sponsored by the National Aeronautics and Space Administration, Washington, DC*, Langley Research Center, VA, April, 1992.

5. Soni, B.K. and Shih, M.-H., "TIGER: Turbomachinery interactive Grid Generation," *Proceedings of the Third International Conference of Numerical Grid Generation in CFD*, Barcelona, Spain, June 1991.

6. Remotigue, M.G., Hart, E.T. and Stokes, M.L., "EAGLEView: A Surface and Grid Generation Program and Its Data Management," *Proceedings of a workshop sponsored by the National Aeronautics and Space Administration, Washington, DC*, Langley Research Center, Hampton, VA, April, 1992.

7. Sorenson, Reese L. and McCann, Karen, "GRAPEVINE: Grids about Anything by Poisson's Equation in a Visually Interactive Networking Environment," *Proceedings of a workshop sponsored by the National Aeronautics and Space Administration, Washington, DC*, Langley Research Center, Hampton, VA, April 1992.

8. Soni, B.K., "Algebraic Methods and CAGD techniques in Structured Grid Generation," *Proceedings of the Fourth International Conference of Numerical Grid Generation in CFD*, Swansea, Wales, April 1994.

9. FArin, Gerald, "Curves and Surfaces for Computer Aided Geiometric Design – A practical Guide," Second Edition, Academic Press, 1990.

10. Boor, Carl de, "A practical Guide to Splines," *Applied Mathematical Sciences*, Volume 27, by Springer-Verlag New York, Inc.

11. Yu, T.Y., "IGES Transformer and NURBS In Grid Generation," Master's Thesis, Mississippi State University, August 1992.

12. Soni, B.K., "Grid Generation for Internal Flow Congiruations," Journal of Computers & Mathematics with Applications, Vol.24, No. 5/6, pp. 191-201, September 1992.

13. Yu, Tzu-Yi and Soni, Bharat K., "Geometry Transformer and NURBS in grid generation", The Fourth International Conference in Numerical Grid Generation and related fields, Swansea, England, April 6-8, 1994.

14. Lyche, T., Markey, K. "Knot Removal for Parametric B-Splinecurves and surfaces", Computer Aided Geometric Design 4 (1987), pp. 217-230, North Holland.

15. Yang, J.C., and Soni, B.K., "Structured Adaptive Grid Generation", Proceedings of the Mississippi State University Annual Conference on Differential Equations and Computational Simulation, MSU, March 19-20, 1993.

Canonical-Variables Multigrid Method for Euler Equations

S. TA'ASAN

Institute for Computer Applications in Science and Engineering
MS 132C, NASA Langley Research Center, Hampton, VA USA

Abstract : In this paper we describe a novel approach for the solution of inviscid flow problems for subsonic compressible flows. The approach is based on canonical forms of the equations, in which subsystems governed by hyperbolic operators are separated from those governed by elliptic ones. The discretizations used as well as the iterative techniques for the different subsystems, are inherently different. Hyperbolic parts, which describe, in general, propagation phenomena, are discretized using upwind schemes and are solved by marching techniques. Elliptic parts, which are directionally unbiased, are discretized using *h-elliptic* central discretizations, and are solved by pointwise relaxations together with coarse grid acceleration. The resulting discretization schemes introduce artificial viscosity only for the hyperbolic parts of the system; thus a smaller total artificial viscosity is used, while the multigrid solvers used are much more efficient. Solutions of the subsonic compressible Euler equations are achieved at the same efficiency as the full potential equation.

1 Introduction

In the past decade a substantial effort has been invested in solving the Euler equations, with multigrid methods playing a central role. The two major directions of research in multigrid solution of the Euler equations are the use of coarse grids to accelerate the convergence of the fine grid relaxations, and the use of defect correction as an outer iteration, coupled with use of multigrid to solve for the low order operator involved [3]. Extensive research has been conducted in both directions. Unfortunately, both approaches can be shown to have limited potential. Methods based on defect correction have h-dependent convergence rates for hyperbolic equations, where h is the mesh spacing; other multigrid methods have p-dependent convergence rates, where p is the order of the scheme involved. This unacceptable situation motivated the research outlined here.

The poor behavior of coarse grid acceleration for hyperbolic equations, which becomes even worse with high order discretizations, leads us to conclude that coarse grids should not be used to accelerate convergence for hyperbolic problems. Rather, the relaxation process alone should converge all components of such problems. This is possible, since hyperbolic problems describe propagation phenomena for which marching in the appropriate direction is a highly effective solver. For elliptic problems, on the other hand, local relaxation with good smoothing properties can be greatly accelerated by coarse grid correction. Moreover, elliptic problems cannot be solved efficiently by local relaxation alone, so that coarse grid acceleration is essential. In short, hyperbolic equations do not need coarse grid acceleration, while elliptic equations require such acceleration.

These observations have motivated a study concerning the separation of the hyperbolic and elliptic parts in steady state inviscid flow calculations. The result is a canonical form for the inviscid equations, where the hyperbolic and elliptic parts reside in different blocks of an upper triangular matrix for the system [5]. The new discretization schemes, which are based on these canonical forms, use upwind discretization only for the hyperbolic variables, and use a central *h-elliptic* discretization for the elliptic variables. This gives a better representation for the physical phenomena, since elliptic problems do not have a bias in any spatial direction, a property that should hold true for the discretization as well, if possible. Upwind discretization for hyperbolic problems, on the other hand, is compatible with the bias of information flow in the physical problem.

The elliptic and hyperbolic parts of the equations are treated differently by the solver. Unlike existing solvers, which use coarse grids to accelerate the hyperbolic part of the system as well, the new method computes the hyperbolic part via relaxations based on the canonical forms. This involves marching in the streamwise direction for the hyperbolic quantities, which are the entropy s and total enthalpy H. The rest of the unknowns, e.g., the velocity components, are relaxed by a Kaczmarz relaxation using preconditioned residuals, yielding a smoothing rate identical to that for the full potential equation.

Numerical results are given for a two dimensional flow around an ellipse. This problem already includes the major difficulties in real problems and serve as a good test for the method proposed. Second order schemes are used for both cases and the solutions are obtained with convergence rates similar to that of the full potential equation (although the work involved here is larger accounting for the multiple equations).

2 Canonical Forms and Discretization

For a two dimensional flow, the canonical form of the compressible Euler equations is [6],

$$\begin{pmatrix} D_1 & D_2 & 0 & 0 \\ qD_y & -qD_x & -\frac{c^2}{\gamma(\gamma-1)}D_0 & \frac{1}{q}D_0 \\ 0 & 0 & -T\rho Q & 0 \\ 0 & 0 & 0 & \rho Q \end{pmatrix} \begin{pmatrix} u \\ v \\ s \\ H \end{pmatrix} = \begin{pmatrix} 0 \\ 0 \\ 0 \\ 0 \end{pmatrix} \quad (1)$$

where:

$$\begin{aligned} D_1 &= \rho/c^2((c^2 - u^2)D_x - uvD_y) \\ D_2 &= \rho/c^2((c^2 - v^2)D_y - uvD_x) \\ D_0 &= vD_x - uD_y \end{aligned} \quad (2)$$

In view of this canonical form we can use the following discretization rule for the invisicd equations:

1. *Use central (unbiased) h-elliptic discretizations for elliptic subsystems*

2. *Use upwind biased schemes for hyperbolic subsystems*

Let a domain $\Omega \in I\!\!R^2$ be divided into arbitrary cells. Let the vertices, edges, and cells be V,E, and C respectively. The well known Euler formula

$$\#V + \#C + \#holes = \#E + 1 \tag{3}$$

suggests several possibilities for discretization of different systems on structured and unstructured meshes.

Rewriting the Euler formula as

$$\#V + \#V + \#C + \#C + \#holes = \#C + \#V + \#E + 1 \tag{4}$$

one obtains the following choice of discretization. Let H be associated with the cell centers, the normal velocity components be at the edges, and the entropy be at the vertices. Quantities other than these are calculated by well known algebraic relations for the thermodynamic quantities and by averaging for the tangential velocity components. With each cell, we associate one continuity equation and one energy equation, while the two momentum equations are associated with each vertex. The following diagram is then obtained:

$$\begin{array}{lcl}
\mathbf{V} \cdot \mathbf{n} & \Longleftrightarrow & \#E \\
s & \Longleftrightarrow & \#V \\
H & \Longleftrightarrow & \#C \\
\mathrm{div}\rho\mathbf{V} = 0 & \Longleftrightarrow & \#C \\
-\mathbf{V} \times \mathrm{curl}\mathbf{V} + \nabla P = 0 & \Longleftrightarrow & 2\#V \\
\mathrm{div}\rho\mathbf{V}H = 0 & \Longleftrightarrow & \#C \\
\int_{\Gamma_{hole}} \mathbf{V} \cdot \mathbf{t} d\sigma & \Longleftrightarrow & \#holes
\end{array} \tag{5}$$

The canonical form for the compressible equations suggests that only the entropy and the total enthalpy will be discretized using upwind biased schemes, and only in the appropriate terms; that is, only those terms in which a derivative in the streamwise direction occurs will be discretized with upwind bias. Other derivatives of these quantities will be discretized using central differencing. More details regarding the discretization can be found in [6].

3 Multigrid Algorithm

The multigrid solver, like the discretization, is based on the canonical forms mentioned in section 2. The main ingredient differing from other multigrid methods is the relaxation scheme. Other elements of the multigrid method are standard and will be mentioned only briefly.

3.1 Relaxation

Let the residual of the compressible Euler equations be denoted by $(R^\rho, R^{\rho u}, R^{\rho v}, R^H)$ and the ones for the canonical form by $(r_c^1, r_c^2, r_c^3, r_c^4)$. Then the following relation prevails

$$\begin{pmatrix} r_c^1 \\ r_c^2 \\ r_c^3 \\ r_c^4 \end{pmatrix} = \begin{pmatrix} 1 & 0 & 0 & 0 \\ -q^2 A & -v & u & 0 \\ -q^2 A & u & v & 0 \\ -H & 0 & 0 & 1 \end{pmatrix} \begin{pmatrix} R^\rho \\ R^{\rho u} \\ R^{\rho v} \\ R^H \end{pmatrix} \qquad (6)$$

where A is a local avaraging operator.

The relaxation for the Euler equations is employed as follows. The total enthalpy is relaxed first using Gauass-Seidel relaxation in the streamwise direction, using the preconditioned residual r_c^4. Next entropy is relaxed with Gauass-Seidel relaxation in the streamwise direction using r_c^3. This is followed by a Kaczmarz relaxation for the equation corresponding to r_c^1, r_c^2, for the velocities.

The coarsening part of the multigrid method used is standard, so its details are omitted. Coarse grids are created by combining neighboring fine grid cells into a coarse grid cell. Linear interpolation of corrections and full weighting of residual and functions are used in an FMG-FAS formulation [1].

4 Numerical Results

We present here numerical results for subsonic cases of flow in a nozzle and around an ellipse. This case already present most of the difficulties encountered in subsonic compressible flows. A body fitted O-grid was used which extended to to a distance of 7 chords. The finest grid involved 128 by 64 cells.

Figure 1 shows results for a flow around an ellipse at a 10 degrees angle of attack, with free stream Mach number of 0.1, with zero circulation at infinity. This is a case of special difficulty. The exact values for the lift and drag coefficient are zero. The calculated values are $C_L = 2.3 \times 10^{-3}, C_D = -7.7 \times 10^{-3}$. A grid refinement study shows that the lift and drag coefficient approach zero with mesh refinement. The full FMG convergence history for this problem is shown in the right picture of figure 1. Note that on the finest grid a reduction of nine orders of magnitudes is obtained in just twelve cycles. Relative errors in entropy are shown in the left picture, and clearly demonstrate the effectiveness of the discretization used.

5 Conclusion

A new approach for the discretization and the solution of the Euler equations have been presented. It is based on a canonical form of the equations which allows for discretization which involve upwind biased discretization only for the physically biased quantities, that is, entropy and total enthalpy. The multigrid method used for that disctetization shows essentially optimal convergence rates for the second order

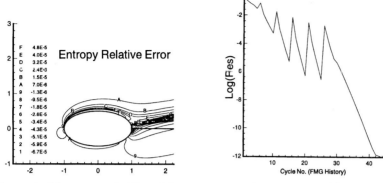

Figure 1: Flow around ellipse:$M_\infty = .1, \alpha = 10, 128 \times 64$, grid, Entropy errors and residuals history

schemes used. Moreover, nonlifting solutions around smooth bodies at angle of attack are obtained. Unlike most other methods, the performance of the method does not degrade as the Mach number approaches zero.

References

[1] A. Brandt: Multigrid Techniques 1984 Guide with Applications to Fluid Dynamics. GMD Studien Nr 85.

[2] A. Brandt and S. Ta'asan: Multigrid Solutions to Quasi-Elliptic Schemes. In *Progress and Supercomputing in Computational Fulid Dynamics* Proceedings of the U.S.-Israel Workshop 1984, Earll. M. Murman and Saul Abarbanel (Eds.), Birkhauser 1985

[3] Hemker, P.W.: Defect Correction and Higher Order Schemes for the Multigrid Solution of the Steady Euler Equations, *Multigrid Methods* II, W. Hackbush and U. Trottenberg (Eds), (*Lecture Notes in Mathematics*) Springer Verlag, Berlin, 149-165.

[4] A. Jameson: Solution of the Euler Equations for Two Dimensional Transonic Flow by Multigrid Method, *Appl. Math. and Computat.*, 13, 327-355.

[5] S. Ta'asan: Canonical forms of Multidimensionl Inviscid Flows. ICASE Report No. 93-34

[6] S. Ta'asan: Canonical-Variables Multigrid Method for Steady-State Euler Equations. ICASE Report No. 94-14

Agglomeration Multigrid for the Euler Equations

V. Venkatakrishnan & D. J. Mavriplis

Institute for Computer Applications in Science and Engineering
MS 132C, NASA Langley Research Center
Hampton, VA, U.S.A.

Abstract : A multigrid procedure that makes use of coarse grids generated by the agglomeration of control volumes is used to solve the two- and three-dimensional Euler equations on unstructured grids about complex configurations. The agglomeration is done as a preprocessing step and runs on a workstation in a time linearly proportional to the number of fine grid points. The agglomeration multigrid technique compares very favorably with existing multigrid procedures both in terms of convergence rates and elapsed times. The main advantage of the present approach is the ease with which coarse grids of any desired degree of coarseness may be generated in three dimensions, without being constrained by considerations of geometry. Inviscid flows over a variety of complex configurations are computed using the agglomeration multigrid strategy.

1 Introduction

Over the last few years, the multigrid method has been demonstrated as an efficient means for obtaining steady-state solutions to the Euler equations on unstructured meshes in two and three dimensions. One popular approach uses unnested unstructured grids as coarse grids and has been shown to be successful in both two- and three-dimensional unstructured grid computations [1, 2, 3, 4]. Grids of varying coarseness are generated independently using any given grid generation strategy. Piecewise linear interpolation operators for the transferring of flow variables, residuals and corrections are derived during a preprocessing step by using efficient search procedures. For complex geometries, especially in three dimensions, constructing coarse grids that faithfully represent the complex geometries can become a difficult proposition.

One approach that circumvents this problem is the generation of coarse grids through cell agglomeration. This method was developed in [5] and independently in [6] and has been pursued by the authors for inviscid and viscous flows past complex configurations in both two and three dimensions [7, 8, 9, 10]. Lallemand et al. use a base scheme that stores the variables at the vertices of the triangular mesh, whereas Smith uses a scheme that stores the variables at the centers of triangles. In the present work, the variables are stored at the vertices of a mesh composed of simplices i.e., triangles in two dimensions and tetrahedra in three dimensions. This paper only briefly reviews the concepts behind agglomeration multigrid. The full details may be found in [8].

2 Governing equations and discretization

The Euler equations in integral form for a control volume Ω with boundary $\partial\Omega$ read

$$\frac{\partial}{\partial t}\int_\Omega \mathbf{u}\ dv + \oint_{\partial\Omega} \mathbf{F}(\mathbf{u},\mathbf{n})\ dS = 0. \tag{1}$$

Here $\mathbf{u}$ is the solution vector comprised of the conservative variables: density, the components of momentum, and total energy. The vector $\mathbf{F}(\mathbf{u},\mathbf{n})$ represents the inviscid flux vector for a surface with normal vector $\mathbf{n}$. The control volumes are nonoverlapping polyhedra which surround the vertices of the mesh. They form the *dual* of the mesh, which is composed of segments of median planes. The contour integrals in Eqn. (1) are replaced by discrete path integrals over the faces of the control volume which are computed by usicng the trapezoidal rule. This non-overlapping control volume formulation can be shown to be equivalent to the overlapping control volume formulation of [11]. For dissipative terms, a blend of Laplacian and biharmonic operators is employed. The Laplacian term acts only in the vicinity of shocks and is inactive elsewhere, while the biharmonic term acts only in regions of smooth flow. A multi-stage Runge-Kutta scheme is used to advance the solution in time. In addition, local time stepping, enthalpy damping and residual averaging are used to accelerate convergence [12]. In previous work [1, 2], as well as in the present work, only the Laplacian dissipative term (with constant coefficients) is used on the coarse grids. Thus the fine grid solution itself is second order accurate, while the solver is only first order accurate on the coarse grids.

The object of agglomeration is to derive coarse grids from a given fine grid. To this end, the median dual to the triangulation is first formed. This comprises of nonoverlapping polyhedral control volumes that surround the vertices of the mesh. The main idea behind agglomeration is to fuse the fine grid control volumes selectively to create coarse polyhedral control volumes. The optimal agglomeration problem falls under the category of intractable problems termed NP-complete in graph theory. Hence, heuristic algorithms are used to derive the coarse grids. The agglomeration is achieved by using greedy-type frontal algorithms. The procedure is applied recursively to create coarser grids. The boundaries between the control volumes on the coarse grids are composed of the edges/faces of the fine grid control volumes. In two dimensions, we have observed that the number of such edges only decreases by a factor of 2 when going from a fine to a coarse grid, even though the number of vertices decreases by a factor of 4. Since the computational load is proportional to the number of edges, this is unacceptable in the context of multigrid. However, if we recognize that the multiple edges separating two control volumes can be replaced by a single edge connecting the end points, then the number of edges does go down by a nearly a factor of 4. A similar construction is made in three dimensions as well. As a result of making this approximation, the work decreases by nearly a factor of 4 in two dimensions and 8 in three dimensions when moving from a fine to a coarse grid. The agglomeration algorithm is efficient and runs in linear time. The output of the agglomeration is a list of fine grid cells belonging to each coarse grid cell and edge coefficients, which represent the projected areas of the planar control-volume faces onto the coordinate planes.

Since the fine grid control volumes comprising a coarse grid control volume are known, the restriction is similar to that used for cell-centered structured grid multigrid algorithms. The residuals are simply summed from the fine grid cells and the variables are interpolated in a volume-weighted manner. For the prolongation operator, we use a simple injection (a piecewise-constant interpolation). This is an unfortunate but an unavoidable consequence of using the agglomeration strategy. A piecewise-linear prolongation operator implies a triangulation, the avoidance of which is the main motivation for the agglomeration. The adverse impact of the injection is minimized by employing additional smoothing steps. This is achieved by applying an implicit smoothing procedure to the injected corrections that uses two Jacobi iterations.

3 Results and discussion

To illustrate the capability of the agglomeration multigrid procedure, we present results from two flow computations. The first case concerns two-dimensional flow over a four-element airfoil. The freestream Mach number is 0.2 and the angle of attack is 5°. The fine grid has 11340 vertices. The coarse grids for use with the non-nested multigrid algorithm contain 2942 and 727 vertices. The four agglomerated coarse grids obtained using the agglomeration algorithm grids contain 3027, 822, 217 and 63 vertices (regions), respectively. Figure 1 illustrates the second level coarse grid containing 822 vertices. The convergence histories of the non-nested and agglomeration multigrid algorithms are shown in Figure 2. The convergence histories with the 3 level multigrid are comparable. Convergence improves considerably with the 5 level multigrid algorithm. The CPU times required on the Cray Y-MP for the 3 level multigrid are are 59 and 60 seconds with the non-nested and the agglomerated algorithms, respectively. The CPU time required for the 5 level agglomeration multigrid is 73 seconds. The convergence rates per cycle with the 3-level and 5-level agglomeration multigrid algorithm are 0.87 and 0.93 and that of the 3-level non-nested multigrid is 0.93. With the non-nested multigrid it was not possible to generate coarser grids for this complex geometry.

The next case concerns the three-dimensional solution of inviscid transonic flow over a low-wing transport configuration. The geometry consists of a half fuselage bounded by a symmetry plane with a wing and a nacelle. The fine mesh for this case contains 804,056 vertices and approximately 4.5 million tetrahedra. The six coarse grids derived by the agglomeration algorithm contain respectively, 132865, 25489, 5678, 1421, 373 and 99 vertices. The freestream conditions are Mach number of 0.77 and and 1.116° incidence. The convergence history using a seven-level agglomerated multigrid strategy is shown in Figure 3. The residuals are reduced by nearly six orders of magnitude in 100 W-cycles, which corresponds to a convergence rate of 0.876 per cycle. This compares favorably with the rate of 0.870 obtained using a 5-level non-nested multigrid strategy for the same case. The agglomerated multigrid computation required 96 MW of memory and 45 minutes of CPU time on one processor of a Cray C-90. The computed Mach contours are depicted in Figure 4 where the upper surface wing shock pattern is evident.

We have shown that the agglomeration multigrid strategy can be made to approximate the efficiency of the unstructured multigrid algorithm using independent, non-nested coarse meshes, both in terms of convergence rates and CPU times. Flows over complex configurations have been computed to illustrate the power and flexibility of the agglomeration multigrid strategy.

References

[1] Mavriplis, D. J., and Jameson, A., " Multigrid Solution of the Two-Dimensional Euler Equations on Unstructured Triangular Meshes", *AIAA Journal*, Vol 26, No. 7, July 1988, pp. 824–831.

[2] Mavriplis,D. J., " Three Dimensional Multigrid for the Euler equations", *AIAA Journal*, Vol. 30, No. 7, July 1992, pp. 1753–1761.

[3] Leclercq,M. P., " Resolution des Equations d'Euler par des Methodes Multigrilles Conditions aux Limites en Regime Hypersonique", *Ph.D Thesis, Applied Math, Universite de Saint-Etienne*, April, 1990.

[4] Peraire, J., Peiro, J., and Morgan, K., " A 3D Finite-Element Multigrid Solver for the Euler Equations", *AIAA paper 92-0449*, January 1992.

[5] Lallemand, M., Steve, H., and Dervieux, A., " Unstructured Multigridding by Volume Agglomeration: Current Status", *Computers and Fluids*, Vol. 21, No. 3, pp. 397–433, 1992.

[6] Smith, W. A., " Multigrid Solution of Transonic Flow on Unstructured Grids", *Recent Advances and Applications in Computational Fluid Dynamics, Proceedings of the ASME Winter Annual Meeting*, Ed. O. Baysal, November, 1990.

[7] Venkatakrishnan, V., Mavriplis, D.J., and Berger,M.J., "Unstructured Multigrid through Agglomeration," *Proceedings of the 6th Copper Mountain Multigrid Conference*, April 1993.

[8] Venkatakrishnan, V. and Mavriplis, D.J., "Agglomeration Multigrid for the Three-dimensional Euler Equations," *AIAA Paper 94-0069*, January 1994; to appear in AIAA Journal.

[9] Mavriplis, D.J. and Venkatakrishnan, V., "Agglomeration Multigrid for Viscous Turbulent Flows," *AIAA Paper 94-2232*, July 1994.

[10] Koobus, B., Lallemand, M. H., and Dervieux, A., "Unstructured Volume Agglomeration: Solution of the Poisson Equation", *INRIA Report 1946*, June 1993.

[11] Jameson, A., Baker, T. J., and Weatherill, N. P., "Calculation of Inviscid Transonic Flow over a Complete aircraft", *AIAA Paper 86-0103*, January 1986.

[12] Jameson, A., "Solution of the Euler Equations by a Multigrid Method", *Applied Mathematics and Computation*, Vol. 13, pp. 327–356, 1983.

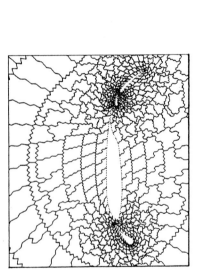

Figure 1. Second-level coarse grid for the four-element test case.

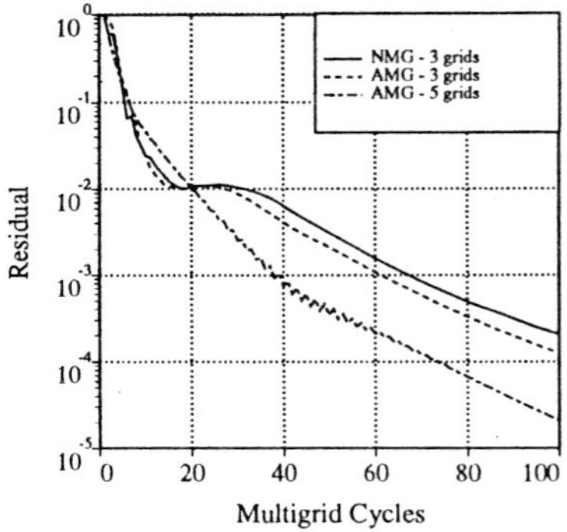

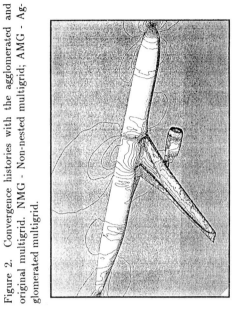

Figure 2. Convergence histories with the agglomerated and original multigrid. NMG - Non-nested multigrid; AMG - Agglomerated multigrid.

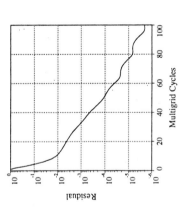

Figure 4. Mach contours on the surface for the low-wing transport.

Figure 3. Convergence history with agglomeration multigrid for flow over a low-wing transport configuration.

Steady Transonic Flow Computations Using Overlapping Grids

Zi-Niu Wu

SINUMEF Laboratory, Ecole Nat. Sup. Arts et Métiers
151 Boulevard de l'Hôpital, 75013 Paris, France

Abstract: An implicit multidomain technique is considered for steady state transonic flow computations using overlapping grids. The convergence to steady state and uniqueness of steady state solution of this technique are studied.

1 Introduction

In previous papers (Lerat and Wu, 1993, 1994), we have studied a GKS-stable, conservative and convergent implicit multidomain procedure for computing steady transonic flows on continuous and patched grids. This procedure is here extended to the case with arbitrarily overlapping grids. We will consider especially the uniqueness of steady state solutions and the convergence to steady state. In our applications, the implicit scheme of Lerat (1979) is chosen for interior point computation and the reconstruction method of Pärt and Sjögreen (1991) is applied for constructing interface conditions.

2 Interface problem in one dimension

Consider the numerical approximation of a hyperbolic system of conservation laws:

$$w_t + h(w)_x = s(w), \quad t \in \mathbf{R}^+, \quad x \in \mathbf{R} \tag{1}$$

with suitable initial data. The computational domain is divided into two overlapping subdomains D_u and D_v having an overlap $a < x < b$ (Fig. 1). The numerical solutions are denoted by $u_j^n = w(b + (j - 0.5)h_u, nk)$, $j \leq 0$ in D_u and $v_j^n = w(a + (j + 0.5)h_v, nk)$, $j \geq 0$ in D_v, where k, h_u and h_v define respectively the time step, the mesh size in D_u and the mesh size in D_v. We assume that $\sigma_u = k/h_u$, $\sigma_v = k/h_v$ are constant. System (1) is approximated by a three-point difference scheme in conservation form in each subdomain:

$$\Delta u_j^n = -\sigma_u(f_{j+\frac{1}{2}}^n - f_{j-\frac{1}{2}}^n), \quad j = -1, -2, \ldots \tag{2}$$

$$\Delta v_j^n = -\sigma_v(g_{j+\frac{1}{2}}^n - g_{j-\frac{1}{2}}^n), \quad j = +1, +2, \ldots \tag{3}$$

where $\Delta u_j^n = u_j^{n+1} - u_j^n$, $\Delta v_j^n = v_j^{n+1} - v_j^n$ denote the time increments and $f_{j+\frac{1}{2}} = f(u_j, u_{j+1})$, $g_{j+\frac{1}{2}} = g(v_j, v_{j+1})$ are numerical fluxes.

The implicit scheme of Lerat (1979) for u_j^n is defined by

$$\Delta u_j - \frac{1}{2}\sigma_u^2[A_{j+\frac{1}{2}}^2(\Delta u_{j+1} - \Delta u_j) - A_{j-\frac{1}{2}}^2(\Delta u_j - \Delta u_{j-1})] = \Delta u_j^{lw}$$

where A is the Jacobian of the flux $h(w)$ and Δu_j^{lw} is given by the classical Lax-Wendroff scheme with $f_{j+\frac{1}{2}}^n = \frac{1}{2}(h(u_{j+1}^n) + h(u_j^n)) - \frac{1}{2}\sigma_u A(h(u_{j+1}^n) - h(u_j^n))$ The scheme for v_j^n is defined similarly. The Lerat's scheme is linearly stable in L_2 and linearly dissipative for any time step.

To define interface conditions, Berger (1987) has developed the flux interpolation method. In this method, the values u_0^n and v_0^n are calculated as in the interior points, the missing numerical fluxes $f_\frac{1}{2}$ and $g_{-\frac{1}{2}}$ being interpolated from the interior points. For linear interpolation, the fluxes $f_\frac{1}{2}$ and $g_{-\frac{1}{2}}$ are given by

$$f_\frac{1}{2} = \alpha g_{p+\ell-\frac{1}{2}} + (1-\alpha)g_{p+\ell+\frac{1}{2}}, \quad g_{-\frac{1}{2}} = \beta f_{-q-m-\frac{1}{2}} + (1-\beta)f_{-q-m+\frac{1}{2}} \quad (4)$$

where $\alpha = 1 - d_l/h_u$ and β is defined in a similar way. The flux interpolation method is conservative but it is only weakly stable (Pärt and Sjögreen, 1991). To have strongly stable (GKS-stable) interface conditions, Pärt and Sjögreen (1991) have proposed the reconstruction method which has only small conservation errors. For piecewise constant reconstruction, the interface condition[1] reads:

$$u_0^{n+1} = h_u^{-1}\sum_{i=0}^{\ell} d_i v_{p+i}^n, \quad v_0^{n+1} = h_v^{-1}\sum_{i=0}^{m} e_i u_{-(q+i)}^n \quad (5)$$

where each d_i for $i = 0, 1, \ldots, \ell$ is the intersection between the cell $j = 0$ of D_u and the cell $j = p + i$ of D_v and $h_u^{-1}\sum_{i=0}^{\ell} d_i = 1$. The e_i for $i = 0, 1, \ldots, m$ are defined similarly.

For purpose of uniqueness and convergence study, we will consider the scalar transport equation $u_t + u_x = 0$ defined on $-1 < x < 1$ and approximated by the difference scheme (2) for $-N_u \leq j \leq -1$, the difference scheme (3) for $1 \leq j \leq N_v$, the interface condition (4) or (5) and the following boundary conditions at $x = -1$ and $x = 1$:

$$u_{-N_u-1}^n = A, \quad v_{N_v+1}^n = v_{N_v}^n \quad (6)$$

This will be called the scalar interface problem on a strip.

3 Uniqueness result for the linearized problem

Let us start with the interface condition (4). There is no difficulty to show that the two interpolation formulae involved in (4) are not linearly independent at steady state. We have thus

Proposition 1 The scalar interface problem on a strip for any pair of three-point schemes and the interface condition (4) has nonunique steady state solutions.

---

[1] As in (Lerat and Wu, 1994), a time lagging of the interface values is made to allow independent solution of the implicit scheme in each subdomain.

For the Lax-Wendroff scheme, one can check that the steady state solution is $u_j = c+(A-c)[(\sigma_u+1)/(\sigma_u-1)]^{j+N_u+1}$, $j \leq 0$ in the left subdomain and $v_j = c$, $j \geq 0$ in the right subdomain. Here c is an arbitrary constant so that the solution is not unique.

For the interface condition (5) obtained by reconstruction, the steady state solution of the scalar interface problem on a strip is unique. But this is not true in the nonlinear case when a shock is caught in the overlap.

Proposition 2 Consider the scalar interface problem on a strip for the Burgers' equation $u_t + uu_x = 0$ linearized about a steady shock in $a < x < b$. The steady state solution is not unique for the Lax-Wendroff scheme and is unique for the implicit scheme of Lerat with high CFL numbers.

This result can be proved by examining the related resolvent problem. Here we check it by computing a quasi-one-dimensional steady shock in a divergent duct whose section is $S(x) = 1.398 + 0.347\tanh(0.8x - 4)$, $0 < x < 10$. The exact shock is at $x = 4.816$. For the explicit Lax-Wendroff scheme with $CFL = 1$, we obtain an unphysical two-shock solution (Fig. 2), that is, the shock position is different in each subdomain. For the implicit scheme of Lerat with CFL=6, we obtain the correct physical solution (Fig. 3).

4 Influence of grid overlapping on the convergence rate

Here we examine the influence of grid overlapping on the convergence to steady state for the scalar interface problem defined by the implicit scheme of Lerat, the boundary conditions (6) and the interface condition (5). The grid system is identical in each subdomain so that we have $N_u = N_v = N$ and $p = q$. The related eigenvalue problem is obtained by letting u_j^n be $z^n \phi_j$ and v_j^n be $z^n \psi_j$ in the scalar interface problem on a strip, yielding $zM_1Y = M_2Y$ where M_1 and M_2 are two real matrices and Y is the unknown vector for ϕ_j and ψ_j. There are in total $N_u + N_v + 4 (= 2N+4)$ eigenvalues z_σ. It is the spectral radius $\rho_o = \max |z_\sigma|$ which dominates the convergence rate. The spectral radius ρ_s for the single domain computation can be obtained in a similar way.

The performance of grid overlapping can be measured by the overlapping efficiency $\epsilon_o = 1 - (2N/N_s)(\ln \rho_s / \ln \rho_o)$ where $N_s = 2N - \frac{1}{2}(p+q)$ is the number of mesh points in the single domain case. The overlapping treatment consumes $100|\epsilon_o|\%$ more (or less) time than the single domain one if $\epsilon_o < 0$ (or $\epsilon_o > 0$).

In Fig. 4 is shown the overlapping efficiency in function of the overlapping length $L_o = (p+q)/2 + 1$ for various CFL numbers. Surprisingly, the overlapping efficiency becomes positive for $L_o \approx CFL$, showing that the overlapping approach may be less expensive than the single domain one (on a non parallel computer!). When $L_o = CFL$, for which there is no more interaction between the two overlapping edges $x = a$ and b, the overlapping efficiency is maximum. This result has also been confirmed by numerical experiments on the linear transport equation. For the nonlinear Euler equations, we find also that the convergence of the overlapping treatment is optimal when $L_o = CFL$ but is never better than that of the single domain one.

5 Two-dimensional transonic flow computations

We have chosen the implicit scheme of Lerat (1979) for interior point computation and the zeroth-order reconstruction of Pärt and Sjögreen (1991) for interface treatment. According to the above theoretical results, this multidomain approximation ensures a unique steady state solution even for shock flows. We have successfully computed various transonic flows over one- and two-element airfoils. In Fig. 5 is displayed the overlapping grid for a staggered Bi-NACA airfoil without incidence and the computed pressure contours for a free-stream Mach number $M_\infty = 0.7$. The pressure distributions on the airfoils are presented in Fig. 6, where the result of a conservative patched grid technique (Lerat and Wu, 1994) is also shown.

Acknowledgements

This work was supported by DRET (French Ministry of Defense) under Contract 92/169 . The author is grateful to Professor A. Lerat (SINUMEF Laboratory) for helpful suggestions during this work.

References

[1] Berger M. (1987), "On conservation at grid interfaces", SIAM J. Numer. Anal., **24**, pp. 967-984.

[2] Lerat A. and Wu Z. N. (1993), "A subdomain matching condition for implicit Euler solvers", In Numerical Methods for Fluid Dynamics IV, Ed. by K. W. Morton and M. J. Baines, Clarendon Press, pp.283-290.

[3] Lerat A. and Wu Z. N. (1994), "A stable conservative multidomain treatment for implicit Euler solvers", submitted to J. Comput. Phys.

[4] Lerat A. (1979), "Une classe de schémas aux différences finies implicites pour les systèmes de loi de conservation", Comptes Rendus Acad. Sci., Paris, **288A**, pp. 1033-1036.

[5] Pärt E. and Sjögreen B. (1991), "Shock waves and overlapping grids", Uppsala Univer., Dept. Sci. Computing, Report No 131, January 1991.

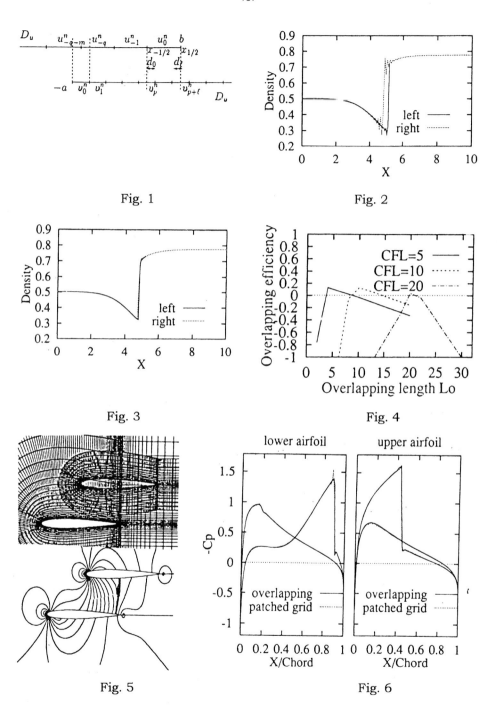

Fig. 1

Fig. 2

Fig. 3

Fig. 4

Fig. 5

Fig. 6

Surface Boundary Conditions for the Numerical Solution of the Euler Equations in Three Dimensions

A. Dadone[1] and B. Grossman[2]

[1]Istituto di Macchine ed Energetica
Politecnico di Bari
via Re David 200, 70125 Bari, Italy
[2]Dept. Aerospace and Ocean Eng.
Virginia Polytechnic Inst. & State Univ.
Blacksburg, Virginia 24061, USA

1 Introduction

In Ref. 1, the present authors considered the implementation of boundary conditions at solid walls in the inviscid Euler solutions by upwind, finite-volume methods. In that paper, the issue of the implementation of surface boundary conditions and the development of characteristic boundary conditions (*e.g.*, see Refs. 2–6) was discussed. We introduced two new boundary-condition procedures, denoted as the symmetry technique (ST) and the curvature-corrected symmetry technique (CCST). Examples of the effects of the various surface boundary conditions considered were presented for several supersonic and subsonic two-dimensional test problems. Dramatic advantages of the curvature-corrected symmetry technique over other methods were shown, with regard to numerical entropy generation, total pressure loss, drag and grid convergence.

In this paper we extend this methodology to three-dimensional geometries. First we review the details of the curvature-corrected symmetry technique for two-dimensional geometries. Then we discuss the local two-dimensional behavior of a three-dimensional flow in the so-called *osculating plane* of an *intrinsic* coordinate system. We then show how a curvature-corrected symmetry technique can be developed for three-dimensional geometries. A few elementary computations are presented for the compressible flow over a sphere, an oblate spheroid and a prolate spheroid, which show the efficacy of this approach. As was found in the two-dimensional case, the curvature-corrected symmetry technique in three dimensions is clearly superior to pressure extrapolation techniques currently in use.

2 Two-Dimensional Formulation

The foundation of the curvature-corrected symmetry technique is based upon utilizing the momentum equation normal to a streamline evaluated on the body surface. For a steady inviscid flow, this equation becomes

$$\left(\frac{\partial p}{\partial n}\right)_w = -\rho q^2/R, \qquad (1)$$

where n is the direction normal to the wall, denoted by subscript w, ρ is the density, q is the magnitude of the velocity and R is the radius of curvature of the wall. In this analysis, we use the symbol $\tilde{u}$ to be the velocity component in the direction of the surface streamline and $\tilde{v}$ to denote the velocity component in the direction normal to the wall. From the impermeability condition at the solid surface $\tilde{v}_w = 0$ and $q_w = \tilde{u}_w$.

An approximation to the integral of Eq. (1) is used to specify the value of pressure p at a system of image cells, illustrated in Fig. 1, such that

$$p_{-1} = p_1 - \rho_w \frac{q_w^2}{R}\Delta n_1, \qquad p_{-2} = p_2 - \rho_w \frac{q_w^2}{R}\Delta n_2, \qquad (2a,b)$$

where Δn_1 represents the distance between cell center (1) and the image-cell center (-1), and likewise Δn_2 is the distance between (2) and (-2).

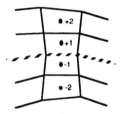

Fig.1 Image cells for CCST.

The other flow quantities may be evaluated by locally modeling the inviscid flow over a solid surface as a vortex flow of constant entropy and total enthalpy. Thus in addition to the pressure condition given by Eqs. (2), we also assume symmetric values of the entropy and total enthalpy. For a perfect gas, it is easily shown that the above conditions lead to

$$\rho_{-1} = \rho_1 \left(\frac{p_{-1}}{p_1}\right)^{1/\gamma}, \qquad \rho_{-2} = \rho_2 \left(\frac{p_{-2}}{p_2}\right)^{1/\gamma}, \qquad (3a,b)$$

and

$$\begin{aligned}\tilde{u}_{-1}^2 &= \tilde{u}_1^2 + \frac{2\gamma}{\gamma-1}\left(\frac{p_1}{\rho_1} - \frac{p_{-1}}{\rho_{-1}}\right) + \tilde{v}_1^2 - \tilde{v}_{-1}^2 \\ \tilde{u}_{-2}^2 &= \tilde{u}_2^2 + \frac{2\gamma}{\gamma-1}\left(\frac{p_2}{\rho_2} - \frac{p_{-2}}{\rho_{-2}}\right) + \tilde{v}_2^2 - \tilde{v}_{-2}^2,\end{aligned} \qquad (4a,b)$$

where γ is the constant ratio of specific heats. The remaining condition stems from the impermeability condition. This condition is enforced by selecting $\tilde{v}_{-1}$ and $\tilde{v}_{-2}$ such that the Roe-averaged value of the normal velocity $\tilde{v}$ is zero at the surface. This is accomplished by setting $\tilde{v}_{-1} = -\tilde{v}_1$ and choosing $\tilde{v}_{-2}$ such that $\tilde{v}_\ell \sqrt{\rho_\ell} + \tilde{v}_r \sqrt{\rho_r} = 0$. In the absence of limiters, using linear extrapolations for the left and right states results in

$$\tilde{v}_{-1} = -\tilde{v}_1,$$
$$\tilde{v}_{-2} = 3\tilde{v}_{-1} + 2[\tilde{v}_1 + (\tilde{v}_1 - \tilde{v}_2)/2]\sqrt{\frac{\rho_1 + (\rho_1 - \rho_2)/2}{\rho_{-1} + (\rho_{-1} - \rho_{-2})/2}}. \qquad (5a,b)$$

The above implementation of the normal-momentum pressure extrapolation is valid for grids which are orthogonal to the surface. It should be noted that for the evaluation of the image points in Eqs. (2) we take $\rho_w = \rho_1$ and $q_w = \tilde{u}_1$. This avoids extrapolations across discontinuities and has a minimal effect on the results. In addition, we generally implement Eqs. (2) neglecting the effect of grid variation by taking $\Delta n_2 = 3\Delta n_1$. With the image points computed as above, the computation of the cell centers nearest to the surface proceeds in the identical manner as the interior cell centers.

In Ref. 1, examples of the effects of various surface boundary conditions were considered for a shock-reflection problem, a supersonic blunt-body problem and for the subcritical compressible flow over a circular cylinder. Dramatic advantages of the curvature-corrected symmetry technique over conventional pressure extrapolation methods were shown, particularly with regard to numerical entropy generation, total pressure loss, drag and grid convergence. In addition, recent evidence indicates that the curvature-corrected symmetry technique is applicable to a variety of other Euler formulations in two dimensions.

3 Three-Dimensional Formulation

For three dimensions we can consider an *intrinsic* coordinate system, *e.g.*, Ref. 7, §20. This local coordinate system is composed of the streamline direction $\hat{s}$, the normal direction $\hat{n}$ and the binormal direction $\hat{b}$. The normal direction is in the direction normal to the streamline direction and parallel to the plane containing ∇p and $\hat{s}$. The binormal direction is perpendicular to both $\hat{s}$ and $\hat{n}$. In this intrinsic system, the inviscid equation of motion for a steady flow are

$$\frac{\partial p}{\partial s} = -\rho q \frac{\partial q}{\partial s}, \qquad \frac{\partial p}{\partial n} = -\rho q^2/R, \qquad \frac{\partial p}{\partial b} = 0. \qquad (6a,b,c)$$

The so-called *osculating plane* is the plane containing $\hat{s}$ and $\hat{n}$. As stated in Ref. 8, "... three-dimensional flow behaves locally like axisymmetric flow in the osculating plane".

A direct extension of Eq. (6b) to determine a curvature-corrected symmetry technique for three dimensions is complicated by the fact that the $\hat{n}$-direction is not, in general, normal to the surface. However, it can be proved that in a plane formed by $\hat{s}$ and $\hat{n}_b$, where $\hat{n}_b$ is the surface normal, that

$$\frac{\partial p}{\partial n_b} = -\rho q^2/R_b, \qquad (7)$$

where R_b is the radius of curvature of the line which is formed by the intersection of the body surface with the plane containing $\hat{s}$ and $\hat{n}_b$. Thus, to develop the CCST, we can adapt the procedures used for two dimensions given in Eqs. (2)-(4). The required changes include the numerical evaluation of the radius of curvature, R_b, of the intersection of the body surface with the plane formed by the tangent to the surface streamline and the surface normal. For the examples presented here, this term is evaluated by approximating the surface streamline direction as the projection of the streamline direction at cell center 1 (of Fig. 1) to the body surface. The R_b term in then evaluated analytically. In addition, we define three velocity components $\tilde{u}$, $\tilde{v}$ and $\tilde{w}$. The component $\tilde{u}$ is in the direction of the streamline projected onto the surface, $\tilde{v}$ is in the direction normal to the surface and $\tilde{w}$ is in the direction normal to $\tilde{u}$ and $\tilde{v}$. We determine these velocity components at the image points by reflecting the surface projection of the streamline direction. Accordingly, at the image points, $\tilde{u}$ are given by Eqs. (4) and $\tilde{v}$ are given by Eqs. (5). Then by definition, $\tilde{w}_{-1} = \tilde{w}_{-2} = 0$.

4 Results

We test the accuracy of three-dimensional CCST boundary condition by comparison with several other boundary condition methods on some simple test cases. The boundary condition procedures will include first and second order pressure extrapolation, which we denote P-I and P-II along with what we call the symmetry technique, ST, and the curvature-corrected symmetry technique, CCST. The ST approach is similar to CCST, but it neglects surface curvature effects in the pressure extrapolations in Eqs. (2). The computations were performed with an explicit cell-centered finite-volume code using Roe's flux-difference splitting with nominal second-order accuracy.

The first test case is the rather trivial case of the subsonic compressible flow over a sphere at $M_\infty = 0.52$. Even though this is an axisymmetric flow, it was computed with a three-dimensional code. The study was performed with three grids denoted *coarse*, *medium* and *fine* in Table 1. The coarse mesh consisted of a grid of $8 \times 16 \times 32$, with 8 cell centers in the radial direction and 16 in the azimuthal direction (from leading to trailing edges) and 32 in the meridional direction. The medium mesh was $16 \times 32 \times 64$ and the fine mesh was $32 \times 64 \times 128$. In Table 1 we present a number of results for each of the three grids, including the maximum surface Mach number, M_{max}, and the trailing-edge surface pressure, p_{te}. As an indication of the accuracy of these approaches, we also show the $\mathcal{L}_2$ norm of the total pressure error on the body surface, $\mathcal{L}_2(p_0 - 1)$ and the $\mathcal{L}_2$ norm of the entropy in the flow field $\mathcal{L}_2(s)$. Finally to indicate how the boundary condition techniques reproduce flow symmetries, we present the $\mathcal{L}_2$ norm of the pressure difference of symmetrically located points on the body surface, $\mathcal{L}_2(p - p_s)$ as well as the density $\mathcal{L}_2(\rho - \rho_s)$. Note that the pressure and density have been normalized by the free-stream total pressure and density, respectively.

Table 1. Subsonic flow over a sphere. Comparison of surface boundary conditions.

	P-I			P-II			ST			CCST		
	coarse	med.	fine	coarse	med.	fine	coarse	med.	fine	coarse	med.	fine
M_{max}	0.768	0.842	0.870	0.809	0.868	0.884	0.799	0.860	0.879	0.838	0.878	0.887
p_{te}	0.917	0.940	0.964	0.962	0.979	0.989	0.951	0.974	0.986	1.013	1.008	1.003
$\mathcal{L}_2(p_0 - 1)$	0.062	0.040	0.023	0.036	0.016	0.007	0.041	0.018	0.008	0.023	0.006	0.0016
$\mathcal{L}_2(s)$	0.013	0.008	0.003	0.012	0.005	0.0017	0.011	0.003	0.0012	0.008	0.002	0.0005
$\mathcal{L}_2(p - p_s)$	0.048	0.029	0.015	0.033	0.013	0.005	0.023	0.012	0.006	0.006	0.002	0.0005
$\mathcal{L}_2(\rho - \rho_s)$	0.043	0.025	0.013	0.028	0.011	0.005	0.019	0.009	0.005	0.006	0.002	0.0009

From the results presented in Table 1, we see that the results obtained with the CCST approach are superior to the other methods. In particular, a good indicator is the trailing-edge pressure, p_{te}, and from the results shown in the table, it appears that all four methods are consistent and approach nearly the same unit pressure in the limit of vanishing cell size. However the CCST results for p_{te} are clearly more converged in grid. Moreover, at the fine grid level, we see that P-I has a 4% error, P-II and ST still present a 1% error, whereas the CCST has an error of 0.3%. The other indicators in Table 1 also confirm this behavior. We note that CCST is significantly more accurate than the P-I technique, which is utilized in many cell-centered finite-volume codes.

Table 2. Subsonic flow over a prolate spheroid. Comparison of surface boundary conditions.

	P-I			P-II			ST			CCST		
	coarse	med.	fine	coarse	med.	fine	coarse	med.	fine	coarse	med.	fine
M_{max}	0.805	0.908	0.938	0.865	0.941	0.955	0.844	0.929	0.948	0.903	0.952	0.958
p_{te}	0.911	0.939	0.964	0.965	0.985	0.992	0.958	0.981	0.990	1.006	1.009	1.004
$\mathcal{L}_2(p_0 - 1)$	0.069	0.043	0.023	0.036	0.014	0.006	0.039	0.015	0.007	0.023	0.005	0.001
$\mathcal{L}_2(s)$	0.017	0.009	0.004	0.013	0.005	0.002	0.011	0.004	0.001	0.008	0.002	0.0004
$\mathcal{L}_2(p - p_s)$	0.050	0.028	0.014	0.029	0.010	0.004	0.027	0.011	0.005	0.008	0.003	0.0007
$\mathcal{L}_2(\rho - \rho_s)$	0.045	0.025	0.013	0.026	0.009	0.004	0.022	0.009	0.004	0.006	0.002	0.0007

The remaining two test cases are flows over ellipsoids of revolution. Case 2 is the $M_\infty = 0.47$ flow over a prolate spheroid. The free-stream is in the x-direction and the equation of the surface is $(x/a)^2 + (y/a)^2 + (z/b)^2 = 1$ with the axis ratio $b/a = 2$. (The free stream is perpendicular to the major axis of the ellipsoid.) The same three grids used in case 1 are used here, namely $8 \times 16 \times 32$, $16 \times 32 \times 64$ and $32 \times 64 \times 128$. The results of case 2 are presented in Table 2. We also present a plot of p_{te} for this case versus $(1/N)^{2/3}$ where N is the number of cell centers. The plot shown in Fig. 2 indicates the degree of grid convergence for the CCST method compared to the other approaches.

Table 3. Subsonic flow over an oblate spheroid. Comparison of surface boundary conditions.

	P-I			P-II			ST			CCST		
	coarse	med.	fine	coarse	med.	fine	coarse	med.	fine	coarse	med.	fine
M_{max}	0.829	0.885	0.901	0.848	0.898	0.908	0.832	0.891	0.906	0.850	0.898	0.908
p_{te}	0.934	0.952	0.971	0.976	0.983	0.991	0.968	0.980	0.988	1.029	1.010	1.003
$\mathcal{L}_2(p_0 - 1)$	0.054	0.031	0.017	0.031	0.013	0.006	0.035	0.015	0.007	0.022	0.006	0.0015
$\mathcal{L}_2(s)$	0.014	0.006	0.002	0.012	0.004	0.002	0.010	0.003	0.0008	0.008	0.002	0.0006
$\mathcal{L}_2(p - p_s)$	0.053	0.029	0.015	0.037	0.013	0.005	0.025	0.013	0.006	0.021	0.005	0.001
$\mathcal{L}_2(\rho - \rho_s)$	0.046	0.024	0.012	0.031	0.010	0.004	0.020	0.009	0.004	0.019	0.005	0.001

Case 3 is the $M_\infty = 0.62$ flow over an oblate spheroid. The free-stream is in the x-direction and the equation of the surface is $(x/a)^2 + (y/a)^2 + (z/b)^2 = 1$ with the axis ratio $b/a = 0.5$. (The free stream is perpendicular to the minor axis of the ellipsoid.) The same three grids used previously are used for this case. The results are presented in Table 3 and the trailing-edge pressure is plotted against $(1/N)^{2/3}$ in Fig. 3. The results for both ellipsoids confirm those for the sphere in establishing the efficacy of the CCST over the other boundary condition procedures. These results indicate that the three-dimensional version of this technique should prove to be as useful as the two-dimensional version.[1] Future work will test the implementation of this method on three-dimensional flows over complicated geometries.

Acknowledgments :

The research of the first author was supported by the Italian agency $MURST$. The second author acknowledges the assistance of the National Aeronautics and Space Administration under NASA Langley grant NAG-1-776.

References

1. Dadone, A. and Grossman, B., "Surface Boundary Conditions for the Numerical Solution of the Euler Equations", *AIAA J.*, **32**, No. 2, 1994, pp. 285–293.

2. Moretti, G., "Importance of Boundary Conditions in the Numerical Treatment of Hyperbolic Equations", in *High-Speed Computing in Fluid Dynamics, Physics of Fluids*, Supplement II, 1969, pp. II-13 – II-20.

3. Kentzer, C. P., "Discretization of Boundary Conditions on Moving Discontinuities", in *Lecture Notes in Physics*, No. 8, Springer-Verlag, N. Y., 1970, pp. 108–113.

4. De Neef, T., "Treatment of Boundaries in Unsteady Inviscid Flow Computations", Delft University of Technology, Dept. Aerospace Eng., Rept. LR-262, Delft, The Netherlands, Feb. 1978.

5. Marcum, D. L. and Hoffman, J. D., "Numerical Boundary Condition Procedures for Euler Solvers", *AIAA J.*, **25**, No. 8, 1987, pp. 1054–1068.

6. Rizzi, A., "Numerical Implementation of Solid Boundary Conditions for the Euler Equations", *Z.A.M.M.*, **58**, No. 7, 1978, pp. T301–T304.

7. Serrin, J., "Mathematical Principles of Classical Fluid Mechanics", *Handbuch Der Physik*, (S. Flügge, C. Truesdell, eds.) **VIII**, 125–262. Springer-Verlag, Berlin, 1959.

8. Frohn, A., "An Analytic Characteristic Method for Steady Three-Dimensional Isentropic Flow", *J. Fluid Mech.*, **63**, part 1, 1974, pp. 81–96.

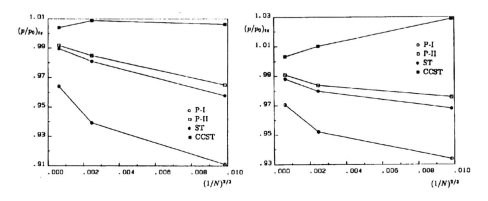

Fig. 2. Trailing-edge pressure - prolate spheroid.

Fig. 3. Trailing-edge pressure - oblate spheroid.

The ψ - q formulation and boundary condition for incompressible flows in multi-connected region

W. Jia[1] and Y. Nakamura[2]

[1]Dept. of Mech. Eng., Yamagata Univ., Yonezawa 992, Japan
[2]Dept. of Aero. Eng., Nagoya Univ., Nagoya 464-01, Japan

1 Introduction

Recently, we proposed a new formulation[1,2] for the incompressible N-S equations by introducing a new variable q, whose curl represents the non-solenoidal terms of the momentum equations. The derived governing equations consist of the linear dynamic equation of $\mathbf{v}$ and the Poisson equation of $\mathbf{q}$ with a non-linear source. These equations theoretically satisfy the divergence free condition. The boundary condition (BC) of $\mathbf{q}$ is the Dirichlet type, so that the Poisson equation can be quickly solved by the conventional point relaxation method. This method has been successfully developed for 2D and 3D problems and applied to both steady and unsteady flows with simple connected region.

However, there are two problems remained to be improved. The first one is that although these equations theoretically satisfy the divergence free condition, numerical errors may accumulate during the time integration process which will influence the resultant velocity field. The second one relates to how to determine the integration constants which appear in the Dirichlet BCs of q for a problem with multi-connected region.

The present paper provide a solution to these problems. It will be shown that by introducing the stream-function ψ along with the new variable q, the first problem can be completely prevented. The integration constants in the Dirichlet BCs of q will be recognized as the time derivative of the stream-function at each boundary. These are determined by satisfying an integration constraint similar to that employed in the conventional stream-function and vorticity method. However, the procedure of inflicting such a constraint is much easier than employed in the conventional method.

2 Derivation of 2D $\psi - q$ formulation

The governing equations for incompressible flows are written as follows:

$$\nabla \cdot \mathbf{v} = 0 \qquad (1)$$

$$\frac{\partial \mathbf{v}}{\partial t} + \omega \mathbf{k} \times \mathbf{v} + \nabla H = -\frac{1}{Re}\nabla \omega \times \mathbf{k} + \mathbf{f} \qquad (2)$$

$$\omega = -\frac{\partial u}{\partial y} + \frac{\partial v}{\partial x}, \quad H = p + \frac{1}{2}\mathbf{v}^2 \qquad (3)$$

where **v** the velocity vector, ω the vorticity, H the total pressure, f the body force, and Re the Reynolds number. k denotes the unit vector normal to the 2D plane.

By taking the divergence of Eq.(2) and substituting Eq.(1) into it, we have

$$\nabla \cdot (\omega \mathbf{k} \times \mathbf{v} + \nabla H - \mathbf{f}) = 0 \qquad (4)$$

To automatically satisfy Eq.(4), a new variable q is introducted as follows:

$$\omega \mathbf{k} \times \mathbf{v} + \nabla H - \mathbf{f} = \nabla q \times \mathbf{k} \qquad (5)$$

By use of this relation, the new dynamic equation of v and the Poisson equation of q are obtained as follows:

$$\frac{\partial \mathbf{v}}{\partial t} + \nabla \left(q + \frac{1}{Re}\omega \right) \times \mathbf{k} = 0 \qquad (6)$$

$$\nabla^2 q = -\nabla \cdot (\mathbf{v}\omega) + \mathbf{k} \cdot \nabla \times \mathbf{f} \qquad (7)$$

These equations were employed as new governing equations in references 1 and 2. The velocity divergence evolves as $\partial \nabla \cdot \mathbf{v}/\partial t = 0$. the divergence free condition is theoretically ensured if the initial field is divergence free. However, owing to the accumulation of numerical errors, the velocity divergence may increase during the time integration process.

Here, we introcuce the stream-function $\psi(\mathbf{v} = \nabla \psi \times \mathbf{k})$ to prevent this problem. It is inserted into Eq.(6), which is spatially integrated as follows:

$$\frac{\partial \psi}{\partial t} + q = \frac{1}{Re}\nabla^2 \psi \qquad (8)$$

The Poisson equation of q becomes

$$\nabla^2 q = \nabla \cdot (\mathbf{v}\nabla^2\psi) + \mathbf{k} \cdot \nabla \times \mathbf{f} \qquad (9)$$

In the present paper, Equations (8) and (9) are employed as governing equations, whcih obviously satisfy the divergence free condition.

3 Boundary Condition of q

We consider a multi-connected region whose boundary Γ is composed of n closed sub-boundaries Γ_i. At each sub-boundary Γ_i, $\psi_i = c_i(t)$. Since q is a derived variable, the BC of which should be derived from the dynamic equation at the boundary. That is $q = Re^{-1}\partial^2\psi/\partial n^2 - a_i$, where $a_i = \partial c_i/\partial t = \partial \psi_i/\partial t$.

The constants a_i are determined from the constraint of pressure be single-valued at each of the sub-boundary Γ_i. Specially, the definition of q (Eq.(5)) is integrated along Γ_i, to satisfy the constraint, we have

$$\oint_{\Gamma_i} \frac{\partial q}{\partial n} ds = \oint_{\Gamma_i} \mathbf{f} \cdot \tau \, ds \qquad (10)$$

where n and τ denote the normal and tangential directions to the boundary, respectively.

Such a constraint is equivalent to $\oint_{\Gamma_i} \partial \omega/\partial n \, ds = -Re \oint_{\Gamma_i} \mathbf{f} \cdot \tau \, ds$, which was imposed on the conventional stream-function and velocity method. The above constraint can easily be embeded in the iterative solution procedure of q as follows.

$$\nabla^2 q = \nabla \cdot (\mathbf{v}\nabla^2 \psi) + \mathbf{k} \cdot \nabla \times \mathbf{f} \quad \text{in the computational region} \tag{11}$$

$$q_1 = \frac{1}{Re}\frac{\partial^2 \psi}{\partial n^2} - \underbrace{\frac{1}{\Gamma_i}\oint_{\Gamma-i}\left(\frac{1}{Re}\frac{\partial^2 \psi}{\partial n^2} + \mathbf{f} \cdot \tau - q_2\right) d\Gamma}_{a_i} \quad \text{on} \quad \Gamma_i \tag{12}$$

where q_1 and q_2 denote the values at and next to the boundary, respectively. The second term on the right-hand side of Eq.(12) becomes equal to a_i when the iterative solution converges.

It is noteworthy to compare the present constraint of Eq.(10) in the $\psi - q$ formulation with the constraint imposed on the conventional stream-function and vorticity method. Because the vorticity transport equation is solved in the latter formulation, the constraint on it can be simply inflicted on the solution procedure. This is the important difference between the two methods.

As a numerical scheme, any existing spatial and temporal discretization techniques employed in the conventional methods can also be applied to the new formulation. Here, the governing equations are spatially discretized by the finite volume method with a regular mesh in generalized coordinates, where the source of the Poisson equation of q is calculated by the QUICK[3] method and the others by the second order approximation. The time integration is performed by the Euler explicit method.

4 Numerical Results

The new formulation and boundary conditions are verified by the flow between two concentric cylinders with zero and unit surface velocities for the outer and inner cylinders, respectively. Figure 1 shows the solution under the BC of Eq.(12), where the constraint of Eq.(10) is satisfied. Figure 1a corresponds to the velocity vectors and Fig.1b the velocity, vorticity as well as $Re \times q$ distributions in the r direction. The velocity distribution is compared with the analytic solution. It is seen that both solutions agree well. In the present problem, the vorticity remains constant in the r direction and equals to $Re \cdot q$.

For comparison, a numerical solution which does not take account of the boundary constants a_i in Eq.(12) in the time integration process is shown in Fig.2. Consequently, the boundary values of the stream-function remain zero at each cylinder surface. The resultant solution is non-physical which has zero flow rate flow section. The proposed boundary conditions are therefore verified by the present problem.

As some more practical examples, the flow between a circular cylinder and an ellipse, the natural convection between two concentric cylinders with different temperatures, and the case between two non-concentric cylinders are conducted. The results are depicted in Figure 3 to 5, respectively.

5 Concluding Remarks

The $\psi - q$ formulation for incompressible viscous flows has been developed which automatically satisfies the divergence free condition in the discretized form.

The boundary conditions of q for the problem with multi-connected region have been derived ans a simple procedure to satisfy these conditions has been proposed

The new formulation and boundary conditions are verified by the flow between two concentric cylinders. The results agree well with the exact solution.

Applications to more complicated flow problems are also conducted where reasonable results are obtained.

References

1. W.Jia and Y.Nakamura: A New Formulation for Incompressible Viscous Flows, in Proc. 1st European CFD Conf., Ch.Hirsch ed., Elsevier Science Publishers B.V., 1(1992), pp. 383-390.

2. W.Jia and Y.Nakamura: The Three Dimensional $v - q$ Method for Incompressible Viscous Flows, AIAA-93-3341-CP.

3. B.P.Leonard: A Stable and Accurate Convective Modeling Procedure Based on Quadratic Upstream Interpolation, Computer Methods in Applied Mech. and Eng., 19(1979), pp. 59-98.

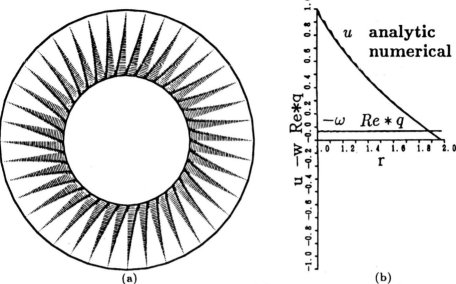

Fig. 1 Flow between two concentric cylinders. Solution under the boundary condition of Eq.(12). (a) Velocity vectors, and (b) velocity, vorticity and $Re \cdot q$ distributions in the r direction.

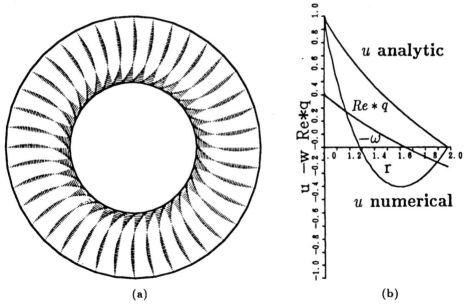

Fig. 2 Flow between two concentric cylinders. Solution which does not take account of the boundary constants a_i in Eq.(12) in the time integration process.

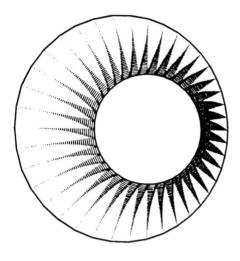

Fig. 3 Velocity vectors of flow between a circular cylinder and an ellipse ($Re = 100$).

Fig. 4 Velocity vectors and temperature contours of natural convection between two concentric cylinders. $Gr = 10^5$ and $Pr = 0.71$. $\theta = 1$ and 0 at the inner and outer cylinders, respectively.

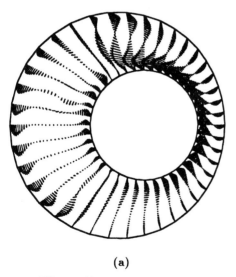

(a)   (b)

Fig. 5 Non-concentric case. (a) velocity vectors and (b) temperature contours. The other conditions are the same as those in Fig.4.

Treatment of Branch Cut Lines in the Computation of Incompressible Flows with C-grids

C. P. Mellen, K. Srinivas

University of Sydney, N.S.W., 2006, Australia

Abstract: It is well recognised that C-grids are ideally suited for computing flows about bodies with a rounded leading edge like aerofoils. A branch cut line is an essential feature of such a grid and is usually treated as one of the boundaries. However, when one computes an incompressible flow using a pressure correction method, the presence of such a feature can be problematic in achieving a stable convergence. This paper describes a novel way of handling the branch cut line which ensures convergence of solution when using SIMPLEC type pressure correction procedures.

1 Introduction

Many reliable methods exist for computing two dimensional incompressible flow through regions with arbitrarily shaped boundaries. One such region which is of practical interest is the flowfield around an aerofoil. If a structured grid is used to discretise the governing equations for a control volume around the aerofoil, it is not uncommon for this grid to be of the 'C' type. A feature of such a grid is the presence of a branch cut line extending from the trailing edge of the aerofoil to the downstream outflow boundary.

In general, the values of flow variables prescribed or set along a boundary determine the flowfield interior to that boundary. Thus, along the branch cut line of the C-grid, the challenge is to assign values which maintain the continuity and smoothness of the flowfield while at the same time satisfying the governing equations. This paper presents a method of treating the branch cut line when using a SIMPLEC type pressure correction procedure. The method is applied to compute the two dimensional flow past a NACA 0012 aerofoil using a non-staggered, non orthogonal, body conforming grid. The effectiveness of the method is brought out.

2 Governing Equations and Numerical Method

In non-orthogonal coordinates, the equations for a steady two-dimensional flow can be written in a general form as :

$$\frac{\partial}{\partial x_1}(C_1\phi + D_{1\phi}) + \frac{\partial}{\partial x_2}(C_2\phi + D_{2\phi}) = JS_\phi \qquad (1)$$

where the C_i terms relate to convection, the D_i terms relate to diffusion, J is the Jacobian of the transformation, and the source term S_ϕ contains (where appropriate) contributions due to pressure gradients and fluid stresses. For a complete description of these terms, the reader is referred to [1]. In this work, the governing equations are solved using the SIMPLEC algorithm [2]. The momentum interpolation technique is used to remove the tendency for the non-staggered grid arrangement to produce non-physical oscillations in the pressure and/or velocity fields.

3   Principles of Branch Cut Treatment

Formulation of a method for treating the branch cut line demands that both the mass flux and the u,v momentum flux across the branch cut are conserved. Further, the equations governing incompressible flow are elliptic, implying that any one region of the flowfield should theoretically be able to be influenced by every other region of the flowfield. Thus, if the ellipticity of the flowfield is to be preserved in the presence of a branch cut line, means must be found of providing strong linkages across it for the pressure field, momentum field and the pressure correction field.

4   Treatment of Pressure Correction Equation

When discretised, the pressure correction equation used in the SIMPLEC algorithm can be expressed in the form

$$A_P p'_p = A_E p'_E + A_W p'_W + A_N p'_N + A_S p'_S - \Delta Q \tag{2}$$

with

$$A_P = A_N + A_S + A_E + A_W \qquad \Delta Q = C_w^* - C_e^* + C_n^* - C_s^* \tag{3}$$

where p' is the pressure correction field and the $C_e^*, C_w^*, C_s^*, C_n^*$ are the mass fluxes through each cell face as calculated using the u^*, v^* velocity field.

When equation (2) is used at a boundary, one needs to impose an appropriate boundary condition. When the pressure in the cell is known, the pressure correction in the cell is set to zero. If the pressure is not known, one needs to know the normal component of velocity or its gradient through the cell face on the boundary. For a branch cut line, where pressures are not known, it appears that the latter boundary condition is the most suitable.

The simplest means of determining the normal component of velocity at the branch cut line is by using the momentum interpolation technique to interpolate from adjoining cell centres. This normal velocity is then used to calculate mass flux across the cell faces along the branch cut. Once the mass flux through a cell face has been explicitly assigned, the pressure correction equation for that cell becomes

$$A_P p'_p = A_E p'_E + A_W p'_W + A_N p'_N - \Delta Q \tag{4}$$

where

$$A_P = A_E + A_W + A_N \qquad \Delta Q = C_w^* + C_e^* + C_n^* - S_{\text{flux}} \tag{5}$$

where the term A_S has to set to zero to denote that the mass flux (S_{flux}) through the south cell face adjoining the branch cut is considered to be a known quantity. However, an important step in obtaining the correct velocity and pressure fields is to correct the velocity through each cell face to account for changes in the pressure gradient arising from the addition of the p' field. For the south face of a typical cell (shown in Fig 1.), this procedure can be represented by

$$v_S = \frac{A_S}{a_S}(p'_S - p'_P) + v^*_S \qquad (6)$$

where a_S is the area of the south cell face. Thus, the implication of setting As to zero and explicitly evaluating the mass flux through the cell face is effectively an assumption of a zero pressure gradient across the C-grid join - an assumption which is usually physically unrealistic.

A practical reason for setting the terms relating to the pressure correction field outside the boundary to zero (the A_S's above) is that the correction field outside the boundary is in fact unknown. However, if the boundary is internal to the computational domain, a pressure correction field 'outside' the boundary does exist. Thus, it should be possible to treat the mass flux across the branch cut in an implicit manner. If the pressure correction equation is written using matrix notation, the $[A]$ coefficient matrix is seen to be of pentadiagonal form. Treating the mass flux across the branch cut in an implicit manner adds an extra diagonal to this matrix, with this new diagonal lying at $90°$ to the original diagonals. In the case where an iterative solver is used to solve the pressure correction equation, the easiest way to account for the terms on this new diagonal is by noting that after each internal iteration of the solver, an 'intermediate' pressure correction field is available throughout the computational domain. Thus, the approach taken is to treat the contributions from the cross diagonal terms as 'pseudo source' terms. After each internal iteration of the solver, the contribution of these terms to the pressure correction equation is updated using the intermediate pressure correction field just calculated. This treatment allows for strong pressure linkages to be maintained across the branch cut. The treatment of the pressure-velocity correction along the branch cut line can is summarised below :

- 1. Use guessed pressure field p^* to obtain guessed velocity field u^*,v^*

- 2. Use momentum interpolation to estimate velocity and mass flux S_{flux} across branch cut.

- 3. Evaluate coefficients from pressure correction equation relating to pressure correction across branch cut - A_S's - in manner identical to that used if join were internal to grid.

- 4. Solve for p' field with A_S's $\neq 0$ along join. After each internal iteration of solver update contribution $A_S p'_S$ to pressure correction equation from interim pressure correction field.

- 5. Update velocity across boundary due to change in pressure gradient - $(p'_P - p'_S)$.

It has been found that the implicit treatment of the mass flow across the C-grid join is crucial to the stability of the method. If this mass flux is treated explicitly, large differences in the p' field arise across the branch cut. These differences cause spurious oscillations in the momentum and velocity fields which eventually lead to divergence of the solution.

5 Treatment of Momentum Equations

The treatment of the momentum equations along the branch cut line follows very closely that described earlier for the pressure correction equation. In the present method, the coefficients of the linearised equations are evaluated as if the branch cut lay along a computational line internal to the grid. Where required, the velocity components are obtained from the cell centres immediately south of the branch cut. It is a simple procedure to verify that this treatment ensures that momentum flux across the C-grid is conserved.

6 Results

A code incorporating the changes outlined above was developed within the framework of [3], and was then used to compute the flow about a NACA 0012 aerofoil. The grid used for these computations consisted of 149 by 80 points, with 93 points distributed over the aerofoil surface. The downstream and freestream boundaries were placed approximately 3 chord lengths away from the aerofoil. Computations were carried out for various Reynold's numbers and angles of attack. Comparisons were made with the measurements in [4], which relate to a NACA 0012 aerofoil at a Reynold's No. of 2.8×10^6 and angle of attack of $6°$. The blockage effects which may have influenced the results in [4] were not simulated - the boundaries of the computational domain were set to be either freestream (if inflow boundaries) or were set to give zero gradient tangential to the flow direction (if outflow boundaries). Adjacent to the aerofoil surface, the average grid size in the direction normal to the surface was approximately 0.025 of the chord length. The surface pressure distribution produced with this grid is shown in Fig 2. Also shown are the results obtained from a mesh which was interactively adapted to better resolve the flow gradients near the aerofoil surface. In general, the pressure distribution shows good agreement with the experimental data, with little difference between the two grids. The contour plots for farfield pressure and velocity contained in Fig's 3 and 4 illustrate that the flowfield is smooth and continuous across the C-grid join.

7 Conclusion

A method for treating the branch cut line on C-grids when using SIMPLEC like pressure correction techniques has been formulated and implemented. The method was verified by computing the flow over a NACA 0012 aerofoil at an angle of attack of

$6°$ and Reynold's No. of 2.8×10^6, with good agreement between the computed and experimental results being observed. Computations for the flow around the above aerofoil were also carried out using an adapted grid designed to more accurately capture regions of high velocity gradient. These computations served to show the flexibility and stability of the present branch cut treatment.

Acknowledgments :

Thanks are due to Professor W. Rodi of the University of Karlsruhe for making available the code which formed the basis of this work.

References

[1] Rodi W., Majumdar S. and Schonung B., 'Finite Volume Methods for Two-Dimensional Incompressible Flows With Complex Boundaries', Computer Methods in Applied Mechanics and Engineering, **75**, pp 369-392, (1989).

[2] Van Doormal J.P. and Raithby G.D., 'Enhancements of the SIMPLE Method for Predicting Incompressible Fluid Flows', Numer. Heat Transfer, **7**, pp 147-163, (1984).

[3] Zhu J., 'FAST-2D: A Computer Program for Numerical Simulation of Two Dimensional Incompressible Flows with Complex Boundaries', Report No. 690, Institute for Hydromechanics, University of Karlsruhe, (1991).

[4] Gregory N. and O'Reilly C.L., 'Low Speed Aerodynamic Characteristics of NACA 0012 Aerofoil Section, Including the Effects of Upper Surface Roughness Simulation Hoarfrost', National Physical Laboratory, Teddington, England, Aero Rept. 1308, (1970).

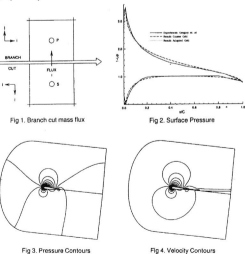

Fig 1. Branch cut mass flux Fig 2. Surface Pressure

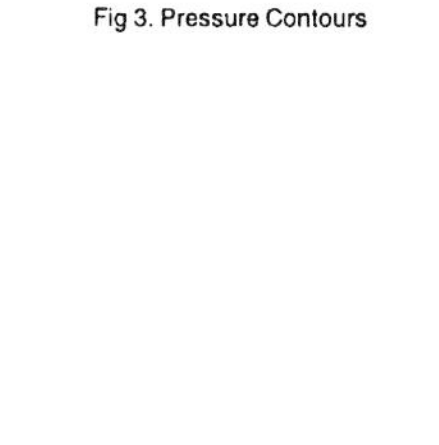

Fig 3. Pressure Contours 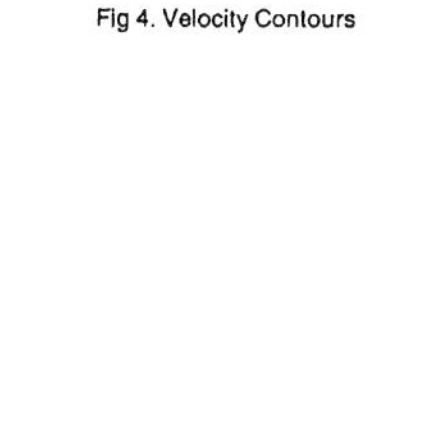 Fig 4. Velocity Contours

Concepts on Boundary Conditions in Numerical Fluid Dynamics

Yo Mizuta

Dept. Engineering Science, Fac. Engineering, Hokkaido
University, Sapporo 060, Japan

1 Introduction

Boundary conditions are important in numerical fluid dynamics – this is a matter of course, though, the methods are not pursued deeply in most problems such as the flow around a body (bodies) or in a cavity, since there seems no more than putting the normal velocity zero on a wall, or putting the tangential velocity zero too, if the wall is non-slip. In some problems, boundary conditions are essential – fluids with free surface or density interface, fluids around a moving or oscillating body (bodies), open boundary, and so on. For successful numerical experiment of these problems, boundary conditions with physical significance are required, instead of expedient ones.

2 Deformable Cell and Laws of Conservation

In the present method of numerical analysis of fluids [1], the fluid in the domain of analysis is divided into deformable cells according to the movable boundaries. Since the shape of each cell differs from one to one, and it deforms, equations for numerical analysis are derived from integral-type laws of conservation here, instead of the hydrodynamic equations finite-differenced. When the fluid is incompressible, the **law of conservation of volume (LCV)**

$$0 = \left(\oint d\mathbf{S} \cdot \mathbf{v} \right)_v^{n+1} \equiv \Delta V(\{\mathbf{q}^{n+1}\}) \tag{1}$$

must be satisfied concerning a small domain around a vertex subscripted as $v = 1, 2, \cdots$ (LCV domain, see Fig. 1) not only within the domain but also adjacent to boundaries and in corners at the time step $n+1$, where $\mathbf{v}$ is the fluid velocity, and the surface integral covers all the surfaces of the LCV domain. In two-dimensional case, this is rewritten as a linear combination of the flux $\mathbf{q} = b\mathbf{v}$ (b:the thickness of the fluid), which $\Delta V(\{\mathbf{q}^{n+1}\})$ shows.

Treatment of boundaries can be described conveniently on the basis of the LCV adjacent to a boundary or in a corner. The flux $\mathbf{q}_A$ on a boundary or at a corner is expressed by the normal flux q_n and the tangential flux q_s as $\mathbf{q}_A = (q_n\mathbf{n} + q_s\mathbf{s})_A$, $\mathbf{q}_A = [(q_n)_{(2)}\mathbf{s}_{(1)} - (q_n)_{(1)}\mathbf{s}_{(2)}]/[(\mathbf{s}_{(1)} \times \mathbf{s}_{(2)}) \cdot \mathbf{z}]$ where $\mathbf{n}$, $\mathbf{s}$ are the normal and the tangential vector on the boundary, (1),(2) denote two boundaries crossing at the corner, and **z** is the unit vector normal to the domain of analysis. The terms without q_n in ΔV are collected into $\Delta V_{B,C}$, and each $(q_s)^{n+1}$ is expressed linear in $\mathbf{q^{n+1}}$ within the domain as $(q'_s)_A^{n+1} = \langle q_s\rangle_A^{n+1} + \mathsf{K}_A \cdot \mathbf{q_D^{n+1}}$ as for most tangential boundary conditions. Then, the LCV is rewritten as

$$0 = (q_n \Delta s)_5^{n+1} + \Delta V_B(\{\mathbf{q^{u+1}}\}), \qquad (2)$$
$$0 = (q_n \Delta s)_{(1)}^{n+1} + (q_n \Delta s)_{(2)}^{n+1} + \Delta V_C(\{\mathbf{q^{n+1}}\}), \qquad (3)$$

where $\Delta s_{5,(1),(2)}$ are the contact length between the LCV domain and the boundary. This formulation removes an instability concerning tangential boundary conditions such as free slip wall when $(q_s)^n$ is used expediently, instead of $(q_s)^{n+1}$.

3 Volume Force and Surface Force

There are fluids which react to an external field: dielectric fluid under an electric field or magnetic fluid under a magnetic field. In the equation of motion considering these fluids also, we find the gravity force, the pressure force, the viscous force and the electric or magnetic force. In physical sense, the forces which tend to zero together with the density are called volume force, and the others are surface force. In numerical analysis of incompressible fluids, however, these terms could be used in another way.

The momentum $M\mathbf{v}$ in a cell $c = A, B, \cdots$ is renewed between the time steps n and $n+1$ with the increment Δt, by adding to it the convective flux including the difference of the fluid velocity $\mathbf{v}$ and the deformation velocity of the surface of the cell $\mathbf{u}$, and the aforementioned forces. This is shown, equivalent to the equation of motion, in two equations as

$$(M\mathbf{v})_\mathbf{c}^{(\mathbf{n})} = (M\mathbf{v})_\mathbf{c}^\mathbf{n} + \left[\oint d\mathbf{S} \cdot \{(\mathbf{u}-\mathbf{v})\rho\mathbf{v} + \mathsf{V}\}\right]_c^n \Delta t + \left(\iiint d V \mathbf{F^*}\right)_c^n \Delta t, \qquad (4)$$

$$(M\mathbf{v})_\mathbf{c}^{\mathbf{n+1}} = (M\mathbf{v})_\mathbf{c}^{(\mathbf{n})} - \left(\iiint dV \nabla p^*\right)_c^{n+1} \Delta t, \qquad (5)$$

where ρ and V are the fluid density and the viscosity stress tensor.

After the momentum in each cell is renewed, the fluxes from these momenta must satisfy the LCV at $n+1$. The procedure to determine the pressure on vertices answering this requirement intervenes between the procedures to add the forces without the pressure force and the pressure force itself. Though the pressure force is evidently the surface force, parts of the gravity force or the electric or magnetic force can be transferred to the pressure force to improve the accuracy of the numerical analysis. Then, the **surface force** $-\nabla p^*$ is newly defined, and the sum of the other forces results in the **volume force** $\mathbf{F^*}$.

4 Equations for Extended Pressure p^* and Normal Flux q_n

Equation (5) is rewritten for the flux **q** as

$$\mathbf{q}_c^{n+1} = \mathbf{q}_c^{(n)} - \mathbf{f}_c^{n+1}\Delta t, \quad \mathbf{q}_c^{(n)} \equiv \left(\frac{b}{M}\right)_c^{n+1}(M\mathbf{v})_c^{(n)}, \quad \mathbf{f}_c^{n+1} \equiv \left(\frac{b}{M}\right)_c^{n+1}\iiint dV \nabla p^*,\quad (6)$$

where the surface force **f** is a linear combination of the **extended pressure** p^*'s on the vertices encircling the cell. With this inserted, the LCV within the domain, (2) and (3) turn to equations for p^*'s and q_n's called here as **pressure equation**, **boundary equation (BE)** and **corner equation**, respectively. When they are solved for p^*'s and q_n's simultaneously (**set of equations**), the numbers of equations and unknowns always agree, and p^*'s and q_n's thus determined assure the LCV's to be satisfied all over the domain of analysis without ambiguity.

5 Categorized Boundary Conditions & their General Treatment

On each boundary, one tangential and one normal boundary condition are specified and must be satisfied together with the LCV, and the BE is the basis for this purpose. One of the tangential boundary conditions such as (SV) q_s is known, (SS) the tangential stress σ_s is known, (SI) both q_s and σ_s are continuous, has been incorporated into the BE. The normal boundary condition is categorized into one of the following, and the BE is used properly.

(NV) As on a wall, q_n is specified, and p^* on the boundary is obtained after the BE is solved simultaneously with other equations in the set of equations. The BE plays the role of the Neumann boundary condition for the Poisson equation of p^* in this case.

(NP) As on a free surface, p^* is specified, and it is used as the Dirichlet boundary condition for the Poisson equation when the set of equations are solved. The normal flux q_n is obtained through the BE or (2) after all $\mathbf{q}^{n+1}$'s within the domain are calculated.

(NI) Both q_n and p^* are unknown but continuous as on a density interface. Two BE's on both sides of the interface and one interface condition for p^* explained in the next section are used to obtain two p^*'s on both sides and q_n on the interface as the solutions of the set of equations.

6 Interface Condition

On the interface across which the density and/or the permeability jump like the free surface of magnetic fluid, the extended pressure p^* jumps according to its definition.

When it is defined as $p^* = p - \bar{\rho}\mathbf{g} \cdot \mathbf{r}$, the **interface condition** used for (NI)

$$[p^*] = -\left[\bar{\rho}\mathbf{g} \cdot \mathbf{r} + \frac{\mu_0}{2}M_n^2 + \mu_0 \int_0^H M dH\right] - p_c \qquad (7)$$

is derived [2] from the continuity of the normal stress across the interface [3], where [···] means the jump quantity across the interface, and p, $\bar{\rho}\mathbf{g} \cdot \mathbf{r}$, $\mu_0 M_n^2/2$, $\mu_0 \int_0^H M dH$, p_c are the (original) pressure, the gravity potential due to the average density $\bar{\rho}$, the magnetic normal traction, the fluid-magnetic pressure and the capillary pressure, respectively. The magnetization $\mathbf{M}$ and its normal component M_n are calculated from the magnetic field $\mathbf{H}$ on the spot which may change in space. On (NP) such as the free surface, $\bar{\rho} = 0$, $\mathbf{M} = 0$ and p =atmospheric pressure on the vacuum side, and (7) reduces to give the value of p^*. Equation (7) can be used naturally for nonmagnetic fluid with no magnetic contributions.

The volume force compatible with the above definition of p^* is $\mathbf{F}^* = (\rho - \bar{\rho})\mathbf{g}$ under a condition that $\mathbf{M}$ is determined only by $\mathbf{H}$ and is always parallel to it [2,3]. Then, calculating $\mathbf{F}^*$ is totally excluded if $\rho = \bar{\rho}$ in addition.

Figure (2) shows an example of forced free surface oscillation of a magnetic fluid leading to parametric instability under an alternating magnetic field.

7 Open Boundary Condition

Open boundary is used to extend the effective range of numerical experiments both spatially and temporally. It is regarded as either (NV) or (NP) here according to whether $(q_n)^{n+1}$ or $(p^*)^{n+1}$ is specified on the boundary. The value of such physical quantity is obtained appropriately, by assuming that its distribution profile travels outward stationary, equivalent to the Sommerfeld radiation condition [5]. The direction of the wave velocity vector is unnecessary to be determined if the distribution profile is confined to be observed in its most descent direction (**local plane wave model**) [4]. When the inward component exists in addition, the rest from the total physical quantity subtracted by the inward component is treated as above.

Figure (3) shows the vertical analysis of a solitary wave in a two-layered fluid, which is reflected at a wall on the left, and travels out from an open boundary on the right. At corners where the open boundary crosses the free surface, the density interface and the bottom, one p^* and two q_n's must be obtained interrelatedly [6].

8 Conclusion

On the basis of the boundary equation derived from the law of conservation of volume, a course to treat several types of boundary conditions generally and related concepts in numerical analysis of incompressible fluids was explained. By the present method, both the problems of a magnetic fluid under an alternating magnetic field and a solitary wave in a two-layered fluid with an open boundary are successfully analyzed, though they are from different fields.

References

[1] Y.Mizuta: Computers & Fluids **19** 377 (1991)
[2] Y.Mizuta: Proc. 5th Int. Symp. on Computational Fluid Dynamics–Sendai, Vol.II, 267 (1993)
[3] R.E.Rosensweig: *Ferrohydrodynamics* (Cambridge University Press, Cambridge) Chap.4, Chap.5 (1985)
[4] Y.Mizuta: Comput. Fluid Dyn. Journal **1** 1 (1992)
[5] I.Orlanski: J. Comput. Phys. **21** 251 (1976)
[6] Y.Mizuta: submitted to Comput. Fluid Dyn. Journal

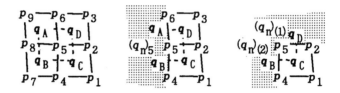

Fig. 1. Cell (solid lines) and LCV domain (broken lines) within the domain of analysis (left), adjacent to a boundary (center) or in a corner (right)

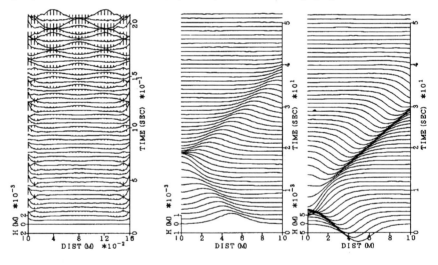

Fig. 2. Forced free surface oscillation of a magnetic fluid leading to parametric instability under an alternating magnetic field

Fig. 3. Solitary wave on a free surface(left) & a density interface(right) in a two-layered fluid traveling out from an open boundary on the right

Far Field Boundary Conditions Based on Characteristic and Bicharacteristic Theory Applied to Transonic Flows

Detlef Schulze

Technical University of Berlin, Germany

Abstract : The influence of two different far boundary conditions on the solution of a transonic profile flow is studied when the distance to the outer grid boundary is reduced. One of the boundary conditions makes use of characteristics in one dimensional flows [1]. The other results from an analysis of bicharacteristic in two dimensional flows [2].
These boundary conditions were combined with a cell-vertex method for solving the Euler equations in two dimensions. The method was used to simulate the transonic flow past a NACA0012 profile set at an angle of attack of one degree and a free stream Machnumber of 0.85.

1 Farfield Boundary Conditions

The purpose of far field boundary conditions is to represent the state of the flow at large distances from the source of disturbance. This is commonly expressed as the flow is being uniform at infinity. However, large outer boundary distances are difficult to model. Either the number of grid points is too large resulting in an unacceptable increase in computing time or the grid cells are highly stretched reducing the accuracy of the computation.

Therefore the outer boundary is always at a more or less small distance to the source of disturbance. Due to the closeness of the outer boundary the requirement arises that signals which impinge on the outer boundary from the interior flow field have to pass through the boundary without being reflected. Otherwise the solution can be deteriorated.

There exist many different proposals of far field boundary conditions which more or less satisfy the requirements stated above. Many of the existing far field boundary conditions are derived by considering characteristics in one dimensional flows.

The far field boundary conditions studied here are based on characteristics in 1D flow or bicharacteristics in 2D flow. The main difference between these two is that the 1D characteristics assume a local variation in one dimension whereas the bicharacteristics allow for a locally varying two dimensional flow at the boundary.

1.1 Characteristic Far Field Boundary Condition

The present scheme adopts a proposal of Salas et al. [1] and utilizes characteristics in one dimension. In the original paper the Riemann invariants are used to match the inner field to a solution of the linearized small-disturbance equation for the outer flow field. However, for reasons of comparison in the present study the inner solution was matched to the freestream conditions in the outer flow field.

To apply this condition a local co-ordinate system, x_n and y_t, is introduced at the outer boundary with x_n normal to and directed outward from the boundary. Assuming that the derivatives along the boundary vanish, the following characteristic equations can be derived, where u_t is tangential velocity, v_n is normal velocity, s is entropy, ρ is density, a is speed of sound, p is pressure and t is time:

$$\frac{ds}{dt} = 0 \qquad \text{along} \qquad \frac{dx_n}{dt} = u_n$$

$$\frac{dv_t}{dt} = 0 \qquad \text{along} \qquad \frac{dx_n}{dt} = u_n \qquad (1)$$

$$\frac{du_n}{dt} \pm \frac{1}{\rho a}\frac{dp}{dt} = 0 \qquad \text{along} \qquad \frac{dx_n^{\pm}}{dt} = u_n \pm a$$

Assuming a locally homentropic flow, the last two equations can be combined to yield

$$\frac{d}{dt}\left(u_n \pm \left(\frac{2a}{\gamma - 1}\right)\right) = 0 \qquad \text{along} \qquad \frac{dx_n^{\pm}}{dt} = u_n \pm a \qquad (2)$$

The term in brackets, the Riemann invariants $R^{\pm}$, can be added and subtracted to yield the local normal velocity and speed of sound at the outer boundary. Inflow and outflow boundaries are distinguished by the sign of the local normal velocity component. For subsonic flow conditions at the boundary R^- can be evaluated from the freestream representation and R^+ can be evaluated from the interior of the domain. For an inflow boundary entropy and v_t are extrapolated from the freestream conditions and for an outflow boundary these quantaties are extrapolated from the interior flow field.

2 Bicharacteristic Far Field Boundary Condition

The present scheme follows a proposal by Roe [2] and makes use of characteristics in two dimensional space. The condition is derived from bicharacteristic equations and involve vorticity, entropy and pressure relationships. Again, the inner solution was matched to the freestream conditions in the outer flow field.

Introducing far field expansions and inserting these into the four component Euler equations written in primitive variable form this boundary condition can be derived.

After some manipulation an equation for the global bicharacteristic is obtained. A local bicharacteristic equation can be identified from this by using a co-ordinate transformation and introducing a wave equation for the pressure. This analysis leads to a boundary condition for the pressure given by

$$\frac{\partial p}{\partial t} + \frac{\beta^2 \alpha_\infty}{\beta R - M_\infty x} \left\{ x\frac{\partial p}{\partial x} + y\frac{\partial p}{\partial y} + \frac{1}{2}(p - p_\infty) \right\} = 0 \qquad (3)$$

and an equation to determine the changes in u, v, $\frac{\partial u}{\partial t}$ and $\frac{\partial v}{\partial t}$ in dependence of p given by

$$(x - \beta M_\infty R)\frac{\partial u}{\partial t} + \beta^3 R\alpha_\infty \frac{\partial u}{\partial x} + \beta^2 y\left(\frac{\partial v}{\partial t} + u_\infty \frac{\partial v}{\partial x}\right) + \alpha_\infty(\beta R - M_\infty x)\frac{\partial v}{\partial y} = \frac{p - p_\infty}{2\rho_\infty} \qquad (4)$$

where β is the Prandtl-Glauert factor, M is Machnumber and R is the distance from the point of disturbance, i.e. the profile, to the boundary point given by

$$R^2 = \frac{x^2}{\beta^2} + y^2 \qquad (5)$$

Equations (4) and (5) are the equations to be implemented as boundary conditions. However, to solve (5) for $\frac{\partial u}{\partial t}$ and $\frac{\partial v}{\partial t}$ a further equation is required. This can be found in the expression for vorticity.

3 Cell-Vertex Scheme

The finite-volume method applied was first introduced by Ni [3] and Hall [4,5] for solving transonic flow fields. The scheme adopted here calculates the steady state solution using a cell-vertex space discretisation and a classical four stage Runge-Kutta time stepping scheme with local time steps and an artificial dissipation operator as proposed by Jameson [6].

4 Numerical Tests and Results

The flow field was discretized by a C-type mesh with a maximum number of 249 times 64 grid points including 189 points on the profile. The largest grid has an outer boundary extension of 50 profile chords. The distance to the outer boundary was reduced by cutting-off the outer cells leaving the rest of the grid unchanged. Thus the outer boundary extension was reduced to 40, 30, 20, 10 and 5 profile chords.

For each grid size and flow condition two simulations of a transonic flow past a NACA0012 profile with the same parameters for the artificial dissipation were performed applying the two different far field boundary conditions. Reducing the outer boundary distance, the influence on the lift and drag coefficient and the shock position on the upper and lower profile side was studied for a freestream Machnumber of 0.85 and an angle of attack of one degree.

Figure 1 and 2 show the Machnumber distribution along the profile when using the characteristic and bicharacteristic boundary conditions, respectively. The pictures clearly show the influence of a reduced outer boundary extension.

Using the characteristic boundary condition the Machnumber distribution is strongly affected by a change of the outer boundary distance. A considerable change in the Machnumber distribution and shock position occurs when the outer distance is reduced to 10 and 5 chords, respectively. Having an outer distance of 5 chords the shock position on the upper profile side is shifted upstream by about 6%. On the lower side the shock is shifted downstream by about 2%. Using the bicharacteristic

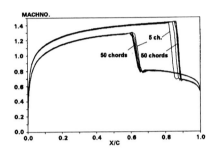

Fig. 1. Effect on M-no. along profile, characteristic boundary condition

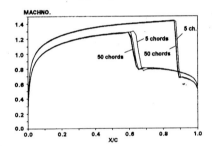

Fig. 2. Effect on M-no. along profile, bicharacteristic boundary condition

boundary condition no severe difference in the Machnumber distribution is visible for outer boundary distances of 10 chords or more. Using an outer boundary distance of 5 chords the shock position is shifted down-stream on the upper profile side by about 1% whereas on the lower side it is shifted by about 3%.

Figure 3 and 4 depict the changes in lift and drag coefficient, respectively. The pictures show that the lift and drag is considerably affected by a reduction of the outer boundary extension when applying the characteristic boundary condition. The calculated lift coefficient is reduced by 24% and the drag coefficient is changed by 12% when the outer boundary distance is reduced from 50 to 5 profile chords.

However, employing the bicharacteristic boundary condition the influence of the reduced boundary distance on the lift and drag is much smaller than before. Reducing the boundary extension from 50 to 5 chords the calculated lift coefficient is reduced by 9% and the drag coefficient by 6%.

Figure 2 and 4 clearly show the benefits assuming a locally varying two dimensional flow field at the outer boundary by applying bicharacteristic theory. It can can be seen that the assumption of a locally varying one dimensional flow field is strongly violated when the outer boundary is closer than 20 chords.

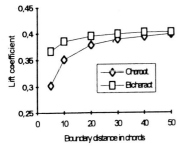
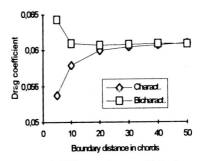

Fig. 3. Effect on lift coefficient   Fig. 4. Effect on drag coefficient

5 Conclusion

The numerical experiments show that the formulation of suitable far field boundary conditions is a crucial point in flow field simulations. The results clearly demonstrate that more elaborated analysis and numerics required for solving the bicharacteristic equations pay-off. The influence of a reduction of the outer boundary extend on lift, drag and Machnumber distribution when applying bicharacteristic based far field boundary conditions is strongly reduced as compared to when using a characteristic based condition.

References

1. Salas, M.D.; Thomas, J.L.: Far-field boundary conditions for transonic lifting solutions to the Euler equations. AIAA Journal, Vol. 24, No. 7, July 1986, pp 1079-1080.

2. Roe, P.L: Remote boundary conditions for unsteady multidimensional aerodynamic computations. ICASE Report 86-75, Nov. 1986.

3. Ni, R.-H.: A Multiple-Grid Scheme for Solving the Euler Equations, AIAA Journal, Vol. 20, No. 11, Nov. 1982.

4. Hall, M.G.: Fast Multigrid Solution of the Euler Equations using a Finite Volume Scheme of Lax-Wendroff Type. RAE Technical Report 84013, 1984.

5. Hall, M.G.: Cell vertex schemes for solution of the Euler equations. Proc. of the Conf. on Num. Methods for Fluid Dynamics, University of Reading, 1985.

6. Jameson, A.: Numerical Solution of the Euler Equations for Compressible Inviscid Fluids, Report MAE 1643, Princeton University.

An Accurate Cubic-Spline Boundary-Integral Formulation for Two-Dimensional Free-Surface Flow Problems

Debabrata Sen

Dept. of Naval Architecture, IIT Kharagpur 721302, India

1 Introduction

One of the most widely used method in numerical free-surface hydrodynamics, particularly for problems involving highly nonlinear free-surface flows in presence of submerged or floating bodies, is the boundary integral equation method (BIEM), e.g. see [1-4]. For a large class of problems where the solution is sought in the time domain, the BIEM used as the field equation solver is required to be applied at every instant of time. For such time-simulation algorithms, the accuracy of the BIEM solution is extremely important since even a small error present may build-up and introduce numerical instability.

It is well known that BIEM performs poorly at the region of discontinuity in geometry and/or boundary conditions. The intersection point between a surface piercing body and the free surface is characterized by both these discontinuities, and thus it is not surprising that earlier studies [4] have encountered numerical problems in this region. Most of these studies utilize either a constant-element version [1-2] or linear element version of BIEM [3-4], although higher order versions have also been used [5].

In this paper, developments of a fully cubic-spline BIEM as Laplace-solver is presented where boundary variables as well as the free-surface part of the geometry are represented by parametric cubic-splines. Here it is possible to enforce the condition of continuous velocity through the body-free surface intersection point in addition to the condition of continuous potential. The merits of the method in comparison to the other versions of BIEM are discussed. Test results show that this method produces significantly improved results compared to its linear or constant element counterpart.

2 Description of the Method

We consider without loss of generality a domain Ω as shown in Fig. 1 bounded by the surface $\partial\Omega$ within which the potential function ϕ satisfies Laplace equation such that $\nabla\phi$ represents velocity of irrotational motion of an ideal fluid in Ω. Application of Green's second identity to ϕ and $G = \frac{1}{2\pi}\ln r$ produces the following well known integral relation:

$$\alpha(\mathbf{x})\phi(\mathbf{x}) + \int_{\partial\Omega} [\phi_n(\xi)G(\mathbf{x},\xi) - \phi(\xi)G_n(\mathbf{x},\xi)]d\partial\Omega = 0 \qquad (1)$$

with ξ located on $\partial\Omega$. In BIEM, the continuous integral in (1) is replaced by a discrete sum in terms of the nodal values of the boundary variables (ϕ, ϕ_n) by making suitable assumptions on their variations. In this work, we approximate $\mathbf{x} = (x, z)$ and (ϕ, ϕ_n) by parametric cubic-splines, e.g. ϕ over $\mathbf{x}_j$ and $\mathbf{x}_{j+1}$ is interpolated by,

$$\phi^j(\zeta) = \{\mathbf{Q}^j\}\{\mathbf{R}^j\} \quad 0 \leq \zeta \leq \zeta_{j+1}, \tag{2}$$

where the elements of $\{\mathbf{Q}^j\}$ are functions of the spline parameter ζ and

$$\{\mathbf{R}^j\}^T = \{\phi_j \quad \phi_{j+1} \quad \phi'_j \quad \phi'_{j+1}\}. \tag{3}$$

In the above, prime denotes differentiation with respect to ζ. It can be shown that integrals in (1) can be written in the following form (for illustration, only the first integral is considered below):

$$\{e\} = [\,[\mathbf{A1}] + [\mathbf{B2}][\mathbf{B4}]^{-1}[\mathbf{B5}]\,]\{a\}, \tag{4}$$

or,

$$\{e\} = [[\mathbf{A1}] + [\mathbf{A2}][\mathbf{A4}]^{-1}[\mathbf{A5}]]\{a\} - [\,[\mathbf{A2}][\mathbf{A4}]^{-1}[\mathbf{A6}] - [\mathbf{A3}]\,]\{c\}, \tag{5}$$

depending on the spline end conditions. Here,

$$e_i = \int_{\partial\Omega} \phi(\zeta)G_n(\mathbf{x}_i, \xi)d\partial\Omega, \tag{6}$$

$$\{a\}^T = \{\phi_1 \quad \phi_2 \quad \ldots \quad \phi_{M-1} \quad \phi_M\}, \tag{7}$$

$$\{c\}^T = \{\phi'_1 \quad \phi'_M\}. \tag{8}$$

Matrices [**A4**], [**A5**], [**A6**], [**B4**] and [**B5**] are simple functions of ζ and [**A1**], [**A2**], [**A3**] and [**B2**] are functions of the so-called influence coefficients:

$$\int_{\partial\Omega_j} \zeta^k G(\mathbf{x}_i, \zeta)d\partial\Omega, \quad \int_{\partial\Omega_j} \zeta^k G_n(\mathbf{x}_i, \zeta)d\partial\Omega; \quad k = 0, 1, 2, 3. \tag{9}$$

Equation (4) is applicable when the end slopes $\{c\}$ is known while (5) is applied when this is not known and a natural or relaxed end condition is used.

Consider the domain shown in Fig.1 and assume that on $\partial\Omega_2$ (the free-surface) Dirichlet condition is prescribed while on the two vertical boundaries $\partial\Omega_1, \partial\Omega_2$ Neumann conditions are given. (The flat impermeable bottom boundary $\partial\Omega_4$ is removed from (1) by augmenting G.) With these specifications the discretized approximation of (1) to be solved for the unknown normal velocity on the free surface (and ϕ on the other two surfaces) turns out to be:

$$\alpha_i\phi_i - [\mathbf{C}^I]\{a^I\} + [\mathbf{E}^I]\{c^I\} + [\mathbf{D}^{II}]\{b^{II}\} - [\mathbf{C}^{III}]\{a^{III}\} + [\mathbf{E}^{III}]\{c^{III}\}$$
$$= [\mathbf{C}^{II}]\{a^{II}\} - [\mathbf{E}^{II}]\{c^{II}\} - [\mathbf{D}^I]\{b^I\} - [\mathbf{D}^{III}]\{b^{III}\}, \tag{10}$$

where

$$[\mathbf{A1}] + [\mathbf{A2}][\mathbf{A4}]^{-1}[\mathbf{A5}] = [\mathbf{C}]$$

$$[\mathbf{A1}] + [\mathbf{B2}][\mathbf{B4}]^{-1}[\mathbf{B5}] = [\mathbf{D}] \qquad (11)$$
$$[\mathbf{A2}][\mathbf{A4}]^{-1}[\mathbf{A6}] - [\mathbf{A3}] = [\mathbf{E}]$$

In the above, the superscripts I, II, III refer to the respective quantities for the boundaries $\partial\Omega_1, \partial\Omega_2, \partial\Omega_3$, and $\{\mathbf{b}\}$ contains the nodal values of ϕ_n. In writing (10), we have regarded the end slopes of ϕ curve over each part of the boundary as known, while the end slopes of ϕ_n are considered as not known. Equation (10) represents a system of N linear algebraic equations with $(N+8)$ unknowns. (N = total number of nodal points on $\partial\Omega$.) Additional 8 auxiliary equations are therefore necessary to render (10) solvable. These relations are obtained by enforcing continuity of potential as well as velocity through the intersections. Continuity of ϕ through $\partial\Omega_1 \cap \partial\Omega_2$ and $\partial\Omega_2 \cap \partial\Omega_3$ give two relations, while a continuous $\nabla\phi$ through the four corners provide the additional six relations: two each from $\partial\Omega_1 \cap \partial\Omega_2, \partial\Omega_2 \cap \partial\Omega_3$ and one each from $\partial\Omega_1 \cap \partial\Omega_4, \partial\Omega_3 \cap \partial\Omega_4$. For example, a continuous $\nabla\phi$ through $\partial\Omega_1 \cap \partial\Omega_2$ gives,

$$\phi'^{I}_{M1} = k_1 \phi^{I}_{n,M1} + k_2 \phi^{II}_{n,1} \ . \qquad (12)$$

$$\phi'^{II}_{1} = k_3 \phi^{I}_{n,M1} + k_4 \phi^{II}_{n,1} \ , \qquad (13)$$

where k_1, k_2 are functions of the Jacobian and the geometric angle at the corner. Equation (10) supplemented by the above 8 relations can be assembled in standard matrix form:

$$[\mathbf{H}]\{\mathbf{Y}\} = \{\mathbf{X}\} \ , \qquad (14)$$

with $\{\mathbf{Y}\}$ containing the unknown nodal values of (ϕ, ϕ_n).

The above developments, although shown for the domain in Fig. 1 with a particular specification of boundary conditions, can be generalised for more complex boundary geometry and other type of boundary conditions. In the linear element BIEM, ϕ can be made continuous through the corner but not $\nabla\phi$ while in the constant element version none of the two conditions can be enforced. Comparing with formulations using higher-order boundary elements, it should be noted that in most such formulations,e.g. [5], velocity compatibility is enforced *after* a solution of the discretized (1) and then modifying the nodal values at the corner by means of auxiliary equations. In contrast, in the present case, the velocity continuity is enforced *within* the Laplace-solver (this condition enters through specification of the end-slopes of ϕ). For algorithms involving time-evolution of nonlinear free surface, spatial derivatives of geometry and boundary data are required. Due to lack of inter-element continuity, the schemes applied to determine these derivatives in the usual higher-order versions of BIEM necessarily involve additional approximations, e.g. [5], while in the present case these can be found from a consistent cubic spline approximation.

3 Results and Discussion

Care is taken to evaluate the influence coefficients (9) such that a consistent level of very high accuracy (e.g. a 6-digit accuracy) is maintained in the computed coefficients. This is achieved by a combination of analytical formulae and quadrature rules of variable order. The performance of the method is demonstrated by considering a propagating Airy wave for which the exact solution is known. Table 1 shows results for a cosine-grid distribution where grids are concentrated near the corner. Vastly improved results from the present spline-formulation compared to linear and constant element BIEM is evident. What should particularly be noted is not just the improvement, but the magnitude of improvement : even for only a moderately fine grid, this improvement ranges by a factor of several hundred to several thousand. Note also that the two lower order versions produce results of comparable accuracy.

We next combine the above Laplace solver with time-integration of the full nonlinear free surface constraints since the final objective is to simulate complex free surface motions in presence of surface-piercing objects. In Lagrangian form, the free surface conditions are,

$$\frac{D\phi}{Dt} = -gz + \frac{1}{2}\nabla\phi^2 \; ; \; \frac{D\mathbf{x}}{Dt} = \nabla\phi \; . \tag{15}$$

The algorithm for following the free surface motion with time t is same as in [1-4] and hence not detailed here.

Figure 2 shows evolution of the free surface for a wave-maker motion where $\partial\Omega_1$ is a translating rigid plate impulsively started at $t = 0^+$ with a constant velocity $\phi_n = 1.0$. Although this implies an infinite initial acceleration of the plate, the fluid motion could be followed for small Δt, and the results show existence of a logarithmic singularity as predicted by analytical theories [6]. For an initially quiescent fluid, the condition of a continuous $\nabla\phi$ through the corner implies that the plate (wave-maker) must be started smoothly with a zero initial velocity. This is given by a sine motion, e.g. ϕ_n on $\partial\Omega_1 = A\omega\sin\omega t$. This however results a finite acceleration, and thus it is better to modulate the initial velocity by a ramp function,

$$\phi_n(t) = 0.5A\sin\omega t(1 - \frac{\pi t}{\sigma}) \quad \text{on } \partial\Omega_1 \quad \text{for } 0 \leq t \leq \sigma, \tag{16}$$

such that $\phi_{n_t} = 0$ at $t = 0$ in addition to a zero initial velocity. Although the algorithm could be advanced even when the initial acceleration is non-zero but finite if for the initial few steps Δt is small, use of (14) is preferred. Due to length restrictions, these results could not be included here but will be reported elsewhere.

References

1. Isaacson, M. and Ng. J.Y.T. (1993), "Time-domain second-order wave radiation in two-dimensions", J. Ship Research, **37** (1), pp.25-33.

2. Sen, D. (1993),"Numerical simulation of motions of two-dimensional floating bodies", J. Ship Research, **37** (4), pp.307-330.

3. Zhao,R. and Faltinsen, O. (1993),"Water-entry of two-dimensional bodies", J. Fluid Mech., **118**, pp.593-612.

4. Dommermuth, D. and Yue, D. (1987),"Numerical simulation of nonlinear axisymmetric flows with the free surface", J. Fluid Mech., **178**, pp.195-219.

5. Grilli, S.T. and Svendsen, I.A. (1990), "Corner problems and global accuracy in the boundary element solution of nonlinear wave flows", Eng. Analysis with Boundary Elements, **7** (4), pp. 178-195.

6. Joo, S.W. , Schultz, W.W. and Messiter, A.F. (1990),"an analysis of the initial value wave-maker problem", J. Fluid Mech., **214**, pp.161-183.

Table 1. $|(\phi_n(\text{computed}) - \phi_n(\text{exact}))/\phi_n(\text{exact})| \times 10^3$ on the free surface (length of domain $= 2\lambda$, $\lambda =$ Airy wave length)

	at the corner			max. excluding at corner		
N	const.	linear	spline	const.	linear	spline
35	31.460908	26.372035	0.452729	21.087910	66.404013	1.117976
70	4.031335	9.989349	0.026881	7.806242	17.419739	0.065509
105	3.766677	5.159402	0.004892	3.764638	7.858501	0.012893
140	7.247903	3.113095	0.001274	2.190854	4.454860	0.004102

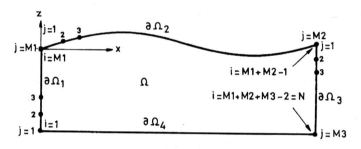

Fig. 1. Definition Diagram.

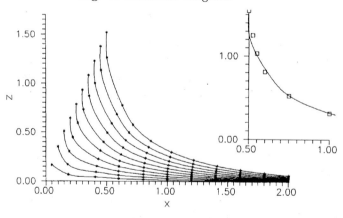

Fig. 2. Free-surface profiles for an impulsive wavemaker from $t = 0.05$ to 0.50 at equal intervals. The inset figure shows comparison of the profile at $t = 0.5$ (solid line) with analytical solution (square symbols).

ELAFINT : A Computational Method for Fluid Flows with Free and Moving Boundaries

H. S. Udaykumar and W. Shyy

Department of Aerospace Engineering,
Mechanics and Engineering Science
University of Florida, Gainesville, FL 32601

1 Background and Formulation

In various materials solidification arrangements, conditions are encountered under which the interface separating the solid and liquid phases experiences a **morphological instability**[1,2] and breaks up into convoluted structures such as **dendrites** and **cells**. Such interfacial instabilities occur in the materials processing industry, and can significantly affect the computational and structural homogeneity of grown crystals, such as high temperature alloys and semiconductors. Furthermore, the instability phenomena are significantly influenced by fluid flow in the melt [4]. In recent years much attention has therefore been devoted to this issue by a wide array of researchers, including fluid mechanicians. Progress has been made in understanding the various aspects of the phenomena since the seminal work of Mullins and Sekerka[5]. While the preliminary analyses were indeed insightful, the highly non-linear nature of the phenomenon has rendered computational or experimental investigation of the fully developed instability regime imperative. Experimental efforts have always been hampered by the extremely small size range, of the order of microns, at which the instabilities occur. Numerical simulations provide the opportunity to delve into the nonlinear stages of growth and to freely vary parameter ranges and growth environments. In the following, we present a method that can simulate the evolution of highly distorted phase fronts, including convective transport in the melt. The method is abbreviated as *ELAFINT* - *E*ulerian *L*agrangian *A*lgorithm *F*or *I*nterface *T*racking due to the combination of a fixed grid and a moving interface. The following equations are solved in the liquid phase:

Continuity:
$$\nabla \cdot \vec{u} = 0 \qquad (1)$$

Momentum:
$$\frac{\partial \vec{u}}{\partial t} + \vec{u} \cdot \nabla \vec{u} = \nu \nabla^2 u - \frac{\nabla p}{\varrho_o} - \beta(T - T_o)\vec{j} \qquad (2)$$

Energy:
$$\frac{\partial T}{\partial t} + \vec{u} \cdot \nabla T = \alpha_1 \nabla^2 T \qquad (3)$$

In the solid phase :
$$\vec{u} = 0 \tag{4}$$
and energy equation reads:
$$\frac{\partial T}{\partial t} = \alpha_s \nabla^2 T \tag{5}$$

The equations in each phase are solved separately subject to the boundary conditions on the interface:

$$T_{\text{interface}} = T_m \left(1 - \frac{\gamma}{L}\aleph\right) \quad (Gibbs - Thomson\ condition) \tag{6}$$

where T_m is the melting temperature, γ is the surface tension, L the latent heat of fusion and $\aleph$ the local curvature of the interface.

The fluid flow velocity boundary conditions are:

$$\vec{u} \cdot \vec{t} = 0 \quad (no - slip) \tag{7}$$

$$\varrho_L \vec{u} \cdot \vec{n} = (\varrho_s - \varrho_L) V_N \quad (no\ penetration) \tag{8}$$

where V_N is the normal velocity of the interface, given by,

$$\varrho_s L V_N = \left(k_s \frac{\partial T_s}{\partial n} - k_l \frac{\partial T_l}{\partial n}\right) \quad (Stefan\ condition) \tag{9}$$

Here subscripts l and s stand for liquid and solid respectively. Appropriate Neumann and Dirichlet conditions are applied at the boundaries of the computational domain.

2 Features of the current computational approach

A numerical solution procedure attacking the problem of interfacial evolution faces two primary challenges.

(a) The interface separating the two phases is a **moving boundary**, which assumes **highly convoluted** forms. Boundary conditions for the equations in **each phase** are applied on this boundary. Eq.(6) involves the curvature of the front. Thus, the slopes and curvatures of the interface are to be computed accurately. Eq.(9) is a statement of heat conservation across the phase discontinuity and inaccuracies in computing heat fluxes can lead to erroneous values of interface velocities.

(b) An efficient field equation solver is to be developed which incorporates regarding interface shape and location. The boundary conditions are to be applied at the **exact location of the interface.**

Conventional Lagrangian approaches, employing moving, boundary conforming grids[6-8] fail when the interface becomes highly contorted and suffers topological changes. The Eulerian methods[9-11], solve for fluid fraction variables on stationary grids, and reconstruct the interface based on the fluid fraction data in each cell, a process that involves several logical operations. In order for the interfaces obtained to be accurate highly refined grids may be required [11]. The computational procedure presented here seeks to surmount the difficulties associated with conventional approaches by combining features of both methods.

A **fixed, Cartesian grid** is employed to perform the computations. The irregularly shaped interface is tracked over the grid and is represented using **marker particles** specified by an index i and coordinates (X_i, Y_i). The markers are connected by piecewise circular arcs to obtain interfacial shape information such as normals and curvatures. The interface is advanced in time by the Lagrangian translation of the marker particles, the normal velocity being computed from Eq.(9). The new interfacial shape is obtained by joining these updated markers. The use of an Eulerian grid facilitates the execution of merger/breakup procedures at the interface, a drawback conventionally attached to pure Lagrangian methods. In the current framework it is possible to associate each marker with the Eulerian grid, a nearest neighbour search is economically performed, and proximal interface segments can be easily identified. Furthermore, simple data structures are employed, and the intensive logical operations are restricted to the one-dimensional arrays associated with the interfacial cells only.

The interface tracking method was first tested independently [12] of the field solver and velocity models were employed to advance the front. Highly contorted interfaces were captured, and even in the highly distorted stage, accurate interfacial curvatures were obtained. The method was then applied to simulate merger and breakup events. Multiple mergers in space and time were performed. Thus, limitations conventionally associated with Lagrangian procedures were overcome.

The interface tracking scheme is next integrated with a method to solve the equations in the solid and liquid phases. The interface now forms an internal, moving boundary in the system and the boundary conditions, Eqs. (6)-(9) need to be imposed at this boundary.

A control volume formulation is designed to solve the field equations (1)-(5). Since the interface cuts through the grid, fragmented control volumes such as shown in Fig.1 may arise. The transport equation for a general ϕ variable (u,v or T):

$$\frac{\partial \phi}{\partial t} + \nabla \cdot \vec{u}\phi = \alpha \nabla^2 \phi + S \tag{10}$$

where alpha is a diffusivity and S is a source term, can be integrated over such irregular shaped control volumes by the application of divergence theorem;

$$\int_A \frac{\partial \phi}{\partial t} dA + \oint_l (\vec{u}\phi) \cdot \vec{n} dl = \alpha \oint_l \frac{\partial \phi}{\partial n} dl + \int_A S dA \tag{11}$$

The time-implicit discretized form of the above conservation statement for an m sided control is then;

$$\frac{\phi_{ij}^{k+1} - \phi_{ij}^k}{\delta t} \delta A_{cv} + \sum_{j=1}^{m} (\vec{u}\phi \cdot \vec{n}) dl_i = \alpha \sum_{i=1}^{m} \frac{\partial \phi}{\partial n} dl_i + S_{ij} \delta A_{cv} \qquad (12)$$

where δA_{cv} is the area and dl is the length of each side of the control volume. Thus, once the convective and diffusive normal flux through each side of the control volume are obtained, the new ϕ field can be iteratively computed. The normal flux through the interface requires special treatment and is obtained via a biquadratic shape function definition of the ϕ field in the vicinity of the interface. The normal gradients at the interface, obtained to supply the fluxes to the control volume are also used to complete interfacial velocity from Eq.(9) and the interface position updates, fully coupled with the field solver. The fixed-grid, time-implicit formulation allows for the application of well-developed pressure based algorithms [13] for the solution of Eqs.(1)-(5). A scaling procedure has been developed to delineate the flow phenomena at the microscopic scales and to identify the interaction with the global flow field [14].

The results obtained demonstrate the effects of surface tension. In the absence of surface tension, singularities quickly develop on the front. Such a situation is shown in Fig 2. For low interfacial tension (Fig.2(a)), the interface is sensitive to perturbation and tends to break down into branched structures. The noise level and resolution of the front features during the breakdown of the interface are strongly dependent on the grid resolution available. For sufficient interface tension however, the results have been demonstrated to converge under grid refinement. The initial perturbation develops in time into long figures (Figs.2(b) and (c)), as in the analogous Saffman-Taylor instability. The finger shapes reached a shape-preserving state, while the tip of the finger accelerated due to the fixed domain size. The results have demonstrated that the predicted physical implications of capillarity are well borne out and the method is capable of tracking evolving interfaces for large times and distortions.

The fluid flow equations were solved for several test cases. A previously tested body-fitted grid was employed to compare with the current method. A driven cavity flow was chosen as a test case. Fig3(a) compares the streamline patterns obtained from *ELAFINT* and the body-fitted computations [16] for a Reynolds number of 1000. Also compared, in Fig.3(b) are the u-velocity component at the centerline of the cavity. Next, the melting of Gallium in a side-wall heated cavity was studied. Natural convection in this case causes the deformation of the phase front. The non-dimensional numbers applied in this case are Rayleigh number=10^4, Pr=0.021, St=0.042. Comparison was made with a purely Eulerian method using an enthalpy formulation[15]. The interface shapes are shown at equal time intervals in Fig 4(a). Fig 4(b) compares the streamline patterns in the two cases.

3 Conclusions

A computational technique called ELAFINT has been developed to handle the existence of highly deformed moving and free boundaries. These boundaries or interfaces cut through an underlying cartesian grid, leading to irregularly shaped control volumes in the vicinity of the interface. A method to reassemble these control volumes to maintain flux conservation has been designed. The accuracy of the methods at each stage of the development have been assessed. Currently highly deformed, moving interfaces with phase change can be tracked in conjunction with an implicit pressure-based Navier-Stokes equation solver. This methodology can be applied to solve fluid flow problems in complex geometrices.

References

1. Langer,J.S., 1980,"Instabilities and pattern formulation in crystal growth," *Rev. Mod.Phys.*, Vol. 52, No. 1, pp.1-56.

2. Kessler,D.A., Koplik,J. and Levine, H., 1988,"Pattern selection in fingered growth phenomena," *Advances in Physics*, Vol.37, No. 11, pp.255-339.

3. Nakaya,U., 1954, *Show Crystals*, Harvard University Press, Cambridge, MA.

4. Mullins,W.W. and Sekerka,R.F., 1964,"Stability of a planar interface during solidification of a dilute binary alloy," *J.Appl. Phys.*, Vol.3, pp.444-451.

5. Glicksman,M.E., Coriell,S.R., and McFadden,G.B., 1986,"Intercation of flows with the crystal melt interface," *Ann. Rev. Fluid Mech.*, Vol. 18, pp307-336.

6. Shyy.W., Udaykumar,H.S., and Liang,S.-J., 1993,"An interface tracking method applied to morphological evolution during phase change," *Int.J.Heat Mass Transf.*, Vol 36, No. 7, pp. 1833-1844.

7. DeGregoria,A.J. and Schwartz,L.W., 1986,"A boundary integral method for two-phase displacement in Hele-Shaw cells," *J.Fluid Mech.*, vol. 164, pp.383-400.

8. Glimm,J., McBryan,O., Melnikoff,R. and Sharp,D.H., 1986,"Front tracking applied to Rayleigh-Taylor instability," *SIAM J.Sci.Stat.Comput.*, Vol.7, No. 1, pp.230-251.

9. Hirt,C.W. and Nichols,B.D., 1981,"Volume of fluid (VOF) methods for the dynamics of free boundaries," *J.Comp.Phys.*, Vol.39,pp.201-225.

10. Sethian,J.A. and Strain.J., 1982,"Crystal growth and dendritic solidification," *J.Comp.Phys*, Vol.98, No.2,pp.231-253.

11. Wheeler,A.A., Murray,B.T. and Schaefer,R.J., 1993,"Computation of dendrites using a phase field model," to appear in *Physics D*.

12. H.S.Udaykumar and Shyy, W., 1993,"Development of a grid-support marker particle scheme for interface tracking," presented at the 11th AIAA Comp. Fluid Dyn. Conf.; Paper No. AIAA-93-3384, Orlando, Florida, USA.

13. Patankar, S.V., 1980, "Numerical Heat Transfer and Fluid Flow", Hemisphere Publicating, New York.

14. Udaykumar, H.S., 1994, PhD Dissertation, Department of Aerospace Engineering, Mechanics and Engineering Science, University of Florida, Gainesville.

15. Shyy, W. and Rao, M.M., 1994, "Enthalpy based formulations for phase change problems with application to g-jitter," Microgravity Sci. and Tech., Vol.7, pp. 41-49.

16. Shyy.,W., 1994, "Computational Modelling for Fluid Flow and Interfacial Transport," Elsevier, Amsterdam, Netherlands.

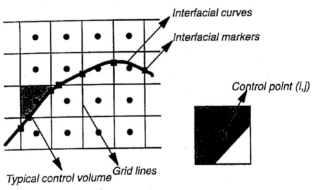

Fig. 1 Illustration of interface grid and control volume

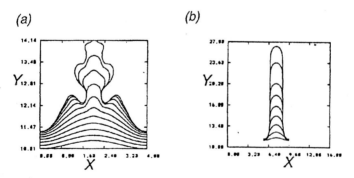

Figure 2

(a) Interface shapes for low surface tension. Successive instabilities lead to branched structure. Each line corresponds to the interface shape at a given instants.
(b) Interface evolution for higher surface tension. A stably propagating finger is obtained.

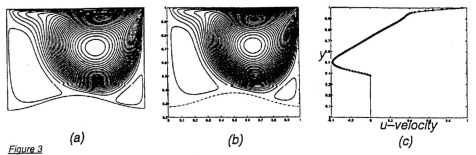

Figure 3
Comparison of solutions from body–fitted computations and ELAFINT

(a) Streamlines for the case of a driven cavity with a deformed base. Re=1000, 121x121 grid. Contours from a boundary–fitted grid computation. (b) On a fixed Cartesian grid using the current method. (c) Comparison of centerline u–velocities. Dots represent solution from a boundary–fitted grid. Full lines from current calculation.

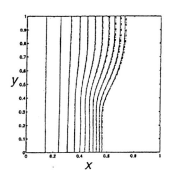

(a) Comparison of interface positions at equal intervals of time. Dotted lines represent the enthalpy–based method, full lines correspond to the present method.

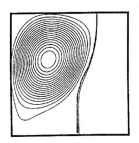
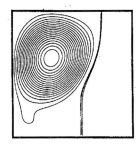

Figure 4
Melting of Gallium, Ra=10^4, Pr=0.021, St=0.042, 81x81 grid. Melting is initiated at the left wall. Streamline contours at t=100. Comparison of the enthalpy method and ELAFINT. (b) Current method, t=100. Enthalpy method, t=100. The interface has been represented by plotting temperature contours from T=–0.005 to 0.005. For ELAFINT, the interface position is actually available explicitly. In the enthalpy method the only information regarding the interface is the contour shown.

An Implicit Spectral Solver for the Compressible Navier-Stokes Equations in Complex Domains

Renzo Arina, Fabrizio Ramella

Dipartimento di Ingegneria Aerospaziale,
Politecnico di Torino, C.so Duca degli Abruzzi,
24, I-10129 Torino, Italy

Abstract : A spectral collocation method for the compressible Navier-Stokes equations written in generalized coordinates is presented. The method is combined with an implicit time integration technique based on nonlinear Krylov methods (GMRES) with a finite volume preconditioning.

1 Introduction

In this work we focus our attention on the numerical simulation of viscous compressible flows in complex configurations. One of the major requirements for viscous flow calculations is the need of a high spatial accuracy, in order to correctly resolve thin shear layers and other flow structures. In the case of turbulence modelling, it is of crucial importance to have the numerical error, induced by the numerical scheme, well below the sensitivity of the turbulence models, otherwise the physical tuning of the turbulence model is grid dependent.

Spectral methods display the best approximation properties with respect to standard finite difference and finite element schemes. For sufficiently smooth flows they display exponential convergence. However their application is strongly limited by the occurrence of flow singularities and complexity of the geometry. A possible way to overcome these difficulties, may consist into the splitting of the domain in different regions depending upon the presence of viscous effects. The regions where the flow is essentially inviscid, with the possible presence of shocks, can be accurately solved by low order schemes (FD, FV or FEM). While the regions where the viscous effects are dominant, and the flow is essentially smooth, can be solved by spectral methods.

A viscous/inviscid domain decomposition strategy is reported in (Arina and Canuto 1993, 1994). In the present work we discuss a spectrally accurate scheme for generally shaped regions. In order to have an efficient technique, a particular attention has been devoted to the time integration scheme. We employ implicit techniques such as Approximate Factorization (AF) and nonlinear Krylov method (GMRES), where the computing time for the matrix inversion is strongly reduced by means of a finite volume preconditioning.

2 Spectral Approximation

Introducing the coordinate transformation $\xi = \xi(x,y)$, $\eta = \eta(x,y)$, the Navier-Stokes equations can be written in the conservation form

$$\frac{1}{J}\frac{\partial Q}{\partial t} + \frac{\partial \hat{F}}{\partial \xi} + \frac{\partial \hat{G}}{\partial \eta} = \frac{1}{Re}\left(\frac{\partial \hat{F}_v}{\partial \xi} + \frac{\partial \hat{G}_v}{\partial \eta}\right) \tag{1}$$

where $J = (x_\xi y_\eta - x_\eta y_\xi)^{-1}$ is the Jacobian of the coordinate transformation, and

$$\hat{F} = y_\eta F - x_\eta G, \quad \hat{G} = -y_\xi F + x_\xi G, \quad \hat{F}_v = y_\eta F_v - x_\eta G_v, \quad \hat{G}_v = -y_\xi F_v + x_\xi G_v$$

are the inviscid and viscous fluxes.

Denoting with

$$\mathcal{L}(Q) = \frac{\partial}{\partial \xi}\left(\hat{F} - \frac{1}{Re}\hat{F}_v\right) + \frac{\partial}{\partial \eta}\left(\hat{G} - \frac{1}{Re}\hat{G}_v\right)$$

the convective and diffusive terms of system (1), the spatial operator $\mathcal{L}(Q)$ can be discretized by finite differences on the computational domain (ξ, η), characterized by an equally spaced grid (e.g. $\Delta\xi = \Delta\eta = 1$). By an appropriate discretization of the metric terms, this approach is equivalent to a finite volume discretization of the governing equations on the physical domain (x,y). On the other hand, if a spectral discretization is employed, the computational domain is no longer equally spaced. In order to have spectral accuracy, the choice of the collocation points for the spectral approximation must follow appropriate quadrature rules (Canuto et al., 1987). Then it is necessary to map the distorted domain of interest into a computational rectangle with the appropriate point distribution dictated by the choice of the interpolating polynomials. Moreover the metric terms must be calculated with spectral accuracy. A collocation method is employed for the discretization of $\mathcal{L}(Q)$. The derivatives are evaluated by the matrix vector multiplication method, knowing the values of $Q(\xi_{i,j}, \eta_{i,j}, t)$ at the collocation points. If the problem involves periodic boundary conditions, Fourier polynomials are employed, otherwise Chebyshev or Legendre polynomials are used.

While the Fourier collocation method automatically enforces periodic boundary conditions, in the other cases it is necessary to impose directly the boundary conditions. On each nonperiodic boundary, the operator $\mathcal{L}(Q)$ is replaced by the operator $\mathcal{L}_B(Q)$. Along the boundaries where the flow is essentially inviscid, $\mathcal{L}_B(Q)$ is obtained by introducing a characteristic decomposition of $\mathcal{L}(Q)$ with respect to the boundary normal direction, then replacing the equations corresponding to incoming signals by appropriate boundary conditions and finally recasting the resulting system in conservative form. Along wall boundaries, $\mathcal{L}_B(Q)$ contains the no-slip and temperature conditions, expressed in conservative variables, in addition to the momentum equation along the wall normal.

3 Time Integration

The time integration scheme can be formulated in the following form

$$\delta Q^n + \theta \Delta t J \mathcal{L}(Q)^{n+1} = -\Delta t (1-\theta) J \mathcal{L}(Q)^n$$

where $\delta Q^n = Q^{n+1} - Q^n$, n being the time index. Varying θ, different well known time integration schemes are recovered.

The fluxes $\hat{F}$, $\hat{G}$, $\hat{F}_v$, $\hat{G}_v$, are non linear. They can be linearized as follows

$$\hat{F}^{n+1} = \hat{F}^n + \hat{A}^n \delta Q^n + O(\Delta t^2)$$

and we obtain the linear system

$$\left(I + \theta \Delta t J \frac{\partial \mathcal{L}(Q)^n}{\partial Q} \right) \delta Q^n = -\Delta t J \mathcal{L}(Q)^n \qquad (2)$$

where the term $\partial \mathcal{L}(Q)/\partial Q$ contains the flux Jacobians

$$\hat{A} = \frac{\partial \hat{F}}{\partial Q}, \quad \hat{B} = \frac{\partial \hat{G}}{\partial Q}, \quad \hat{R} = \frac{\partial \hat{F}_v}{\partial Q}, \quad \hat{S} = \frac{\partial \hat{G}_v}{\partial Q}.$$

The algorithm is formulated in *delta form*, updating δQ. Then in the case of steady flows, the left hand side, being multiplied by δQ, tends to zero and the spatial precision of the solution depends only upon the spatial discretization of the term $\mathcal{L}(Q)^n$. Spectral accuracy is obtained by introducing the spectral approximation of the operator $\mathcal{L}(Q)^n$.

The left hand side could also be discretized with spectral accuracy, but the resulting matrix would be full. It can be convenient to introduce a suitable preconditioning technique consisting in discretizing the left hand side with a low order scheme, such as a finite difference scheme, equivalent to a cell centered finite volume scheme. By exploiting the band structured form of the resulting matrix, its inversion is by far less expensive than the inversion of a full matrix.

The calculation of the flux Jacobian matrices $\hat{A}$, $\hat{B}$, $\hat{R}$ and $\hat{S}$ can be very expensive. Moreover using upwind schemes, there are two Jacobians for each convective flux, precisely $\hat{A}^{\pm}$ and $\hat{B}^{\pm}$. Consequently it may be appropriate to introduce an approximation of the type

$$\hat{A}^{\pm} = \frac{1}{2} \left(\hat{A} \pm \rho(\hat{A}) \mathcal{I} \right), \quad \hat{B}^{\pm} = \frac{1}{2} \left(\hat{B} \pm \rho(\hat{B}) \mathcal{I} \right)$$

for the inviscid Jacobian fluxes, being $\rho(\hat{A})$ and $\rho(\hat{B})$ the spectral radii of $\hat{A}$ and $\hat{B}$. Similarly the Jacobian of the viscous fluxes are approximated as follows

$$\hat{R} = \rho(\hat{R}) \mathcal{I}, \quad \rho(\hat{R}) = 2J(x_\eta^2 + y_\eta^2)\frac{\mu}{\rho}$$

$$\hat{S} = \rho(\hat{S}) \mathcal{I}, \quad \rho(\hat{S}) = 2J(x_\xi^2 + y_\xi^2)\frac{\mu}{\rho}$$

For multidimensional problems, system (2) can be solved by an AF technique. However, from numerical experiments we have noted slow convergence in some cases. In order to improve the efficiency of the implicit solver we have focused our attention on the construction of a Newton-like scheme based on a nonlinear Krylov subspace projection method for solving system (2). The particular Krylov method we have considered is the Generalized Minimun RESidual Method (GMRES) for solving

$$J_\mathcal{L} \delta Q = -\mathcal{L}(Q) \qquad (3)$$

with $J_\mathcal{L} = \frac{\partial \mathcal{L}(Q)}{\partial Q}^n$. System (3) corresponds to (2) for steady problems.

A linesearch backtracking procedure and a Jacobian-free technique, leading to reduction on storage, are also employed (Brown and Saad, 1990). In order to improve the convergence properties a left preconditioning is introduced, and precisely

$$M^{-1} J_\mathcal{L} \delta Q = -M^{-1} \mathcal{L}(Q) \qquad (4)$$

The Jacobian-free approach does not require an explicit evaluation of $J_\mathcal{L}$, therefore we can adopt as preconditioner the finite volume discretization of the left hand side of (2), which, in the multidimensional case, is also factored by the AF technique. Then we have

$$M^{-1} = \left(\frac{I}{\theta \Delta t J} + \tilde{J}_\mathcal{L} \right)^{-1}$$

where $\tilde{J}_\mathcal{L}$ is the Jacobian formed with the approximated Jacobian fluxes. The preconditioned problem to be solved is

$$J_\mathcal{G} \delta Q = \mathcal{G}(Q)$$

with $J_\mathcal{G} = M^{-1} J_\mathcal{L}$ and $\mathcal{G}(Q) = -M^{-1} \mathcal{L}(Q)$. The evaluation of the initial residual $r_0 = -M^{-1} \mathcal{L}(Q)$ as well as the next steps of the Arnoldi process are reduced to the solution of finite volume systems with AF.

4 Results

The previous scheme has been tested for a one-dimensional case, consisting into the subsonic inviscid flow in a nozzle. Characteristic boundary conditions are imposed at the inlet and outlet. The space discretization is obtained by Legendre polynomials. Figure 1 show the numerical solution obtained with 16 modes, compared with the exact solution. An almost comparable accuracy is obtained with a finite difference discretization on a grid with not less than 100 intervals. Figure 2 show the convergence history of the calculation compared with a finite difference solution (101 points) using a Flux-Vector splitting method for the convective terms.

Figure 3 compare the results of the simulation of the flow past a cylinder at $Re = 40$, on a 24×32 grid, with a Fourier-Chebyshev collocation discretization. The length of the steady separation bubble is compared with experimental data and a finite difference simulation with a 101 × 101 grid, Flux-Vector splitting upwinding for the convective terms and centered differences for the diffusive ones.

References

1. Arina R., Canuto C. (1993): "A Self-Adaptive Domain Decomposition for the Viscous/Inviscid Coupling. I - Burgers Equation", J. Comput. Phys., Vol. 105, pp. 290-300

2. Arina R., Canuto C. (1994): "A χ-Formulation of the Viscous-Inviscid Domain Decomposition for the Euler/Navier-Stokes Equations", in Proc. of the 7th Int. Conf. Domain Decomposition Methods in Scientific and Engineering Computing, ed. by D. Keyes and J. Xu (Contemporary Mathematics, AMS)

3. Canuto C., Hussaini M.Y., Quarteroni A., Zang T.A. (1987): "Spectral Methods in Fluid Dynamics", Springer-Verlag

4. Coutanceau M., Bouard R. (1977): "Experimental Determination of the Main Features of the Viscous Flow in the Wake of a Circular Cylinder in Uniform Translation. Part I. Steady Flow", J. Fluid Mech., Vol. 79-2, pp. 231-256

5. Brown P.N., Saad Y. (1990): "Hybrid Krylov Methods for Nonlinear Systems of Equations", SIAM J. Sci. Stat. Comput., Vol.11, 3, pp 450-481

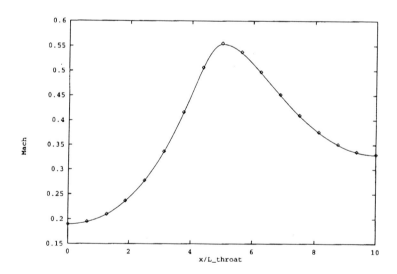

Fig. 1. Inviscid nozzle flow, Mach distribution, — exact solution, ◇ Legendre polynomials

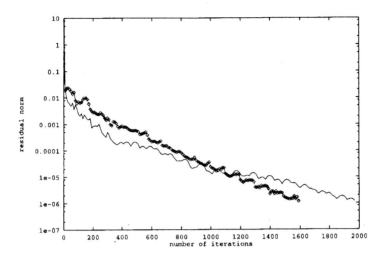

Fig. 2. Inviscid nozzle flow, Convergence hystory of $\| \mathcal{L}(Q) \|_\infty$ — Legendre collocation, □ finite differences

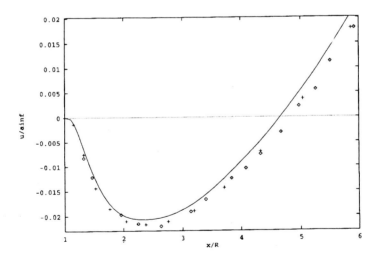

Fig. 3. Viscous flow past a circular cylinder, Re=40, Velocity distribution $\frac{u}{a_\infty}$ on the flow axis behind the cylinder, — Finite Differences, + Legendre Polynomials, ◇ Experimental Data (Coutanceau and Bouard, 1977)

Accurate 2D Euler Computations by means of a High Order Discontinuous Finite Element Method

F. Bassi[1] and S. Rebay[2]

[1]Dipartimento di Energetica
Politecnico di Milano
Piazza Leonardo da Vinci 32
20133 Milano, Italy

[2]Dipartimento di Ingegneria Meccanica
Università degli studi di Brescia
Via Branze 38
25121 Brescia, Italy

Abstract: This work describes a high order accurate discontinuous finite element method for the numerical solution of the equations governing compressible inviscid flows. Our investigation has focused on the problem of correctly prescribing the boundary conditions along curved boundaries. We show by numerical testing that, in the presence of curved boundaries, a high order approximation of the solution requires a corresponding high-order approximation of the geometry of the domain. Numerical solutions of transonic flows are presented which illustrate the versatility and the accuracy of the proposed method.

1 Introduction

The discontinuous Galerkin method is a finite element method based on a discontinuous approximation of the solution. In this respect the method is not dissimilar from a classical finite volume method where the solution is regarded as a piecewise constant function inside each element (cell). In the discontinuous finite element method, however, higher accuracy is obtained by means of high-order polynomial approximation within an element rather than by means of wide stencils as is the case of finite volume numerical schemes. The strict local nature of the high order approximation used in the discontinuous Galerkin method is perfectly suited to deal with complex computational domains possibly (and more conveniently) discretized by unstructured meshes.

To the authors' knowledge, much of the theoretical work which has been published on the discontinuous Galerkin method has focused on devising effective smoothing techniques to control the numerical oscillations generated by the numerical scheme when the discontinuities of the solution are located in the interior of the elements [4, 5, 6, 3]. The methods developed have been successfully used to compute several test case solutions of one dimensional and of simple two dimensional hyperbolic problems. However, no theoretical analysis nor numerical testing has yet been performed in order to assess the performance of the discontinuous Galerkin method in geometrically more complex 2-dimensional calculations, namely those involving curved boundaries. In fact, the authors have found an unexpected inaccuracy of the method in earlier attempts to compute the numerical solution of rather standard transonic calculations over airfoils [1, 2]. In these earlier calculations the strightforward assumption was made that the domain could be approximated by means of geometrically linear elements. In this work we show that, in order to cure the poor performance of the method in the presence of curved boundaries, it is necessary to adopt a high order accurate geometrical representation of the boundary of the domain.

We present several numerical solutions of internal and external transonic 2 dimensional flows which illustrate the geometrical versatility and the high accuracy (4-th order) of the method. In our computations, we control the numerical oscillations across the discontinuities by using a "discontinuity capturing" operator similar to that described by Hansbo and Johnson [8] in the streamline diffusion method. We should mention however that no special attention has been given to obtain a strictly monotone solution in the presence of discontinuitues. In fact, our primary goal in this investigation is to show the feasibility of high order accurate transonic computations and to enlight the importance of the corresponding high order geometrical approximation of curved boundaries.

2 The Numerical Method

Consider the weak form of the Euler equations in conservation variables

$$\int_\Omega v \frac{\partial u}{\partial t}\, d\Omega + \int_\Omega v \nabla \cdot \mathbf{F}(u)\, d\Omega = 0. \tag{1}$$

A discrete analogue of eq. (1) is obtained by considering functions u_h and v_h which are polynomials P^k of degree less or equal to k inside each element of a triangulation of the domain Ω, i.e.,

$$u_h = \sum_i U_i(t)\phi_i(\mathbf{x}), \quad v_h = \sum_j V_j \phi_j(\mathbf{x}), \tag{2}$$

where U_i denote the degrees of freedom of the unknown solution, V_j the degrees of freedom of the test function and ϕ the shape functions. By substituting u and v by u_h and v_h in equation (1) and integrating by parts, the semidiscrete equations for a generic element E can be written as

$$\frac{d}{dt}\int_E v_h u_h\, d\Omega + \int_e v_h \mathbf{F}(u_h)\cdot \mathbf{n}\, d\sigma - \int_E \nabla v_h \cdot \mathbf{F}(u_h)\, d\Omega = 0, \tag{3}$$

where e denotes the boundary of element E.

Due to the discontinuous function approximation, flux terms are not uniquely defined at element interfaces. The flux function term $\mathbf{F}(u)\cdot \mathbf{n}$ is therefore replaced by a numerical flux function $H(u_l, u_r)$ which depends on the two interface states u_l and u_r and which is determined, as in finite volume schemes, by an exact (as done in this work) or by an approximate Riemann solver. For elements having an edge on the boundary, the flux function must be consistent with the boundary conditions.

As stated in the introduction, an accurate geometrical approximation of the elements having edges on the boundary is crucial for the solution accuracy. For example, as will be shown in the next section, a piecewise linear geometrical approximation of the boundary results in very high spurious entropy production near to the solid wall. To the authors' experience, the level of spurious entropy production decreases very slowly as the grid is refined. However, a dramatic improvement of the solution accuracy is obtained if the boundary is represented by means of piecewise quadratic functions. We have also tested a cubic representation of the boundary, but, at least for the test cases attempted, the gain in accuracy is much lower than from the linear to the quadratic case. In the case of curved boundaries, the integrals appearing in equation (3) have been performed by means of Gauss numerical quadrature formulae using so called iso- and/or super-parametric elements. In practice this amounts 1) to map the real elements to a reference element by means of linear, quadratic or cubic coordinate transformations and 2) to evaluate the integrals numerically on the reference element. In the following we will denote the interpolation function and the geometrical transformation functions of order k by P^k and by Q^k, respectively.

In order to obtain a nonoscillatory scheme, Cockburn (e.g. [4]) and Bey and Oden [3] introduce limiters in the formulation of the discontinuous Galerkin method, similar in nature to those developed for finite volume TVD schemes. A second possible approach consists in supplementing the discontinuous Galerkin formulation with a "discontinuity capturing" term similar to that originally introduced by Hughes (e.g. [9]) and Johnson (e.g. [10]) in the context of the SUPG or the SD finite element methods. In this work we have used the latter limiting strategy, since it retains the strict local nature which characterizes the original formulation of the discontinuous Galerkin method. The discontinuity capturing operator to be added on the left hand side of equation (3) is written as

$$\int_E (\nabla v_h)^T \epsilon \nabla u_h \, d\Omega. \tag{4}$$

The scalar coefficient ϵ depends on the residual of the finite element solution and on mesh spacing, and is given by

$$\epsilon = Ch^2 \left\{ \sum_i [(|u_{h_i}| + c)^{-1} \nabla \cdot \mathbf{F}_i(u_h)]^2 \right\}^{-1/2}, \tag{5}$$

where C and c are positive parameters, and h is the mesh spacing. The parameter C controls the amount of dissipation added to the scheme, while c is a small constant, here put equal to 0.01.

The scheme is extremely robust, and the artificial viscosity operator does not need fine problem dependent tuning. In addition, the artificial viscosity operator has proven

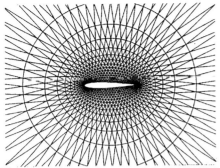

Figure 1: Triangulation around the NACA0012 airfoil

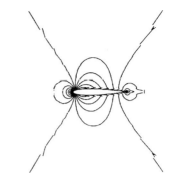

Figure 2: Mach isolines: $M_\infty = 0.5$, $\alpha = 0°$, P^1-Q^1 elements

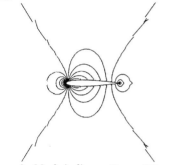

Figure 3: Mach isolines: $M_\infty = 0.5$, $\alpha = 0°$, P^1-Q^2 elements

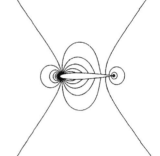

Figure 4: Mach isolines: $M_\infty = 0.5$, $\alpha = 0°$, P^2-Q^3 elements

to be effective also with high-order elements (tests have been performed with cubic elements).

1 Examples

In this section we present the results computed with increasing order of accuracy for the classical test cases of the subsonic and transonic flow around a NACA0012 airfoil and around the bi-NACA0012 airfoil configuration [7]. We have also performed simpler test calculations such as the simple shock reflection and the supersonic shear wave problems, which, for reasons of space, will be presented only in the full paper. In all the following Figures, we plot the solution inside each element according to the polynomial order used in the computations.

The first case considered is the subsonic flow at $M_\infty = 0.5$ and $\alpha = 0°$ about the

NACA0012 airfoil. We have used the relatively coarse triangulation containing 1088 points shown in Figure 1 which extends 8 chords far away from the airfoil. In this computation the value of the viscosity coefficient C is 0.01. Figure 2 shows the Mach isolines obtained with P^1–Q^1 isoparametric elements. It appears clearly that a very high spurious entropy production is generated at the leading edge and is convected along the entire airfoil surface. In Figures 3 and 4 we show the solutions computed by means of P^1–Q^2 and P^2–Q^3 super-parametric elements. Notice that, even for P^1–Q^2 elements, there is a remarkable improvement of the solution accuracy.

Next, in Figures 5 through 8 we show the computed transonic flow at $M_\infty = 0.85$ and $\alpha = 1°$ around the same airfoil. This is a challenging test case since the scheme must resolve accurately both the shocks and the strong gradients at the leading edge. The value of the viscosity coefficient C is 0.03. In this case we find that the P^1–Q^2 computation results in a rather strong entropy production near the leading edge, which is considerably reduced in the more accurate P^2–Q^3 and P^3–Q^3 numerical solutions but not completly eliminated. Notice however that the viscosity parameter used in this case is too low to completely suppress the spikes in the discontinuities, and this nonmonotone resolution is responsible for the waving shape of the Mach isolines behind the shocks.

Finally, we present the results of the transonic flow at $M_\infty = 0.55$ and $\alpha = 6°$ around the geometrically more complex bi-NACA0012 configuration. Also in this case the value of the viscosity coefficient C is 0.03. Figure 9 shows the triangulation containing 3459 nodes which has been generated by means of the grid generation algorithm described by Rebay in [11]. Figures 10 and 11 show the Mach and the pressure isolines obtained with P^2–Q^3 elements. From Figure 12, showing a detail of the Mach isolines and of the mesh around the leading edge of the top airfoil, we can appreciate the accuracy of the solution offered by the quadratic interpolation functions.

4 Conclusions

In this work we have shown the feasibility of high order accurate Euler computations by means of a discontinuous Galerkin method. We have put in evidence the importance of an accurate representation of the boundaries when the solution is computed with high order accurate elements. Because of the locality of both the basic discontinuous Galerkin scheme and of the operator employed to control the oscillations across shocks, the described method is particularly suited to be used with unstructured meshes. The viscosity coefficient ϵ depends on the residual of the finite element solution and on mesh spacing but does not take into account features such as the flow direction or the stretching of the elements. Our choice has been mainly dictated by simplicity. In the future we plan to test more sophisticated, and hopefully more accurate, alternative formulations available in the literature.

References

[1] Bassi, F., Rebay, S., Savini, M. (1992). "A High Resolution Discontinuous Galerkin Method for Hyperbolic Problems on Unstructured Grids". In proceedings, *3-rd ICFD Conference, Oxford University Press*.

[2] Bassi, F., Rebay, S., Savini, M. (1992). "Discontinuous Finite Element Euler Solutions on Unstructured Meshes". In proceedings, *13-th ICNMFD, Lecture Notes in Physics, Vol 414*, 245–249.

[3] Bey, K. S. and Oden, J. T. (1991). "A Runge-Kutta Discontinuous Finite Element Method for High Speed Flows". *AIAA paper no. 91-1575-CP*, 541–555.

[4] Cockburn, B. and Shu, C.-W. (1989). "TVB Runge-Kutta Local Projection Discontinuous Galerkin Finite Element Method for Conservation Laws II: General Framework". *Math. Comp., 52*, 411–435.

[5] Cockburn, B. and Shu, C.-W. (1989). "TVB Runge-Kutta Local Projection Discontinuous Galerkin Finite Element Method for Conservation Laws III: One Dimensional Systems". *J. Comput. Phys., 84*, 90–113.

[6] Cockburn, B., Hou, S. and Shu, C.-W. (1990). "The Runge-Kutta Local Projection Discontinuous Galerkin Finite Element Method for Conservation Laws IV: The Multidimensional Case". *Math. Comp., 54*, 545–581.

[7] Dervieux, A., Van Leer, B., Periaux, J., Rizzi, A., (1989). "Numerical Simulation of Compressible Euler Flows". *Notes on Numerical Fluid Dynamics, Vol. 26*.

[8] Hansbo, P. and Johnson C. (1991). "Adaptive Streamline Diffusion Methods for Compressible Low Using Conservation Variables". *Comp. Meth. in Appl. Mech and Eng., 87*, 267–280.

[9] Shakib F., Hughes T. J. R., Johan Z. (1991). "A New Finite Element Formulation for Computational Fluid Dynamics: X. The Compressible Euler and Navier-Stokes Equations". *Comp. Meth. in Appl. Mech and Eng., 89*, 141–219.

[10] Johnson, C. (1992). "Finite Element Methods for Flow Problems". In *Unstructured Grid Methods for Advection Dominated Flows, AGARD Report 787 — VKI Numerical Grid Genaration 1992*.

[11] Rebay, S. (1993) "Efficient Unstructured Mesh Generation by means of Delaunay Triangulation and Bowyer-Watson Algorithm". *J. Comput. Phys., 106*, 125–137.

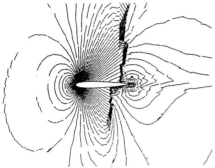

Figure 5: Mach isolines: $M_\infty = 0.85$, $\alpha = 1°$, P^1-Q^2 elements

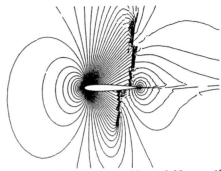

Figure 6: Mach isolines: $M_\infty = 0.85$, $\alpha = 1°$, P^2-Q^3 elements

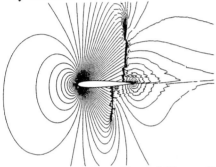

Figure 7: Mach isolines: $M_\infty = 0.85$, $\alpha = 1°$, P^3-Q^3 elements

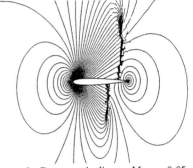

Figure 8: Pressure isolines: $M_\infty = 0.85$, $\alpha = 1°$, P^3-Q^3 elements

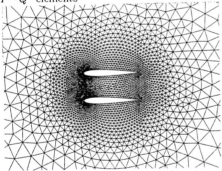

Figure 9: Unstructured triangulation around the bi-NACA0012 airfoil configuration

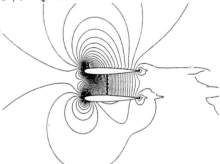

Figure 10: Mach isolines: $M_\infty = 0.55$, $\alpha = 6°$, P^2-Q^3 elements

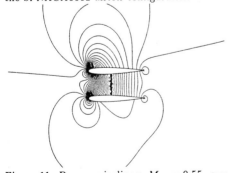

Figure 11: Pressure isolines: $M_\infty = 0.55$, $\alpha = 6°$, P^2-Q^3 elements

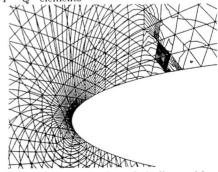

Figure 12: Detail of Mach isolines: $M_\infty = 0.55$, $\alpha = 6°$, P^2-Q^3 elements

A Higher-Order Multidimensional Upwind Solution-Adaptive Multigrid Solver for Compressible Flows

L. A. Catalano, P. De Palma, G. Pascazio, M. Napolitano

Istituto di Macchine ed Energetica, Politecnico di Bari
Via Re David 200, 70125 Bari, Italy

1 Introduction

In the last years, a genuinely multidimensional upwind methodology for compressible inviscid flows has been developed in the framework of a Brite/Euram programme [1]. The Euler system is decomposed either into a set of five-to-six simple-wave equations [2], or into a system of four optimally decoupled characteristic equations [3]. A compact-stencil conservative discretization is achieved using a cell-vertex triangular mesh: each simple-wave or characteristic contribution to the flux balance over each cell is sent to its downstream nodes, according to one of the Fluctuation Splitting (FS) schemes proposed in [4]. For the case of the more robust six-wave (6W) model, combined with the linear first-order-accurate N-scheme, a standard FAS multigrid strategy [5] has been combined with an optimal explicit multi-stage Runge-Kutta (RK) smoother for scalar advection equations [6], to provide a very efficient genuinely multidimensional upwind Euler solver [7]. An adaptive local-refinement strategy has been finally developed to obtain accurate solutions to complex subsonic, transonic and supersonic flow problems at very reasonable computer costs [8,9]. On the other hand, an excessive entropy generation at solid boundaries makes the methodology developed so far unsuitable for viscous flow computations. Moreover, also the low-diffusion-A (LDA) and the positive-streamwise-invariant (PSI) FS schemes [4], which achieve second-order accuracy for homogeneous scalar advection equations, are only first-order-accurate when combined with a simple-wave model of the Euler equations employing more than three waves, as shown in [10,11]. For such a reason, the characteristic decomposition has been recently reconsidered and its lack of robustness, experienced when choosing gradient-dependent propagation directions, has been overcome by abandoning the requirement of decoupling the compatibility equations [12]. In this paper, the FAS multigrid and adaptive local refinement strategies developed by the authors are extended to such a more accurate multidimensional upwind solution procedure. It is anticipated that the improved accuracy with respect to the previous approach, based on the 6W decomposition, will be somewhat counterbalanced by a reduced efficiency, due to the higher-order-accuracy of the scheme and to the presence of coupling terms in the compatibility equations, which forbids the optimization of the smoother.

2 Characteristic decomposition method

The 2D Euler equations are decomposed into a set of four compatibility equations with three degrees of freedom, represented by the unit vectors $\mathbf{k}^{(i)}$, $i = 2, 3, 4$, namely:

$$\frac{\partial W}{\partial t} + \Lambda \cdot \nabla \mathbf{W} + \mathbf{q} = 0. \tag{1}$$

In eq. (1), Λ is the diagonal matrix containing the bicharacteristic vectors $\lambda^{1,2} = \mathbf{u}$, $\lambda^{3,4} = \mathbf{u} + a\mathbf{k}^{(3,4)}$, W is the vector of the characteristic variables and q is the vector of the coupling terms:

$$\partial W = \begin{pmatrix} \partial \rho - \partial p/a^2 \\ \mathbf{s}^{(2)} \cdot \partial \mathbf{u} \\ \mathbf{k}^{(3)} \cdot \partial \mathbf{u} + \partial p/\rho a \\ \mathbf{k}^{(4)} \cdot \partial \mathbf{u} + \partial p/\rho a \end{pmatrix}, \quad q = \begin{pmatrix} 0 \\ \mathbf{s}^{(2)} \cdot \nabla p/\rho \\ a\mathbf{s}^{(3)} \cdot (\mathbf{s}^{(3)} \cdot \nabla)\mathbf{u} \\ a\mathbf{s}^{(4)} \cdot (\mathbf{s}^{(4)} \cdot \nabla)\mathbf{u} \end{pmatrix}, \tag{2a, b}$$

$\mathbf{s}^{(i)}$ being a unit vector orthogonal to $\mathbf{k}^{(i)}$. In 1986, Deconinck et al. [1] proposed to choose the $\mathbf{k}^{(i)}$ directions according to the flow variable gradients so as to minimize q. The resulting model, combined with the FS approach of [4], has been capable of providing results only for supersonic flows and experienced convergence difficulties [1]. Therefore, the idea of minimizing the coupling terms has been abandoned and two more robust models, depending on the local flow direction and Mach number, have been recently proposed [12]. The first model (CS) takes $\mathbf{k}^{(3)}$ aligned with the velocity vector ($\theta_3 = \theta = arctan(v/u)$), whereas the second more costly one (CMA) averages the characteristic contributions calculated using two different directions θ_3, $\theta_3 = \theta \pm (\pi/2 + arctan(1/\sqrt{|M^2 - 1|}))$. It is noteworthy that these two directions are related to the Mach angle for supersonic flows and that this decomposition is continuous at sonic point. In both models, $\mathbf{k}^{(4)} = -\mathbf{k}^{(3)}$ and $\mathbf{k}^{(2)} = \mathbf{k}^{(3)}$.

The compatibility equations are discretized in space on a cell-vertex triangular grid following the multidimensional upwind FS approach of [4]. In particular, in order to obtain higher-order-accurate solutions, the two linearity preserving PSI (non-linear, positive) and LDA (linear, non-positive) schemes are used to discretize the convective and coupling terms, respectively, see [4,12] for more details.

3 Multigrid adaptive strategy

A solution-adaptive local-refinement strategy for cell-vertex residual distribution methods, recently developed by the authors and applied to the six-wave decomposition model [8,9], is combined here with the characteristic decomposition method described in the previous section. Starting from a regular structured grid, nested levels of local refinement are created and managed by a quad-tree data-stucture [13]. The standard FAS multigrid V-cycle [5] using full-weighting collection and bilinear prolongation is employed in combination with an explicit multi-stage RK scheme. For the case of the simple-wave decomposition model, it is possible to optimize the smoothing properties of the RK scheme by using a 2D scalar Fourier analysis [6,1], so as to achieve a very good multigrid efficiency [7,8]. Unfortunately, the approach of [6] cannot be extended to the present characteristic decomposition, due to the presence of the coupling term q. Moreover, the stability range of RK schemes is severely reduced so that a higher number of stages needs to be employed, with coefficients and CFL number determined empirically.

4 Results

The subsonic flow through a channel with a cosine shaped wall and inlet Mach number equal to 0.5 has been considered at first to validate the present methodology. Figure 1 provides the Mach number contours obtained using the CS decomposition on a 64 × 16 uniform grid. Tables 1 and 2 show the L_1 and L_∞ norms of the entropy error for the simple-wave and characteristic decomposition models. These are seen to provide second-order accuracy while generating a very low amount of numerical entropy with respect to the 6W model. Concerning the efficiency, figure 2 shows the convergence histories, on a 64 × 16 uniform grid with five multigrid levels, of the 6W model (using the optimal three-stage RK scheme [1,8]) and of the CMA (CFL=0.6) and CS (CFL=0.3) characteristic decompositions (using a six-stage scheme with coefficients: .07, .12, .20, .35, .65, 1, and CFL numbers equal to .6 and .3, respectively). The convergence rate of the optimized 6W approach is clearly superior. Furthermore, the CMA decomposition, providing the same accuracy as the CS one [12], enjoys a faster convergence rate, due to its higher cross-wind diffusion. It must be remarked that the three models required slightly different CPU times per work unit.

A more complex test-case, namely, the subsonic flow through a cascade of VKI gas turbine rotor blades (VKI LS-59), has been then computed using a C-type non-periodic grid [14]. Figure 3 shows the Mach number distribution along the profile obtained with the 6W and the characteristic models on a 256 × 8 uniform grid. The optimal three-stage and the empirical six-stage RK schemes have been employed for the 6W, CMA (CFL=.5) and CS (CFL=.1) models, respectively. The solution obtained with the characteristic approach is closer to the experimental data of [15], as anticipated. However, some wiggles are present on the front part of the suction side of the blade and the acceleration after the stagnation point on the pressure side is underestimated. This latter feature disappears when locally refining the mesh, as shown in figure 4, which provides the solution calculated on a composite grid obtained starting from a 128 × 4 uniform one with two refinement cycles. The persistence of the wiggles, which are present only very near the blade surface, seems to indicate that they are caused by the local definition of the blade itself. Concerning the efficiency, figure 5 shows the convergence histories for the 6W, CMA and CS decompositions employing a three level multigrid on the uniform grid: the conclusions drawn for the previous test case are confirmed, with a further worsening of the convergence rate for the CS model.

5 Conclusions

A solution-adaptive local-refinement multigrid strategy previously developed for a simple-wave model of the Euler equations has been extended to two newly developed higher-order-accurate characteristic decomposition methods for the solution of steady inviscid compressible flows. Numerical results show that the new methodology provides a significant accuracy improvement with respect to the simple-wave approach, at the expense of a reduced multigrid convergence rate. While further improvements are warranted, the new method, thanks to its second-order spatial accuracy and very low entropy generation, appears to be well suitable for computing viscous flows.

Acknowledgements

This work has been supported by the EEC under BRITE/EURAM contract No. AER CT92-0040, and by MURST, 40%.

References

[1] H. Deconinck et al.: CWI-Quarterly 6-1 1 (1993)

[2] P. L. Roe: J. of Comp. Phys. **63** 458 (1986)

[3] H. Deconinck et al.: Lecture Notes in Physics **264** 216 (1986)

[4] R. Struijs et al.: VKl LS 1991-01 (1991)

[5] A. Brandt: Lecture Notes in Mathematics **960** 220 (1982)

[6] L. A. Catalano, H. Deconinck: TN-173, von Karman Institute (1990)

[7] L. A. Catalano et al.: Notes on Numerical Fluid Mechanics **35** 69 (1992)

[8] L. A. Catalano et al.: Lecture Notes in Physics **414** 90 (1993)

[9] L. A. Catalano et al.: 5th European Multigrid Conference, CWI (1993)

[10] L. A. Catalano et al.: 5th ISCFD, Sendai (1993)

[11] H. Paillére et al.: AIAA Paper 93-3301 (1993)

[12] H. Deconinck et al.: VKI LS 1994-05 (1994)

[13] P. W. Hemker et al.: Report NM-R9014 (1990)

[14] L. A. Catalano et al.: AIAA Paper 94-0076 (1994)

[15] R. Kiock et al.: J. Eng. for Gas Turbines and Power **108** 277 (1986)

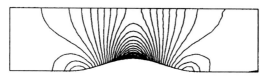

Fig. 1. Channel flow: Mach contours ($\Delta M = 0.02$)

Table 1. Accuracy study for the six wave decomposition with N-scheme

grid	$L_1(S)$	$L_\infty(S)$	M_{max}
16 × 4	1.00(-2)	5.60(-2)	.717
32 × 8	6.41(-3)	4.42(-2)	.803
64 × 16	3.28(-3)	2.51(-2)	.873

Table 2. Accuracy study for the characteristic decomposition with PSI-scheme

grid	$L_1(S)$	$L_\infty(S)$	M_{max}
16 × 4	9.76(-4)	6.57(-3)	.862
32 × 8	3.55(-4)	3.73(-3)	.927
64 × 16	8.29(-5)	1.06(-3)	.943

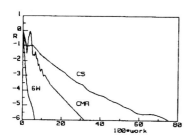

Fig. 2. Channel flow: convergence histories

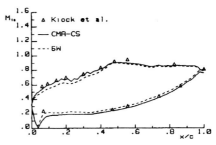

Fig. 3. Cascade flow: Mach number distribution (uniform)

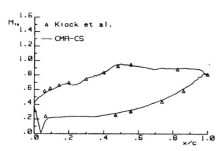

Fig. 4. Cascade flow: Mach number distribution (adaptive)

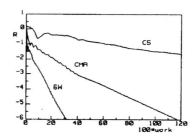

Fig. 5. Cascade flow: convergence histories (uniform)

On some advances in Fluid Flow Modelling through coupling of Newton method and new implicit PPV schemes

Valery V. Dedesh, Ivan V. Yegorov

Central Aerohydrodynamics Institute (TsAGI),
Zhukovsky-3, 140160 Moscow Region, Russia
Telex: 412573 ZAKAT SU, Fax: (095) 271.00.19

1 Introduction

During recent 20 years there has been a considerable effort devoted to new TVD schemes refinement and large scale applications. Sophisticated TVD difference schemes are constantly undergoing simplifications [1,2], while implicit numerical methods to solve discretized equations of fluid motion are becoming more and more refined. Newly introduced Godunov-type schemes of the PPV (Polynomially Presented Viscosity [3]) method in their implicit formulation, are used to demonstrate that for the flows considered extremely simple PPV schemes work and converge faster than and yield solutions as fine as the most sophisticated high resolution schemes [1].

For the problems of viscous aerodynamics, at high Reynolds numbers, simplified PPV schemes converge more readily than reference [1] scheme and, if needed, supply a perfect initial guess for the latter converging calculations. Newton method is used in this study because the code [4], implementing it, offers maximum flexibility of flow type physics variation. The code involves quick in-home sparse matrix technology [5] and is compact w.r.t. computer memory and CPU operation count.

The point of the report is the PPV method to construct Godunov-type schemes [3], aimed at a proper construction of numerical viscosity matrices, associated with Godunov-type schemes. The ground idea of the method is to express a numerical viscosity matrix of a scheme as a quadratic (full) or linear (simplified) polynomial of a gas dynamics Jacobian matrix. Simplified PPV schemes is a subject of the present study. No characteristic transformations, no matrix manipulations: they are fast and easily vectorized. By construction these schemes, however simple, work as good as full PPV schemes [1] in many cases and do the same job at a reduced computer effort price.

2 PPV schemes in 1-D

Modelling fluid and gas flows, governed by Euler and Navier-Stokes equations $w_t + f_x = 0$ by means of difference schemes $w^0 = w_0 - \lambda(f_{\frac{1}{2}} - f_{-\frac{1}{2}})$ focuses attention on the numerical viscosity matrix **Q**, determining the specific appearance a Godunov-type scheme flux

$$f_{\frac{1}{2}} = \frac{1}{2}(f_1 + f_0 - \mathbf{Q}_{\frac{1}{2}}) \tag{1}$$

Matrix **Q** is assumed to depend on the spectrum of the problem in hand in the form $\mathbf{Q} = q_0\mathbf{I} + q_1\mathbf{A} + q_2\mathbf{A}^2$, where $\mathbf{A} = \frac{\partial f}{\partial w}$. Obviously, **Q** and **A** are simultaneously diagonalized by the same set of eigenfunctions. **Q**-matrix eigenvalues $\lambda_-, \lambda_0, \lambda_+$ are related to the three distinct eigenvalues a_-, a_0, a_+ of **A** via: $\lambda_{-,0,+} = q_0 + q_1 a_{-,0,+} + q_2(a_{-,0,+}^2)$ which leads to expressions for

$$q_0 = \lambda_0 - d_1 a_0 + d_2 a_0^2, \quad q_1 = d_1 - 2d_2 a_0, \quad q_2 = d_2 \tag{2}$$

where $d_1 = \frac{1}{2}(\lambda_+ - \lambda_-)$ and $d_2 = \frac{1}{2}(\lambda_+ - 2\lambda_0 + \lambda_-)$. When $\lambda_{-,0,+}$ are chosen to be $|a_{-,0,+}|$, the result is the Roe scheme [1]. Account for all the eigenvalues, determining solution of the problem gives us the right to call the scheme

$$f_{\frac{1}{2}} = \frac{1}{2}(f_1 + f_0 - (q_0)_{\frac{1}{2}}\Delta_{\frac{1}{2}}w - (q_1)_{\frac{1}{2}}\Delta_{\frac{1}{2}}f - (q_2)_{\frac{1}{2}}\Delta_{\frac{1}{2}}\mathbf{A}_{\frac{1}{2}}f) \tag{3}$$

full or F-scheme. Here, the exact equality $\Delta_{\frac{1}{2}}f = \mathbf{A}_{\frac{1}{2}}\Delta_{\frac{1}{2}}w$ is used, $\mathbf{A}_{\frac{1}{2}}$ and $(q_{0,1,2})_{\frac{1}{2}}$ are computed using the corresponding Roe averaged [2] values.

If one takes $\lambda_{-,+} = |a_{-,+}|, \lambda_0 = \frac{1}{2}(|a_-| + |a_+|)$ then (3) becomes essentially simpler, since $q_2 = 0$. Account for the two limiting characteristics a_+, a_- influence gives grounds to call the corresponding algorithm (4) simplified or S-scheme. Order of accuracy is raised from unity to two within the MUSCL approach [3].

3 2-D Problems

Steady-state 2-D problems $f_x + g_y = 0$ in curvilinear coordinate systems $\hat{f}_{\frac{1}{2},0} - \hat{f}_{-\frac{1}{2},0} + \hat{g}_{0,\frac{1}{2}} - \hat{g}_{0,\frac{1}{2}} = 0$ are solved using local rotation of the coordinate system, where $\hat{f}_{\frac{1}{2},0} = \mathbf{T}^{-1}f_{\frac{1}{2}}(\mathbf{T}w)$, and $\hat{g}_{0,\frac{1}{2}} = \mathbf{T}g_{\frac{1}{2}}(\mathbf{T}^{-1}w)$, where $f_{\frac{1}{2}}, g_{\frac{1}{2}}$ are usual numerical fluxes, computed according to (4) and **T** is the local rotation matrix.

The resulting computational stencil in node (i,j)=(0,0) of a second - order PPV scheme to compute viscous flows consists of 13 points. Contributions from all over the structured grid result in a vector of equations $\Phi(q) = 0$, to be solved for q by means of the fully implicit Newton method of consecutive approximations $q_{n+1} = q_n - (\frac{\partial \Phi}{\partial q})_n^{-1}\Phi_n$, where n is an iteration index and $\frac{\partial \Phi}{\partial q}$ is a sparse matrix, inverted by means of a nested dissection aided direct solver [5]. Quadratic convergence of Newton method is usually realized after 7-10 iterations. All the results presented in Ch.4 refer to the direct sparse matrix solver, however iterative solvers are also available in [4].

4 Applications

4.1 Viscous and inviscid flows of an ideal gas

Substitution of $a_- = u - c, a_0 = u, a_+ = u + c$ in (4), where u and c are gas velocity and $\gamma = 1.4$ ideal gas sound speed, yields the F-scheme with $q_0 = c|M|(1 - M^2)$, $q_1 = M(2|M| - 1)$, $q_2 = \frac{(1-|M|)}{c}$, whereas for the S-scheme $q_0 = c(1 - M^2), q_1 = M, q_2 = 0$, for $|u| < c$, and $q_{0,2} = 0, q_1 = sign(M)$ otherwise (for both S and F schemes), where $M = \frac{u}{c}$. Viscous terms in Navier-Stokes equations are approximated by central differences.

4.2 Viscous flows of incompressible fluid (pseudo - compressibility approach, with $\beta = 1$.)

Here again $a_- = u - c, a_0 = u, a_+ = u + c$, yet c is now defined as $c = \sqrt{u^2 + 1}$. Thus $q_0 = \frac{|M|}{c}$, $q_1 = M(2|M| - 1)$, $q_2 = \frac{(1-|M|)}{c}$ for scheme F and $q_0 = \frac{1}{c}, q_1 = M, q_2 = 0$ for scheme S.

4.3 Turbulent flows of viscous compressible gas (Reynolds - averaged Navier-Stokes equations and 2-equation Coakley $q - \omega$ turbulence model)

Here $q_{0,1,2}$ are identical to those in case a. Noteworthy is that $q_{\frac{1}{2}}$, $\omega_{\frac{1}{2}}$, entering $\mathbf{A}_{\frac{1}{2}}$, should also be Roe-averaged for the F and are not needed for the S scheme.

5 Numerical examples

The first problem, solved using the above numerical technique is the 2D inviscid transonic flow past NACA0012 airfoil ($\alpha = 1.25^0, M = 0.8$, Fig.1). No visible difference between "S" and "F" scheme results is seen in $-Cp$ value distribution along the airfoil surface (Fig.2). Implicit Newton method convergence history and time account the for both schemes make one conclude that CPU time savings are of order of 30% with no deterioration of numerical results on the background.

The second case, considered in the report is laminar Navier-Stokes supersonic 2D flow past a cylinder ($M = 8.0, Re = 1000$, Fig.3). Here the "F" scheme failed to converge in the Newton method framework, while the "S" scheme did converge and supplied a successful initial guess for the scheme "F", now convergent also (see cylinder surface pressure distribution in Fig.4). Computer time savings, now less significant: about 25% in favor of simplified PPV schemes, and excellent quality of the results, again suggest that the simplified PPV schemes be good candidates for further research and practical utilization.

The third result relates to the incompressible fluid flow past a cylinder at Re number equal to 40. Essentially separated flow (Fig.5) is modelled almost identically by both schemes, after 8 iterations. See for example $-Cp$ distribution on the cylinder surface, Fig.6. To be noted here is the outstanding simplicity of the simplified scheme for incompressible fluid flow simulation, allowing to save up to 35 % of CPU operation count.

Finally, Figures 7 and 8, coming from the solution of Reynolds-averaged Navier-Stokes equations ($M = 0.5, Re = 200000$) using Coakley $q - \omega$ turbulence model again shows an excellent agreement between scheme "F" and "S" results ($-Cp$ and Cf surface distributions). "F"-scheme needs about twice the number of Newton iterations until convergence, as compared to the "S" algorithm. Besides, "S" is itself almost by 30% cheaper, than scheme "F".

Acknowledgements :

It gives us pleasure to thank Prof. V.Ya.Neiland, Prof. R.Narasimha, Dr. S.S.Desai, Prof. P.J.L.Zandbergen, for their cooperation. We also acknowledge valuable financial assistance from TsAGI, NAL and Russian Basic Research Fund.

References

[1] P. L. Roe, J.Comput.Phys.,v.**41**,p.357,(1981)

[2] M. S. Liou, Lecture Notes in Phys.,v.**414**,p.115-119,(1993)

[3] V. V. Dedesh, C.R.Acad.Sci.Paris,t.**316**,Sér.II,p.1357-1362,(1993)

[4] I. V. Yegorov, O. L. Zaycev, Proc. 5th Int. Symposium on CFD, Sendai, Japan, v.**3**, p.393-400, (1993)

[5] J. A. George, SIAM J.Numer.Anal.,v.**10**, N.2, p.347-363, (1973)

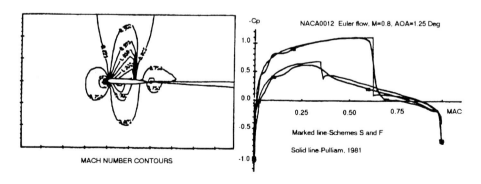

Fig. 1   Fig. 2

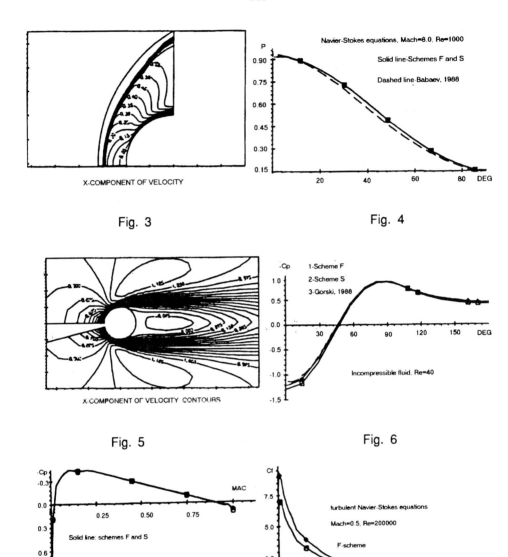

An improved approximate Riemann Solver for Hypersonic Bidimensional flows

Laurence Flandrin and Pierre Charrier

CeReMab Universitè Bordeaux 1, 33405,
France, Bruno Dubroca
CEA/CESTA Le Barp, France

Abstract : We propose a new version of HLLEM approximate Riemann solver where local antidiffusive coefficients are calculated to have robustness and accuracy in the obtained scheme. We introduce the concept of Riemann-stability, property satisfied by our method, which guarantees its good behaviour in known difficult situations. Numerical results are presented to demonstrate the efficiency of our approach.

1 Introduction

Nowadays a great number of finite volume schemes based on approximate Riemann solvers are used to simulate complex flows, well capturing shocks and discontinuities. Nevertheless, some of them, often the less dissipative, if they are known for their accuracy could present embarrassing failures in peculiar situations: negative densities, carbuncle phenomenon [1] ... It is well-known that adding numerical viscosity in the approximate fluxes could prevent these problems. It is quite natural with some methods, like Roe's scheme modified with Harten's entropy fix in which viscosity could be well adjusted by means of a viscosity parameter, or like HLLE scheme, very dissipative by construction [2]. However, the robustness of these methods is only insured to the detriment of their accuracy, which could be embarrassing, for the boundary layers approximations, for example. When HLLE is used, the lack of accuracy is particularly noticeable on linearly degenerate fields. Hence a modification of this scheme has been proposed introducing antidiffusive coefficients on these fields [3], [4]. If this method, called HLLEM, seems to be really efficient for one dimensional flows computations, we have found again the classical difficulties tied with the lack of dissipation computing two-dimensional problems. A delicate study of Roe's scheme and HLLEM allowed to show important role played by the tangential velocities approximations for multidimensional calculations. Indeed, when they reach too small or too large values compared to the initial ones, the carbuncle could appear on the bow shock. More generally we have noted that it is of importance that the intermediate states calculated by the approximate Riemann solver rely on the physical ones. Consequently, we introduce the concept of Riemann-stability (or R-stability) which insures this property: a R-stable solver will more take into account the exact solution than solvers which do not get the property, and it is a guarantee of its robustness.

Numerically, we have brought into question the values of the HLLEM antidiffusive coefficients and we have proposed a way to compute tham to have a R-stable solver that we have called HLLEMR. We note that this one is not less accurate than classical schemes, which is essential for the treatment of viscous flows and the capture of contact discontinuities.

In the following section a short background about HLLEM and the notations used is given. Section 3 will be devoted to the Riemann-stability property. In section 4, HLLEMR is built and analyzed. This solver is in fact used for the simulation of two-dimensional flows on unstructured hybrid meshes. In the last section numerical experiments demonstrate the efficiency of HLLEMR to compute with a minimum amount of dissipation, flows in situations where classical solvers fail without additional viscosity. Navier-Stokes computations substantiate its accuracy.

2 HLLEM Presentation

The following Riemann problem posed in the (n_x, n_y) direction is considered:

$$W_t + \overline{F}(W)_\xi = 0 \qquad W(\xi, 0) = \begin{cases} W^l & \text{if } (\xi < 0) \\ W^r & \text{if } (\xi > 0) \end{cases} \tag{1}$$

where $\xi = x n_x + y n_y$, is the normal variable. For the Euler equations, the vector of conservative variables is $W = (\rho, \rho V_n, \rho V_\tau, E)$ ρ denoting the density, V_n and V_τ the normal and tangential components of the velocity and E the total energy. The flux in the $\overline{n}$ direction is given by $\overline{F}(W) = (\rho V_n, p + \rho V_n^L, \rho V_n V_\tau, V_n(E + p))$.

Using HLLEM leads to the following formulation of the approximate solution:

$$W^*\left(\frac{\xi}{t}; W^l, W^r\right) = \begin{cases} W^l & \xi/t < b^l \\ W^{HLLE} +(\xi - \tilde{u}t)(\delta_m \alpha_2 \overline{R}_2 + \delta_\tau \alpha_3 \overline{R}_3) b^l < & \xi/t < b^r \\ W^r & \xi/t > b^r \end{cases} \tag{2}$$

Quantities noted $\overline{q}$ refer to Roe's averaged values. $\overline{R}_i$ denotes the ith right eigenvector of the Roe's matrix and $\alpha_i = \overline{R}_i^{-1}(W^r - W^l)$. b^l and b^r are approximations for the largest and the smallest physical signal velocities: $b^l = \min\left(V_n^l - c^l, \overline{V}_n - \overline{c}\right)$ and $b^r = \text{Max}\left(V_n^r + c^r, \overline{V}_n + \overline{c}\right)$ are taken and $\overline{u} = \left(b^l + b^r\right)/2$. The constant state W^{HLLE} is given by the equality of consistency with the integral form of the conservation law.

δ_m and δ_τ are the antidiffusive coefficients between 0 and 1 and acting on the linearly degenerate fields numbered 2 and 3. The former is found in the one dimensional problem while the latter corresponds to the purely bidimensional field. If these coefficients are equal to zero, then **HLLE** is obtained. To define **HLLEM**, they take the same value, calculated so that Roe's scheme is recovered if b^l and b^r are equal to the smallest and the largest eigenvalues of the Roe-average matrix.

3 The Riemann-Stability property

Let P the function which permits to deduce from the conservative variables the primitive ones:

$$P(W) = P(\rho, \rho V_n, \rho V_\tau, E) = (\rho, V_n, V_\tau, p) \tag{3}$$

Definition 1: *The set of physically acceptable states for the Riemann problem (1) is* :

$$E(W^l, W^r) = \{W / \exists \eta \in \mathcal{R}, W = W^e(\eta; W^l, W^r)\} \tag{4}$$

with $W^e(\eta; W^l W^r)$ referring to the exact solution of the problem.
We note $\overline{C}_v(E(W^l, W^r))$ the closure of the convex hull of $E(W^l, W^r)$ and we define the set:

$$E_R(W^l, W^r) = \left\{ \begin{array}{c} W^q \text{ so that } \forall i = 1,4 \\ \min\{P(W)\}_i \leq \{P(W^q)\}_i \leq \max\{P(W)\}_i \\ W \in \overline{C}_v(E(W^l, W^r)) \end{array} \right\} \tag{5}$$

Definition 2: *We say that an approximate Riemann solver is Riemann-stable (or R-stable) if the set $A(W^l, W^r) = \{W / \exists \eta \in \mathcal{R}, W = W^a(\eta; W^l, W^r)\}$ is enclosed in $E_R(W^l, W^r)$.*

Then the primitive variables computed by a R-stable solver satisfy a maximum principle according to the states of the exact solution. Moreover a R-stable scheme is positively conservative. Gudunov's scheme is R-stable. Similarly, HLLE, introduces the following constant state:

$$W^{HLLE} = \frac{1}{b^l - b^r} \int_{b^l}^{b^r} W^e(\eta; W^l, W^r) d\eta \tag{6}$$

which obviously belongs to $\overline{C}_v(E(W^l, W^r))$ and so to $E_R(W^l, W^r)$. Conversely, Roe's sheme based on a linearization of the Riemann problem could generate non physical states [3] and so is not R-stable. It is the same thing for HLLEM which sometimes coincides with Roe's solver.

4 A R-Stable Scheme: HLLEMR

In this section we propose a way to calculate the antidiffusive coefficients of HLLEM, to obtain a R-stable solver [5]. Two main steps lead to the determination of the new pair δ_m, δ_τ .

4.1 Determination of an admissible convex

We look for a convex, noted C such that if $(\delta_m, \delta_\tau) \in C$, then the approximate Riemann solver is R-stable. to simplify the implementation of the method, we do not require the inclusion given in definition 2 but we impose,

$$A(W^l, W^r) \subset \tilde{H}(W^l, W^r) \tag{7}$$

where we have set $H(W^l, W^r) = \{W/ \exists \eta \in \mathcal{R}, W = W^{HLLE}(\eta; W^l, W^r)\}$ and,

$$\tilde{H}(W^l, W^r) = \left\{ \begin{array}{c} W^q \text{ so that } \forall i = 1, 4 \\ \text{Min } \{P(W)\}_i \leq \{P(W^q)\}_i \leq \max \{P(W)\}_i, \ W \in H(W^l, W^r) \end{array} \right\} \tag{8}$$

Obviously $\tilde{H}(W^l, W^r) \subset E_R(W^l, W^r)$ and if 97) is satisfied, the scheme is R-stable. Considering the density, we now give an example of how the primitive variables are limited to have the previous inclusion. the intermediate density given by a HLLEM type solver is

$$\rho^*(\xi/t) = \rho^{HLLE} + \delta_m \alpha_2 (\xi - \tilde{u}t) \qquad \frac{\xi}{t} \in [b^l, b^r] \tag{9}$$

Because it is a linear function, we just have to limit it at the ends b^l and b^r. If $\rho^l \leq \rho^{HLLE}$ and $\rho^r \leq \rho^{HLLE}$ then the same inequalities are imposed for ρ^* that is to say, $\text{Max}(\rho^l, \rho^r) \leq \rho^*(\xi/t) \leq \rho^{HLLE}$. Similar expressions but with reversed inequalities are obtained if $\rho^l \geq \rho^{HLLE}$ and $\rho^r \geq \rho^{HLLE}$. In these two cases, the conditions are satisfied if and only if $\delta_m = 0$. If $\rho^l \leq \rho^{HLLE} \leq \rho^r$, then to have $\rho^l \leq \rho^*(\xi/t) \leq \rho^r$ it is easy to show that δ_m must satisfy the following condition:

$$\delta_m \leq \text{Min} \left\{ \frac{\rho^{HLLE} - \rho^l}{|\alpha_2|}, \frac{\rho^r - \rho^{HLLE}}{|\alpha_2|} \right\} \tag{10}$$

Such a reasoning is done for all other primitive variables as easily as previously for the two components of the velocities but the treatment of the pressure which is not a monotone function, is more complicated. Indeed we have

$$p^*(\xi/t) =$$

$$\rho^{HLLE} + \frac{1}{2\rho^*(\xi/t)} \left[\rho^{HLLE} \alpha_2 \delta_m t(\xi/t - \tilde{u}) \left(\left[V_n^{HLLE} - \overline{V}_n\right]^2 + \left[V_\tau^{HLLE} - \overline{V}_\tau\right]^2 \right) \right.$$

$$\left. + 2\rho^{HLLE} \alpha^3 \delta_\tau t(\xi/t - \tilde{u}) \left(\overline{V}_\tau - V_\tau^{HLLE} \right) - (\alpha^3 \delta_\tau t(\xi/t - \tilde{u}))^2 \right] \tag{11}$$

An efficient way to overcome this difficulty is to split p^* in two párts, and to compare each of them to $\rho^l/2$, $\rho^{HLLE}/2$, $\rho^r/2$. We first take,

$$p1(\xi/t) =$$

$$\rho^{HLLE} + \frac{1}{2\rho^*(\xi/t)} \left[\rho^{HLLE} \alpha_2 \delta_m t(\xi/t - \tilde{u}) \left(\left[V_n^{HLLE} - \overline{V}_n\right]^2 + \left[V_\tau^{HLLE} - \overline{V}_\tau\right]^2 \right) \right.$$

$$\left. + 2\rho^{HLLE} \alpha^3 \delta_\tau t(\xi/t - \tilde{u}) \left(\overline{V}_\tau - V_\tau^{HLLE} \right) \right] \tag{12}$$

which is a monotone function and so it is only necessary to limit it at the ends of $[b^l, b^r]$. Secondly we consider

$$p2(\xi/t) = p^{HLLE}/2 - \frac{(\alpha^3 \delta_\tau t(\xi/t - \tilde{u}))^2}{2\left(\rho^{HLLE} + \alpha_2 \delta_m t(\xi/t - \tilde{u})\right)} \quad (13)$$

which has an extremum at $\tilde{u}$ equal to $p^{HLLE}/2$ and here again the limitations only done from its values at b^l and b^r which leads to the following inequality.

$$p^{HLLE}(p^g - p^{HLLE}) \leq -\mid \alpha_2 \mid \hat{\delta}_m(p^g - p^{HLLE}) \quad (14)$$

$$-\alpha_3^2 \mid \hat{\delta}_\tau^2 \leq -\mid \alpha_2 \mid \hat{\delta}_m(p^g - p^{HLLE}) - \alpha_3^2 \mid \hat{\delta}_\tau^2$$

By this way, linear inequalities issued from the limitation of the primitive variables are obtained. They are completed with the definition constraints $0 \leq \delta_m \leq 1$ and $0 \leq \delta_\tau \leq 1$ and all these conditions form a convex, C, in which (δ_m, δ_τ) must be taken to have a R-stable scheme.

4.2 Final choice of the antidiffusive coefficients

Among all the admissible pairs we select the one which minimizes the diffusion in the scheme. This could be done taking it in viscous form, and minimizing the eigenvalues related to linearly degenerate fields in the viscosity matrix, that is to say,

$$q^2 = \frac{b_+^r + b_-^l}{b_+^r - b_-^l}\overline{\lambda}^2 - 2(1-\delta_m)\frac{b_+^r b_-^l}{b_+^r - b_-^l} \quad q^3 = \frac{b_+^r + b_-^l}{b_+^r - b_-^l}\overline{\lambda}_3 - 2(1-\delta_\tau)\frac{b_+^r b_-^l}{b_+^r - b_-^l} \quad (15)$$

$$b_+ = \max(b, 0), b_- = \min(b, 0)$$

Finally the selected pair (δ_m, δ_τ) will be the one which minimizes in the convex C the function $J(\delta_m, \delta_\tau) = 2 - \delta_m - \delta_\tau$.

4.3 Comments on HLLEMR

4.3.1 Differences with HLLEM

Two points make HLLEMR very different from HLLEM. First, we have a completely local approach; one pair (δ_m, δ_τ) is computed by edge, depending on the regularity of the flow. Moreover, the coefficients have generally different values, because of the different role they play in the simulation. δ_m relies on the densities and the heat transfer coefficient while δ_τ limits the tangential velocities and so avoids, for example, the carbuncle.

4.3.2 Other applications of HLLEMR

This solver, here described for the two-dimensional Euler equations can be applied to other problems where the treatment of linearly degerate fields is essential. Indeed, such a method can be used to compute reactive flows: to insure the positivity of the mass fractions, one antidiffusive coefficient by chemical species could be introduced. Another example is given by the second order turbulent model involving the Reynold tensor, governed by a transport equation which must be kept positive definite. Computing a unique and appropriate antidiffusive coefficient for all tensor component guarantees the positivity.

In [6] another modification of HLLE scheme is given to keep the positively conservative property. In [7] another way to limit the tangential velocities and to insure the positivity is proposed, improving Roe or Osher schemes.

5 Numerical Experiments

HLLEMR has been implemented in a code, CHRONOS, which computes Euler and Navier-Stokes bidimensional equations using unstructured hybrid meshes. A second order extension of the solver has been developed based on a MUSCL approach.

The first examples demonstrate the robustness of the scheme. the carbuncle phenomenon has been caused computing an inviscid flow around an incident cylinder with a 25 freestream mach number, with Roe's scheme or HLLEM scheme. Fig.1 shows this anomaly and nearby we have the same flow computed with HLLEMR without any problems. The carbuncle is explained by a wrong calculation of the tangential velocities in the intermediate state given by the solver. Indeed a numerical study showed us how these values were very far from the initial ones issued from the considered Riemann problem. Moreover we have found a recirculation area in the carbuncle, due to wrong sign changes velocities. We note that limiting δ_τ is sufficient to remove this problem.

Similarly HLLEMR enables us to compute the inviscid flow over the double ellipsoid with no negative densities problems, which are often numerically created because of the strong rarefaction areas encountered in this flow. The amount of viscosity which has to be added to keep the densities always positive is intrinsically and locally calculated (fig.2).

We conclude this section with the presentation of Navier-Stokes calculations which substantiate the accuracy of the scheme. CHRONOS, with HLLEMR has computed the viscous hypersonic flow over a blunt body with a 28 × 92 coarsed grid. The same calculation has been done with a reference code using a 90 × 150 grid. The representations of the results given by the both codes are presented in (fig.3). The heat transfer coefficients and the skin friction coefficients near the wall are compared. We can note the good quality of these results.

6 Conclusion

We have proposed a new approximate Riemann solver, HLLEMR, particularly robust which could compute flows in some situations where classical solvers fail without additional viscosity. This solver is R-stable, a property which insures that the primitive variables in the intermediate states given by the solver satisfy a maximum principle according to the exact solution. This property explains its robustness for computing two dimensional hypersonic flows without loss of accuracy towards the classical schemes. this advantage may reveal many useful applications especially for problems involving linearly degenerate fields treatments (tridimensional flows, reactive flows, second order turbulence model).

References

1. J.J. Quirk, A Contribution to the Great Riemann Solver Debate, Icase Report 92-64, (1992)

2. A. Harten, P.D. Lax, B. Van Leer, On Upstream Differencing and Godunov Type Schemes for Hyperbolid Conservation Laws, Icase Report 82-5, (1982)

3. B. Einfeldt, C.D. Munz, P.L. Roe, B. Sjogreen, On Godunov Type Methods near Low Densities, Journal of Computational Physics, 92, 273-294, (1991)

4. B. Einfeldt, On Godunov Type Methds for Gaz Dynamics, SIAM Journal of Numerical Analysis, 25, 294-317. (1988)

5. P. Charrier, B. Dubroca, L. Flandrin, An Appriximate Riemann Solver for Hypersonic Bidimensional Flows, C.R. Acad. Sci. Paris. t. 317 serie 1, 1083-1086, (1993)

6. S. Obayashi, Y. Wada, Practical Formulation of a Positively Conservative Scheme, AIAA Journal, Vol.32, 1093-1095, (1994)

7. B. Larrouturou, How to Preserve the Mass Fractions Positivity when Computing Compressible Multi-Component Flow, Journal of Computational Physics, 95, 59-84, (1991)

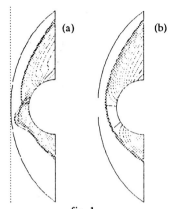

fig. 1
densities contours computed
(a) with Roe'scheme, (b) with HLLEMR

fig. 2
Densities contours over the double ellipsoid
computed with HLLEMR

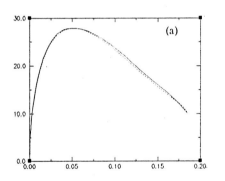
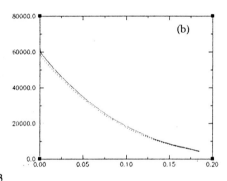

fig. 3
(a) Skin friction coefficients and (b) heat transfert coefficients near the wall
computed with CHRONOS and a reference code

An Implicit Lambda Method for 2-D Viscous Compressible Flows

Bernardo Fortunato[1] and Vinicio Magi[2]

[1]Istituto di Macchine ed
Energetica - Politecnico di Bari
Via Re David, 200 – 70125 – Bari, Italy
[2]Dipartimento di Meccanica dei Fluidi ed
Ingegneria Offshore Via E. Cuzzocrea,
48 – 89128 – Reggio Calabria, Italy

1 Introduction

The present paper provides a Fast Navier-Stokes Solver for the computations of compressible flows. This methodology is an extension of the classical Lambda formulation [1-3], and is based on the integration of the compatibility conditions along the bicharacteristic lines, thus reducing multi-dimensional flow problems to a sequence of simple quasi one-dimensional problems. The classical Lambda formulation is an accurate, non conservative, upwind technique which recasts the time dependent Euler equations in terms of generalized Riemann variables (four compatibility conditions) and integrates them by using upwind differences. The Lambda formulation has undergone several improvements. In particular Moretti [1] has introduced a methodology, named Fast Solver, which improves the computational efficiency of the explicit Lambda scheme; Dadone et al. [2] have extended the 2-D version of Fast Solver to handle 3-D, subsonic internal flows of simple geometry and Fortunato [3] has shown the accuracy of 3-D results in a real geometry (the Stanitz elbow). In the present work the 2-D compressible viscous governing equations are obtained as follows: the convective contributions are treated as the four compatibility conditions along the bicharacteristic lines, like in the inviscid formulation, and the diffusive contributions are treated as corrective terms. The present model has been applied to solve subsonic flows around the NACA 0012 airfoil. The numerical results have been compared with those obtained by several authors in [4], in order to confirm the accuracy and the reliability of the proposed algorithm.

2 Governing equations and numerical method

The non-dimensional continuity, momentum and energy equations for compressible, viscous flows are given in vector form as

$$\frac{a_t}{\delta} + \frac{\mathbf{q} \cdot \nabla a}{\delta} + a \nabla \cdot \mathbf{q} - a(S_t + \mathbf{q} \cdot \nabla S) = 0 \tag{1}$$

$$\mathbf{q}_t + \mathbf{q} \cdot \nabla \mathbf{q} + \frac{a}{\delta} \nabla a - \frac{a^2}{\gamma} \nabla S = \frac{\sqrt{\gamma} M_\infty}{\rho Re_\infty} \nabla \cdot \tau \tag{2}$$

$$S_t + \mathbf{q} \cdot \nabla S = -\frac{\sqrt{\gamma} M_\infty}{Re_\infty} \frac{\gamma^2}{(\gamma-1)\rho a^2 Pr_\infty} \nabla \cdot \Phi + \frac{\sqrt{\gamma} M_\infty}{Re_\infty} \frac{\gamma}{\rho a^2} \tau : \nabla \mathbf{q} \tag{3}$$

where a is the speed of sound, $\mathbf{q}$ the velocity vector, τ the stress tensor, ρ the density, M_∞, Re_∞ and Pr_∞ are Mach, Reynolds and Prandtl numbers respectively, Φ is the heat flux vector and the subscript t indicates time partial derivatives. Moreover $\delta = (\gamma - 1)/2$, where γ is the specific heat ratio and S is the entropy, defined as

$$S = \frac{\gamma}{\gamma - 1} \ln \frac{a^2}{\gamma} - \ln p \tag{4}$$

where p is the pressure. The scalar governing equations are then written in orthogonal curvilinear coordinates (ξ, η). Let us introduce the Riemann variables

$$\begin{aligned} R_1^x &= \frac{a}{\delta} + u & R_2^x &= \frac{a}{\delta} - u \\ R_1^y &= \frac{a}{\delta} + v & R_2^y &= \frac{a}{\delta} - v \end{aligned} \tag{5}$$

and the corresponding characteristic slopes

$$\begin{aligned} \lambda_1^x &= \frac{u+a}{h_1} & \lambda_2^x &= \frac{u-a}{h_1} & \lambda_3^x &= \frac{u}{h_1} \\ \lambda_1^y &= \frac{v+a}{h_2} & \lambda_2^y &= \frac{v-a}{h_2} & \lambda_3^y &= \frac{v}{h_2} \end{aligned} \tag{6}$$

where u and v are the velocity components along the coordinate lines and h_1 and h_2 the metric scale factors. Adding and subtracting the two components of the momentum equations (2) to the continuity equation (1), the following compatibility conditions along the four bicharacteristic lines can be obtained:

$$\begin{aligned} R_{1_t}^x &= aS_t + f_1^x + f_{31}^y + f_1^L - D + V_1 \\ R_{2_t}^x &= aS_t + f_2^x + f_{32}^y + f_2^L - D - V_1 \\ R_{1_t}^y &= aS_t + f_1^y + f_{31}^x + f_3^L - E + V_2 \\ R_{2_t}^y &= aS_t + f_2^y + f_{32}^x + f_4^L - E - V_2 \end{aligned} \tag{7}$$

and the entropy equation can be written as

$$S_t = f_3^x + f_3^y + V_3 + V_4 \tag{8}$$

In the equations (7) and (8) f_i^x and f_i^y represent the convective contributions written in terms of Riemann variable derivatives, entropy derivatives and the corresponding characteristic slopes. The f_i^L terms depend on the curvilinear coordinate system, and the V_i terms derives from the viscous contributions. Moreover D and E are given by

$$D = \frac{a}{h_2} \frac{\partial v}{\partial \eta} \qquad E = \frac{a}{h_1} \frac{\partial u}{\partial \xi} \qquad (9)$$

The Fast Solver methodology integrates implicitly equations (7) and (8) as uncoupled equations so that a multidimensional problem is reduced to a sequence of quasi 1-D problems. Moreover the four equations (7) define only three physical variables, so that a redundancy may occur; indeed the first two equations define u and a, the third and forth define v and a once more.

Hence a is defined twice in different ways. In order to avoid the redundancy, the terms D and E must be computed such that the following identity is satisfied exacly [3]:

$$R^x_{1_t} + R^x_{2_t} = R^y_{1_t} + R^y_{2_t} \qquad (10)$$

The implicit numerical technique employed is first order in time and second order in space. The resulting linearized equations, written in an incremental delta-form, are discretized in space in the spirit of the Lambda formulation: all derivatives of the old time level terms are approximated by three-point second order windward differences, and the derivatives of the unknown delta variables are approximated by first order windward differences. The viscous terms are approximated by second order central differences. The entropy equation and the first two Riemann variables are integrated by sweeping along the streamwise direction, and the other two Riemann variables are integrated by sweeping along the transversal direction. As far as the boundary conditions are concerned, the pressure is prescribed at the outlet, the total speed of sound and the direction of the velocity vector are assigned at the inlet. At the wall, the temperature is assigned equal to the free stream total one and the no-slip condition is used for the velocity.

3 Results

The present model has been applied to the solution of the well-known test cases presented at the Gamm Workshop [4]. In order to show the reliability and accuracy of the present model three significant cases have been considered. The first test case refers to the flow over a NACA 0012 airfoil with a free stream Mach number $M_\infty = 0.85$, with an angle of attack $\alpha = 0°$ and a Reynolds number $Re_\infty = 500$. The mesh used is a body fitted C-mesh obtained by using the Theodorsen conformal mapping transformation. The computational domain is discretized by 256x64 meshes; an exponential stretching is used in the transversal coordinate such that the first meshline is located at 0.0009 chord lenght away from the wall. The number of points on the airfoil is 120 and the farfield and the outflow boundaries are located at a distance of 15 chord lengths away from the body. The Mach number contours, the pressure coefficient c_p and the skin friction coefficient c_f distributions are reported respectively in Figs. 1,2, and 3. The second test case refers to the flow over the same airfoil with a free stream Mach number $M_\infty = 0.8$, with an angle of attack $\alpha = 10°$

and a Reynolds number $Re_\infty = 500$. The Mach number contours and c_p and c_f distributions are reported respectively in Figs. 4,5, and 6. Finally, Figs. 7,8 and 9 show the results for the flow around the same airfoil with $M_\infty = 0.8$, $\alpha = 10°$ and $Re_\infty = 73$. In all cases a very good agreement with [4] can be observed, confirming the accuracy and the reliability of the proposed methodology.

Acknowledgements :

The present paper has been supported by MURST and CNR, Italy.

References

1. Moretti, G., "A Fast Euler Solver for Steady Flows," AIAA 6th CFD Conference, Danvers, MA, 1983.

2. Dadone, A., Fortunato, B. and Lippolis, A., "A Fast Euler Solver for Two and Three Dimensional Internal Flows," Computers and Fluids, Vol. 17, pp. 25-37, 1989.

3. Fortunato, B. "Three Dimensional Rotational Compressible Flow Computations by means of a Fast Euler Solver," AIAA paper 91-2471, 27th Joint Propulsion Conf., June 1991.

4. Proceedings of the GAMM-Workshop on Numerical Simulation of Compressible Navier-Stokes Flows, INRIA, Sophia-Antipolis, Notes on Numerical Fluid Mechanics, Vol. 18, Vieweg-Verlag, 1986.

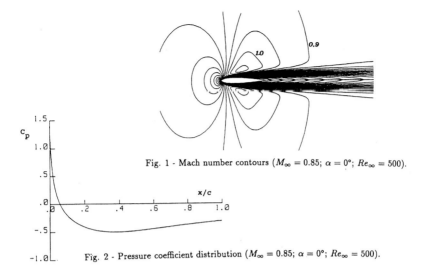

Fig. 1 - Mach number contours ($M_\infty = 0.85$; $\alpha = 0°$; $Re_\infty = 500$).

Fig. 2 - Pressure coefficient distribution ($M_\infty = 0.85$; $\alpha = 0°$; $Re_\infty = 500$).

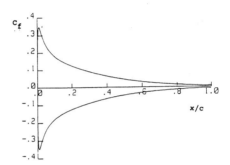

Fig. 3 - Skin friction coefficient distribution ($M_\infty = 0.85$; $\alpha = 0°$; $Re_\infty = 500$).

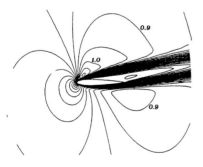

Fig.4 - Mach number contours ($M_\infty = 0.8$; $\alpha = 10°$; $Re_\infty = 500$).

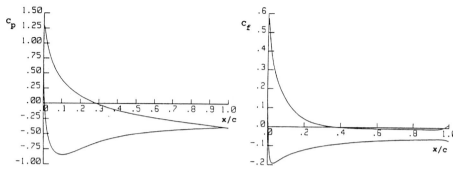

Fig.5 - Pressure coefficient distribution ($M_\infty = 0.8$; $\alpha = 10°$; $Re_\infty = 500$).

Fig.6 - Skin friction coefficient distribution ($M_\infty = 0.8$; $\alpha = 10°$; $Re_\infty = 500$).

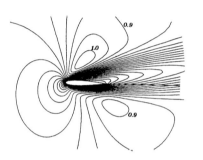

Fig.7 - Mach number contours ($M_\infty = 0.8$; $\alpha = 10°$; $Re_\infty = 73$).

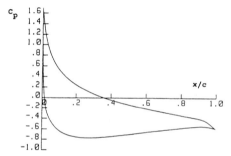

Fig.8 - Pressure coefficient distribution ($M_\infty = 0.8$; $\alpha = 10°$; $Re_\infty = 73$).

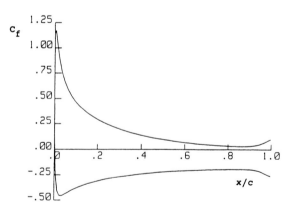

Fig.9 - Skin friction coefficient distribution ($M_\infty = 0.8$; $\alpha = 10°$; $Re_\infty = 73$).

Numerical Methods for Simulating Supersonic Combustion

J. Groenner, E. von Lavante, M.Hilgenstock

University of Essen, Lehrstuhl fuer Stroemungsmaschinen,
Essen, Germany

Abstract : An efficient computer program was developed for the computation of supersonic combustion problems. Several test cases showed the capabilities of implementations of different upwind schemes in calculating chemical reactions. The FAS multigrid procedure accelerated in some cases the convergence to steady state. The numerical results were validated by the corresponding experiments for the more demanding cases. A reasonable qualitative as well as quantitative agreement was achieved.

1 Introduction

The research and development of several hypersonic flight vehicles in the USA and Europe has brought many new computational methods for the prediction of hypersonic flow fields. The consideration of new physical phenomena in this regime, like, for example, chemical reactions and vibrational excitation, leads to the solutions of inhomogenous Navier-Stokes equations with a source term and additional equation of mass conservation for each chemical species. The difficulties in numerically solving the resulting system of governing equations arise from the extreme stiffness of the equation system, due to the very short characteristic time scales associated with the chemistry, and the uncertainties about the material, physical and chemical properties of the participating species.

While the main attention was paid to the analysis of hypersonic flows, the problem of simulation of supersonic flows in typical combustion chambers of high speed flight vehicles was treated less extensively. The main goal of the present research was to investigate the possibilities of efficient simulation of complex flows in typical combustion chambers of propulsion units for high supersonic and hypersonic flight regimes. One of the goals of the present work was to address various numerical issues using a few well published test cases. From the many possibilities, recently published in open literature, following problems were selected:

- shock induced ignition in a supersonic diffusor (Bussing and Murmannn [1])

- transverse hydrogen injection in a supersonic airstream

- 2-D scramjet model (Wada et al [10])

The second test case has also been extensively experimentally investigated by the ISL (German-French Research Institute, Saint Louis), allowing a direct comparison between the experimental data and the numerical results. Three different schemes for the spatial discretization were worked out and tested. Several numerical treatments of the source terms, arising due to the chemical species production, were investigated. The different possibilities of the temporal discretization, leading to stable numerical algorithms, were compared as well, and will be discussed.

2 Algorithm

In this work, the flow was assumed to be compressible, viscous, and a mixture of thermally perfect species. Due to the relatively low temperature and high pressure in the present configurations, the gas mixture can be treated as in vibrational equilibrium. The governing equations were in this case the compressible Navier-Stokes equations for n_s species:

$$\frac{\partial \hat{Q}}{\partial t} + \frac{\partial \hat{F}}{\partial \xi} + \frac{\partial \hat{G}}{\partial \eta} = \frac{S}{J} \qquad (1)$$

where $\hat{F}$ and $\hat{G}$ are the flux vectors in the corresponding ξ and η directions, Q is the vector of the dependent variables and J is the jacobian of the transformation of coordinates. S is the vector of the chemical source terms. The details of the governing equations are given in [11].

A simple model according to the Fick's law for the binary diffusion coefficient was used, along with the Sutherland equation for the viscous coefficient. The chemical reactions were realized with an 8-reaction model of Evans and Schexnayder [2], for the H_2-air combustion.

Three different upwind methods were extended to chemical reactions systems. These methods are based on the work of Roe (flux-difference splitting) [8], van Leer (flux-vector-splitting) [5] and Liou (AUSM Advection-Upstream-Splitting-Method) [6].

2.1 Roe Scheme

The present research originally started with a numerical scheme based on Roe's Flux Difference Splitting in finite volume form. In previous numerical predictions, this scheme was highly effective in providing accurate viscous results at a wide range of Mach numbers. In the present version, the reconstruction of the cell-centered variables to the cell-interface locations was done using a monotone interpolation as introduced by Grossmann and Cinella in [3]. The interpolation slope was limited by an appropriate limiter, according to the previously published MUSCL type procedure (see, for example, [11]).

2.2 Liou Scheme

The AUSM is relatively new. It separates the corresponding flux into three parts, the convective part, the pressure part and the viscous part. As usual, the viscous part is evaluated with central differences, where as the other parts are upwind differenced. The damping behaviour is proportional to the signal speed, in our case the Mach number in normal direction to the cell interface, and, therefore, the numerical damping decreases in the case of vanishing signal velocity. This behaviour causes pressure oscillations, especially in shear layers. In order to improve the damping characteristics of this scheme in the case of vanishing signal velocities, von Lavante and Yao [7] have proposed a modification of the original Liou scheme. In this modified form the signal velocity is the product of the Mach number and the density at the cell interface. The damping of the continuity equation is now equal to the van Leer scheme. The details of this scheme, with the corresponding modifications, are given by Hilgenstock et. al. [12]. This scheme is automatically positivity preserving in the sence of Larrouturou [4].

2.3 van Leer Flux Vector Splitting

The split fluxes are constructed according to the formulation given by Shuen in [9], and will not be repeated here. The fluxes are constructed from the variables Q_L and Q_R, extrapolated from the left or right of the cell interface, depending on the sign of the split fluxes $F^{\pm}$, using the same MUSCL interpolation as in the Roe Scheme.

2.4 Temporal Integration

The governing equations were integrated by a semi-implicit method, with different multi-stage Runge-Kutta type schemes used for the explicit operator of the fluid-dynamics part. Following an idea of Bussing and Murmann [1], only the chemical source terms were treated implicitely,

$$\left\{ I - \Delta t \, \Theta \, \frac{\partial S^n}{\partial Q^n} \right\} \frac{\partial Q^n}{\partial t \, J} = \hat{S}^n - \frac{\partial \hat{F}^n}{\partial \xi} - \frac{\partial \hat{G}^n}{\partial \eta} \qquad (2)$$

with the relaxation parameter Θ. In most of the present computations, a two-stage Runge-Kutta procedure with $\Theta = 1$ seemed to be the best choice.

The numerical effort to invert the Matrix $B = I - \Delta t \, \Theta \, \frac{\partial S^n}{\partial Q^n}$ depends on the formulation of the jacobian of the chemical source terms. Several different forms of the jacobian matrix, with increasing complexity and accuracy, were implemented and compared. The most obvious choice is to invert the full $n_s \times n_s$ matrix B. This, however, is a problem from the numerical point of view, since the inversion is CPU time consuming, and the matrix B usually illconditioned. This approach worked, but was rather inefficient. The next possibility to simplify the matrix B consists of dropping all the off diagonal terms, while keeping only the diagonal terms. In our case of eight reactions with seven species, this turned out to be an effective means of accelerating the convergence, with stability limits given by the acoustic wave speeds.

Using a multi-block grid structure resulted in a flexible code with the possibilty of working with different chemical models (nonequilibrium, equilibrium, frozen) in different blocks. Besides, some of the blocks were selectively refined, depending on the evolving results. The present geometrical treatment of the computational domain was simple, yet flexible enough.

3 Multigrid Acceleration

The still relatively slow rate of convergence was accelerated using the standard FAS multigrid procedure. The residuals and the chemical production terms were restricted using simple summation; in the restriction of the dependent variables Q, a volume weighted averaging was employed. An bilinear interpolation was used in the prolongation. A detailed description of the FAS scheme can be found in [11]. A comparison of the rates of convergence with and without the MG is presented in Fig.3 for steady flow simulations in a supersonic diffusor.

4 Results

4.1 Supersonic Diffusor

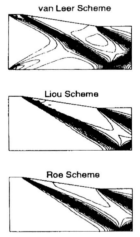

Figure 1: Pressure contours in supersonic diffusor

In this configuration, the supersonic flow of a premixed hydrogen-air mixture enters the diffusor at a Mach number $M = 2.5$, and a temperature of $T = 900K$. The equivalence ratio was set to $\Phi = 0.1$. The pressure contours for the three different schemes are shown in Fig. 1. Clearly, the best results were obtainded for the Roe flux-difference-splitting, although the AUSM scheme of Liou gave similar results. The only difference are slight oscillations at the shock. The scheme of van Leer is the most dissipative one and gives thicker boundary layers, causing significant leading edge shocks. These shocks lead to unphysical solutions for the chemical reaction calculations, Fig. 2. The

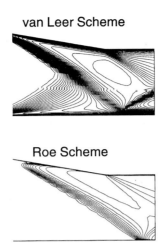

Figure 2: H_2O contours in supersonic diffusor

resulting H_2O contours, given by the Roe- and Liou-schemes, are practically identical, so that only the results from the Roe scheme are shown in Fig. 2. They indicate that the combustible mixture is ignited by the oblique shock. The convergence histories for this configuration are given in Fig. 3, including the cases of explicit and implicit multigrid computations.

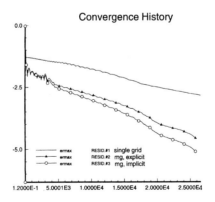

Figure 3: Convergence histories with and without MG, diffusor

4.2 Transverse H_2 jet in a Supersonic Airstream

The geometry and the boundary conditions of this configuration are decribed in detail by von Lavante et. al. [11] . Here, a hydrogen jet is injected into a two-dimensional channel with parallel walls. At the inflow, the Mach number was $M = 4.0$, the pressure was $p = 0.1 MPa$ and the temperature was $T = 1000 K$. The hydrogen jet enters at sonic conditions through a slot of 1.3 mm width; its temperature was $T = 600 K$. The Reynolds number was in this case $Re_h = 9 \times 10^6$,based on the channel height.The

Figure 4: Pressure contours, transverse H_2 Jet, Roe scheme

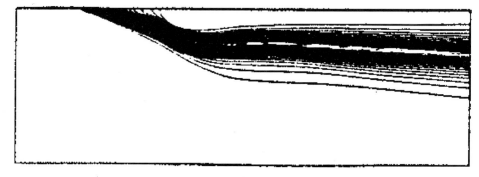

Figure 5: H_2O contours, transverse H_2 jet, Roe scheme

computational grid of 131 × 131 points allowed a reasonable resolution of the important regions. This case is of particular interest, since it has been frequently used in numerical simulations by several authors and was (and is still being) extensively experimentally investigated.

As the numerical experiments, discussed above, indicated, the van Leer's FVS scheme was too dissipative for accurate predictions of the chemical reactions in the regions where the viscous effects were dominant, so that only the Roe scheme was applied in this test case. The position of the H_2 jet is also visible in Fig. 7, where the pressure contours are shown. Clearly visible are the weak leading edge shock, the strong separation shock, the bow shock, the barrel shock and the recompression shock. The separation shock and the bow shock merge and reflect at the lower wall. This reflected shock then reflects again off the shear layer between the air and the product gases at the upper wall. The H_2O contours in Fig. 8 show the position of the reaction zone. In accordance with the experimental measurements, published in [11], the H_2 is carried upstream of the injection opening by recirculating fluid in the boundary layer. Downstream of the region where most of the H_2O production occures, the flow is basically chemically frozen, with H_2O being convected. A small part of the produced water is convected into the boundary layer ahead of the jet by the recirculation, present at this location. The velocity vectors are shown in detail in Fig. 9.

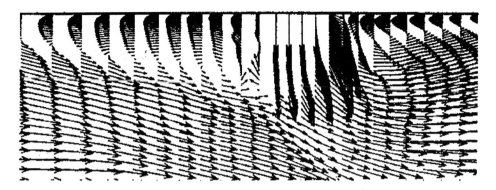

Figure 6: Velocity vectors in detail, transverse H_2 jet, Roe scheme

The flow at the shear layer-shock interaction was in this case weakly unsteady, with the shedding of small vortices. These are consequently convected downstream. The computed velocities fall well within the limits of the experimentally determined values.

4.3 Two Dimensional Scramjet

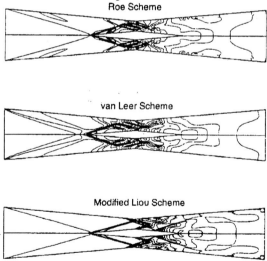

Figure 7: Pressure contours, scramjet

The flow in a two-dimensional model scramjet configuration, shown by Wada et al [10], is presented last. The inflow Mach number was 4, and the Reynolds number based on the maximum height was $Re_h = 3 \; 10^6$. The resulting pressure contours for the three schemes are given in Fig. 10. The concentrations of H_2O can be seen in Fig. 11. These results support the conclusions drawn in the previous sections.

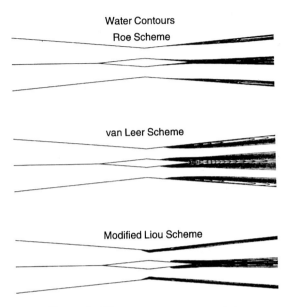

Figure 8: Water contours, scramjet

5 Conclusions

In the present work, aimed at finding a simple numerical method for simulations of compressible, viscous flows in chemical nonequilibrium, three different spatial discretization schemes were tested on a few supersonic configurations with increasing complexity. The van Leer FVS scheme, known to be very dissipative, was too inaccurate to be of practical use. The prolific Roe scheme gave consistently good results on adequate computational grids, but displayed a distinct lack of robustness. The relatively untested AUSM scheme, with the proper modifications, gave some promising results, while being simple and comparatively reliable. The CPU time savings due to its simplicity are, however, negligible compared with the effort spent on the modelling of the other real gas effects.

The simplest possibility of removing the exponential stiffness of the governing equations by implicit formulation was the most effective one. The diagonal form of the jacobian matrix $\frac{\partial S}{\partial Q}$ worked well, making CFL numbers, based on the acoustic wave speeds, of $O(1)$ possible.

The FAS multigrid procedure effectively accelerated the rate of convergence in the steady case, but was not suited for unsteady computations in the present form.

References

[1] Bussing, T.R.A. and Murmann, E.M., " Finite Volume Method for the Calcultion of Compressible Chemically Reacting Flows ", AIAA Paper 85-0331, Jan. 1985

[2] Evans, J.S., Schexnayder, C.J., " Influence of Chemical Kinetics and Unmixedness on Burning in Supersonic Hydrogen Flames", AIAA-Journal, Febr. 1980, pp. 188-193

[3] Grossmann, B., and Cinella, P., " Flux-Split Algorithms for Flows with Nonequilibrium Chemistry and Vibrational Relaxation ", J. Comp. Phys., vol. 88, pp. 131-168, 1990

[4] Larrouturou, B., and Fezoui, L. " On the Equations of Multi-Component Perfect or Real Gas Inviscid Flow ", Nonlinear Hyperbolic Problems, Lecture Notes in Mathematics, 1402, Springer Verlag, Heidelberg 1989

[5] van Leer, B., " Flux-Vector-Splitting for the Euler Equations ", Institue for Computer Applications in Science and Engineering, Hampton, VA, Rept. 82-30, Sept. 1982

[6] Liou, M.S., " On a New Class of Flux Splitting ",Proceedings , 13th International Conference on Numerical Methods in Fluid Dynamics, Rome 1992, pp. 115-119

[7] von Lavante, E. and Yao, J. " Simulation of Flow in Exhaust Manifold of an Reciprocating Engine ", AIAA-93-2954

[8] Roe, P.L., Pike, J., " Efficient Construction and Utilisation of Approximate Riemann Solutions ", Computing Methods in Applied Sciences and Engineering, VI, pp. 499-516, INRIA, 1984

[9] Shuen, Jiang-Shun, " Upwind Differencing and LU Factorization for Chemical Non-equilibrium Navier-Stokes Equations ", Journal of Comp. Phys., vol. 99, pp. 213-250, 1992

[10] Wada, Y. Ogawa, S. and Ishiguro, T., " A Generalized Roe's Approximate Riemann Solver for Chemically Reacting Flows ", AIAA 89-0202

[11] von Lavante, E., Hilgenstock, M. and Groenner, J., " Simple Numerical Method for Simulating Supersonic Combustion ", AIAA Paper 94-3179

[12] Hilgenstock, M., von Lavante, E. and Groenner, J.," Efficient Computations of Navier-Stokes Equations with Nonequilibrium Chemistry ", ASME Paper 94-GT-251

SHAPE OPTIMIZATION GOVERNED BY THE QUASI 1D EULER EQUATIONS USING AN ADJOINT METHOD

Angelo Iollo[1,2], Manuel D. Salas[3], Shlomo Ta'asan[2,4]

[1] Dip. Ingegneria Aeronautica e Spaziale - Politecnico di Torino; 10129 Torino, Italy
[2] Institute for Computer Applications in Science and Engineering;
23681 Hampton, Va USA
[3] NASA Langley Research Center; 23681 Hampton, Va USA
[4] The Weizman Institute of Science; 76100 Rehovot, Israel

Abstract : In this paper we discuss a numerical approach for the treatment of optimal shape problems governed by the quasi one-dimensional Euler equations. In particular, we focus on flows with embedded shocks. We introduce a cost function and a set of Lagrange multipliers to achieve the minimum. Some results are given to illustrate the effectiveness of the method.

1 Introduction

The physical structure of the complex flows that occur in aerodynamic design can be predicted by reliable numerical simulations. If one can predict performance, it is fundamental to know which modification of an aerodynamic configuration improves performance. A broad category of techniques to design aerodynamically better aircraft is known as inverse design methods. The inverse design method, pioneered by Lighthill [5], requires *a priori* knowledge of a desirable pressure or velocity distribution. The *quality* of the shape obtained strongly depends on this choice. Therefore a weakness of this approach is its dependence on the experience of the designer. In addition the existence of the solution is not assured for a general pressure or velocity distribution. An original example of such an approach is found in [3].

The numerical approach that we will use in this paper, lift the dependence on heuristic choices of the desirable distribution, allowing the imposition of constraints to be satisfied by the solution found. The numerical simulation of the flow and a numerical optimization code are coupled. The optimization code calculates the best perturbation to the geometry to decrease a cost function. The geometry itself is described by a set of shape coefficients.

The optimization code can be devised in one of several ways. A common approach is to perturb one shape coefficient at a time and compute the derivative of the cost function with respect to this coefficient as a finite difference. Although such codes are simple to devise, the procedure is costly and can introduce large errors. In a further evolution of this approach, an equation is first calculated for the derivative of the cost function with respect to the shape coefficient and then solved numerically. An equation must be solved for each shape coefficient.

The approach presented in this paper is a classical optimal control method. In [2] this technique is applied to design by conformal mapping. In [7], this method has been applied for the design of an airfoil in a subsonic potential flow.

Here, we consider a flow with embedded shocks where the governing equations are the Euler equations. We show how to derive an analytical expression of the cost function derivatives with respect to the shape coefficients. For this purpose, we introduce costate variables (Lagrange multiplier) and solve only one set of costate equations. In the present approach an optimal shape can be found for problems with embedded shocks, without additional complications. A careful examination of the structure of the costate equations suggests a method for integrating them with a robust algorithm developed for fluid dynamics purposes.

2 Problem Statement

Let the extension of the Laval nozzle in physical space be $\Omega = [0, l]$. An energylike functional denoted by $\mathcal{E}$, is the cost function we want to minimize. An optimal shape of the nozzle is reached when we meet the necessary conditions for a minimum of $\mathcal{E}$. Let

$$\mathcal{E} = \frac{1}{2} \int_0^l (p - p^*)^2 dx$$

where p^* is a target pressure distribution along the abscissa x and p is the pressure field for the present geometry.

In the case of a quasi-one-dimensional flow, if we denote ρ = density, u = velocity, e = specific total energy, p = pressure, a = speed of sound, h = height of the channel, γ = specific heat ratio, $\kappa = \frac{\gamma-1}{2}$, then the Euler equations for unsteady flows reduce to

$$\mathbf{U}_t + \mathbf{F}_x + \mathbf{Q} = 0 \tag{1}$$

where

$$\mathbf{U} = \begin{pmatrix} \rho \\ \rho u \\ \rho e \end{pmatrix} \quad \mathbf{F} = \begin{pmatrix} \rho u \\ p + \rho u^2 \\ u(\rho e + p) \end{pmatrix} \quad \mathbf{Q} = \begin{pmatrix} \rho u \beta \\ \rho u^2 \beta \\ u(\rho e + p)\beta \end{pmatrix}$$

$\beta = h_x/h$, and $p = \kappa \rho(2e - u^2)$. In the following derivation, we use the homogeneity property

$$\mathbf{F} = \frac{\partial \mathbf{F}}{\partial \mathbf{U}} \mathbf{U} = \mathbf{A}(\mathbf{U})\mathbf{U} \tag{2}$$

The source term $\mathbf{Q}$ can be written to display its dependence on $\mathbf{U}$; in fact, the multiplication can be carried out to show that

$$\mathbf{Q} = \frac{\partial \mathbf{Q}}{\partial \mathbf{U}} \mathbf{U} = \mathbf{S}(\mathbf{U})\mathbf{U} \tag{3}$$

We refer to the solution of the above equations as the analysis problem. We will consider the inlet flow with a constant total pressure and entropy, i.e., $p_{in}^o = p_{in}(1 + \kappa M_{in}^2)^{\gamma/\gamma-1} = $ Constant, $p_{in}/\rho_{in}^\gamma = $ Constant. At the outlet, if the flow is subsonic, the static pressure is fixed as $p_{out} = $ Constant.

3 Adjoint Formulation

The design problem can be thought of as a search for a minimum of a functional under constraints. Let

$$\mathcal{L}(\alpha_i, \mathbf{\Lambda}, \mathbf{U}) = \mathcal{E} + \int_\Omega \mathbf{\Lambda}^T (\mathbf{F}_x + \mathbf{Q}) dx \qquad (4)$$

where $\mathbf{\Lambda}^T$ is an arbitrary vector with components $(\lambda_1(x), \lambda_2(x), \lambda_3(x))$, Ω is the domain $[0, l]$, and α_i are shape coefficients that define the geometry of the nozzle, for example by $h(\alpha_i, x) = \sum_i \alpha_i f_i(x)$ with $f_i(x)$ a generic function of x. If we increase the shape coefficients by $\varepsilon \widetilde{\alpha_i}$, the multipliers by $\varepsilon \mathbf{\Lambda}$ and the flow variables by $\varepsilon \mathbf{U}$, then the latter functional will change by an amount, say, $\varepsilon \delta \mathcal{L}$. As proposed in [1], taking $\mathbf{\Lambda}$ such that

$$-\mathbf{A}^T \mathbf{\Lambda}_x + \mathbf{S}^T \mathbf{\Lambda} + \frac{\partial p}{\partial \mathbf{U}} (p - p^*) = 0 \qquad (5)$$

we have

$$\frac{\partial \mathcal{E}}{\partial \alpha_i} = \frac{\partial \mathcal{L}}{\partial \alpha_i} = \int_0^l \frac{c_i \mathbf{\Lambda}^T \mathbf{S} \mathbf{U}}{\beta} dx \qquad (6)$$

where $c_i(x) = (h f_{i_x} - h_x f_i)/h^2$. Suppose that the flow field is known, such that all the variables dependent upon $\mathbf{U}$ are fixed. If we solve eq.(5) with the appropriate boundary conditions for $\mathbf{\Lambda}$ and substitute the results in eq.(6), then we obtain a formulation for the gradient of the Lagrangian. To actually determine a solution of this system, we embed eq.(5) in time as

$$\mathbf{\Lambda}_t - \mathbf{A}^T \mathbf{\Lambda}_x + \mathbf{S}^T \mathbf{\Lambda} + \frac{\partial p}{\partial \mathbf{U}} (p - p^*) = 0 \qquad (7)$$

Until now, we have limited our investigation to shockless nozzles to avoid certain difficulties that we will discuss here. One problem is that eq.(1) and, therefore, eq.(7) are not defined at the shock. This problem is overcome by extending the solution space of $\mathbf{U}(x)$ to a set of generalized functions, such that eq.(1) will reduce to the Rankine-Hugoniot jumps at the shock. A more subtle shortcoming is presented in [1]. The characteristics pattern of the costate equation shows the necessity of some boundary conditions on both sides of the the flow-field discontinuity to ensure the existence of a steady-state solution. A suitable choice for the Lagrange multiplier is to take $\mathbf{\Lambda} = 0$ on both sides of the shock such that $\mathbf{\Lambda}$ is continuous. This selection frees us from imposing a condition on $\widetilde{\mathbf{U}}$.

4 Numerical Approach

The flow-field solution is obtained by introducing a discrete grid $(x_n, t_k) = (x_0 + n\Delta x, t_0 + \sum_k \Delta t_k)$, where Δx is constant and Δt_k changes to satisfy the CFL condition. The conservative variables $\mathbf{U}(x)$ are computed at the cell centers and integrated in time. In this implementation, we interpolate $\mathbf{U}(x)$ to the cell faces by using characteristic differences and a minmod limiter. The flux derivative in eq.(1) is then computed using an approximate Riemann solver. See [4]. Away from discontinuities, the scheme is second order accurate.

Depending on the case considered, the solutions of the costate eq.(5), are sought as steady results of eq.(7) with proper boundary conditions. In the present formulation we put in evidence the signals that propagate along the characteristics; therefore finite-differences are one-sided depending on the sign of the corresponding propagation speed. Note that for this equation it would be impossible to use a conservative scheme since no conservation law exists to satisfy. The integration in time is made by explicit time stepping. The scheme is first-order accurate.

In the results that follow, we used a representation of the nozzle geometry defined by $h = \alpha_1 X + \alpha_2/X + \alpha_3$, where $X = x + 10^{-3}$; this representation allows two independent design variables because $\beta = h_x/h$. The strategy used to achieve the minimum of the functional is straightforward:

1. Start with a first guess for the shape coefficients.

2. Solve the flow equation.

3. Solve the costate equation with the values computed in step 2 for the flow field.

4. Update the shape coefficient with a gradient-based criterion.

5. Restart the procedure from step 2 until the gradient is zero.

To update the shape coefficients a BFGS algorithm was used. See [6].

5 Discussion of the Results and Conclusions

We computed the functional $\mathcal{E}$ by an analytical solution of eq.(1) and integrated by trapezoidal approximation. The discrete functional, see fig. 1, shows some discontinuities and a local minimum that disappears as the number of grid points increases. As the mesh is refined, the number of the discontinuities increases, while the jumps become smaller.

A method that derives a formulation for the gradient of $\mathcal{E}$ from a discrete approximation of the functional will obtain meaningless solutions as a result of the discontinuities of the discrete functional, such that no optimization algorithm alone could anyway get to the minimum. In the present formulation, an approximate representation of the analytical gradient of the functional is derived. For this reason, the approximation of the analytical gradient will be, at most, affected by discontinuities.

In fig. 2 we present a model problem in which the target pressure distribution can be exactly recovered by this procedure. The target pressure distribution is obtained starting from a subsonic first guess. This result shows the effectiveness of the method.

In conclusion, a method has been presented to calculate the gradient components of a generic functional, in which (regardless of the number of the shape coefficients) only one linear costate equation must be solved. The minimum computed in this way differs from the minimum of the discrete functional; however these minima indefinitely approach as the grid is refined.

References

[1] Iollo A. and Salas M.D. and Ta'asan S., "Shape Optimization Governed by the Euler Equations Using an Adjoint Method", ICASE, 93-78, 1993

[2] Jameson A.,"Aerodynamic Design via Control Theory", ICASE, 88-64, 1988

[3] Zannetti L.,"A Natural Forulation for the Solution of Two-Dimensional or Axisymmetric Inverse Problems", Int. J. Numerical Methods in Engineering, vol. 22, pp. 451-463, 1986

[4] Pandolfi M., "On the flux-difference splitting formulation", vol. 26, Notes on Numerical Fluid Mechanics, Vieweg Verlag, 1989

[5] Lighthill M.J., "A New Method of Two Dimensional Aerodynamic Design", ARC Rand M 2112, 1945

[6] Fletcher R., "Practical Methods of Optimization", John Wiley & Sons, vol.I, 1980

[7] Ta'asan S. and Kuruvila G. and Salas M.D.,"Aerodynamic Design and Optimization in One Shot", 30th Aerospace Sciences Meeting and Exhibit, AIAA 92-005, Jan., 1992

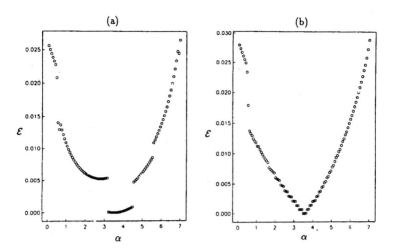

Figure 1: Analytical solution calculated with shape function $h(x) = \alpha(x^3 - x^2) + 1.05$; functional has been calculated with trapezoidal approximation. (a) Solution distributed over 20 grid points. (b) Solution distributed over 80 grid points.

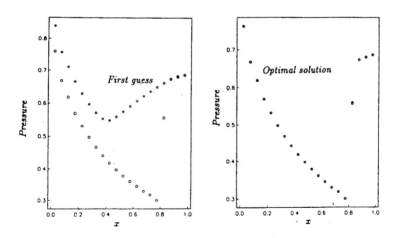

Figure 2: Target solution is $h(x) = 2.5X + 0.3/X + 5$. First guess is $h(x) = 2X + 0.52/X + 5$. Shape function for which minimum is sought is $h(x) = \alpha_1 X + \alpha_2/X + 5$, and optimization algorithm used is BFGS. Starting value of the functional is of order 10^{-3}. At minimum it is of order 10^{-9}. Gradient components are of order 10^{-12} at minimum.

Relaxation Method for 3-D Problems in Supersonic Aerodynamics

Vladimir Karamyshev, Victor Kovenya, Sergei Cherny

Institute of Computational Technologies SD RAS,
Prospect Lavrentyeva 6, Novosibirsk, 630090, Russia

Abstract : A version of global iterations method is considered for the numerical solution of stationary 3-D problems of supersonic aerodynamics with the accelerated convergence of internal iteration processes in marching cross-sections by the least square method.

1 Introduction

Many stationary flows of viscous gas with insignificant gradient changes of gas dynamic parameters along the predominant flow direction can be sufficiently accurately studied in the approximation of the simplified Navier-Stokes equations whose viscous terms include second derivatives only with respect to coordinates lateral to the main flow. A resource-sparing method of global iterations is suggested for the numerical solution of these equations in [1]. Via it, numerical modeling of a wide class of three-dimensional supersonic flows is possible in the presence of subsonic areas and in the absence of local areas of reversed flows.

The implementation of the method requires organizing internal iteration processes for obtaining the solutions of systems of non-linear difference equations in the cross-sections lateral to the flows. Note that as a rule the solutions have singularities such as the presence of boundary and internal layers (narrow regions of abrupt changes of the functions gradients), and traditional stationary methods on non-uniform grids have too low rate of convergence, therefore the question of iterations acceleration arises.

The present paper deals with the possibility to accelerate convergence of internal iterational processes by means of algorithm [2] constructed by the least square method for the linear iterations acceleration. Convergence acceleration in [2] is achieved through the construction of the error vector projection on the subspace of vector residuals obtained on the previous iterations and its correction by means of the current approximation. Unlike the common algorithms acceleration method [2] does not require a priori information on the limits of the matrices range, does not include operations on matrices and makes use of discrepancies vectors only which makes it readily applicable to already programmed iteration schemes. In [3] as examples of numerical solution of two-dimensional stationary problems of aerodynamics by relaxation methods, algorithm [2] was shown efficiently to suppress residual also in non-linear iteration processes.

2 Basic Equations and Numerical Method

Let us consider stationary Navier–Stokes equations in a generalized coordinate system (q^1, q^2, q^3) and truncate in their viscous terms repeated and combined derivatives with respect to q^1. As a result we shall obtain simplified equations which we shall write in a convenient form

$$A^1 \frac{\partial \vec{f}}{\partial q^1} = -A^2 \frac{\partial \vec{f}}{\partial q^2} - A^3 \frac{\partial \vec{f}}{\partial q^3} - D\vec{f} + \vec{F}, \qquad (1)$$

where

$$\vec{f} = \begin{pmatrix} u \\ v \\ w \\ p \\ \rho \end{pmatrix}, \quad D = \begin{pmatrix} 0 & 0 & 0 & 0 & 0 \\ 0 & 0 & v/r & 0 & 0 \\ 0 & -v/r & 0 & 0 & 0 \\ 0 & 0 & 0 & \gamma w/r & 0 \\ 0 & 0 & 0 & 0 & \delta \end{pmatrix}$$

$$A^i = \begin{pmatrix} a^i & 0 & 0 & z_1^i/\rho & 0 \\ 0 & a^i & 0 & z_2^i/(\rho r) & 0 \\ 0 & 0 & a^i & z_3^i/\rho & 0 \\ \gamma p z_1^i & \gamma p z_2^i/r & \gamma p z_3^i & a^i & 0 \\ 0 & 0 & 0 & 0 & a^i \end{pmatrix},$$

$$a^i = u z_1^i + v \frac{z_2^i}{r} + w z_3^i, \quad \delta = \frac{w}{r} + z_1^i \frac{\partial u}{\partial q^i} + \frac{z_2^i}{r} \frac{\partial v}{\partial q^i} + z_3^i \frac{\partial w}{\partial q^i},$$

$$z_1^i = \frac{\partial q^i}{\partial z}, \quad z_2^i = \frac{\partial q^i}{\partial \theta}, \quad z_3^i = \frac{\partial q^i}{\partial r},$$

p is the pressure, ρ designates the density, u, v, w are the components of the velocity vector in the original cylindrical coordinate system (z, θ, r). The components of the vector $\vec{F}$ include the viscous terms.

If there no return flows along the marching direction q^1 in the flux the following scheme could be employed for the solution of such problems:

$$(A_1^1)^n \frac{\vec{f}^n - \vec{f}^{n-1}}{\Delta q^1} + (A_2^1)^n \frac{\vec{f}^{n+1} - \vec{f}^n}{\Delta q^1} = \vec{RHS}^n, \qquad (2)$$

where

$$A_1^1 = \begin{pmatrix} a^1 & 0 & 0 & 0 & 0 \\ 0 & a^1 & 0 & 0 & 0 \\ 0 & 0 & a^1 & 0 & 0 \\ \gamma p z_1^1 & \gamma p z_2^1/r & \gamma p z_3^1 & a^1 & 0 \\ 0 & 0 & 0 & 0 & a^1 \end{pmatrix}, \quad A_2^1 = A^1 - A_1^1.$$

The approximation of the right-hand side ($\vec{RHS}$) of Equation (1) is constructed with the first order of approximation on the basis of splitting-up differential operators into physical processes [1].

In order to solve the system of grid Equations (2) an efficient method of global iterations was suggested in [1] which is a non-linear version of Seidel method (block relaxation with respect to the planes $q^1 = (q^1)^n$). From the kind of matrix A it follows that only one 3-D array is sufficient for the pressure field approximation obtained on the previous Seidel iteration. Therefore, the method of global iterations is almost as economic with respect to computer resources as the marching schemes but, unlike them, it is applicable to the calculation of a wider class of aerodynamics problems.

Since the system of grid equations is non-linear, the solution in marching cross-section is obtained making use of an implicit iteration process. The realization of the iterational process is reduced to economic scalar sweeps through approximated factorization of a two-dimensional stabilizing operator on the basis of the method of splitting up into physical processes and coordinate directions in the marching cross-section [1].

The calculation of gas dynamic parameter in the marching cross-section is resumed on each current Seidel iteration. As can be seen from calculations, the number of global pressure iterations required for the acceptable convergence level is not big, therefore main computation costs are associated with the realization of internal iteration process. Evidently, its accelerated convergence will result in the increased efficiency of the entire method of global iterations.

3 Method of Iterations Acceleration

Let us consider method [2] using by way of example the solution of a system of linear equations

$$A\vec{x} = \vec{b} \tag{3}$$

by means of an iteration process

$$\vec{x}_{k+1} = T\vec{x}_k + \vec{f}. \tag{4}$$

Let us introduce the following designations for residual $\vec{z}_{k+1} = T\vec{x}_k + \vec{f} - \vec{x}_k$ and for the error $\vec{y}_k = \vec{x} - \vec{x}_k = \vec{x} - \vec{x}_{k+1} + \vec{z}_k = \vec{y}_{k+1} + \vec{z}_k$, where $\vec{x}$ stands for the exact solution of the system of Equations (3). Obviously, the following relations hold true: $\vec{z}_{k+1} = T\vec{z}_k$, $\vec{y}_{k+1} = T\vec{y}_k$, $\vec{y}_k = T^{-1}\vec{y}_{k+1}$ (on assumption that the operator T is reversible), hence $(I - T)\vec{y}_k = \vec{z}_k$ and also

$$(T^{-1} - I)\vec{y}_{k+1} = \vec{z}_k, \tag{5}$$

where I is the identical operator.

Note that

$$\vec{x} = \vec{x}_k + \vec{y}_k. \tag{6}$$

Then the convergence acceleration of the iteration process (4) can be attained through correction of the solution $\vec{k}_n$ from (6) if we are able to construct good approximation for $\vec{y}_k$

Let $m+2$ approximation be calculated for $\vec{x}$ from (4): $\vec{x}_i$, $i = k, k+1, \ldots, k+m+1$. The corrected value of $\vec{x}^*_{k+m+1}$ will be calculated from the following formula:

$$\vec{x}^*_{k+m+1} = \vec{x}_{k+m+1} + \vec{y}^*_{k+m+1},$$

where $\vec{y}^*_{k+m+1}$ is written as a linear combination of residual vectors (on the assumption of their linear independence):

$$\vec{y}^*_{k\mid m+1} = \sum_{i=1}^{m} \alpha_i \vec{z}_{k+i}.$$

Coefficients α_i are obtained from the condition of the least square of residual norm for (5), i.e. they are the solution of functional minimization problem:

$$F(\alpha_1, \alpha_2, \ldots, \alpha_m) = \|(T^{-1} - I)\vec{y}^*_{k+m+1} - \vec{z}_{k+m}\|^2$$

The square of residual norm is obtained from a scalar product: $\| * \|^2 = (*, *)$. This results in the following system of equations:

$$\sum_{i=1}^{m} (\vec{z}_{k+i+1} - \vec{z}_{k+i}, \vec{z}_{k+j-1} - \vec{z}_{k+j})\alpha_i = (\vec{z}_{k+m}, \vec{z}_{k+j-1} - \vec{z}_{k+j}),$$

$$j = 1, 2, \ldots, m.$$

The algorithm of iterations acceleration successfully suppresses the residual and for the non-linear iteration process as well as in the small vicinity of the solution the errors of the approximations of non-linear iteration process obey almost the same laws as the iteration method errors on solving systems of linear equations.

4 Calculation Results

A calculation series of supersonic flow past a circular cone at the angle of attack under varying problem parameters has been performed. The calculation domain was limited by the body surfaces, bow shock wave and the symmetry plane of the problem. The numerical solution in the vicinity of the bow part were obtained from the complete Navier–Stokes equations or from quasicone approximation. Two- and three-step algorithm of iterations acceleration were employed in calculations.

Note several specific features of iterations convergence. In spite of non-monotonous behavior of the discrepancy of the accelerated iteration process the error decreases monotonically. Hence we can conclude that the algorithm of the convergence acceleration minimizes the error in the vicinity of the solution and in the case of non-linear iterations. At in all, 9-12 global iterations are required for the convergence of the numerical solution. The typical calculation time of a single global iteration on the grid 31x31x31 for the input algorithm is 3000 s on PC AT 486 (50 MHz) and 2000 s for the accelerated algorithm at $m = 2$. General computational time consumption for the problem solution is 5 hr 43 min making use of the traditional method and 3 hr 40 min when the convergence is accelerated using the least squares method.

We also used following method for monotonization of the iteration process:

$$\vec{x}^*_{k+m+1} = \vec{x}_{k+m} + \vec{r}^*_{k+m+1}, \tag{7}$$

where

$$\vec{r}^*_{k+m+1} = \begin{cases} \vec{z}_{k+m} & \text{, if } (\vec{z}_{k+m}, \vec{r}_{k+m+1}) < 0, \\ \vec{r}_{k+m+1} & \text{, if } (\vec{z}_{k+m}, \vec{r}_{k+m+1}) \geq 0, \end{cases}$$

$$\vec{r}_{k+m+1} = \sum_{i=1}^{m-1} \alpha_i \vec{z}_{k+i} + (1 + \alpha_m) \vec{z}_{k+m}.$$

The CPU time is reduced about 40% by using (7) for case with $m = 2$.

Some three-dimensional computed results from solving the simplified Navier–Stokes equations for supersonic flows past wing-body configurations suggested that the computational time can be reduced about 50% by using the modified method of convergence acceleration at $m = 3$.

5 Conclusion

Non-linear relaxation method with variational optimization of iteration processes has been developed for 3-D simplified Navier–Stokes equations. The calculation results enable us to conclude that the algorithm of convergence acceleration [2] significantly increases the efficiency of the method of global iterations with the minimum additional consumption of computer storage.

These studies were supported by Grant 03-013-16362 from Russian Fundamental Research Fund.

References

[1] V.M.Kovenya,A.G.Tarnavskii,S.G. Cherny: *Splitting Method in the Problems of Aerodynamics* (Nauka, Novosibirsk, 1990)

[2] A.G. Sleptsov: *Russian J. of Theoretical and Applied Mechanics* **1** (1) 73 (Elsevier, New York, 1991)

[3] V.B. Karamyshev: *Computational Technologies* **2** (6) 278 (Institute of Computational Technologies, Novosibirsk, 1993)

Virtual Zone Navier-Stokes Computations for Oscillating Control Surfaces

G.H. Klopfer and S. Obayashi

Senior Research Scientists
MCAT Institute
NASA Ames Research Center,
Moffett Field, CA 94035-1000

A new zoning method called "virtual zones" has been developed for applications to an unsteady finite difference Navier-Stokes code. The virtual zoning method simplifies the zoning and gridding of complex configurations for use with patched multi-zone flow codes. An existing interpolation method has been extensively modified to bring the run time for the interpolation procedure down to the same level as for the flow solver. Unsteady Navier-Stokes computations have been performed for transonic flow over a clipped delta wing with an oscillating control surface. The computed unsteady pressure and response characteristics of the control-surface motion compare well with experimental data.

Present transport aircraft as well as highly maneuverable fighter aircraft are often subject to unsteady aerodynamics. In this unsteady environment aircraft designers utilize active controls to achieve controllability and safety to the aircraft. Active control can also be used to suppress transonic flutter characteristics of high aspect ratio wings and thus reduce the structural weight to achieve more efficient flight conditions.

The physics of unsteady transonic flow around a control surface has been simulated with small disturbance theory [1,2]. Unsteady Navier-Stokes simulations have been performed in two dimensions [3,4]. A more recent study [5] conducted a three dimensional simulation of the unsteady thin-layer Navier-Stokes equations on the flow field surrounding a wing with a forced oscillating control surface. In that study an unsteady Navier-Stokes code, ENSAERO, was extended to simulate unsteady flows over a rigid wing with an oscillating trailing-edge flap.

The numerical simulation of the unsteady Navier-Stokes equations about complex and realistic aerodynamic configurations requires the use of zonal methods. In this method the overall flow field domain is subdivided into smaller blocks or zones. In each of these zones the flow field is solved independently of the other zones. The boundary data for each zone is provided by the neighbouring zones. A major difficulty of the zonal methods applied to oscillating control surfaces has been how to account for the variable exposure of the ends of the control surfaces to the flow field.

The virtual zoning method, first implemented in a multizone finite volume code, CNSFV [6,7], has been modified for application to the unsteady finite difference code,

ENSAERO. The main purpose of these zones is to convert, for example, a solid wall boundary condition into an interface condition. The interface condition. The interface conditions are required for the interzonal communication. In a multi-zonal code the virtual zones are treated like real zones as far as boundary and interface conditions are concerned, however, no flow field computations are done within these zones. Hence, the name " virtual" zone is appropriate.

In addition to the introduction of the virtual zones, it is necessary to speed up the process of determining the interpolation coefficients required for the interzonal communication if the unsteady Navier-Stokes simulation is to be practical.

The present study considers the transonic vortical flow over a clipped delta wing.

References

1. Ballhaus, W., Goorjian, P.M., and Yoshihara, H.;"Unsteady Force and Moment Alleviation in Transonic Flow," AGARD Conference Proceeding No. 227, May 1981.

2. Guruswamy, G.P. and Tu, E.L.; "Transonic Aeroelasticity of Fighter Wings with Active Control Surface," Journal of Aircraft, Vol. 26, No. 7, July 1980, pp. 682-684.

3. Steger, J.L. and Bailey, H.E.; "Calculation of Transonic Aileron Buzz," AIAA Journal, Vol. 18, March 1980, pp. 249-255.

4. Horiuti, K., Chyu, W.J., and Buell, D.A.; "Unsteady Transonic Flow Computation for an Airfoil with an Oscillating Flap," AIAA Paper 84-1562, June 1984.

5. Obayashi, S., and Guruswamy, G.; "Navier-Stokes Computations for Oscillating Control Surfaces," AIAA Paper No. 92-4431, August 1992.

6. Klopfer, G.H. and Molvik, G.A.; "Conservative Multizonal Interface Algorithm for the 3-D Navier-Stokes Equations," AIAA Paper No. 91-1601CP, June 1991.

7. Chaussee, D.S. and Klopfer, G.H.; "The Numerical Study of 3-D Flow Past Control Surfaces," AIAA Paper No. 92-4650, August 1992.

The Domain Decomposition Method and Compact Discretization for the Navier-Stokes Equations

Jacek Rokicki[1] and Jerzy M. Floryan[2]

[1]Institute of Aeronautics and Applied Mechanics,
Warsaw University of Technology, Nowowiejska 22/24,
00-665 Warsaw, Poland,
[2]Department of Mechanical Engineering
The University of Western Ontario, London,
Ontario, N6A 5B9, Canada,

Abstract : The domain decomposition method is considered as means for handling nonstandard geometries and for speeding up of calculations via multiprocessing. Different variants of this method are investigated analytically for the model problem and verified numerically for the full Navier- Stokes equations. The proposed algorithm is based on the fourth-order compact discretization schemes for the Navier-Stokes equations in streamfunction-vorticity formulation expressed in terms of a general orthogonal curvilinear coordinate system.

1 Introduction

The main objective of the present work is to develop a fast and accurate algorithm, capable of accurately predicting medium to large Reynolds number flows in nonstandard geometries. Second-order discretization schemes (eg. classic [1]) require so many grid points for accurate calculations that their application to realistic flow problems becomes impractical (cf. [3]). Spectral methods provide high accuracy but have difficulties in handling non-standard geometries. The present work focuses on higher-order compact finite-difference methods which are sufficiently flexible to deal with complicated geometries and can provide the required accuracy at an acceptable computational cost. The streamfunction-vorticity formulation is selected to bypass direct evaluation of pressure.

Typical, realistic problems in fluid dynamics require enormous amounts of computer work (as measured by the execution time). With the fixed processor speed, significant acceleration can be obtained only by using several processors working in parallel. The full advantage of this approach can be achieved only if the numerical algorithm itself is suitable for parallelisation. The domain decomposition method is well suited for parallelisation of a very broad class of problems. The method consists of dividing the computational domain into overlapping subdomains and solving the original problem on each subdomain separately, with the appropriate transfer of boundary information between the neighbouring subdomains, with each subdomain being served by a different processor.

The algorithm described in this paper takes advantage of the domain decomposition principles and thus is suitable for implementation on several processors working concurrently. Various practical and theoretical problems arising from multiprocessor implementation are analyzed for a model problem. The acceleration of computations is tested in numerical experiments with the multiprocessor-computer being simulated using a conventional one.

2 Flow Problem

The Navier-Stokes equations written in the streamfunction-vorticity ($\psi - \zeta$) formulation for a plane, steady, incompressible, two-dimensional flow have the form:

$$\Delta \zeta - Re(\mathbf{V} \cdot \nabla \zeta - curl \mathbf{f}_e) = 0 \tag{1}$$

$$\Delta \psi = -\zeta \tag{2}$$

$$\mathbf{V} = \mathbf{curl}\,\psi \tag{3}$$

with ψ, $\frac{\partial \psi}{\partial \mathbf{n}}$ known at the boundary. In the above, $\mathbf{f}_e$ is the external body force, $\mathbf{V}$ is the velocity vector, $\frac{\partial}{\partial \mathbf{n}}$ stands for the derivative normal to the boundary and Re denotes the Reynolds number. The field equations (1), (2) when expressed in terms of a curvilinear, orthogonal reference system ($\xi = f(x,y)$, $\eta = g(x,y)$) take the generic form

$$a\Phi_{\xi\xi} + b\Phi_{\eta\eta} - \tilde{q}\Phi_\xi - \tilde{s}\Phi_\eta = R \tag{4}$$

where $a(\xi,\eta) = f_x^2 + f_y^2$, $b(\xi,\eta) = g_x^2 + g_y^2$, $\tilde{q}(\xi,\eta) = q_* - (f_{xx} + f_{yy})$, $\tilde{s}(\xi,\eta) = s_* - (g_{xx} + g_{yy})$ and subscripts denote respective derivatives ((x,y) are the cartesian coordinates of the physical domain). In the case of the vorticity transport equation (1), one should substitute $\Phi = \zeta$, $R = -Re \cdot curl \mathbf{f}_e$, $q_* = Re \cdot \psi_\eta(f_x g_y - f_y g_x)$, $s_* = -Re \cdot \psi_\xi(f_x g_y - f_y g_x)$. In the case of the equation describing streamfunction (2), one should substitute into (4) $\Phi = \psi$, $R = -\zeta$, $q_* = s_* \equiv 0$.

The proposed fourth-order, compact discretization scheme for the generic equation (4) has a functional form similar to the one obtained by Dennis and Hudson [1] for a cartesian reference system, i.e.,

$$-d_0\Phi_0 + (d_1\Phi_1 + \cdots + d_8\Phi_8) + B_0 = 0 \tag{5}$$

and was derived by an analogous procedure (lower numerical subscripts refer to nine points of the computational molecule shown in Fig. 1). The actual functional dependence of $B_0, d_0, \ldots, d_9$ on a, b, $\tilde{q}$, $\tilde{s}$, R and h can be found in the Appendix (h denotes the grid step size in the computational plane). The system of linear equations resulting from (5) is solved by the simple Gauss-Seidel relaxation procedure with the nonlinear terms being updated every few iteration cycles [3].

In order to impose the boundary condition for the normal derivative of ψ, an algebraic boundary formula for vorticity has to be employed. Contrary to the statement of Gupta [2], low order formulas, e.g. of second- or first-order accuracy, cannot be used without negatively affecting the overall accuracy as shown in [3]. Fourth-order, implicit boundary formula for vorticity was developed for cartesian reference system by Rokicki and Floryan in [3]. Generalization of this formula to the curvilinear coordinate is given in [4]. The description here is limited to the case of a impermeable, no-slip boundary corresponding to a line $\eta = const$, with the computational domain located (see Fig. 2) above ($\lambda = 1$) or below ($\lambda = -1$) this line. In this case the formula can be conveniently written using $Z = \zeta/b$ instead of ζ resulting in

$$\mu_{0+}Z_{0+} + \mu_0 Z_0 + \mu_{0-}Z_{0-} = \mu_1 Z_1 + \mu_2 Z_2 + \mu_3 Z_3 + \mu_{1+}Z_{1+} + \mu_{1-}Z_{1-} + 15 P_* \quad (6)$$

where Z_{0+}, Z_0, Z_{0-}, which are related to the vorticity at the boundary, are considered unknown (lower subscript refers to the points in the computational molecule shown in Fig. 2). The coefficients μ and P_* are given by

$$\mu_0 = 23 - 4A - 9Sh/2 - (5S^2 - 10S_\eta)h^2, \quad \mu_{0\pm} = 2A \pm 5Qh/2, \quad \mu_{1\pm} = -3A,$$

$$\mu_1 = -16 + 6A - 4Sh, \quad \mu_2 = 11 - Sh/2, \quad \mu_3 = -2, \quad P_* = -(8\psi_1 - 7\psi_0 - \psi_2)/h^2$$

where $A = a/b$, $Q = -\tilde{q}/b$, $S = -\tilde{s}/b$ and subscript 0 on the right hand side is omitted in order to simplify notation. In the above formula, S_η has to be evaluated using standard second-order finite-difference formula. In the special case of a rectangular grid ($A = 1$, $Q = S = 0$) the formulas above reduce to those derived in [3].

Equation (6) written for each grid point along the boundary (except corners) results in a system of linear equations with a tridiagonal matrix of coefficients. This matrix remains diagonally dominant if (i) $23/8 > a/b$ and (ii) the step size h is sufficiently small. The numerical cost of obtaining the solution is negligibly small in comparison with the cost of a single iteration of the discretized field equations.

3 Pressure Problem

After obtaining vorticity and velocity fields it is possible to calculate pressure p, by solving the Poisson's equation

$$\Delta w = \zeta^2 + \mathbf{V} - \nabla \zeta, \quad w = p - \frac{\mathbf{V}^2}{2}$$

with the Dirichlet boundary conditions obtained by projecting suitable version of the momentum equation onto the tangent to the boundary and integrating it along the boundary (density is taken to be one, $\mathbf{f}_e \equiv 0$) [4]. This procedure was used because Dirichlet boundary conditions are easier to implement in the higher-order schemes studied here.

4 Numerical Testing

We employed two different methods to verify accuracy of the proposed algorithm. In the first method an artificial exact solution is compared with a corresponding numerical solution [3]. In the second method, the accuracy is estimated by repeating calculations on a sequence of grids (with a diminishing grid size). The second approach is presented here. The test problem consists of flow between eccentric cylinders. The inner cylinder (of radius 1) rotates with the unit angular speed, the outer cylinder (of radius 2) remains at rest. The typical vorticity pattern is presented in Fig. 3 and the estimated relative error of vorticity is displayed in Fig. 4 for a wide range of Reynolds numbers (here Reynolds number is defined as reciprocal of the kinematic viscosity). The results shown in Fig. 4 confirm that the finite-difference algorithm introduced here is fourth-order accurate.

5 Domain Decomposition

The domain decomposition methodology consists of dividing the computational domain into overlapping subdomains and solving the original problem on each subdomain separately, with the appropriate transfer of boundary information between the neighbouring subdomains. Since calculations on each subdomain can be carried out simultaneously on different processors, this procedure offers an opportunity for tremendous acceleration of calculations.

Success of the domain decomposition in accelerating the overall calculations crucially depends on the strategy for information transfer between different subdomains. Two of the obvious issues are (i) what type of information should be transferred and (ii) what should be the size of the overlap domain. Some of these questions can be answered by identifying and analyzing transfer matrix for a simple model problem. Our analytical results show that convergence rate for a linear problem with two subdomains is $\sim(1-\delta^3)$ for transfer of $(\psi, \frac{\partial \psi}{\partial \mathbf{n}})$ and $\sim(1-\delta)$ for transfer of (ψ, ζ), where δ denotes the size of overlap domain. These results suggest that the former type of information transfer may be impractical. Numerical experiments with the full Navier-Stokes equations confirm these predictions and show lack of convergence in the case of transfer of $(\psi, \frac{\partial \psi}{\partial \mathbf{n}})$.

Acceleration of calculations due to multiprocessing was investigated using computational examples discussed in [3]. The solution domain was subdivided into k equal size subdomains and multiprocessing was simulated on a single processor with the appropriate information transfer. Results shown in Fig. 5 demonstrate that it is possible to come very close to the maximum theoretically possible acceleration rate. These results also show that the advantage of multiprocessing increases for the increasing problem size.

6 Appendix

The fourth-order discretization scheme for the generic equation (4) is not restricted to equations of the Navier-Stokes type. Indeed this discretization is valid for all PDE of generic form (4) provided $a > 0$, $b > 0$ and all u, b, $\tilde{q}$, $\tilde{s}$ are sufficiently regular functions of ξ and η. The free term B_0 in (4) is expressed as

$$B_0 = -h^2[8R_0 + R'_1 a_0(1 - \frac{q_0 h}{2}) + R''_2 b_0(1 - \frac{s_0 h}{2}) + R'_3 a_0(1 + \frac{q_0 h}{2}) + R''_4 b_0(1 + \frac{s_0 h}{2})]$$

with $R' = R/a$, $R'' = R/b$, $s = \tilde{s}/b$, $q = \tilde{q}/a$, and the subscripts refer again to points shown in Fig. 1. The coefficients d_j in (4) are expressed as

$$d_0 = D_{00} + D_{02} h^2,$$

$$d_{1/3} = D_{10} \mp D_{11} h + D_{12} h^2 \pm D_{13} \frac{h^3}{2}, \quad d_{5/6} = D_{50} + (D_{511} \pm D_{512})\frac{h}{2} \pm D_{52}\frac{h^2}{4},$$

$$d_{2/4} = D_{20} \mp D_{21} h + D_{22} h^2 \pm D_{23} \frac{h^3}{2}, \quad d_{7/8} = D_{50} - (D_{511} \pm D_{512})\frac{h}{2} \pm D_{52}\frac{h^2}{4},$$

and the coefficients $D_{00}, \cdots, D_{52}$ have the form

$$D_{00} = 20 D_{50} = 20(a_0 + b_0), \quad D_{10} = 10 a_0 - 2 b_0, \quad D_{20} = 10 b_0 - 2 a_0,$$

$$D_{02} = 2[a_0(q_0^2 - 2q_{\xi 0} + b'_{\xi\xi 0} - s'_0 a'_{\eta 0}) + b_0(s_0^2 - 2 s_{\eta 0} + a'_{\eta\eta 0} - q'_0 b'_{\xi 0})],$$

$$D_{11} = \tilde{q}_0(5 - b'_0) + 2 a_0 b'_{\xi 0}, \quad D_{12} = a_0(q_0^2 - 2 q_{\xi 0}) - a'_{\eta 0} \tilde{s}_0 + a'_{\eta\eta 0} b_0,$$

$$D_{21} = \tilde{s}_0(5 - a'_0) + 2 b_0 a'_{\eta 0}, \quad D_{22} = b_0(s_0^2 - 2 s_{\eta 0}) - b'_{\xi 0} \tilde{q}_0 + b'_{\xi\xi 0} a_0,$$

$$D_{13} = a_0 q_0 q_{\xi 0} + \tilde{s}_0 q'_{\eta 0} - (a_0 q_{\xi\xi 0} + b_0 q'_{\eta\eta 0}), \quad D_{511} = 2 b_0 a'_{\eta 0} - \tilde{s}_0(a'_0 + 1),$$

$$D_{23} = \tilde{q}_0 s'_{\xi 0} + b_0 s_0 s_{\eta 0} - (a_0 s'_{\xi\xi 0} + b_0 s_{\eta\eta 0}), \quad D_{512} = 2 a_0 b'_{\xi 0} - \tilde{q}_0(b'_0 + 1),$$

$$D_{52} = \tilde{q}_0 s'_0 + \tilde{s}_0 q'_0 - 2(a_0 s'_{\xi 0} + b_0 q'_{\eta 0})$$

where $b' = b/a$, $a' = a/b$, $s' = \tilde{s}/a$, $q' = \tilde{q}/b$, lower indices ξ and η denote differentiation in the corresponding directions and subscript 0 on the right hand side corresponds to the point 0 in Fig. 1. If $\xi = x$, $\eta = y$, we have $q_* = \tilde{q} = q = q'$, $s_* = \tilde{s} = s = s'$, $a = b \equiv 1$, and all derivatives of a and b vanish. In this case the above formulas for d_j, $(j = 0, \cdots, 8)$ reduce to those derived by Dennis and Hudson [1]. The special case of $\xi = f(x)$ and $\eta = g(y)$ is equivalent to generation of a non-uniform rectangular grid in the original flow domain (x, y) with a uniform square grid in the computational domain (ξ, η).

All first and second derivatives of coefficients a, b, a', b', s, s', q, q', necessary to evaluate coefficients d_j, can be calculated with the usual second-order finite-difference formulas without compromising the fourth-order accuracy of the scheme (cf. [3]). All coefficients, however, have to be evaluated with the fourth-order accuracy (cf. [1] [3]). This means that fourth-order formulas for the first derivatives of streamfunction are required to calculate q_* and s_* in (4). This accuracy can be obtained by using both the streamfunction and the vorticity grid data.

Acknowledgements

This work was supported by the NSERC of Canada.

References

[1] S.C.R. Dennis, J.D. Hudson: J. Comput. Phys., **85**, 390 (1989).
[2] M.M. Gupta, R.P. Manohar: J. Comput. Phys., **31**, 265, (1979).
[3] J. Rokicki, J.M. Floryan: Report ESFD - 3/93, Department of Mechanical Engineering, The University of Western Ontario, London, Ontario, Canada.
[4] J. Rokicki, J.M. Floryan: Report ESFD - 5/93, ibid.
[5] J. Rokicki, J.M. Floryan: Report ESFD - 1/94, ibid.

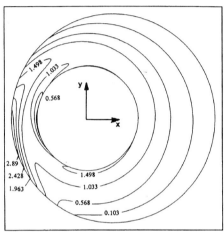

Fig. 1. Distribution of vorticity for Re = 200.

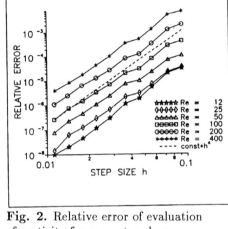

Fig. 2. Relative error of evaluation of vorticity for geometry shown in Fig. 1.

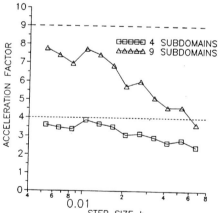

Fig. 3. Acceleration of calculations due to use of several processor, with each of them serving a different subdomain.

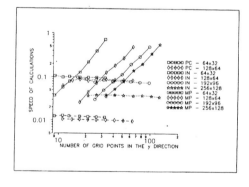

Fig. 4. Speed of calculations for different grid sizes with storage matrices defined as either 64 × 32, or 128 × 64, or 192 × 96, or 256 × 128. PC, IN & MP denote personal computer, Indigo workstation and Maspar computer, respectively

Viscous flow modelling using unstructured meshes for aeronautical applications

J. Szmelter and A. Pagano

British Aerospace Airbus Limited, Filton, Technology and
Computing, PO BOX 77, Bristol, BS99 7AR, U.K.

1 Introduction

The study of complex configurations and complex viscous flows is receiving increasing attention by the aerospace industry. Due to recent progress in computational methods, the numerical solution of the Euler equations over whole aircraft configurations has become possible. The flexibility and generality offered by unstructured meshes (tetrahedral or triangular) to model geometrically very complex configurations have been well recognised particularly as the method offers relatively short lead times to generate new meshes. In addition many forms of adaptivity such as mesh enrichment/derefinement, mesh regeneration and mesh movement are easily introduced within unstructured meshes methods and do not require changes within the flow solver. The most obvious advantages in the application of adaptivity techniques come from enhancing the accuracy of the flow solution, and relaxing the demand on the quality of the initial mesh, which in an industrial environment, often is not generated by a CFD expert.

The successful application of unstructured meshes techniques within the aircraft design process is still however limited to the solution of the Euler equations since generation of highly stretched meshes suitable for the solution of the Navier Stokes equations over three dimensional complex geometries is not readily available. Also at present a solution of the three dimensional Navier Stokes equations over whole aircraft configurations poses high CPU and memory requirements even for structured Multi-block meshes and is not routinely performed. Alternatively viscous-inviscid interaction techniques can be used in which the viscous layer (boundary layer and wake) and the external inviscid flow are calculated separately and then coupled together through an iterative, interative process. Although they can based on the simplified description of the fluid flow viscous coupled techniques can be used for many aeronautical applications. These techniques have been proven to be very effective in predicting the high speed performance of civil transport aircraft and they are one of the mainstay of the CFD prediction/design environment at British Aerospace.

In this paper two dimensional flows are discussed, as a necessary initial study leading towards three dimensional generalisations. A new inviscid coupling method has been developed which is combination of steady state Euler solvers coupled to an integral boundary layer code.

2 Method Description

The inviscid solution is obtained on unstructured triangular meshes, which are ideally suited to model complex geometry. The unstructured meshes suited, originated at University College of Swansea, it includes mesh generator based on the advancing front technique [1] and Euler solvers, and has moving meshes and local refinement capabilities using local remeshing, thus allowing for adaptive capturing of flow features such as shocks. The mesh generator and solvers have also two dimensional capabilities to solve inviscid time dependent flows with moving bodies, [2]. The generalisation of the presented viscous coupling to such applications is also well advanced and is very revelant to the study of flutter and buffet problems. The basic space discretisation is accomplished by employing a finite element (cell vertex finite volume equivalent). Two versions of solver based on the Runge Kutta time stepping and on Taylor-Galerkin discretisation in time [2] are used and compared.

The semi-inverse viscous-inviscid interaction technique [3] has been introuced, and is shown in TABLE 1. Other forms of coupling are under investigation, particularly for unsteady aerodynamics applications.

The integral laminar/turbulent boundary layer solver pioneered at DRA uses either prescribed pressure distribution (forward mode) or the predicted growth in displacement thickness (inverse mode) to drive the turbulent boundary layer equations. The approach is based on 'lag-entrainment' method [4,5], which has been further developed to incorporate curvature effects on turbulence structure [6], compressibility effects [7], Low Reynolds Number correction, normal shear stress terms [8] as well as wake calculations [9]. Laminar flow is calculated by the Thwaites compressible method, with natural transition predicted by the Granville correlation criterion. With forced transition, the laminar separation bubble is calculated by Horton's method.

For applications to realistic civil aircraft configurations it is important that the theoretical design/prediction tools used are able not only to calculate well behaved attached laminar and turbulent flows but also to handle 'off design' behaviour such as for instance: shock induced boundary layer interaction and separation, trailing edge separation and interaction between shock induced and trailing edge separation. Both the integral and the finite difference method are able to handle limited regions of separation. This was demonstrated first by DRA [10], by coupling the lag-entrainment to a full potential method and later by BAe [11] - the first in europe of couple the lag-entrainment to Euler method. Le Balleur has extended the semi-inverse coupling method to deal with massively separated flows [12].

3 Numerical Results

To illustrate the potential capability of the method, preliminary calculations were performed for two well known RAE2822 test cases [13]:

Case 3 : $M = 0.600$ alpha $= 2.57$ Re $= 6.3 \times 10^6$ - Subcritical

Case 6 : $M = 0.725$ alpha $= 2.92$ Re $= 6.5 \times 10^6$ - Supercritical

Tunnel effect corrections were applied.

The calculations were carried out with and without wake treatment within the boundary layer code. Adapted mesh and corresponding pressure contours for Case 6. are shown in Fig.1. The concept of wake treatment using non-overlapping 'Euler' and 'boundary layer' mesh is illustrated in Fig.2. The numerical inviscid and viscous results are compared with experiment in Fig.3. The quality of the solution is good and illustrates the importance of proper trailing edge and wake treatment.

4 Conclusion

The novel application of viscous coupling to unstructured meshes has been proposed and developed. The method allows for viscous flows modelling and avoids the difficulty of generating highly stretched tetrahedral in 3D or triangular in 2D elements required for Navier-Stokes solvers. The time step allowed by the explicit euler solver is limited by the size of the 'Euler' mesh, resulting in faster algorithms than standard explicit Navier-Stokes solvers.

The generalisation of the presented method to three dimensional and unsteady flows is already in progress.

References

1. J.Peraire, J.Peiro, L.formaggia, K.Morgan, O.C. Zienkiewicz, 'Finite Element Euler Computations in three Dimensional.' Int. J. Num. Meth. Engng., Vol 26, pp 2125-2159, 1988.

2. O.Hassan, E.J. Probert, K. Morgan, J. peraire, 'Domain Decomposition Combined with Adaptive Remeshing for Problems of Transient Compressible Flow', Proc. 13-th International Conference on Numerical Methods in Fluid Dynamics, Rome 1992.

3. J. C. Le Balleur, 'Calcul par Couplage Fort des Ecoulements Visqueux Transsoniques incluant Sillages et Decollements. Profil d'Ailes Porlant. La Recherche Aerospatiale, May-June 1981.

4. J.E. Green, D.J. Weeks, J.W.F. Brooman, 'Prediction of Turbulent Boundary Layers and Wakes in Compressible Flow by a Lag-Entrainment Method'. RAE Technical Report TR72231, 1973.

5. J.E. Green, 'Application of Head's Entrainment Method to the Prediction of Turbulent Boundary Layers and Wakes in Compressible Flows'. RAE Technical Report TR72079, 1972.

6. P. Bradshaw, 'The Analogy Between Streamline Curvature and Buayancy in Turbulent Shear Flow'. J. Fluid Mech, Vol 36, pp 177-191, 1969.

7. J.E. Green, 'The prediction of Turbulent Boundary Layer Development in Compressible Flow'. J. Fluid Mech, Vol 31, part 4 pp. 753-778, 1968.

8 P.R. Ashill, DRA Bedford - Private Communication.

9 H.b. Squire, A.D. Young, 'The Calculation of the Profile Drag of Aerofoils'. A.R.C.R. & M. 1838, 1937.

10 B.R. Williams, 'The prediction of Separated Flows Using a Viscous - Inviscid Interaction Method'. RAE Tech Memo Aero 2010, 1984.

11 A. Pagano, 'A Viscous Coupled Euler Method'., Presentation at the AGCFM Workshop on Separated Flows, RAE Farnborough, 1985.

12 J.C. Le Balleur, 'Viscous-Inviscid Calculation on High-Lift Separated Compressible Flows over Airfoils and Wings.', Proceedings AGARD Cp-415, 1992.

13 P.H. Cook, M.A. Mc Donald, M.C.P. Firmin, 'Aerofoil RAE1822: Pressure Distribution and Boundary Layer and Wake Measurements AGARD AR 138, Paper A6 1979.

VISCOUS-INVISCID COUPLING TECHNIQUES

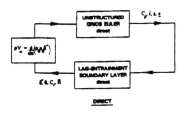
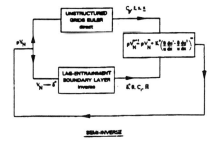

TABLE 1

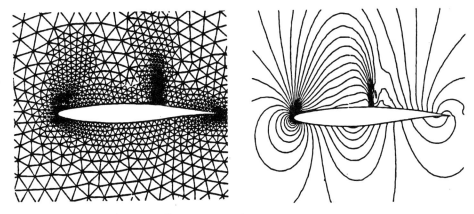

Fig 1. Adapted mesh and corresponding pressure contours
TEST CASE 6

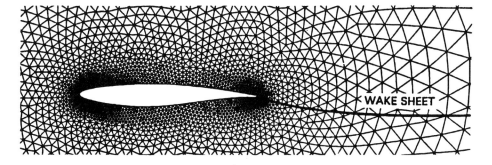

Fig 2. Wake treatment, non-overlapping meshes

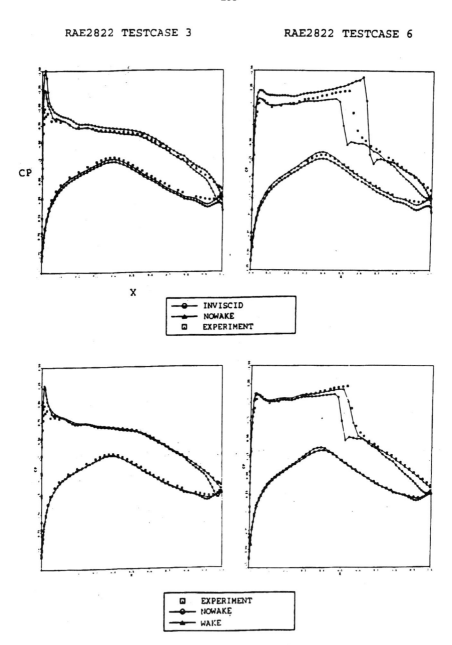

Fig 3. Cp plots
Comparison of flow code and experimental results.

Upwind Numerical Scheme for a Two-Fluid Two-Phase Flow Model

I. Toumi and P. Raymond

French Atomic Energy Commission
CE Saclay. DMT/SERMA
91191 Gif sur Yvette Cedex, France.
Tel.: (331) 69 08 21 61 Fax: (331) 69 08 23 81

1 Introduction

The modelling and numerical simulation of two-phase flow phenomena keeps causing complex problems for the development of computer codes dedicated to design and safety studies of nuclear reactors. Generally, engineering two-phase flow calculations require the determination of the velocity, the pressure and the energy fields for each phase. The usual way to establish physical modelling for two-phase flows is to start from a single continuous description for each phase given by Navier-Stokes equations. The presence of topologically complicated interfaces suggest the need for a multifield model of the flow. Various averaging techniques have been used to obtain practical and computable two-fluid models [2]. However, the averaging procedure leads to an ill-posed initial value problem and many differential terms as added mass and interfacial pressure are proposed in order to make the models hyperbolic.

The model considered here, is a first order equal pressure six equation two-fluid model. Except the interfacial pressure term, which contains partial derivatives, the other terms of mass and momentum transfer between phases are assumed to be absent. They will appear as source terms, and will be added to the numerical scheme. The resulting model is a nonconservative hyperbolic one.

The most common approach to solving these equations is a numerical method based on staggered grids and donor-cell differencing [3]. This method, now almost universal in two-fluid codes like TRAC [10], RELAP [9], CATHARE [11], introduces a large amount of numerical diffusion. Moreover, Ransom [4] has shown that high frequency oscillations appear with a large number of cells. The non-convergence of the results may be due either to the ill-posed character of the model or to the numerical scheme.

In this paper, we present a numerical method based upon a generalized Roe's approximate Riemann solver. Such numerical schemes have been widely used for hyperbolic systems of conservation laws. The linearized approximate Riemann solver of Roe [1] was proposed in 1981 for the numerical solution of the Euler equations governing the flow of an ideal gas. These last years, generalized Riemann solvers for the Euler equations with real gases have motivated many. Several extensions of

Roe's linearization to an arbitrary equation of state have been proposed [6], [8]. A weak formulation of Roe's approximate Riemann solver, based on the choice of a path Φ in the states space, has been introduced in [7]. This weak formulation was applied in order to build a Roe-averaged matrix for a conservative system governing a homogeneous equilibrium two-phase flow. We seek here to extend this formulation to a hyperbolic nonconservative system modelling a two-component two-phase flow.

In practice this new numerical scheme has proved to be stable and capable of generating non-oscillating solutions in problems where other schemes have failed.

2 Two-Phase Flow Model

We consider the first order differential equations of mass, momentum and energy conservation of the two-fluid model which might describe two-component two-phase flow in a straight pipe.

$$\begin{aligned}
&\partial_t \left(\rho_k \alpha_k\right) + \partial_x \left(\rho_k \alpha_k u_k\right) = 0 \\
&\partial_t \left(\rho_k \alpha_k u_k\right) + \partial_x \left(\rho_k \alpha_k u_k^2\right) + \alpha_k \partial_x p + (p - p_i) \partial_x \alpha_k = 0 \\
&\partial_t \left(\rho_k \alpha_k H_k\right) - \alpha_k \partial_t p + \partial_x \left(\rho_k \alpha_k u_k H_k\right) = 0
\end{aligned} \qquad (1)$$

with $H = h + \frac{u_k^2}{2}$ and $\alpha_v + \alpha_l = 1$. Here the subscript k refers to the vapor phase ($k = v$) or the liquid phase ($k = l$); ρ_k, u_k, h_k and α_k are the mass density, the velocity, the enthalpy and the void fraction of the k-phase. P is the pressure assumed to be equal in the two phases. The expression for p_i is chosen to provide the hyperbolicity of the system. To close the system the liquid phase is assumed to be incompressible with constant mass density ρ_l, while the vapour mass density is given by a state equation $\rho_v = \rho_v(p, h_v)$. The results, however, can be generalized to a compressible liquid phase.

In order to study the hyperbolicity of the system (1) we are looking for the eigenvalues of this system. To determine these eigenvalues we must find the six roots of a polynomial of degree six. We prefer to assume the relative velocity between the two phases much lower than the speed of sound of the two-phase mixture c_m. This is the case in many physically interesting configurations, for example for steam and water. Then, we introduce the following small parameter

$$\xi = \frac{(u_v - u_l)}{c_m} \qquad (2)$$

and we use a perturbation method around $\xi = 0$ in order to know the behaviour of the eigenvalues $\lambda_i(\mathbf{u})$, $i = 1, 6$. We will show that using a pressure correction p_i given by

$$p_i = \frac{\alpha_v \alpha_l \rho_v \rho_l}{\alpha_v \rho_l + \alpha_l \rho_v} (u_v - u_l)^2, \qquad (3)$$

all the eigenvalues of the system (1) are real.

3 Numerical Method

To solve the nonlinear Riemann problem for hyperbolic systems of conservation laws

$$\partial_t U + \partial_x f(U) = 0$$
$$U(x,0) = U_i^t \quad (x < x_{ij}) \qquad U(x,0) = U_j^t \quad (x > x_{ij}) \tag{4}$$

Roe [1] introduces a local linearization

$$\partial_t U + A\left(U_i^t, U_j^t\right)\partial_x U = 0$$
$$U(x,0) = U_i^t \quad (x < x_{ij}) \qquad U(x,0) = U_j^t \quad (x > x_{ij}) \tag{5}$$

where the matrix $A\left(U_i^t, U_j^t\right)$, known as the Roe averaged matrix, is some Jacobian matrix depending on the initial states U_i^t and U_j^t.

The Roe averaged matrix $A\left(U_i^t, U_j^t\right)$ is constructed to have the property

$$F\left(U_j^t\right) - F\left(U_i^t\right) = A\left(U_i^t, U_j^t\right)\left(U_j^t - U_i^t\right) \tag{6}$$

This property guarantees that, when U_i^t and U_j^t are connected by a single shock wave, the approximate Riemann solution agrees with the exact solution. In this case, the Rankine-Hugoniot condition is satisfied for some shock speed σ (equal to an eigenvalue of $A\left(U_i^t, U_j^t\right)$)

$$F\left(U_j^t\right) - F\left(U_i^t\right) = \sigma\left(U_j^t - U_i^t\right) \tag{7}$$

In Reference [1], Roe gives a method to construct the matrix for Euler equations with perfect gases. Several extensions of Roe's method to real gases have been proposed [8], [6].

However, the various solutions suggested for real gas extensions are not applicable to thermal-hydraulic two-fluid models. We have additional difficulties due to the nonconservative form of the hyperbolic system (1) since its jacobian matrix is not the derivative of a flux function $f(U)$. To overcome these difficulties, we use a weak formulation of Roe's approximate Riemann solver [7], which has been applied to construct a Roe averaged Jacobian matrix for the Euler equations with arbitrary equations of state. This weak formulation has also been used to build approximate Riemann solvers for an homogeneous two-phase flow model [7], and for an isentropic two-fluid model [5].

Precisely, we consider approximate solutions to the Riemann problem for the system (1) which are exact solutions to the approximate linear problem:

$$\partial_t U + A\left(U_i^t, U_j^t\right)_\Phi \partial_x U = 0$$
$$U(x,0) = U_i^t \quad (x < x_{ij}) \qquad U(x,0) = U_j^t \quad (x > x_{ij}) \tag{8}$$

Now, the average matrix $A\left(U_i^t, U_j^t\right)_\Phi$ is also depending on a smooth path $\Phi\left(s, U_i^t, U_j^t\right)$ linking the two states U_i^t and U_j^t in the states space. This generalized Roe averaged matrix must satisfy the jump condition

$$A\left(U_i^t, U_j^t\right)_\Phi \left(U_j^t - U_i^t\right) = \int_0^1 A\left(\Phi\left(s, U_i^t, U_j^t\right)\right) \frac{\partial \Phi}{\partial s}\left(s, U_i^t, U_j^t\right) ds \qquad (9)$$

which is a generalization of Roe's condition (6). We refer the reader to [7] for more details on this formulation of Roe's approximate Riemann solver and its application to nonlinear hyperbolic systems.

To construct such matrix we follow the method introduced in [7], where the main feature is the choice of the canonical path for a parameter vector W:

$$\Phi\left(s, U_i^t, U_j^t\right) = \psi_0\left(W_i + s(W_j - W_i)\right) \qquad (10)$$

where Ψ_0 is a smooth function such that $\Psi_0(W) = U$ and $A_0(W) = \frac{\partial \Psi_0}{\partial W}$ is a regular matrix for every state W. Here, using the above weak formulation, with the parameter vector chosen as follow

$$W = \left(\sqrt{\rho c}, \sqrt{\rho(1-c)}, \sqrt{\rho c}u_v, \sqrt{\rho(1-c)}u_l, \sqrt{\rho c}H_v, \sqrt{\rho(1-c)}H_l,\right) \qquad (11)$$

we will build a hyperbolic approximate Riemann solver for the two-fluid model.

4 Numerical Results

In this section we extend Roe's numerical scheme for the calculation of one-dimensional two-phase flow based on the two-fluid model. The resulting first-order numerical scheme may be written:

$$U_i^{t+\Delta t} = U_i^t + \frac{\Delta t}{\Delta x}\left(F^+\left(U_{i-1}^t, U_i^t\right) + F^-\left(U_i^t, U_{i+1}^t\right)\right) \qquad (12)$$

with the positive and the negative part of the flux given by

$$F^+\left(U_{i-1}^t, U_i^t\right) = A^+\left(U_{i-1}^t, U_i^t\right)_\Phi \left(U_i^t - U_{i-1}^t\right) \qquad (13)$$

and

$$F^-\left(U_i^t, U_{i+1}^t\right) = A^-\left(U_i^t, U_{i+1}^t\right)_\Phi \left(U_{i+1}^t - U_i^t\right) \qquad (14)$$

- **Problem 1** : Shock-tube problem
  This problem consists in a Riemann problem for the two-fluid model where the left and right states are given by

$$\begin{array}{lllll} U_L: & P_L = 25 MPa & \alpha_L = 0.25 & u_v^L = 0.ms^{-1} & u_l^L = 0.ms^{-1} \\ U_R: & P_R = 10 MPa & \alpha_R = 0.1 & u_v^R = 0.ms^{-1} & u_l^R = 0.ms^{-1} \end{array} \qquad (15)$$

The vapour phase is assumed to be an ideal isentropic gas. The computations have been done with 300 nodes. The solution is composed by seven constant states separated by rarefaction waves or shock waves. Figure 1 gives the total energy profile.

Riemann Problem

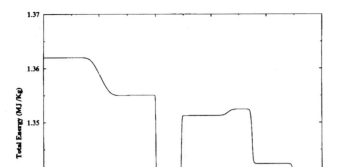

Figure 1: Riemann Problem

- **Problem 2** : Water faucet problem

    This test, proposed by Ransom [4] consists in a vertical water jet, contained within a cylindrical channel, that is accelerated under the action of gravity. At the initial state, the pipe is filled with a uniform column of water surrounded by stagnant vapour, such that the void fraction is 0.2 and the column has a uniform velocity of 10m/s and a uniform pressure of $10^5 Pa$. The boundary conditions are specified velocities of 10m/s for the liquid and 0m/s for the vapour at the inlet, and constant pressure at the outlet.

    The calculation was carried out until a steady-state is reached, with 100 nodes and a constant CFL number equals to 0.9. Figure 2 shows the vapour void fraction profile at various time. These results clearly demonstrate the ability of the numerical scheme to capture discontinuities.

    In order to test the convergence and the stability character of the scheme, computations have been made using discretization with 12 cells to 768 cells, but constant CFL numbers equal to 0.9. The Figure 3 gives the void fraction profile for the various discretization. An interesting feature of the results shown is that there is no oscillations at the discontinuity of the void fraction.

- **Problem 3** : Edwards pipe blowdown experiment

    This standard test for transient two-phase codes concerns the prediction of the blowdown of an initially subcooled liquid from a pipe of 4m length. The water in the pipe has an initial pressure of 7.0MPa. The transient is initiated by opening the right side of the pipe. Figure 4 gives the pressure history at the closed end for various rate τ of the interfacial heat transfer process.

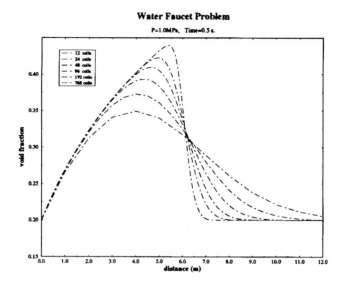

Figure 2: Void fraction profile (transient calculations)

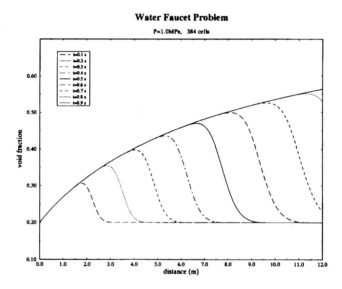

Figure 3: Void fraction profile (convergence)

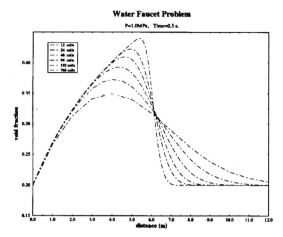

Figure 4: Pressure profile for the Edwards pipe blowdown

References

[1] P. L. Roe, " Approximate Riemann Solvers, Parameter Vectors and Difference Schemes", "Journal of Computational Physics", 1981, Vol. 43, pp. 357–372

[2] H. B. Stewart and B. Wendroff, "Journal of Computational Physics", 1984, Vol. 56, pp. 363

[3] D. R. Liles and W. H. Reed, "Journal of Computational Physics" 1978, Vol. 26, pp. 77

[4] V. H. Ransom and V. Mousseau, "Convergence and Accuracy of the Relap5 Two-Phase Flow Model", "Proceedings of the ANS International Topical Meeting on Advances in Mathematics, Computations, and Reactor Physics", 1991, Pittsburgh, Pennsylvania,

[5] Imad Toumi and A. Kumbaro, "A Linearized Approximate Riemann Solver for a Two-Fluid Model", "Sub. for pub. to J. Comput. Phys."

[6] P. Glaister, "Journal of Computational Physics" Vol. 74, pp. 382, 1988

[7] Imad Toumi, " A Weak Formulation of Roe's Approximate Riemann Solver", " Journal of Computational Physics", Vol. 102, pp. 360–373, 1992

[8] H. Yee, "A class of High-Resolution Explicit and Implicit Shock-Capturing Methods", "Von Karmann Institute for Fluid Dynamics", Lecture Series, no. 1989-04, 1989,

[9] V. H. Ransom, "RELAP5/MOD1 Code Manual Volum 1 : Code Structure, System Models and Numerical Method", NUREG/CR-1826, 1982

[10] " TRAC-PF1 / MOD1. An Advanced Best-Estimate Computer Program for Pressurized Water Reactor Thermal-Hydraulic Analysis", NUREG/CR-3858, LA-10157-MS, 1986

[11] J. C. Micaelli, " CATHARE An Advanced Best-Estimate Code for PWR Safety Analysis", SETH/LEML-EM, no. 87-58, 1987

A Compact Formalism to Design Numerical Techniques for Three Dimensional Internal Flows

Mauro Valorani and Bernardo Favini

Dipartimento di Meccanica e Aeronautica,
Università di Roma "La Sapienza"
Via Eudossiana 18, 00184 Roma

1 Outline

The paper illustrates the application of a compact matricial formulation [1] to the numerical solution of three-dimensional, compressible, inviscid, non-reactive, internal flows. The proposed formulation is aimed at developing numerical techniques to integrate hyperbolic system of conservation laws. It provides an analytic framework to derive algorithms for different model equations, different numerical techniques together with a clear and flexible coding. The present paper is confined to the analysis of smooth flows. The treatment of shock waves and contact discontinuities by means of a front tracking approach [2] will be considered elsewhere. On the basis of the proposed formulation, a non-oscillatory, second-order numerical technique is developed. The technique, valid for non-orthogonal systems of curvilinear co-ordinates, belongs to the family of λ schemes [3, 5]. A diversified component-wise representation of the unknowns in the time and the space derivatives allows a simple and accurate enforcement of the boundary conditions, and provides a form of the equations free of local terms originated by the domain transformation [4, 5]. Thus, a consistent upwind discretization of the flux jacobians can be defined, since all metric terms appear exclusively as coefficients of spatial derivatives. Following [4, 6, 7], a uniform treatment of the integration at the boundaries, when the boundary conditions can be expressed in terms of combinations of either time or space derivatives of the unknowns is also present.

2 Description of the formulation

The overall accuracy and efficiency of a numerical method for hyperbolic systems of equations mainly depend on how the wave propagation and the geometry are described and how the boundary conditions are enforced. The proposed method adopts the wave front model [4,5], which can be considered as an extension of the λ scheme approach to non-orthogonal grids. According to the wave front model, the component-wise form of the governing equations written in a non-orthogonal curvilinear system of co-ordinates can be cast into a compact matrix form as

$$w_{,\tau} + \sum_{j=1,N_d} A_j w\,|_j = 0 \qquad N_d \text{ number of space dimensions} \qquad (1)$$

In the equations above, the symbol w denotes the algebraic vector of the dependent variables when the velocity vector $\mathbf{q}$ is projected over the co-variant base, whereas the notation $\hat{w}$ is used when $\hat{\mathbf{q}}$ is projected over the Cartesian base. The matrices A_j are the flux jacobian of the conservation laws projected along the curvilinear co-ordinates ξ_j. The spatial co-variant derivatives of w can be evaluated by means of standard partial derivatives of $\hat{w}$ [5] according to the rule

$$w\,|_j = T\hat{w}_{,j} \qquad (2)$$

where T is a transformation matrix defined as a function of the jacobian of the curvilinear mapping. Being system (1) hyperbolic, each matrix A_j is similar, by definition, to the diagonal matrix Λ_j formed by its eigen-values λ_j^m through the corresponding right R_j and left L_j eigen-vector matrices. System (1) can thus be written

$$w_{,\tau} + \sum_{j=1}^{N_d} R_j\,\Lambda_j\,L_j T\hat{w}_{,j} = 0 \qquad (3)$$

The numerical integration of system (3) is obtained by means of an explicit two-step characteristic-biased scheme [3,5,10], which is second-order accurate both in time and space. At each step, it uses a two point stencil for each wave direction, and for three-dimensional problems is stable for a CFL number less than 2/3. Relations (2) allow to express the co-variant derivatives without the need of introducing the undifferentiated Christoffel symbols, whose characteristic-biased discretization is computationally expensive. A TVD modification presented in [11] by one of the authors of this paper makes the scheme oscillation free.

3 Integration at the boundaries

The adoption of the wave front model and of the co-variant base to represent vectors and differential operators allows a simple and accurate enforcement of the boundary conditions. When system (3) is evaluated at a boundary point, some of the qualities

$$\mathcal{L}_j^m = \lambda_j^m l_j^m \cdot w\,|_j \quad \text{or} \quad \mathcal{L}_j^m = \lambda_j^m l_j^m \cdot T\hat{w}_{,j} \qquad j = 1, N_d;\, m = 1, N_d + 2 \qquad (4)$$

will correspond to waves carrying information inwards the computational domain. Each of these missing pieces of information has to be replaced by relations modelling the outside world in order to obtain a well-posed problem at that boundary point [6,8]. The unknown term $\mathcal{L}_j$ can be evidentiated in system (3) by writing

$$w_{,j} + R_j \mathcal{L}_j + c_{ik} = 0 \qquad (5)$$

where all valid information carried by outgoing and transversal waves are summed up in vector c_{ik}. System (5) provides the general form of the equations at the boundary. The new $w_{,\tau}$ is found once the term $\mathcal{L}_j$ is determined by solving the system (10) formed by combining waves coming from the interior of the domain and boundary conditions. Following [4,6,7], these boundary conditions can be formulated in differential form. Thus, a condition expressing the time invariance of some integral quantity can be written as a, generally non-linear, combinations of the time derivative of w

$$b \cdot w_{,\tau} = B \qquad (6)$$

or, similarly, a relation on spatial gradients can be cast as a combinations of the co-variant derivatives of w, or partial derivatives of $\hat{w}$

$$s_j \cdot w \mid_j = S_j \qquad s \cdot \hat{w}_{,j} = S_j \qquad (7)$$

These relations must be rewritten as conditions upon $\mathcal{L}_j$. Equation (5) and (6) readily provide

$$b \cdot (-R_j \mathcal{L}_j - c_{ik}) = B \qquad (8)$$

whereas equations (7) combined with the inverse relations of (4) give

$$s_j \cdot \left[(r_j^m \cdot \mathcal{L}_j) / \lambda_j^m \right] = S_j \qquad s_j \cdot \left[T^{-1} \left(r_j^m \cdot \mathcal{L}_j \right) / \lambda_j^m \right] = S_j \qquad (9)$$

The information carried by the outgoing waves, equations (4), and those related to the boundary conditions, equations (8) and (9), form a generally non-linear system in the variable $\mathcal{L}_j$, whose canonical form is

$$D_{j,j} \mathcal{L}_j = d_j \qquad (10)$$

where $D_{j,j}$ is an $(N_d + 2) \times (N_d + 2)$ matrix, and d_j is an $(N_d + 2)$ vector.

A similar approach can be followed to derive relations at edges and corners. Lack of space forces us to postpone the presentation of the complete analysis of the integration at the boundaries to a forthcoming paper. This part of the work will complete the analysis presented in [9].

4 Results

The axi-symmetric, transonic flow through an annular nozzle with a central plug was chosen to test the methodology presented above. This test is particularly severe for the following reasons. The contours defining the computational domain in the meridional planes display both sharp convex corners and zones to strong curvature (fig.1 and 2). These effects are enhanced by the axi-symmetry of the geometry. Among other consequences, the complex nozzle geometry produces an S-shaped sonic locus surface, whose area in turn defines the actual mass flow elaborated by the nozzle. The shape of the sonic locus is highly sensitive with respect to the mesh quality, and this can lead to strong inaccuracies in the prediction of the mass flow and thus of the overall flow.

The zone seeming to have the strongest influence on the sonic locus shape is the axi-symmetric stagnation point flow encountered moving along the central plug. There, the flow is at first slowly brought to rest, and then is strongly accelerated along the conical wedge. The flow undergoes large accelerations passing over the round corners having a strong curvature. These are also zones difficult to resolve. A further prominent feature of the flow is the weak oblique embedded shock forming downstream the sonic locus on the cowl side. The formation adopted, although non conservative, can provide a satisfactory capture of this weak shock. The portion of the flow affected by the approximate treatment of the shock is confined to the recompression region following the oblique shock along the cowl.

To assess the accuracy and the consistency of the numerical solution, we compared results obtained both by orthogonal and H-type grids (fig.1 and 2). Exponential stretching in the radial direction provide a uniform slenderness of the computational cell in planes orthogonal to the axis of simmetry (fig.2). The orthogonal grid, generated by conformal mapping [12], provides a satisfactory distribution of the grid nodes at the concave portions of the contours, whereas the description at the convex regions is poor. This reduces the accuracy of the solution at the stagnation point on the central plug. On the contrary, the H-type grid is able to allocate a sufficient number of nodes at the stagnation point region, but performs very poorly around the concave regions (the rounded corners). Moreover, the contour discontinuity produced by the conical wedge exercises its influence throughout the whole plane which is normal to the axis and passes through the wedge corner (note the kink on the contour lines of fig.1).

The solutions obtained with the two different types of grids, both having the same number of nodes ($128 \times 64 \times 4$), are compared in fig.1. It is only at this relatively high number of grid points that the solutions agree with each other at least as far as the sonic locus shape, the 'isentropically captured' shock shape and the overall flow feature are concerned. Differences persist at and downstream the stagnation points, and, although less marked, downstream the rounded corners. The Mach number distribution along the center body and at the cowl side obtained by the conformal and H type grids are compared in fig.3. The comparison puts into evidence a discrepancy after the rounded corner at the central body. The percent error on the total temperature, again evaluated along the two contours (fig.4), shows that the solution obtained by the conformal grid is globally the most accurate of the two. The most important causes of error are the wedge corner ($x \approx 0.85$), the sonic transition at the wall ($x \approx 1.15$), and the acceleration after the rounded corner ($1.15 \leq x \leq 1.3$).

To assess the effect on the solution at steady state of the TVD limiter, we compared the solutions obtained, on a conformal grid, turning on and off the TVD limiter (fig.5). The largest differences are found in the region along the conical wedge, and behind the oblique shock. This may be attributed to the lowered order of accuracy of the scheme in these zone of steep gradients because of the limiter action. The smoother flow away from the boundaries agrees rather well in the two solutions.

The Mach distribution relative to this last test (fig.6) shows again that the largest errors are found downstream the rounded corner, and that globally the scheme without limiter provides a more accurate solution (see the error distribution in fig.7). However, the limiter realizes a less oscillatory transition through the oblique shock, as conformed by the sound speed distribution along two different grid lines in fig.8.

References

1 M.Valorani, B.Favini, XII Congresso Nazionale AIDAA, Como (1993)

2 G. Moretti, AIAA J., 26 (1988)

3 G. Moretti, Comput. Fluids, 15 (1987)

4 L. Zannetti, B.Favini, Comput. Fluids, 17 (1989)

5 B.Favini, R.Marsilio, L.Zannetti, ISCFD, Nagoya (1989)

6 G.Moretti, NASA CP-2201 (1981)

7 C.P.Kentzer, Lecture Notes in Physics, 8 (1971)

8 T.J.Poinsot, S.K.Lele, J. Comput. Phys., 101 (1992)

9 K.Thompson, J. Comput. Phys., 89 (1990)

10 Gabutti, Comput. Fluids, 11 (1983)

11 A.Di Mascio, B.Favini, Meccanica, 26 (1991)

12 F.Nasuti, M.Onofri, M.Valorani, ICNMLTF, Swansea (Pineridge Press, 1993)

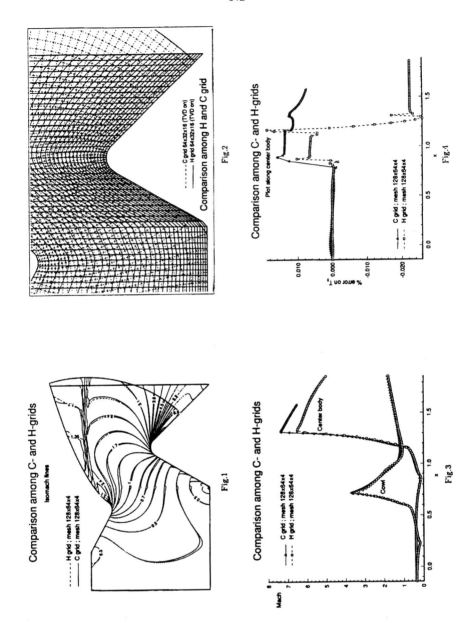

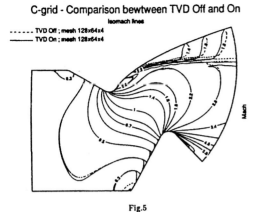

Fig.5

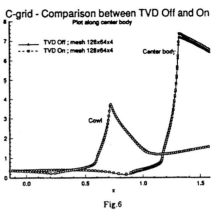

Fig.6

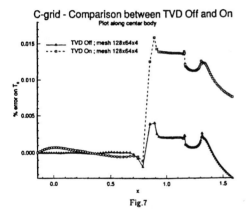

Fig.7

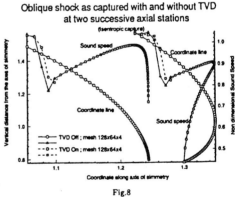

Fig.8

An Implicit-Explicit Flux Vector Splitting Scheme for Hypersonic Thermochemical Nonequilibrium Flows

Satoru Yamamoto, Hideo Nagatomo and Hisaaki Daiguji

Dept. of Aeronautics and Space Engineering,
Tohoku University, Sendai 980, Japan

1 Introduction

Hypersonic flows associated with chemical and thermal nonequilibrium flows usually occurred around re-entry vehicles must be calculated using not only ordinary mass, momentum and energy conservation equations based on the Navier-Stokes equations but mass equations of chemical species related to chemical reactions. Exactly considering, additional equations such as vibrational energy equations of each species taken account of the thermal nonequilibrium phenomena and an electronic energy equation for ionizations might be solved. Many of calculations for those flows have been already performed to design hypersonic vehicles[1]-[4]. Most of them only take mass equations for each chemical species, momentum equations, an energy equation for the translational-rotational temperature and an additional equation for the vibrational-electron temperature into consideration to get reasonable results under some approximations. In this paper, we solve same set of equations.

On the other hand, from a numerical point of view, many numerical approaches have been tried to get results fast and economically. The physical time scale of chemical reactions is extremely smaller than that of the fluid motion. If the flow is calculated explicitly, a very small time step must be taken to obtain a physical result. It seems to be necessary to employ an implicit unconditionally stable scheme based on the relaxation method such as the Gauss-Seidal method which is able to take a very large CFL number. Moreover, a robust shock capturing scheme must be also taken to get very strong bow shocks. Recently, several TVD-like schemes have been approached. But, in some research cases still the first-order upwind scheme into the sweep normal to the wall across the shock is used. It suggests that the existing TVD scheme may not always be robust on simulating hypersonic flows including very strong shocks.

The basic concept of making the present new flux vector splitting scheme for thermochemical nonequilibrium flows is not only in the points of fast, economical and robust calculations such as the above existing approaches, but those of easy and simple calculations, because most of existing approaches have relatively complex forms.

The present flux vector splitting scheme can be applied for the implicit calculation as well as the explicit calculation. It means that the same flux vector form in the explicit calculation can be used also in the implicit calculation only replacing the vector of unknown variables into the vector of the time derivatives. Therefore, most of complicated procedures, such as to calculate the eigenvectors, Jacobian matrices and so on, can be neglected. Since this scheme is based on the existing scheme in the Euler and Navier-Stokes solvers already developed by us [5][6], some of useful algorithms in the solvers can be also available. The Fourth-order Compact MUSCL TVD(FCMT) scheme, which can capture weak discontinuities and vortices as well as shocks more accurately than most of the existing shock capturing scheme, is one of them. The detail is explained in Ref.[6].

In the following sections, we will briefly explain the derivation of the present scheme, that is, an implicit-explicit flux vector splitting scheme, and some calculated results will be shown.

2 Fundamental Equations

The fundamental equations used here to calculate hypersonic thermochemical non-equilibrium flows are composed of continuity equations for chemical species (N_2, O_2, NO, NO^+, N, O and electron e), momentum equations, a total energy equation and a vibrational-electron energy equation in curvilinear coordinates, that is

$$\partial Q/\partial t + L(Q) \equiv \frac{\partial Q}{\partial t} + \frac{\partial F_i}{\partial \xi_i} + S + H = 0 \quad (1)$$

where $Q = J(\rho_s, \rho u_1, \rho u_2, E, E_v)$. Actually, Eq.(1) is composed of 11 equations. The values ρ_s, u_i, E, E_v, U_i, p in Q and F_i are the density of chemical species s, velocity components, total energy, vibrational-electron energy, contravariant velocity components and the pressure. Density ρ is calculated as $\sum_{s=1}^{n} \rho_s$. S and H mean the diffusion term and the source term, respectively.

The source terms for chemical reactions and for vibrational-electron energy are included in the vector H. In this paper, we use the most typical formulation such as employed in Ref.[4]. The details of them are also written in Ref.[2].

3 Numerical Methods

Some efficient implicit methods for the hypersonic flow problem have been proposed Refs.[2]-[4]. All of them are based on the Gauss-Seidal relaxation method. The present method used here is almost similar to the method in Ref.[4]. We briefly explain the procedure.

$A_i^{\pm}(i = 1, 2)$ are the Jacobian matrices composed of only positive or negative eigenvalues, respectively. $Q^{L(R)}$ is the vector of unknown variables, which is composed of $(q_s\ q_{u_1}\ q_{u_2}\ q_e\ q_{e_v}) = (\rho_s\ \rho u_1\ \rho u_2\ E\ E_v)$. $\delta Q^{L(R)}$ means the time derivative of $Q^{L(R)}$. L, R mean the weighted extrapolation from the left and right by the MUSCL approach.

The fundamental equations Eq.(1) are implicitly discretized as

$$\delta Q + \Delta t\{\Delta_i \delta F_i^n + \delta S^n + \delta H^n\} = RHS$$
$$RHS = -\Delta t(\Delta_i F_i^n + S^n + H^n) \tag{2}$$

The vectors of δS^n and δH^n are diagonally approximated as

$$\delta S^n \simeq \text{diag}(\alpha_j)\delta Q, \quad \delta H^n \simeq \text{diag}(\frac{1}{\tau})\delta Q \tag{3}$$

where $1/\tau \equiv 1/\tau_s$ ($s = 1, \cdots, n$ and E_v) have been proposed in Ref.[4] and α_j is set to $2\mu g_{jj}/(Re\rho\Delta\xi_j)$. If some approximations are performed such as $\delta Q_{\ell-1/2+k}^L \simeq \delta Q_{\ell-1+k}$, $\delta Q_{\ell-1/2+k}^R \simeq \delta Q_{\ell+k}$ ($k = 0, 1$), $\Delta\xi_1 = \Delta\xi_2 = 1$ and an efficient diagonal approximation proposed by Yoon[3], then Eq.(2) is finally simplified as

$$[I + \Delta t\{\beta\sigma(A_i) + \text{diag}(\frac{1}{\tau}) + \text{diag}(\alpha_j)\}]\delta Q$$
$$= RHS - \Delta t\{(A_j^-)_{\ell+1/2}\delta Q_{\ell+1} - (A_j^+)_{\ell-1/2}\delta Q_{\ell-1}\} \tag{4}$$

$\sigma(A_i)$ is the spectral radius of A_i. In this paper, the maximum of absolute eigenvalues for A_i is used. β is a relaxation parameter. The left hand of Eq.(4) is completely diagonalized. Consequently, Eq.(4) is algebraically calculated by the Gauss-Seidal method as

$$\delta Q = D^{-1}[RHS - \Delta t\{(A_j^-)_{\ell+1/2}\delta Q_{\ell+1} - (A_j^+)_{\ell-1/2}\delta Q_{\ell-1}\}] \tag{5}$$
$$D = [I + \Delta t\{\beta\sigma(A_i) + \text{diag}(\frac{1}{\tau}) + \text{diag}(\alpha_j)\}] \tag{6}$$

Eq.(5) can be also solved by LU-decomposition like LU-SGS method [3] to reduce the computational efforts.

In the present paper, a very useful form is derived, that is, an implicit-explicit flux vector splitting scheme which can be used in both explicit and implicit calculations. It means that not only the flux $A_i^\pm Q$, but the flux $A_i^\pm \delta Q$ can be calculated from a same flux vector splitting form. This form can be written in the vector form composed of sub-vectors as

$$A_i^\pm \tilde{Q} = J \begin{bmatrix} \tilde{q}_s \\ \tilde{q}_{u_1} \\ \tilde{q}_{u_2} \\ \tilde{q}_e \\ \tilde{q}_{e_v} \end{bmatrix} \lambda_{i1}^\pm + \frac{J}{c\sqrt{g_{ii}}} \begin{bmatrix} 0 & + & \bar{q}_s/\bar{q}_0 \cdot \Delta\bar{U}_i \\ \xi_{i,1}\bar{p} & + & \bar{q}_{u_1}/\bar{q}_0 \cdot \Delta\bar{U}_i \\ \xi_{i,2}\bar{p} & + & \bar{q}_{u_2}/\bar{q}_0 \cdot \Delta\bar{U}_i \\ \bar{U}_i\bar{p} & + & (\bar{\chi}^2 + c^2)/\tilde{\gamma} \cdot \Delta\bar{U}_i \\ 0 & + & \bar{q}_{e_v}/\bar{q}_0 \cdot \Delta\bar{U}_i \end{bmatrix} \lambda_{ia}^\pm$$

$$+ \frac{J}{c^2} \begin{bmatrix} \bar{q}_s/\bar{q}_0 \cdot \bar{p} & + & 0 \\ \bar{q}_{u_1}/\bar{q}_0 \cdot \bar{p} & + & \xi_{i,1}c^2/g_{ii} \cdot \Delta\bar{U}_i \\ \bar{q}_{u_2}/\bar{q}_0 \cdot \bar{p} & + & \xi_{i,2}c^2/g_{ii} \cdot \Delta\bar{U}_i \\ (\bar{\chi}^2 + c^2)/\tilde{\gamma} \cdot \bar{p} & + & \bar{U}_i c^2/g_{ii} \cdot \Delta\bar{U}_i \\ \bar{q}_{e_v}/\bar{q}_0 \cdot \bar{p} & + & 0 \end{bmatrix} \lambda_{ib}^\pm \tag{7}$$

where

$$\bar{p} = \tilde{q}_0\bar{\phi}^2 - \tilde{\gamma}(\bar{q}_{u_i}\tilde{q}_{u_i} - \bar{q}_0\tilde{q}_e - \bar{q}_0\tilde{q}_{e_v})/\bar{q}_0$$

$$\bar{\phi}^2 = \tilde{\gamma}(\bar{q}_{u_i}\bar{q}_{u_i}/2\bar{q}_0^2 - \sum_{s \neq c}^{n} \bar{q}_s h_s^0/\bar{q}_0 - \sum_{s \neq e}^{n} \bar{q}_s e_{els}/\bar{q}_0)$$

$$\bar{\chi}^2 = \tilde{\gamma}(\bar{q}_{u_i}\bar{q}_{u_i}/2\bar{q}_0^2 + \bar{q}_{e_v}/\bar{q}_0 + \sum_{s \neq e}^{n} \bar{q}_s h_s^0/\bar{q}_0 + \sum_{s \neq e}^{n} \bar{q}_s e_{els}/\bar{q}_0)$$

$$\Delta \bar{U}_i = \xi_{i,j}\tilde{q}_{u_j} - \tilde{q}_0 \xi_{i,j}\bar{q}_{u_j}/\bar{q}_0 \qquad (8)$$

$$\lambda_{ij}^\pm = (\lambda_{ij} \pm |\lambda_{ij}|)/2 \qquad (j = 1, 3, 4)$$

$$\lambda_{ia}^\pm = (\lambda_{i3}^\pm - \lambda_{i4}^\pm)/2, \quad \lambda_{ib}^\pm = (\lambda_{i3}^\pm + \lambda_{i4}^\pm)/2 - \lambda_{i1}^\pm$$

$\bar{Q}$ is calculated from the values at ℓ and $\ell+1$ points by an averaging technique, which avoids excessive numerical viscosity in the boundary layer. If we use Eq.(7) in the explicit calculation, then $\tilde{Q} = (\tilde{q}_s \ \tilde{q}_{u_1} \ \tilde{q}_{u_2} \ \tilde{q}_e \ \tilde{q}_{e_v})$ is specified to Q.

On the other hand, $\tilde{Q}$ is set to δQ if it is for the implicit calculation. That is

$$A_i^\pm \delta Q = J \begin{bmatrix} \delta q_s \\ \delta q_{u_1} \\ \delta q_{u_2} \\ \delta q_e \\ \delta q_{e_v} \end{bmatrix} \lambda_{i1}^\pm + \frac{J}{c\sqrt{g_{ii}}} \begin{bmatrix} 0 & + & \bar{q}_s/\bar{q}_0 \cdot \Delta\bar{U}_i \\ \xi_{i,1}\bar{p} & + & \bar{q}_{u_1}/\bar{q}_0 \cdot \Delta\bar{U}_i \\ \xi_{i,2}\bar{p} & + & \bar{q}_{u_2}/\bar{q}_0 \cdot \Delta\bar{U}_i \\ \bar{U}_i\bar{p} & + & (\bar{\chi}^2+c^2)/\tilde{\gamma} \cdot \Delta\bar{U}_i \\ 0 & + & \bar{q}_{e_v}/\bar{q}_0 \cdot \Delta\bar{U}_i \end{bmatrix} \lambda_{ia}^\pm$$

$$+ \frac{J}{c^2} \begin{bmatrix} \bar{q}_s/\bar{q}_0 \cdot \bar{p} & + & 0 \\ \bar{q}_{u_1}/\bar{q}_0 \cdot \bar{p} & + & \xi_{i,1}c^2/g_{ii} \cdot \Delta\bar{U}_i \\ \bar{q}_{u_2}/\bar{q}_0 \cdot \bar{p} & + & \xi_{i,2}c^2/g_{ii} \cdot \Delta\bar{U}_i \\ (\bar{\chi}^2+c^2)/\tilde{\gamma} \cdot \bar{p} & + & \bar{U}_i c^2/g_{ii} \cdot \Delta\bar{U}_i \\ \bar{q}_{e_v}/\bar{q}_0 \cdot \bar{p} & + & 0 \end{bmatrix} \lambda_{ib}^\pm \qquad (9)$$

$$\bar{p} = \delta q_0 \bar{\phi}^2 - \tilde{\gamma}(\bar{q}_{u_i}\delta q_{u_i} - \bar{q}_0\delta q_e - \bar{q}_0\delta q_{e_v})/\bar{q}_0$$

$$\Delta\bar{U}_i = \xi_{i,j}\delta q_{u_j} - \delta q_0 \xi_{i,j}\bar{q}_{u_j}/\bar{q}_0$$

We need no additional calculations such as matrices for eigenvectors.

4 Numerical Results

A typical test case for the hypersonic thermochemical nonequilibrium flow problem is calculated [2]. The computational grid has 39 × 31 grid points generated around a quarter of sphere by the axisymmetric assumption.

The initial conditions are specified as $M_\infty = 15.3$, $u_\infty = 5280$[m/s], $p_\infty = 664$[Pa] and $T_\infty = 293$[K]. The wall conditions are treated as non-catalytic and $T_{\text{wall}} = (T_v)_{\text{wall}} = 1000$[K].

Figure 1 shows the calculated distributions of T and T_v along the stagnation line in which the solid lines show the results by the 3rd-order MUSCL TVD scheme and the broken lines show those by the 1st-order upwind scheme. Figure 2 shows the temperature contours of T_v. Same problem has been already calculated by Candler[2] and Imlay[4]. Being compared to the results, the maximum temperature of T in Fig.1 is a little bit higher than that by Candler's but almost same with that by Imlay about the 1st-order results. The shock distance from the wall along the stagnation line is almost same with each other. It means that our results also agree with the experiment by Lobb's[4].

Figures 3 and 4 show the Mach number contours and the contour of the O_2 mass concentration. It can be seen that O_2 is gradually dissociated to O from just behind the shock and finally is completely dissociated near the mid-point between the shock and the wall.

5 Conclusions

An implicit-explicit flux vector splitting scheme which can be used in both explicit and implicit calculations is proposed and applied to a hypersonic thermochemical nonequilibrium flow problem. Since the consistency of these calculations must be remarkably good, the effort to make the present computational code has been greatly reduced. In this paper, consequently the numerical algorithm of the present scheme is weightedly explained.

References

[1] Gnoffo, P.A., McCandless, R.S. and Yee, H.C., "Enhancements to Program Laura for Computation of Three-Dimensional Hypersonic Flow," AIAA Peper 87-0280 (1987).

[2] Candler, G., 'The Computation of Weakly Ionized Hypersonic Flows in Thermochemical Nonequilibrium," Ph.D Thesis, Stanford University, (1988).

[3] Park, C. and Yoon, S., "Fully Coupled Implicit Method for Thermochemical Nonequilibrium Air at Suborbital Flight Speeds.," *J. Spacecraft and Rockets*, Vol.28, No.1 (1991), pp.31-39.

[4] Imlay, S.T. and Eberhardt, S., "Nonequilibrium Thermo-Chemical Calculations using a Diagonal Implicit Scheme," AIAA Peper 91-0468 (1991).

[5] Daiguji, S., Yamamoto, S. and Ishizaka, K., "Numerical Simulation of Supersonic Mixing Layers Using a Fourth-Order Accurate Shock Capturing Scheme," *Proc. of the 13th International Conf. on Numerical Methods in Fluid Dynamics-Oxford, Lecture Notes in Phys.*, **414** (1992), pp.315-319, Springer-Verlag.

[6] Yamamoto, S. and Daiguji, S., "Higher-Order-Accurate Upwind Schemes for Solving the Compressible Euler and Navier-Stokes Equations," *J. Compu. & Fluids*, Vol.22, No.2/3(1993), pp.259-270.

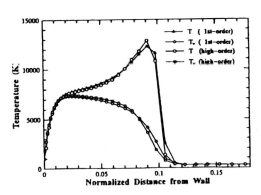

Fig.1 Temperature distributions

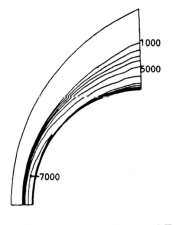

Fig.2 Temperature contours of T_v

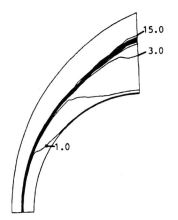

Fig.3 Mach number contours

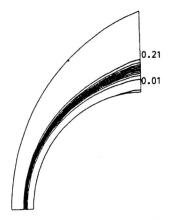

Fig.4 O_2 mass concentration

Computation of Incompressible Flows and Direct Simulation of Transition to Turbulence

Ch.-H. Bruneau[1] and P. Fabrie[2]

[1]CeReMaB - URA 226, Université Bordeaux I - 33405 Talence - France
[2]CeReMaB - URA 226, et LEPT - ENSAM URA 873

Abstract : The aim of this paper is to build a computational method efficient enough to carry out a direct simulation of transition to turbulence. This requires to have an accurate time and space scheme, as well as a fast solver as large time behaviours are needed. The flow is computed behind a cylinder in a channel and new open boundary conditions are set downstream. Numerical tests show various solutions from stable steady state to unstructured flows. To complete this study, a spectrum analysis is performed via Fourier transform.

1 Introduction

These last ten years, many authors have dealt with direct simulation of transition to turbulence for various geomertices [5], [6]. They find good qualitative scenario of transition in agreement with theoretical studies of dynamical systems. Our purpose is to build a robust and precise method able to capture realistic solutions at a given time. Indeed, in most cases, the complex solutions of incompressible Navier-Stokes equations have only a qualitative meaning due to numerical viscosity and time discretization. In this paper we study the flow behind a cylinder in a channel. We focus our attention on space discretization to take into account small eddies, on time discretization to follow well the solution during a large time and finally on the open boundary conditions to convey properly the vortices downstream [2]. For efficiency, we use a multigrid method with a cell by cell relaxation procedure as smoother like in [4].

Numerical tests establish the robustness of the method which is able to explore a large range of Reynolds numbers and to deal with unstructured flows. Moreover, we choose small time and space steps such that the frequency of a periodic solution reaches a stable value, independent of extra refinement.

2 Governing equations and boundary conditions

The dimensionless Navier-Stokes equations for incompressible flows are written in the domain $\Omega \times (0,T)$ (figure 1) as:

(1) $$\frac{\partial U}{\partial t} + (U \cdot \nabla).U - \text{div } \sigma(U,p) = 0$$

(2) $$\text{div } U = 0$$

where $U = (u_1, u_2)$ is the velocity, p the pressure, Re the Reynolds number and $\sigma(U,p)$ the stress tensor.

(3) $$\sigma(U,p) = \frac{2}{Re} D(U) - p\, I$$

where $$D(U)_{ij} = \frac{1}{2}\left(\frac{\partial u_i}{\partial x_j} + \frac{\partial u_j}{\partial x_i}\right).$$

The equations (1), (2) are associated with the initial condition

(4) $$U(x,0) U_0 \text{ in } \Omega$$

and boundary conditions

(5) $$U(x_1, x_2) = U_p \text{ on } \Gamma_{D,1} \times (0,T)$$

(6) $$U(x_1, x_2) = (0,0) \text{ on } \Gamma_{D,0} \times (0,T)$$

(7) $$\sigma(U,p).n + \frac{1}{2}(U.n)^-(U - U_p) = \sigma(U_p, p_0).n \text{ on } \Gamma_N \times (0,T)$$

Where $(U_p, p_0) = ((6x_3(1-x_2), 0), p_0)$ is Poiseuille solution with pressure normalized to zero downstream, n is the outward unit normal vector on Γ_N and any real number α is written $\alpha = \alpha^+ - \alpha^-$.

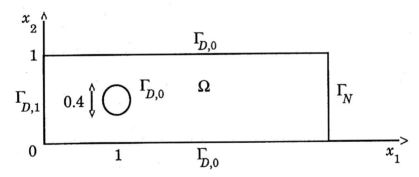

Figure 1 - Domain and notations

The condition (7) yields a well-posed problem as it is shown in [3]; it gives two equations and is coupled to the continuity equation to determine the three unknowns (u_1, u_2, p) in one or two fictitious cells downstream.

$$\frac{\partial u_2}{\partial x_1} = -\frac{\partial u_1}{\partial x_2} - \frac{Re}{2}(u_1)^- u_2 + 6(1 - 2x_2).$$

$$\frac{\partial u_1}{\partial x_1} = -\frac{\partial u_2}{\partial x_2}$$

$$p = \frac{2}{Re}\frac{\partial u_1}{\partial x_1} + \frac{1}{2}(u_1)^-(u_1 - 6x_2(1-x_2))$$

3 Outline of the method and numerical results

To follow well unsteady solutions, the key point is the time discretization. For accuracy, we choose a third order Gear-like scheme with implicit treatment of diffusion terms and explicit treatment of convection terms. Indeed, as it is pointed out by one dimensional tests on Burgers equation and on advection diffusion equation, the implicit treatment of advection terms adds artificial diffusion.

Then, at each time step, we solve a Stokes problem by means of a multigrid algorithm with a cell by cell relaxation procedure. The discretization is achieved on staggered grids. So, on each cell, we solve a Stokes problem for the five unknowns of the cell, four velocities at the middle of the sides and the pressure.

The space discretization is made by centered finite differences except for convection terms for which an upwind anti-diffusion scheme is used [1]. Consequently, the convection terms are computed on the right hand side as follows:

$$\left(u_1 \frac{\partial u_1}{\partial x_1}\right)_i = (u_1)_{i-1/2} \left(4(u_1)_i - 5(u_1)_{i-1} + (u_1)_{i-2}\right)/3\delta x \quad \text{if } (u_1)_{i-1/2} > 0$$

$$- (u_1)_{i+1/2} \left(4(u_1)_i - 5(u_1)_{i+1} + (u_1)_{i+2}\right)/3\delta x \quad \text{if } (u_1)_{i+1/2} < 0$$

The numerical tests are performed for large Reynolds numbers up to several thousands. For $Re = 100$, we get a symmetric steady state solution. Then this steady solution looses its stability and at $Re = 200$ we obtain a purely periodic solution the frequency of which f is close to one. In this solution, the vortices are confined to a small area behind the cylinder whereas for $Re = 100$, the solution exhibits alternate vortices that cross the open boundary downstream (figure 2). However, we still observe a purely periodic solution until $Re = 2100$. From $Re = 2200$ to 3600 the solution remains periodic with two sub-harmonics of the fundamental frequency $f \cong 1$ (figure 3). At $Re = 3700$, there is again a periodic solution with seven sub-harmonics. Then we observe quasi-periodic flows that evolve to chaos as the Reynolds number increases. Consequently, the phase portrait becomes more and more dense and the spectrum becomes almost continuous (figure 3).

Moreover, the downstream boundary condition does not affect the flow as it is shown for a test on a shorter channel (figure 2) [2].

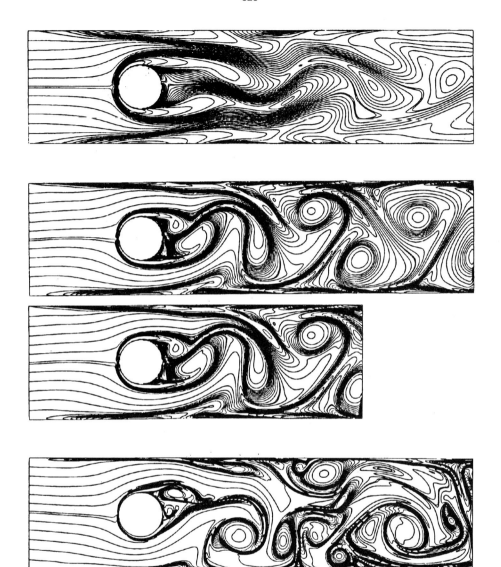

Figure 2 - Vorticity lines for $Re=200$, $Re=1000$ and $Re=10000$.

4 Conclusions

In the paper, we propose an efficient and accurate method to solve unsteady Navier-Stokes equations in an open domain. We demonstrate the quality of the method by numerous numerical tests for high Reynolds numbers. Moreover, we believe that solutions computed on grids as fine as 1024×256 have a quantitative meaning.

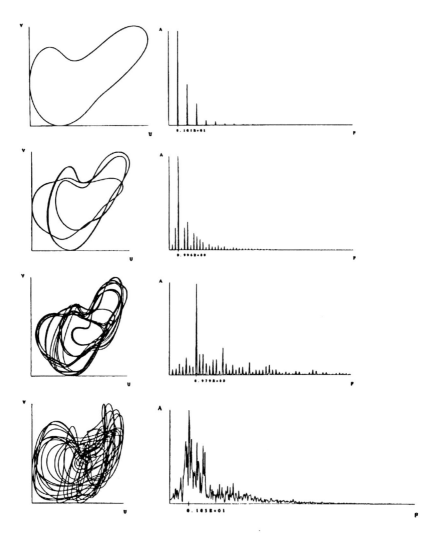

Figure 3 - Phase portraits and spectra for $Re=1000$, $Re=2500$, $Re=3700$ and $Re=10000$

References

1. Ch.-H. Bruneau, Thése d'Etat, Orsay (1989)
2. Ch.-H. Bruneau and P. Fabrie, *to appear in Int. J. Num. Meth. Fluids*
3. Ch.-H. Bruneau and P. Fabrie, *submitted to SIAM J. Math. Anal.*
4. Ch.-H. Bruneau and C. Jouron, *J. Comp. Physics* 89 (1990)
5. A. Fortin, M. Fortin and J.J. Gervais, *private comminication*
6. J.W. Goodrich, K. Gistafson and K. Halasi, J. Comp. *Physics* 90(1990)

Incomplete LU Decomposition in Nodal Integral Methods for Laminar Flow

W. J. Decker and J. Dorning[1]

Engineering Physics Program, University of Virginia,
Charlottesville, VA 22903-2442, USA
[1] Also Dept. of Appl. Math. and Dept. of Mech. Aero. and Nucl. Engr.

1 Introduction

An efficient iterative method using incomplete lower- and upper-triangular matrix decomposition (LUD) for a nodal integral method (NIM) discretization of the incompressible Navier-Stokes (N-S) equations and the Navier-Stokes-Boussinesq (N-S-B) equations has been developed. The composite method was implemented and tested for the N-S equations describing mechanically forced convection and the N-S-B equations describing thermal convection – for steady-state and time-dependent 2-D and 3-D flows. The extension of the formulation of the NIM discrete-variable equations from 2-D to 3-D flows is straightforward; however, the resulting system presents some practical problems because of the real limitations of computer storage when using direct methods to invert the matrices involved in the solution of large problems. In this composite method, an extremely memory-efficient overlapping-block iterative technique using the incomplete LU decomposition (ILUD) scheme of Radicati and Robert (Radicati and Robert, 1987) was employed to invert the NIM Jacobian matrix within each of the Newton iterations used to solve the final nonlinear discrete-variable equations. Comparisons have been made among various ILUDs of the Jacobian to determine which is optimal with respect to speed and memory use. Also, the Sherman-Morrison formula (Gollub and Van Loan, 1989) was examined in order to exploit the sparseness of the matrix when possible and thereby eliminate a portion of the floating point operations needed to invert the Jacobian. The resulting scheme greatly reduces memory requirements to the point of making it possible to solve 3-D fluid dynamics problems on workstations and even on PCs. The power and efficiency of the composite overlapping-block iterative nodal integral method was demonstrated by using it to solve steady-state and time-dependent versions of the classic driven-cavity and double-glazing problems of computational fluid dynamics on coarse 2-D and 3-D computational grids.

2 The Iterative Method and Numerical Results

The set of nonlinear equations for the NIM discrete-variable unknowns (velocity, temperature and normal stress) derived from the N-S or N-S-B equations normally is solved via Newton's method (Azmy and Dorning, 1983). Therefore, the Jacobian must be inverted at each Newton iteration. To reduce the amount of memory required for this procedure, we have identified an iterative technique appropriate for the structure of the NIM Jacobian which obviates the need to store the whole (global) matrix. In a different context, Radicati and Robert (Radicati and Robert, 1987) recognized that a block diagonal matrix with blocks that overlap can be inverted efficiently via ILUD. Based on their ideas, we have used a similar ILUD as the basis of a Gauss-Seidel-like iterative procedure to invert the NIM Jacobian. We have identified the overlapping blocks in it as "local" or "template" Jacobian matrices, each a "global" Jacobian corresponding to the equations on a subgrid of the global grid with local boundary conditions temporarily taken as known. This subgrid may be chosen for 2-D (3-D) problems from grids ranging from a 2×2 ($\times 2$) subgrid to the entire global grid.

In an attempt to further reduce the number of floating point operations, we also have used the fact that the subgrid template Jacobian matrices have the form $J_s = A_s + T_s$, where A_s is the Jacobian of the linear terms in the discrete-variable equations and T_s is the Jacobian of the nonlinear terms. Because only a few of the terms are truly nonlinear (those that result from the advection terms), T_s is sparse and it might be advantageous to use the Sherman-Morrison formula (Gollub and Van Loan, 1989) to reduce the number of floating point operations, and therefore the amount of time needed to invert the matrix. We accomplished this by first noting that A_s need be inverted only once, since for a uniform grid the linear terms do not depend on the location of the subgrid within the global grid, and then the Sherman-Morrison formula is used to construct the inverse of $A_s + T_s$ for each subgrid template at each iteration.

3 Numerical Results

We have studied the amount of CPU time it takes to solve the flow problems using different ILUDs based on the inverse found via LUD of templates of different sizes. The results for the technique applied to time-dependent flows are omitted because they lead to the same conclusions as the steady-state results presented below. We also have compared the iterative technique using LUD to invert the subgrid template Jacobian with the iterative technique using the Sherman-Morrison formula to invert it. The computations were done on an IBM RS6000/340, with 32 megabytes of memory. The four steady-state problems, all starting with a zero initial guess, were the following: a 2-D driven-cavity problem, a 2-D double-glazing problem, a 3-D driven-cavity problem, and a 3-D double-glazing problem.

The process of inverting the 367×367 matrix formed when solving the 2-D steady-state N-S equations for the driven-cavity problem in a square cavity with Re = 100 on an 8×8 grid was examined using a range of templates, from 2×2 to 8×8 subgrids (19×19 to 367×367 matrices), which corresponds to six different ILUDs and a direct LUD method. In all cases solving the problem required six Newton iterations to reach the ℓ_1-norm convergence criterion $\epsilon < 10^{-6}$. The ℓ_1-norm convergence criterion for the iterative technique was $\delta < 10^{-12}$. The CPU times for various templates, corresponding to different ILUDs, are given in Table 1. As expected, direct inversion took the least time. However, since memory often is an important consideration, it is useful to consider seperately just the three templates that used the least memory; then the shortest time was taken by the decomposition using the 3 × 3 subgrid template (47 × 47 matrix). Further, because of the sparseness of T_s, using the technique based on the Sherman-Morrison formula significantly reduces the CPU time, as seen by comparing columns two and three in Table 1. We also examined the process of inverting the 495 × 495 matrix formed when solving the 2-D steady-state N-S-B equations for the double-glazing problem in a square cavity with Ra = 1000 and Pr = 0.71 on an 8 × 8 grid. The templates used ranged from 2 × 2 to 8 × 8 subgrids (23 × 23 or 25 × 25 – depending on the template's proximity to the insulated boundary – to 495 × 495 matrices). For all templates five Newton iterations were required to reach the $\epsilon < 10^{-6}$ convergence criterion. The CPU times for various templates are given in Table 2. Among the three templates that required the least memory, the shortest CPU time was taken by the decomposition using the 3 × 3 template. Also, because T_s is less sparse – owing to the addition of the temperature terms present in thermal convective flow – using the Sherman-Morrison formula does **not** reduce the CPU time, as seen by comparing columns three and two of Table 2.

Three-dimensional flows also were studied. Based on the 2-D calculations and some representative 3-D calculations it was assumed that T_s was sparse enough in the 3-D N-S calculations to take advantage of the Sherman-Morrison formula and that it was not in the 3-D N-S-B calculations. Further, because of memory limitations on the RS6000/340, we only compared the CPU times based on 2 × 2 × 2 templates and 3 × 3 × 3 templates. We examined the process of inverting the 2375 × 2375 matrix formed when solving the 3-D steady-state N-S driven-cavity problem in a cubic cavity with Re = 100 on a 6 × 6 × 6 grid using two templates, 2 × 2 × 2 and 3 × 3 × 3 subgrids (71 × 71 and 161 × 161 matrices, respectively), which corresponds to two different ILUDs. Each required six Newton iterations to reach $\epsilon < 10^{-6}$. The CPU times for the two templates are given in Table 3. The shorter time was taken by the decomposition using the 2 × 2 × 2 template. We also studied the process of inverting the 3059 × 3059 matrix formed when solving the 3-D steady-state N-S-B double-glazing problem in a cubic cavity with Ra = 1000 and Pr = 0.71 on an 6 × 6 × 6 grid using two templates: 2 × 2 × 2 and 3 × 3 × 3 subgrids (99 × 99, 107 × 107 or 115 × 115 and 359 × 359, 368 × 368 or 377 × 377 matrices, depending on the proximity to the boundary). Five Newton iterations were required to reach $\epsilon < 10^{-6}$. The CPU times are given in Table 4. As in the driven cavity problem, the shorter time was taken by the iterative technique based on the 2 × 2 × 2 template.

Finally, we studied the CPU times for the fastest iterative technique for each of the four test problems studied (using 8 × 8 grids for the 2-D flows, but only with 4 × 4 × 4 grids for the 3-D flows) on the RS6000/340, and compared them with the CPU times on a PC with a 25 megahertz Intel i486 central processor and eight megabytes of memory. Table 5 shows the comparison between these two computing platforms and demonstrates that this iterative technique makes it possible to solve 3-D flow problems on a personal computer. (The large amount of time it took the PC to solve the 3-D double-glazing problem was due to the extensive swapping to disk.)

As mentioned above, the results for the time-dependent versions of the four problems are not included here because they led to the same conclusions as those for the steady-state problems. The flow fields that resulted from a typical time-dependent 3-D calculation for the double-glazing problem is shown in Fig. 1.

4 Conclusion

We have developed and studied an iterative procedure that uses ILUD of the Jacobian that arise in NIMs for the N-S and N-S-B equations. This procedure makes possible inversion of large matrices while meeting the goal of dramatically reducing memory requirements, so that using the composite method we now can compute 3-D flows on platforms ranging from PCs to massively parallel computers. (The same ILUDs that make it possible to solve these problems on PCs also make possible their parallelization.) We have examined how different ILUDs, resulting in Jacobians for different size templates, affect the amount of CPU time needed to carry out a calculation on an IBM RS6000/340 and on an i486, 25 megahertz PC, and we found the optimal decompositions for typical classes of 2-D and 3-D CFD problems. This research was supported by NASA under Grants No. NAGW-3021 and No. NGT-51063.

Table 1.

Template Size	CPU Time (LU) (sec)	CPU Time (S-M) (sec)
2 × 2	1202.3	618.8
3 × 3	794.0	561.6
4 × 4	1013.9	757.6
5 × 5	1050.8	798.6
6 × 6	982.2	721.6
7 × 7	684.5	502.8
8 × 8	91.3	60.0

Table 2.

Template Size	CPU Time (LU) (sec)	CPU Time (S-M) (sec)
2 × 2	1443.6	1523.5
3 × 3	1116.0	1958.8
4 × 4	1411.4	2654.4
5 × 5	1548.4	3406.0
6 × 6	1644.1	4015.4
7 × 7	1996.8	4103.8
8 × 8	278.6	696.4

Table 3.

Templ. Size	CPU Time (sec)
$2 \times 2 \times 2$	11268.3
$3 \times 3 \times 3$	14143.3

Table 4.

Templ. Size	CPU Time (sec)
$2 \times 2 \times 2$	12611.1
$3 \times 3 \times 3$	14951.3

Table 5.

Flow Prob. (Templ.)	Time (340) (sec)	Time (PC) (sec)
2-D D. C. (3×3)	561.6	3420.4
2-D D. G. (3×3)	1116.0	8678.5
3-D D. C. (2×2×2)	574.8	5951.5
3-D D. G. (2×2×2)	1177.6	53680.9

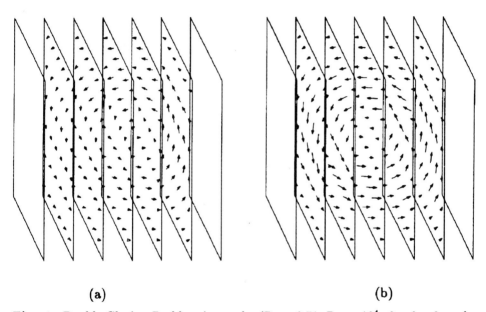

(a) (b)

Fig. 1. Double-Glazing Problem in a cube (Pr = 0.71, Ra = 10^4, 6 × 6 × 6 mesh, Δ t = 0.01 [dimensionless units]); (a) t = 0.04, (b) t = 4.0 (asymptotic)

References

1. Y. Y. Azmy and J. J. Dorning, A Nodal Integral Approach to the Numerical Solution of Partial Differential Equations, Advances in Reactor Computations, Volume II, 1983, pp. 894 – 909.

2. Gene Gollub and Charles Van Loan, Matrix Computations, Second Edition, The Johns Hopkins University Press, Baltimore, MD, 1989.

3. G. Radicati and Y. Robert, Vector and Parallel CG-like Algorithms for Sparse Non-Symmetric Systems, IMAG/TIM3 Technical Report, Grenoble, France, 1987.

On Turbulent Flow in a Sudden Pipe Expansion and its Reverse Transfer of Subgrid-Scale Energy

Rainer Friedrich and Claus Wagner

Lehrstuhl für Fluidmechanik, TU München
Arcisstr. 21, D-80333 München, Fed. Rep. of Germany

Abstract : Results of a direct numerical simulation (DNS) of turbulent sudden pipe expansion flow at a Reynolds number based on upstream centerline velocity and diameter of 6950 are presented. They are obtained with a central finite-difference scheme of second order accuracy in space and time. It is shown that the scheme is well suited for DNS as long as the smallest turbulent scales are properly resolved. One-dimensional spatial sharp cut-off filtering of the data is performed in order to assess the energy transfer between resolved and unresolved motions in large-eddy simulations (LES). Backward transfer of kinetic energy from subgrid-scale (SGS) to grid-scale turbulent motions is found in the mean which cannot be described by simple eddy-viscosity type SGS models.

1 Numerical Method and Computational Details

The turbulent flow in a sudden pipe expansion involves several complex flow phenomena the accurate prediction of which is difficult not only for statistical modeling but also for large-eddy simulation. There is a mixing layer that rapidly develops downstream of the expansion and hits the wall forming a broad reattachment region. It drives the primary recirculation zone. DNS is a tool capable of describing these phenomena in all details. The geometry of the flow makes the use of spectral numerical methods very complicated. A finite-difference approach is therefore adopted.

We use Schumann's (1975) volume balance procedure to generate a second order accurate finite-difference form of the incompressible, non-dimensional Navier-Stokes equations in cylindrical (z, φ, r)-coordinates, *viz*:

$$\sum_{\alpha} \left(\Delta A_\alpha \, \bar{u}_\alpha \, |_{\xi_\alpha^+} - \Delta A_\alpha \, \bar{u}_\alpha \, |_{\xi_\alpha^-} \right) = 0, \qquad (1)$$

$$\Delta V \frac{\partial \bar{u}_\alpha}{\partial t} +$$
$$\sum_{\beta} \left[\Delta A_\beta \left(\bar{u}_\beta \bar{u}_\alpha + \bar{p} \delta_{\alpha\beta} - \bar{\tau}_{\alpha\beta} \right) |_{\xi_\beta^+} - \Delta A_\beta \left(\bar{u}_\beta \bar{u}_\alpha + \bar{p} \delta_{\alpha\beta} - \bar{\tau}_{\alpha\beta} \right) |_{\xi_\beta^-} \right]$$
$$- (\text{Term})_\alpha = 0. \qquad (2)$$

(Term)$_\alpha$ contains all curvature terms (See Wagner and Friedrich 1994a). $\xi_\alpha^+, \xi_\alpha^-$ indicate positions $\xi_\alpha \pm \Delta\xi_\alpha/2$.
The shear stress is defined as

$$\overline{\tau}_{\alpha\beta} = \overline{S}_{\alpha\beta}/Re_\tau. \qquad (3)$$

The Reynolds number $Re_\tau = u_\tau D/\nu$ is based on friction velocity and pipe diameter of the incoming flow. A typical strain rate component is

$$\overline{S}_{r\varphi} = r\delta_r(\overline{u}_\varphi/r) + \frac{1}{r}\delta_\varphi \overline{u}_r. \qquad (4)$$

δ_r, δ_φ represent central difference operators. The flux differences across the mesh volume ΔV in (1), (2) correspond to central differences as well. The overbars over velocity components denote averages taken over mesh surfaces, like $\Delta A_z = r\Delta r \Delta \varphi$, while $\overline{p}$ represents a volume average which is stored in the cell-center. Equation (2) is integrated in time on staggered grids using a semi-implicit scheme in which all convection and diffusion terms containing derivatives in the circumferential direction are treated implicitly. A leapfrog-scheme which is second-order accurate serves to integrate the remaining convection terms in time. An averaging step all the 50 time steps avoids possible $2\Delta t$-oscillations. Diffusive terms with derivatives in (r,z)-directions are treated with a first-order Euler backward step. The time-step is chosen following a linear stability analysis. A fractional step method provides the coupling between pressure and velocity fields. It leads to a Poisson equation for the pressure which is solved using a Fast Fourier Transform in φ-direction. The remaining set of 2D Helmholtz problems in irregular domains is treated with a cyclic reduction algorithm in combination with the influence matrix technique (Schumann 1980).

The geometric singularity of the basic partial differential equations on the pipe centerline does not pose serious problems in the finite-difference formulation (1-4). It requires an approximation concerning the radial velocity component only. This component is extrapolated from neighbouring mesh points.

1.1 Boundary, Initial Conditions and Simulation Data

Due to the elliptic nature of the problem considered, boundary conditions are required for all boundaries of the computational domain. At walls the wall-normal velocity component is taken as zero (impermeable walls). Since tangential velocity components are not defined there, the aligned shear stresses are prescribed instead.

The specification of inflow boundary conditions is a complicated task. We have chosen an accurate way of generating these conditions by performing a DNS of fully developed pipe flow separately. It provides the 3D time-dependent velocity vector for each cell of the inflow plane at each time step and guarantees the correct physical flow behaviour there. In the outflow plane the mean velocity vector is linearly extrapolated, whereas its fluctuating part is computed from a pure convection equation with the local mean velocity as convective speed.

The DNS of fully developed pipe flow is started with velocities consisting of experimental mean values plus random fluctuations satisfying the constraints of incompressibility and typical rms profiles. The Reynolds number is initially increased by a factor of 5 and then gradually decreased during one eddy-turnover time (D/u_τ) in order to avoid the decay of initial fluctuations. Non-zero constant velocities are chosen as initial conditions for the DNS of turbulent pipe expansion flow. Due to the use of exact Dirichlet boundary conditions in the inflow plane these unphysical initial values are swept out of the computational domain in the course of six eddy-turnover times.

The computational domain consists of an upstream pipe section of diameter D and length 1.48 D. The downstream section has the diameter 1.2 D and the length 2.03 D. An equidistant cylindrical grid system with 180 × 128 × 115 cells in (z, φ, r)-directions is used. The grid spacing in terms of wall units (of the incoming flow) is:

$$\Delta z^+ = 7.03, \quad (r\Delta\varphi)^+ = 0.046, \ldots, 8.78, \quad \Delta r^+ = 1.87. \tag{5}$$

It is comparable to that of Kim et al. (1987) for plane channel flow. Computation of the minimal Kolmogorov length in the upstream pipe section provides the value 1.56. A proper resolution of all turbulent scales is thus guaranteed.

A great number of statistically independent realizations of the flow must be provided to ensure stable averages. Out of 247 000 time steps 780 samples have been taken for averaging. These samples correspond to 800 GB of data. Between each sample there is a time lag of 250 time steps or 1/20 th of a eddy-turnover time. Such a simulation requires 745 CPU hours on a CRAY-YMP 8/864 and 29 MW of memory. The CPU-time per time step and grid point is $3.9 \cdot 10^{-6}$ sec.

The DNS of fully developed pipe flow in a domain which is 5 D long and contains 256 × 128 × 96 cells needs around 500 CPU hours and 29.7 MW of memory. This is indeed a very expensive way of generating inflow boundary conditions.

1.2 Comparison Between Pipe and Channel Flow

The Reynolds number $Re_\tau = u_\tau D/\nu$ for fully developed pipe flow has been chosen 360 in order to allow for a comparison with the spectral DNS data of Kim, Moin & Moser (1987) for channel flow. Their Reynolds number based on channel width was $u_\tau h/\nu = 360$. Figure 1 shows profiles of rms-vorticity fluctuations. The indices φ and r indicate the spanwise and wall normal directions in the case of channel flow. Although pipe and channel data are very close, there are small systematic differences. They are due to transverse curvature (or Reynolds number) effects, see Eggels et al. (1994).

2 DNS Results of Pipe Expansion Flow

The complicated instantaneous flow structure in the recirculation zone and in the mixing layer at $z/D = 2.2$ is demonstrated in Fig. 2. The mixing layer is extremely corrugated by high speed fluid which comes from the core region and penetrates the recirculation zone causing local patches of positive axial momentum although the mean flow is reversed.

Figure 3 shows fluctuating velocity vectors in cylinder surfaces ($y^+ = 14$) of both pipe sections, unrolled into the plane. The incoming flow is characterized by streaky structures which have a mean spacing of 120 wall units. Such structures will develop again, but only far downstream of the mean reattachment line. Around this line the mean turbulent transport of fluctuating kinetic energy towards the wall is strongly enhanced. It in turn leads to a redistribution of energy from the radial to the axial and circumferential velocity components, as evident from the vector plot.

The flow is axially symmetric in the mean. The primary separation zone is 10.2 step heights or 1.02 D long. Contour lines of $<u_z>$ are plotted in Fig. 4. Dotted lines indicate reversed flow. In the corner appears a very weak and small secondary recirculation zone. The highest turbulence activity occurs in the mixing layer around $r = D/2$, $2.3 < z/D < 2.5$ from where it is convected and diffused towards the axis. The turbulence intensities exceed those of the incoming flow by factors of 2 roughly. Correspondingly the rate of dissipation of turbulent kinetic energy reaches much higher values than upstream (cf. Wagner, Friedrich 1994).

3 Reverse Transfer of Subgrid-Scale Energy

The fact that the turbulent flow in a sudden pipe expansion contains regions of fast downstream development will certainly be reflected in the small scale motions. Thus, intelligent subgrid-scale (SGS) models are needed in large-eddy simulations of separated flows which account for the energy transfer among different scales in the right way. The question arises whether the presently available models have this potential.

In a LES the velocity fields are decomposed as

$$u_i = \tilde{u}_i + u_i''. \tag{6}$$

$\tilde{u}_i$ represents the resolved or grid-scale (GS) part and u_i'' the unresolved (SGS) part. Mostly $\tilde{u}_i$ is based on spatial low-pass filtering (see Aldama 1990). In the present case we apply a one-dimensional ideal filter in the circumferential direction. Its space Fourier transform is

$$\hat{G}(k_\varphi) = \begin{cases} 1 & for \mid k_\varphi \mid \leq K_\varphi^c = \pi/(r\Delta\varphi) \\ 0 & otherwise. \end{cases} \tag{7}$$

The non-dimensional cut-off wavenumber $\left(K_\varphi^c\right)^+$ has been assumed 0.065 which corresponds to a grid spacing of $(r\Delta\varphi)^+ = 48$ at $r^+ = 170$. Thus, 22 grid points are retained in φ-direction.

The space Fourier transform of $\tilde{u}_i$ is obtained from

$$\hat{\tilde{u}}_i(z, k_\varphi, r, t) = \hat{G}(k_\varphi) \cdot \hat{u}_i(z, k_\varphi, r, t). \tag{8}$$

The inverse transform provides the filtered velocity $\tilde{u}_i$. u_i'' follows from (6). Now, the stress tensor ϕ_{ij}, unknown in LES, can be computed. It consists of a pure SGS part and a cross term exchanging energy among unresolved and resolved scales, viz:

$$\phi_{ij} = \widetilde{u_i'' u_j''} + \widetilde{\tilde{u}_i u_j''} + \widetilde{u_i'' \tilde{u}_j}. \tag{9}$$

Usually the traceless SGS stress tensor τ_{ij}, namely

$$\tau_{ij} = \phi_{ij} - \frac{1}{3}\phi_{kk}\delta_{ij} \tag{10}$$

is computed and the isotropic part is added to the pressure. The product $-\tau_{ij}\tilde{S}_{ij}$ is responsible for the generation of SGS energy, since it appears in the balance equation of the SGS kinetic energy. We consider its statistical average and follow Härtel et al. (1994) who split the SGS stress and the GS rate of strain tensors into mean and fluctuating parts, to obtain:

$$< \tau_{ij}\tilde{S}_{ij} > = < \tau_{ij} >< \tilde{S}_{ij} > + < \tau_{ij}'\tilde{S}_{ij}' > . \tag{11}$$

The first term on the r.h.s. of (11) can be interpreted as the SGS contribution to the total turbulence production, whereas the second term accounts for a redistribution of kinetic energy within the turbulence spectrum. Figure 5 contains profiles of the mean SGS production rate $< \tau_{ij}\tilde{S}_{ij} >$ and of the sum of covariances between fluctuating stresses and rates of strain at two axial positions. The dashed curves correspond to the upstream position $z/D = 0.58$ and thus, with good accuracy, to the case of fully developed pipe flow. The solid curves are taken at $z/D = 2.04$, i.e. within the recirculation zone.

The mean rate of production is positive throughout, not only in simple shear layers ($z/D = 0.58$), but also within the separation zone, which means that on the average energy is transferred from resolved to unresolved scales although locally and instantaneously the product $\tau_{ij}\tilde{S}_{ij}$ strongly fluctuates and changes sign. For pipe and channel flow a priori tests of Härtel (1994) show that all presently known eddy-viscosity SGS models overestimate $< \tau_{ij}\tilde{S}_{ij} >$ because they are incapable of predicting the reverse transfer of energy in the buffer layer, which is reflected by the minimum in the $< \tau_{ij}'\tilde{S}_{ij}' >$ profile. Since similar effects of energy backscatter occur within the mixing layer of the present separated flow, it can be concluded that LES cannot be made as accurate as desired, unless SGS models are improved. For higher cut-off wavenumbers very similar effects are found. Additional filtering in axial direction for fully developed pipe flow led to results qualitatively identical to those for $z/D = 0.58$.

References

[1] Aldama, A.A. (1990): Filtering techniques for turbulent flow simulation, Lecture Notes in Engineering, Vol. 56, Springer-Verlag, Berlin. New York

[2] Eggels, J.G.M., Unger, F., Weiss, M.H., Westerweel, J., Adrian, R.J., Friedrich, R., Nieuwstadt, F.T.M. (1994): Fully developed turbulent pipe flow: A comparison between direct numerical simulation and experiment. J. Fluid Mech. 268, pp. 175-209.

[3] Härtel, C., Kleiser, L., Unger, F., Friedrich, R. (1994): Subgrid-scale energy transfer in the near-wall region of turbulent flows. To appear in: Phys. Fluids A.

[4] Härtel, C. (1994): Analyse und Modellierung der Feinstruktur im wandnahen Bereich turbulenter Scherströmungen, Ph.D.thesis, Munich Univ. of Technology.

[5] Kim, J., Moin, P., Moser, R.(1987): Turbulence statistics in fully developed channel flow at low Reynolds number, J. Fluid Mech., 177, pp.133-166.

[6] Schumann, U. (1975): Subgrid scale model for finite difference simulations of turbulent flows in plane channels and annuli, J. Comp. Phys., 18,pp. 376-404.

[7] Schumann, U. (1980): Fast elliptic solvers and their application in fluid dynamics.-In: Kollmann, W. (ed.): Computational Fluid Dynamics. Washington, Hemishere, pp. 402-430.

[8] Wagner, C., Friedrich, R.(1994): Direct numerical simulation of turbulent flow in a sudden pipe expansion, Proc. of the 74th AGARD Fluid Dynamics Panel Meeting and Symposium on "Application of Direct and Large Eddy Simulation to Transition and Turbulence", Chania, Crete, April 18-21, 1994, AGARD CP 551.

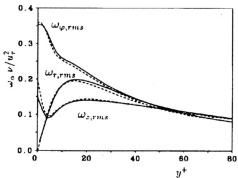

Fig. 1. Rms vorticity fluctuations. DNS pipe: ———, DNS Kim et al.(1987): - - - .

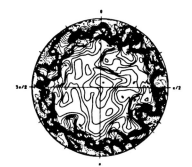

Fig. 2. Contour lines of the instantaneous axial velocity component in the plane $z/D = 2.2$. Solid/dashed lines represent positive/negative values ranging from $-4.81 < u_z/u_{\tau 0} < 20.35$.

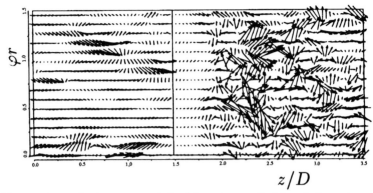

Fig. 3. Fluctuating velocity vectors (u'_z, u'_φ) upstream (left) and downstream of the step at equal distances from the walls ($y^+ = y\dot{u}_{\tau 0}/\nu = 14$). In φ-direction each 2nd vector is plotted only.

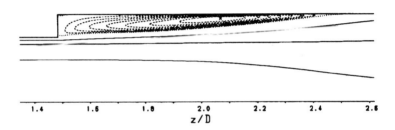

Fig. 4. Contour lines of the mean axial velocity component. Dotted lines separate foreward from backward flow (dashed lines).

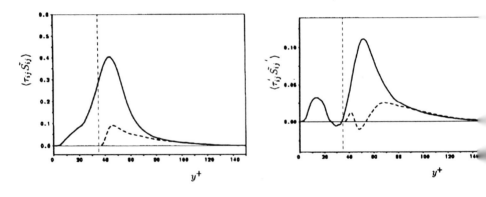

Fig. 5. Mean production of SGS kinetic energy (left) and sum of covariances between fluctuating SGS stresses and GS rates of strain (right) at two axial positions: - - - $z/D = 0.58$, ——— $z/D = 2.04$.

Analysis of 3-D Unsteady Separated Turbulent Flow using a Vorticity-Based approach and Large-Eddy Simulation[1]

K. S. Muralirajan, K. Ghia, U. Ghia* and B. Brandes-Duncan

Computational Fluid Dynamics Research Laboratory
Department of Aerospace Engineering and Engineering Mechanics
*Department of Mechanical, Industrial and Nuclear Engineering
University of Cincinnati, Cincinnati, OH 45221, USA

1 Introduction

The long-term goal of this investigation is to numerically simulate complex high-Reynolds number (Re) flows by solving the Velocity-Vorticity formulation of the Navier-Stokes Equations ($\overline{V} - \overline{\omega}$ NSE) using the Large-Eddy Simulation (LES) approach. The first step towards achieving this goal, and hence, the primary objective of this paper, is the development of a numerical method for the $\overline{V} - \overline{\omega}$ NSE, for the simulation of unsteady, turbulent flows in three-dimensional (3-D) geometries using LES. In LES, a spatial filter is employed to separate the small-scale, i.e., the subgrid scale (SGS) eddies from the large-scale (LS) eddies, as the latter vary less rapidly than their total instantaneous counterparts; the LS eddies are computed while the SGS eddies are modeled; this permits simulation with a relatively coarser grid as compared to that required if all the turbulent scales are resolved.

Extensive LES work has been done using the Velocity-Pressure formulation of the Navier-Stokes Equations ($\overline{V} - p$ NSE). Moin and Kim [11] and Piomelli [16] studied straight channel flows; Kaiktsis, Karniadakis and Orszag [9] and Akselvoll and Moin [1] simulated backstep channel flows; Yang and Ferziger [17] investigated flows in a channel with an obstacle. In these studies, the velocity-velocity correlations for the SGS eddies were represented using the Smagorinsky model.

It has been shown earlier by Osswald, K. Ghia and U. Ghia [13, 14, 15], Huang, U. Ghia, Osswald and K. Ghia [5, 6, 8], Noll, K. Ghia and Muralirajan [12], and Huang and U. Ghia [7] that the $\overline{V} - \overline{\omega}$ formulation has some advantages over the $\overline{V} - p$ formulation in a variety of applications, including maneuvering bodies. All of these studies were carried out using the Direct Numerical Simulation (DNS) technique using coarse grids. The DNS is computationally resource intensive and therefore limited to simple geometries and/or low-Re flows. Hence, to simulate higher Re flows and/or flows involving complex geometries, a LES technique for the $\overline{V} - \overline{\omega}$ formulation is necessary.

---

[1] This research is supported, in part, by AFOSR Grant No. F49620-93-1-0393, and Ohio Supercomputer Center Grant No. PES070-5.

In the present study, the unsteady $(\overline{V} - \overline{\omega})$ DNS analysis developed by Ding, K. Ghia, Osswald and U. Ghia [3] is modified to use LES for simulating turbulent flows. A vorticity SGS model is developed from the Smagorinsky SGS model for velocity correlations to close the filtered NSE. Generalized body-fitted coordinates are employed, so that the software is not restricted to simple geometries. In this paper, only essential details will be provided due to space limitation.

2 Large-Eddy Simulation: SGS Modeling

In large-eddy simulation, the filtered NSE are solved for the large-scale flow, while, a model is employed for the subgrid scales. The eddy-viscosity model developed by Smagorinsky and its variants are widely used in large-eddy simulation research. However, these models have been developed for use in $\overline{V} - p$ NSE formulations, and currently do not incorporate generalized coordinates. The remainder of this Section describes how, with appropriate modifications, these models can equivalently be used in $\overline{V} - \overline{\omega}$ formulations in generalized coordinates. This is shown by beginning with the filtered Vorticity Transport Equation (VTE), presented as (Note: single overbar denotes filtered variable vector; double overbar denotes a filtered variable tensor; prime denotes a SGS variable):

$$\frac{\partial \overline{\omega}}{\partial t} + \nabla \times (\overline{\omega} \times \overline{V}) = \frac{1}{Re} \nabla \times (\nabla \times \overline{\omega}) - \nabla \times (\overline{\omega' \times V'}) \quad (1)$$

Here, the $(\overline{\omega' \times V'})$ term needs to be modeled. In the filtered linear-momentum equation (LME),

$$\frac{\partial \overline{V}}{\partial t} + \nabla \bullet (\overline{V}\,\overline{V}) = \frac{1}{Re} \nabla^2 \overline{V} - (\nabla \bullet \overline{V'V'}) \quad , \quad (2)$$

the $(\overline{V'V'})$ term needs to be modeled.

Comparing the terms, and noting that $\nabla \times$ (LME) = VTE, it can be shown that $\nabla \times (\overline{\omega' \times V'}) = \nabla \times (\nabla \bullet \overline{V'V'}) = \nabla \times (\nabla \bullet \overline{\overline{\tau}}) = \nabla \times (\overline{M}) = \overline{N}$, where $\overline{\overline{\tau}}$ is given by the Smagorinsky model for the SGS terms in the $\overline{V} - p$ NSE. In other words, the divergence of the Smagorinsky model can be used as a model for the $(\overline{\omega' \times V'})$ term in VTE. For later reference, this model will be termed the vorticity model $(\overline{M})$.

The Smagorinsky model can be written as: $\overline{\overline{\tau}} = \overline{V'V'} = -2\nu_T \overline{\overline{S}}$ where the eddy viscosity $\nu_T = l^2 (2\overline{\overline{S}} : \overline{\overline{S}})^{\frac{1}{2}}$, the strain-rate tensor $\overline{\overline{S}} = \frac{1}{2}(\nabla \overline{V} + \nabla \overline{V}^T)$, the length scale $l = C_s \Delta$, with $\Delta = (\Delta_1 \Delta_2 \Delta_3)^{\frac{1}{3}}$, and Δ_i is the grid step in coordinate direction i (i.e., $\Delta_1 = \Delta x$, etc.). Furthermore, C_s is the Smagorinsky coefficient, and can be assumed to be constant or can be computed dynamically as given by Germano et al. [4]. If C_s is not computed dynamically, then it must be multiplied by a wall damping function such as $(1 - exp(-\frac{y^+}{26})^3)^{\frac{1}{2}}$, since the Smagorinsky model is based on homogenous isotropic turbulence. This helps the model to be turned off at a wall, where $y^+ = 0$.

The vorticity model development starts with $\overline{M} = \nabla \bullet \overline{\overline{\tau}}$, where $\overline{\overline{\tau}}$ contains velocity terms that are known when the VTE is being solved. The Curl of $\overline{M}$, (i.e., $\overline{N} =$

$\nabla \times \overline{M}$), can be added as a source term in the filtered VTE, thereby closing the filtered $\overline{V} - \overline{\omega}$ system.

In generalized coordinates ξ^1, ξ^2, ξ^3, the strain-rate tensor becomes, $\overline{\overline{S}} = S_{ik}\overline{e}^i\overline{e}^k$, with $S_{ik} = \frac{1}{2}(\frac{\partial v_k}{\partial \xi^i} + \frac{\partial v_i}{\partial \xi^k} - 2v_j\Gamma^k_{ij})$, and $\overline{\overline{S}} : \overline{\overline{S}} = S_{ik}S_{ik}g^{ii}g^{kk}$. Therefore, $\overline{M} = \nabla \bullet \overline{\overline{\tau}} = m_i\overline{e}^i$, where $m_i = g^{jj}(\frac{\partial t_{ij}}{\partial \xi^j} - t_{ik}\Gamma^k_{jj} - t_{jk}\Gamma^k_{ij})$, and $\overline{N} = \nabla \times \overline{M} = n_k\overline{e}^k$, where $n_i = \frac{g_{ii}}{\sqrt{g}}(\frac{\partial m_k}{\partial \xi^j} - \frac{\partial m_j}{\partial \xi^k})$, etc. Here, g_{ij}, g^{ij} are components of the contravariant and covariant metric tensors, and t_{ij} are components of $\overline{\overline{\tau}}$. It is interesting to note that the Christoffel symbols given as $\Gamma^k_{ij} = \frac{\partial \overline{e}_i}{\partial \xi^j} \bullet \overline{e}^k$, represent only seven distinct components, due to the orthogonality of the grid.

3 Problem Configuration and Numerical Approach

A backstep duct, shown in Fig. 1, which is geometrically simple but belies the complex physics of the separated flow, is chosen for this study. The x_1, x_2, x_3 directions refer to the spanwise, normal and streamwise directions, respectively. The expansion ratio, defined as the ratio of the outlet plane height to the inlet plane height, is 1.94. All lengths are normalized by the outlet plane height.

A staggered, generalized, orthogonal coordinate mesh with cubic spline clustering is used to discretize the unknown variables. A central-difference scheme is used for all spatial derivatives except the vorticity convection term, for which a third-order, biased-upwind-difference scheme is used. A hybrid method, which makes use of the Block-Gaussian-Elimination (BGE) direct solver and a Gauss-Seidel iterative solver, is employed to solve the velocity problem. A multigrid scheme is implemented to accelerate convergence. The VTE is solved using a modified Douglas-Gunn alternating-direction implicit (ADI) scheme. No-slip boundary conditions are specified along the duct walls at the spanwise and normal boundaries. A turbulent mean velocity profile, with superimposed random perturbations filtered in space, is used to provide time-dependent inlet profiles. At the outlet, second-order extrapolation, combined with the buffer domain technique of C. Liu and Z. Liu [10], is employed for smooth passage of eddies. The solution evolves from an initial condition, given as fully-developed turbulent mean flow at the inlet and outlet ducts, while in the region of the backstep these are scaled by the conformal tranformation scale factor. This procedure avoids discontinuities in the initial conditions at the backstep.

4 Discussion of Results and conclusions

The software developed has been validated with test functions using *Mathematica*, for LES subroutines; i.e., test functions are defined for the input variables in the software, and the results are compared with the analytical solutions generated by *Mathematica* using the same input test functions. Note that this only validates the software and not the theory. To validate the theory itself, the results obtained by the software were compared with experimental data; in particular, results for a square duct configuration were compared with experimental results. The backstep duct results presented here are from coarse-grid DNS (i.e., LES with no model for the SGS eddies) computations, and these are only of preliminary nature, to the extent that they have not reached the time-asymptotic state. The Reynolds number chosen for this is $Re = 324$ based on outlet height and velocity (or $Re = 648$ based on inlet height and velocity). There is general qualitative agreement between the present computational results and the experimental results of Armaly et al. [2]. The reattachment length was found to be 8.8, compared to 12 from experiments. The discrepencies can be attributed to SGS eddies that are not being accounted for, and the fact that, even though the ratio of inlet and outlet height are same, the spanwise and streamwise lengths of the duct considered here are smaller than those used in the experiment by Armaly et al. [2], because of computational resource limitations.

Figure 2 shows the streamwise velocity profile on a lateral mid-plane, above the reattachment point. The flow can be seen to be almost two-dimensional near the center plane (plane of symmetry), but, in the vicinity of side walls, it is highly three-dimensional even at this Reynolds number. Figure 3 shows the instantaneous streamfunction contours on the longitudinal plane of symmetry. The primary recirculation region consists of two vortices, one large primary vortex and a small secondary vortex. Two small recirculation regions are also found at the top wall. Figure 4 shows the instantaneous velocity vectors on the longitudinal plane of symmetry. The flow, which is fully developed at the inlet, becomes separated in the recirculation zone and then grows into a fully developed profile far downstream of the step. The absence of small eddying structures indicate that the flow is laminar. Computations are currently being carried out for transitional and turbulent Reynolds number regimes. The subgrid scale model in LES will be turned on for the latter computations, and the LES methodology will be fully validated.

Acknowledgment :

The authors sincerely acknowledge Mr. S. Siravuri, for his assistance on many aspects of the work reported in this study.

References

1. Akselvoll, K. and Moin, P., (1993), ASME FED, vol. 162, pp. 1-6.

2. Armaly, B. F., Durst, F., Pereira, J. C. F. and Schönung, B., (1983), JFM, vol. 127, pp. 473-496.

3. Ding, Z., Ghia, K. N., Osswald, G. A. and Ghia, U., (1992) Bulletin of A. P. S., November.

4. Germano, M., Piomelli, U., Moin, P. and Cabot, W. H., (1991), Phys. Fluids A, vol. 3, pp. 1760-1765.

5. Huang, Y., Ghia, U., Osswald, G. A. and Ghia, K. N., (1991), AIAA CP 914, June.

6 Huang, Y., Ghia, U., Osswald, G. A. and Ghia, K. N., (1992), NMFM, vol. 36, pp. 54-66.

7 Huang, Y. and Ghia, U., (1992), SIAM J. Comm. Appl. Numerical Mathematics, vol. 8, pp. 707-719.

8 Huang, Y., Ghia, U., Osswald, G. A. and Ghia, K. N., (1993), AIAA Paper No. 93-0682.

9 Kaiktsis, L., Karniadakis, G. and Orszag, S., (1991), JFM, vol. 231, pp. 501-528.

10 Liu, C. and Liu, Z., (1993), JCP, Vol. 106, No. 1, pp. 92-100.

11 Moin, P. and Kim, J., (1982), JFM, vol. 118, pp. 341-377.

12 Noll, C., Ghia, K. N. and Muralirajan, K. S., (1992), ASME FED, vol. 133, pp. 221-222.

13 Osswald, G. A., Ghia, K. N. and Ghia, U., (1987), AIAA-CP 874, pp. 408-421.

14 Osswald, G. A., Ghia, K. N. and Ghia, U., (1988), Lecture Notes in Physics, vol. 323, pp. 454-461.

15 Osswald, G. A., Ghia, K. N. and Ghia, U., (1990), Lecture Notes in Physics, vol. 371, pp. 215-218.

16 Piomelli, U., (1987), Ph.D. Thesis, Stanford University.

17 Yang, K-S. and Ferziger, J., (1993), AIAA J., vol. 31, No. 8, pp. 1406-1413, August.

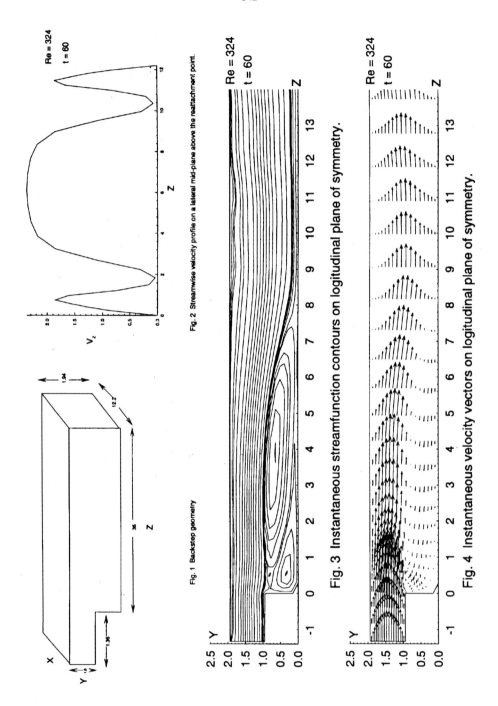

Penalty Finite Element Method for Transient Free Convective Laminar Flow

K. Ravi and K. Ramamurthi

Indian Space Research Organisation, Trivandrum, India

Abstract: The influence of the penalty parameter, grid size and relaxation parameter in the penalty finite element method is studied for Rayleigh number between 10^3 and 10^8. The relaxation parameter is shown to exert a pronounced influence on convergence and values near unity are required for Rayleigh numbers exceeding 10^6. The flow currents and temperature distributions are shown to be dominated by the outer wall heating as Rayleigh number increases.

1 Introduction

The Penalty Finite Element Method is applied for the prediction of the transient buoyancy-driven flows of incompressible viscous liquids in cylindrical annuli. The method is first demonstrated for Rayleigh number greater than 10^6 for a two dimensional cavity comprising of alternate hot and cold walls for which bench-mark solutions using finite difference schemes are available. After validating the computations, the method is applied for the cylindrical annulus.

2 Governing Equations and Solution Procedure

The governing equations for continuity, momentum and energy are written in the axisymmetric coordinate system in terms of the nondimensional pressure P, velocities U and V and temperature T as given by Pelletier[1]. The boundaries of the problem at the inner and outer cylindrical walls of the annulus are assumed as isothermal surfaces whereas the upper and lower boundaries are taken as insulated.

The pressure terms in the momentum equation is expressed as a multiple of the continuity equation using the penalty parameter λ as given below:

$$P = -\lambda \left(\frac{\partial U}{\partial r} + \frac{U}{r} + \frac{\partial V}{\partial y} \right) \tag{1}$$

Here r denotes the nondimensional radial coordinate and y is the nondimensional axial coordinate.

The differential equations for the momentum and energy are integrated and expressed in an algebraic form at each element using the Galerkin's weighted residual technique[2]. Four-noded rectangular elements are used for discretizing the variational formulations of the equations with linear interpolations for U,V and T. A backward difference scheme is adopted for the transient terms so that the algorithm is fully implicit and stable.

The initial values of U,V and T are taken as zero at all nodes except at the boundaries where they are specified. Picard's iterative technique is employed for the solution. The energy equation is solved for temperature after which the momentum equations are solved for the velocity fields.

The computations are carried out in double precision mode on a 33 MHz i860 computer having 64 MB RAM and 780 MB hard disk. The results are validated for a cavity problem using the bench-mark solutions of Davis[3].The cavity consisted of alternate hot and cold walls on the left and right sides with the upper and lower surfaces being insulated.

3 Choice of Penalty Parameter

In order to solve the global matrix equation, the penalty parameter(λ) needs to be specified 'a priori'. The value of λ needs to be large enough to enforce continuity but not too large that other terms are calculated inaccurately. Different values of λ between 10^3 and 10^{13} were tried at different values of Rayleigh numbers(Ra). Figure 1 shows the number of iterations required for convergence as λ is varied. It is seen that the convergence is very weakly dependent on λ for all Ra $< 10^5$. The magnitudes of temperatures and velocities determined in the computation also did not change with changes in λ (Fig.2). Beyond a λ of 10^{13} convergence becomes problematic.Below a λ of 10^4 the incompressibility constraint is not satisfied and convergence is not possible.When Ra $> 10^6$ problems in convergence are encountered for all values of λ and this aspect is considered subsequently.

4 Relaxation Parameter

The weightage for the previous computed values(ϕ_{r-1}) and the present computed values (ϕ_r) in the iterations are related to the current value(ϕ_n) by,

$$\phi_n = (1 - R)\phi_r + R\phi_{r-1} \qquad (2)$$

where R is the relaxation parameter. A small value of R is seen to give poor convergence characteristics. Thus in Fig.3 and 4 the convergence behaviour of the V component at Ra $= 10^6$ is shown as a function of the number of iterations for R of 0.7 and 0.99 respectively. It is seen that at R $= 0.7$ the solution oscillates as iteration proceeds(Fig.3) whereas at R $= 0.99$ monotonic convergence is achieved(Fig.4). The need for a higher R as Ra increases is evident. While for a Ra $= 10^6$ adequate convergence is obtained with R $= 0.99$ in 865 iterations,it is found necessary to increase R to a value of 0.999 for Ra $= 10^7$ to achieve convergence. The convergence is achieved in 1615 iterations with a relaxed tolerance. For Ra $= 10^8$ a minimum R $= 0.9999$ is required to achieve convergence of the results.

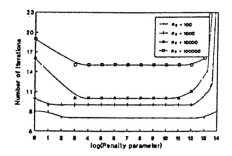

Fig. 1. Convergence characteristics for different values of λ and Ra

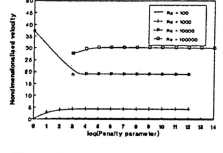

Fig. 2. Predicted values of vertical velocities for different values of λ

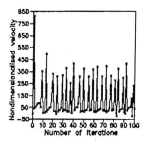

Fig. 3. Convergence behaviour of V with $R = 0.70$ for $Ra = 10^6$

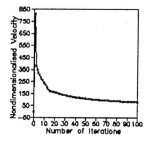

Fig. 4. Convergence behaviour of V with $R = 0.99$ for $Ra = 10^6$

The demarkation of the zones of converged and oscillatory solutions in the domain of R and Ra is shown in Fig.5. It is to be noted that convergence does not imply a reduced number of iterations. The number of iterations for convergence when Ra $< 10^5$ is given as a function of R in Fig.6. For Ra $> 10^5$ the number of iterations are much higher as discussed earlier.

5 Grid Sizes

Different sizes of grids do not influence the convergence behaviour though the accuracy of the results is poorer for the larger grid sizes. The number of grids considered are 17x17, 21x21, 39x39 and 61x61. Robert's transformation which makes use of the stretching parameter $\beta = (1 - \delta/h)^{-1/2}$ [4] where δ is the characteristic boundary layer thickness and h is the height of the grid is used to generate fine grids at the wall region. No improvement in the convergence is possible with the finer grids for all values of Ra. Only when R is increased better convergence is achieved.

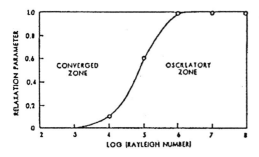 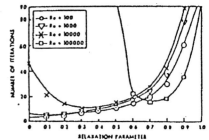

Fig. 5. Converged and oscillatory zones   Fig. 6. Influence of R on convergence

6   Flow Pattern and Temperature Distribution in Annulus

Figures 7 and 8 give typical plots of flow velocities and isotherms at a nondimensional time of 0.001 and 0.05 for a $Ra = 10^4$, radius ratio of 2 and Prandtl number of 7.The computations show the evolution of secondary circulation and formations of cellular flow with pronounced longitudinal velocities in the mid region of the annulus. The flow pattern becomes skewed at smaller radius ratios with the longitudinal currents progressively shifting towards the outer walls. The symmetry of the flow velocities and temperatures is lost with significantly increased diffusion from the outer walls. The convection currents are found to become stronger at smaller times with increased Ra.

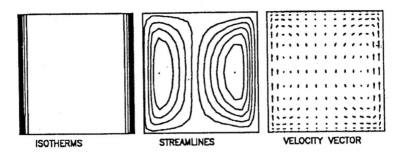

Fig. 7. Isotherms, streamlines and velocity vectors at a nondimensional time of 0.001

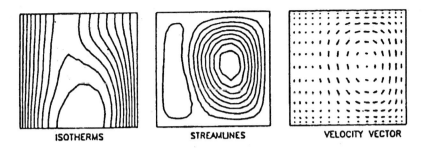

Fig. 8. Isotherms, streamlines and velocity vectors at a nondimensional time of 0.05

Changes to Prandtl number did not significantly influence the transient temperatures or the velocity distribution.

7 Concluding Remarks

The parametric study carried out with the penalty finite element method shows that convergence of the method is possible by increasing the relaxation parameter as Rayleigh number increases. The method is applied to annulus flow and shows the evolution of secondary currents. Diffusion of heat from outer wall dominates the free convective flow velocities and temperature distribution as Rayleigh number increases.

References

[1] D.H. Pelletier, J.N. Reddy, J.A. Schetz: Annual Rev. of Num. Fluid Mech. and Heat Trans. **2** 39 (1989)

[2] J.N. Reddy, A. Satake: Trans. of ASME, Int. Jl. of Heat Trans. **102** 659 (1980);

[3] G. de vahl Davis: Int. Jl. of Num. Meth. in Fluids **3** 249 (1983);

[4] D.A. Anderson, J.C. Tannehill, R.H. Pletcher: Comp. Fluid Mech. and Heat Trans. Chapter 5.6 Hemisphere Publishing Corp. 247 (1984)

Numerical Simulation of Three Dimensional Turbulent Underexpanded Jets

M.Si Ameur and J.P.Chollet

Laboratoire des Ecoulements Géophysiques et Industriels
CNRS-UJF-INPG BP.53X-38041 Grenoble Cedex, France

Abstract : The exhaust jet of a high-speed aircraft engine expands into a low pressure atmosphere. The gas is a mixture of nitric oxydes NO_x with CO_2 and air. Three dimensional underexpanded jet with low pressure ratio is computed by large eddy simulation. Hydrodynamic instabilies arise in the shear layer inducing the release of turbulence. The mechanism which is responsible for transition to turbulence in low underexpanded jet is identifyed. The turbulent zone of jet is well reproduced thanks to high-order Godunov method which dissipates the small scale energy fluctuations while providing good accuracy.

1 Introduction

Underexpanded jets are of central interest within the framework of supersonic propulsion systems. Combustion efficiency and pollutant emission are crucial issues for engine builders. As test benches are expensive and measurements in real flight conditions are difficult, numerical simulations are used as a complement to experiments. Therefore, the computation of shear flows is then important not only in fundamental research but also for industrial applications. Most of flows of practical interest are turbulent, involving a wide range of length scales and requiring a three-dimensional computation of the flow field. Simulations give detailed information about the structure and dynamics of turbulent scales. In this work, we present a numerical simulation of an underexpanded jet. We give a complete description of this underexpended sonic jet by fully solving the compressible Navier-Stockes equations. Because of the size of the jet nozzle, the simulation is definitely a large eddy simulation(LES). The numerical method is based on the resolution in time and in space of the physical process using piecewise parabolic method(PPM) combined with linearized Riemann solver. This scheme introduces non-linear dissipation just where needed by approximating the evolution of any variable across the grid-mesh through high-order interpolation followed by monotonicity constraints.

2 The Numerical Method

The full compressible Navier-Stokes equations for ideal gases are considered. For a flow of a mixture made of different chemical species. Each species is modelled as an ideal gas. The non-dimensionalized compressible Navier-Stokes equation is written under conservative form:

$$\frac{\partial U}{\partial t} + \frac{\partial F(U)}{\partial x} + \frac{\partial G(U)}{\partial y} + \frac{\partial H(U)}{\partial z} = S(U) \qquad (1)$$

S(U) represents the source terms of reaction and momentum and species diffusion. F(U), H(U), G(U) regroups inviscid terms.

$$U = \{\rho, \rho u, \rho v, \rho w, \rho E, \rho Y_i\}^T$$

E is the total energy, Y_i the mass fraction of species. Eq. (1) is solved by decomposition into three one-dimensional hyperbolic problems and one three-dimensional parabolic problem.

$$L_x : \frac{\partial U}{\partial t} + \frac{\partial F(U)}{\partial x} = 0, \quad L_y : \frac{\partial U}{\partial t} + \frac{\partial G(U)}{\partial y} = 0 \quad L_z : \frac{\partial U}{\partial t} + \frac{\partial H(U)}{\partial z} = 0$$

$$\Psi : \frac{\partial U}{\partial t} = S(U)$$

The combined solution is computed as follows :

$$U^{n+2} = \Psi^{\frac{1}{2}} L_z L_x L_y \Psi L_y L_x L_z \Psi^{\frac{1}{2}} U^n \qquad (2)$$

The code was developed by Gathmann[1], it is based on the piecewise parabolic method(PPM[3])in combination with the linearized Riemann solver of Roe[5]. The PPM scheme is an higher-order extension of Godunov's method. Its distinctive characteristics are -High-order spatial interpolation, which allows to resolve sharp features in the flow; -Performs well in advection dominated problems and handles waves in a very precise way due to a characteristic decomposition of the flow; -Acts as a subgrid scale by filtering unresolved fluctuations, thanks to monotonicity constraints; -Each characteristics decomposed by the PPM scheme is integrated by upwind scheme.

We use a non-reflective boundary conditions on both inflow and outflow boundaries. In cross-stream and spanwise directions we impose free outflow boundary conditions, with information carried along all incoming characteristics set to zero.

3 Results

Low underexpanded jet in homogeneous ambiant flow develops two flow regions (Fig1). In region 1 , it develops a periodic expansion and compression waves. The jet centerline pressure oscillates around the pressure of the coflow region 2. This pressure oscillation is damped by the turbulence in the jet mixing layer.

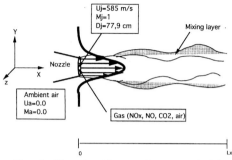

Fig. 1. Sketch of low underexpanded jet

We investigate jets with expansion ratio $p_j/p_a = 1.51$. The jet at the nozzle exit is made of hot gas(NO, NO_x, CO_2, air) at Mach number $M_j = 1$, the ambiant flow is composed of an homogeneous base state environment. The inflow profile is a hyperbolic tangent profile with two length parameters: the initial momentum thickness and the jet radius R. All length scales are non-dimensionalized by D_j. The basic profile is that studied by Michalke and Hermann[2]

$$u(r) = U_a + \frac{U_j - U_a}{2}\left(1 - \tanh\left[\frac{R}{4\Theta}(\frac{r}{R} - \frac{R}{r})\right]\right)$$

index j assigns jet values and index a coflow values. R is the jet radius defined in such a way that $u(R) = (u_j + u_a)/2$. Θ is the momentum thickness:

$$\Theta = \int_0^\infty \left[\frac{u(r) - U_a}{U_j - U_a}\right]\left[\frac{U_j - u(r)}{U_j - U_a}\right] dr$$

A weak white noise was superimposed to the inflow profile in the shear region of the jet. The amplitude of the noise was 1% of the jet pressure. This noise models the turbulent fluctuations of the jet near the wall of the nozzle. Calculations are made on a $125 \times 125 \times 125$ cartesian grid, which is sufficient to resolve the large scales of the jets containing the wavelength of the most unstable modes. Three dimensional small-scale turbulence develops in the flow from nonlinear interactions, a subgrid scale model has to be used in order to model the unresolved scales. In the jet presented in this work, the dominant instability is fully resolved and the dissipation of small scale fluctuations is done through filtering by the numerical algorithm. The effective Reynolds number $Re = U_j D_j^3/\nu_1$ is between $10^5 < Re < 10^6$. Where μ_1 is a hyper-viscosity. The jet has a radius $R = 2.5\delta_i$ and the domain size is $L_x = 20D_j, L_y = 5D_j, L_z = 5D_j$. The parameters of flow are given in table 1.a. The chemical composition of the jet at the nozzle exit was obtained from measurements on the Concorde engine Olympus (Table 1.b).

	Mach	density(g/cm^3)	pressure(bar)	speed(m/s)
Jet	1.0	0.624	1.53	585.7
Ambiant flow	0.0	1.249	1.013	0.0

Tab. 1a

	$Y(NO)$	$Y(NO_x)$	$Y(CO2)$	air
Jet	4.3E-04	4.3E-04	3.9E-02	96E-02
Ambiant flow	0.0	0.0	0.0	1.0

Tab. 1b

Table 1. Parameters of flow

Fig. 2. is an isosurface of vorticity modulus, it shows the three stages of the developement of a jet(laminar, transition and turbulence). We see helical large scale structures observed in laboratory experiment by Gutmark and al.[4]. The growth of very elongated Λ-shaped structures can be seen. Fig 3 displays zoomed views on Λ structures, these structures are formed by a pair of helicoidal Kelvin-Helmholtz vortex rolls, they pump ambiant fluid to the jet flow and appear when the flow mechanism turns nonlinear.

Fig. 2. Three-dimensional isosurface of vorticity modulus ($\omega = 3U_j/D_j$ at $t = 20D_j/U_j$. Note the development of helical large structures and the formation of downstream directed Λ structures. The flow is coming from left to right. The domain is shown for $4D_j < x < 20D_j$

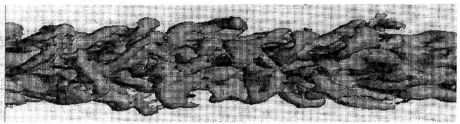

Fig. 3. Zoom of $15D_j < x < 20D_j$ region: Λ structures on three dimensional isosurface of vorticity modulus($\omega = 3U_j/D_j$) at $t = 20D_j/U_j$. These Λ-shaped structures are formed by pairs of helicoidal two Kelvin-Helmholtz vortex rolls

Acknowledgements

To Snecma for making olympus engine data available and to IDRISS-CNRS computing center for crau computing time.

References

[1] Gathmann, R.J., Baudoin, C., Chollet J.P.,1992, Direct numerical simulation of nitric oxide evolution in underexpanded jets. ASME-paper 92 - GT - 372

[2] Michalke, A., Hermann, G., 1982, On the inviscid instability of a circulair jet with external flow, JFM 114, pp.343 - 359

[3] Colella, P., Woodward, P.R. 1984 The piecewise parabolic method (PPM) for gas-dynamical simulations. JCP 54, 174 - 201

[4] Gutmark, E., Schadow, K.C, Wilson, K.J., Bicker, C.J., 1988, Near-field pressure radiation and flow charactersitics in low supersonic circular and elliptic jets, Phys.Fluids A **31** (9)

[5] Roe P.L. 1981 Approximate Riemann Solvers, parameter vectors and differnce schemes, JCP 43 , 357 - 372

[6] Chollet J.P., Si Ameur M., Gathmann., R.J., Spatial development of supersonic mixing layer between rigid walls, Ninth Symposuim on Turbulent Shear Flows, Kyoto, august 1993, proceedings of Turbulent shear flows, 19 (2) : 1 - 4

[7] Gathmann, R.J., Si Ameur, M., Mathey F.,1993,"Num. simulations of natural transition in the 3-D supersonic shear layer", Phys.Fluids A 5(11), 2946 - 2968

[8] Si Ameur, M., Gathmann, R.J., Chollet J.P., Mathey F, 1992 "Simulations numériques de transition naturelle et forcée d'une couche de mélange supersonique" Colloque: Groupement de recherche hypersonique, Garchy 5-7.oct, Actes du colloque publiés par le CNRS, Garchy5. -7.oct, p 213 - 220

Conditioned Navier-Stokes Equations Combined with the K-E Model for By-pass Transitional Flows

Johan Steelant and Erik Dick

Department of Mechanical and Thermal Engineering,
Universiteit Gent, Sint Pietersnieuwstraat 41,
9000 Gent, Belgium.

1 Introduction

Turbomachinery flows are characterized by a very high intensity turbulent mean part. As a consequence, laminar flow in boundary layer regions undergoes transition through direct excitation of turbulence. This is the so-called bypass transition. Regions form that are intermittently laminar and turbulent. In particular in accelerating flows, as on the suction side of a turbine blade, this intermittent flow can extend over a very large part of the boundary layer. Classical turbulence modelling based on global time averaging is not valid in intermittent flows. To take correctly account of the intermittency, conditioned averages are necessary. These are averages taken during the fraction of time the flow is turbulent or laminar respectively. Starting from the Navier-Stokes equations, conditioned continuity, momentum and energy equations are derived for the laminar and turbulent parts of an intermittent flow. The turbulence is described by the classical k-ε model. The supplementary parameter introduced by the conditioned averaging is the intermittency factor. In the calculations, this factor is prescribed in an algebraic way.

2 Conditioned Averaging

We define an intermittency function $I(x, y, z, t)$ that takes the value 1 in a turbulent region and the value 0 in a non-turbulent, say laminar, region. The time-averaged value of this function during some time interval T is defined as the intermittency factor γ. The time interval T is chosen to be large with respect to the time scales of the turbulence, but still small with respect to the time scales of the mean flow. For Reynolds-averaging, the velocity component in x-direction u can be decomposed in mean and fluctuating components as follows:

$u = \overline{u} + u'$ where $\overline{u}$ is a global time average;
$u = \overline{u}_t + u'_t$ for I=1 where $\overline{u}_t$ is an average during the turbulent state;
$u = u_l$ for I=0 where u_l is the value of u during the laminar state.

The turbulent mean value and fluctuation satisfy

$$\overline{Iu_t''} = 0 \qquad \overline{Iu} = \gamma \bar{u}_t = \frac{1}{T}\int_0^T Iu\,dt.$$

The laminar mean value and the global mean value satisfy

$$\overline{(1-I)u} = (1-\gamma)u_l \quad \text{and} \quad \bar{u} = \gamma \bar{u}_t + (1-\gamma)u_l.$$

Further, we derive the equations for the conditioned averages of a space or time derivative quantity. For this purpose, we analyze the contributions coming from the fronts between turbulent and laminar zones. We assume that a front partly belongs to the laminar phase and partly to the turbulent phase. This results in an extra term for both phases. For a space derivative in the turbulent phase, we find

$$\overline{I\frac{\partial u}{\partial x}} = \gamma \frac{\partial \bar{u}_t}{\partial x} + \frac{1}{2}(\bar{u}_t - u_l)\frac{\partial \gamma}{\partial x}.$$

The laminar conditioned mean value is

$$\overline{(1-I)\frac{\partial u}{\partial x}} = (1-\gamma)\frac{\partial u_l}{\partial x} + \frac{1}{2}(\bar{u}_t - u_l)\frac{\partial \gamma}{\partial x}.$$

The above rules apply to any other variable and is valid for every other space direction or time.

3 Conditioned Flow Equations

The rules for conditioned mean values and derivatives go over to Favre-averages. We define mean and fluctuating parts of density:

$$\rho = \bar{\rho}_t + \rho_t' \quad \text{for I = 1, where} \quad \overline{I\rho} = \gamma \bar{\rho}_t;$$

$$\rho = \rho_l \quad \text{for I = 0}.$$

Hence $\bar{\rho} = \gamma \bar{\rho}_t + (1-\gamma)\rho_l$. Further, a turbulent Favre-average for velocity is defined by

$$\overline{I\rho u} = \gamma \bar{\rho}_t \tilde{u}_t.$$

The global Favre-average follows from $\overline{\rho u} = \bar{\rho}\tilde{u} = \gamma \bar{\rho}_t \tilde{u}_t + (1-\gamma)\rho_l u_l$. We derive now the conditioned turbulent mean mass equation. The unaveraged equation is

$$\frac{\partial \rho}{\partial t} + \frac{\partial \rho u}{\partial x} + \frac{\partial \rho v}{\partial y} = 0.$$

According to the rules for derivatives, we obtain as turbulent conditioned mean equation

$$\gamma\frac{\partial \bar{\rho}_t}{\partial t} + \frac{1}{2}(\bar{\rho}_t - \rho_l)\frac{\partial \gamma}{\partial t} + \gamma\frac{\partial \bar{\rho}_t \tilde{u}_t}{\partial x} + \frac{1}{2}(\bar{\rho}_t \tilde{u}_t - \rho_l u_l)\frac{\partial \gamma}{\partial x} + \gamma\frac{\partial \bar{\rho}_t \tilde{v}_t}{\partial y}$$

$$+\frac{1}{2}(\bar{\rho}_t \tilde{v}_t - \rho_l v_l)\frac{\partial \gamma}{\partial y} = 0.$$

The laminar equation is

$$(1-\gamma)\frac{\partial \rho_l}{\partial t} - \frac{1}{2}(\bar{\rho}_t - \rho_l)\frac{\partial \gamma}{\partial t} + (1-\gamma)\frac{\partial \rho_l \tilde{u}_t}{\partial x} + \frac{1}{2}(\bar{\rho}_t \tilde{u}_t - \rho_l u_l)\frac{\partial \gamma}{\partial x}$$

$$+(1-\gamma)\frac{\partial \rho_l \tilde{v}_t}{\partial y} + \frac{1}{2}(\bar{\rho}_t \tilde{v}_t - \rho_l v_l)\frac{\partial \gamma}{\partial y} = 0.$$

The laminar equations are analogous to the turbulent ones and will be left out in the sequel. The conditioned turbulent mass equation used in the calculation is

$$\frac{\partial \bar{\rho}_t}{\partial t} + \frac{\partial \bar{\rho}_t \tilde{u}_t}{\partial x} + \frac{\partial \bar{\rho}_t \tilde{v}_t}{\partial y} =$$

$$\frac{1}{2}(\rho_l - \bar{\rho}_t)\frac{1}{\gamma}\frac{\partial \gamma}{\partial t} + \frac{1}{2}(\rho_l u_l - \bar{\rho}_t \tilde{u}_t)\frac{1}{\gamma}\frac{\partial \gamma}{\partial x} + \frac{1}{2}(\rho_l v_l - \bar{\rho}_t \tilde{v}_t)\frac{1}{\gamma}\frac{\partial \gamma}{\partial y}$$

Similarly, the momentum equations can be treated. We write the momentum equations in compact form as

$$\frac{\partial \rho u_i}{\partial t} + \frac{\partial \rho u_i u_j}{\partial x_j} + \frac{\partial p}{\partial x_i} = \frac{\partial \tau_{ij}}{\partial x_j},$$

where the summation convention is used. During the turbulent phase, the Favre- and Reynolds-decompositions are

$$\rho u_i = \rho(\tilde{u}_{ti} + u''_{ti}); \quad \rho u_i u_j = \rho(\tilde{u}_{ti} + u''_{ti})(\tilde{u}_{tj} + u''_{tj}); \quad p = \bar{p} + p'; \quad \tau_{ij} = \bar{\tau}_{ij} + \tau'_{ij}.$$

The turbulent conditioned equations are

$$\gamma\frac{\partial \bar{\rho}_t \tilde{u}_{ti}}{\partial t} + \frac{1}{2}(\bar{\rho}_t \tilde{u}_{ti} - \rho_l u_{li})\frac{\partial \gamma}{\partial t} + \gamma\frac{\partial \bar{\rho}_t \tilde{u}_{ti} \tilde{u}_{tj}}{\partial x_j} + \gamma\frac{\partial \overline{\rho u''_{ti} u''_{tj}}}{\partial x_j}$$

$$+\frac{1}{2}(\bar{\rho}_t \tilde{u}_{ti}\tilde{u}_{tj} + \overline{\rho u''_{ti} u''_{tj}} - \rho_l u_{li}u_{lj})\frac{\partial \gamma}{\partial x_j} + \gamma\frac{\partial \bar{p}_t}{\partial x_i} + \frac{1}{2}(\bar{p}_t - p_l)\frac{\partial \gamma}{\partial x_i}$$

$$= \gamma\frac{\partial \overline{\tau_{tij}}}{\partial x_j} + \frac{1}{2}(\overline{\tau_{tij}} - \tau_{lij})\frac{\partial \gamma}{\partial x_j}.$$

The usual eddy viscosity modelling approximations are now introduced:

$$\overline{\rho u''_{ti} u''_{tj}} = \frac{2}{3}\bar{\rho}_t \tilde{k}_t + \mu_t \tilde{S}_{tij}; \quad \overline{\tau_{tij}} = \tilde{\mu}\tilde{S}_{tij},$$

where $\tilde{k}_t$ is the turbulence kinetic energy during the turbulent phase, μ_t is the eddy viscosity and $\tilde{S}_{tij}$ is the shear tensor based on Favre-averages during the turbulent phase. We neglect here front contributions to the turbulent mean shear stress.

For instance, the resulting momentum-x equation is:

$$\frac{\partial(\bar{\rho}_t \tilde{u}_t)}{\partial t} + \frac{\partial(\bar{\rho}_t \tilde{u}_t \tilde{u}_t)}{\partial x} + \frac{\partial(\bar{\rho}_t \tilde{u}_t \tilde{v}_t)}{\partial y} + \frac{\partial(\bar{p}_t + \frac{2}{3}\bar{\rho}_t \tilde{k}_t)}{\partial x} = \frac{\partial(\tilde{\mu} + \mu_t)\tilde{S}_{txx}}{\partial x}$$

$$+\frac{\partial(\tilde{\mu}+\mu_t)\tilde{S}_{txy}}{\partial y} + \frac{1}{2}(\rho_l u_l - \overline{\rho}_t \tilde{u}_t)\frac{1}{\gamma}\frac{\partial \gamma}{\partial t} + \frac{1}{2}(\rho_l u_l u_l - \overline{\rho}_t \tilde{u}_t \tilde{u}_t)\frac{1}{\gamma}\frac{\partial \gamma}{\partial x}$$

$$+\frac{1}{2}(\rho_l u_l v_l - \overline{\rho}_t \tilde{u}_t \tilde{v}_t)\frac{1}{\gamma}\frac{\partial \gamma}{\partial y} + \frac{1}{2}(p_l - \overline{p}_t - \frac{2}{3}\overline{\rho}_t \tilde{k}_t)\frac{1}{\gamma}\frac{\partial \gamma}{\partial x}$$

$$-\frac{1}{2}\left[\mu_l S_{lxx} - (\tilde{\mu}+\mu_t)\tilde{S}_{txx}\right]\frac{1}{\gamma}\frac{\partial \gamma}{\partial x} - \frac{1}{2}\left[\mu_l S_{lxy} - (\tilde{\mu}+\mu_t)\tilde{S}_{txy}\right]\frac{1}{\gamma}\frac{\partial \gamma}{\partial y}.$$

The energy equation and the Reynolds stress equation, which form the basis of the k- and ε-equations, can be treated in the same way. The result is, in each case, an equation which is similar to the global averaged equation supplemented with source terms due to the front passages. For the Reynolds stress equations, the source terms largely compensate each other. For the present research, we decided therefore to neglect these source terms. So, we model the turbulence by the classical (low Reynolds number) k- and ε-equations, but written for the turbulent conditioned averaged values. We employ the Yang-Shih variant [1].

4 Results

The above equations have been used to calculate intermittent flows for the T3A test case ($U_e = 5.2m/s; Tu = 3\%$) [2]. The intermittency γ can be prescribed algebraically according to Dhawan et Narasimha [3] by

$$\gamma(x) = 1 - \exp\left[-\hat{n}\sigma(Re_x - Re_{x_t})^2\right],$$

where they assume a concentrated breakdown at x_t which results in a linear spot growth. According to Mayle [4] and the measurements of Gostelow and Walker [5] a Gaussian distribution of the spot production is more appropriate. The position of x_t is then somewhere midway $\gamma = 1\%$ and 20%. In our test case we assumed a zero spot growth at the start of transition ($\gamma = 1\%$) which tends to the linear spot growth once γ reaches the value of 20%. This growth is modelled by a rational function. Start of transition is given by $Re_\theta = 460Tu^{-.65}$ [6] and $\hat{n}\sigma = 1.5 \times 10^{-11} Tu^{7/4}$ [4]. Figures 1a and 1b show respectively the skin friction coefficient and the shape factor H as function of Re_x. The transition is spread over the correct distance (prescribed here). As is well known, the transition obtained by the k-ε turbulence model without intermittency occurs too early and is too rapid. The velocity profile for $\gamma = 60\%$ is given in Fig. 1c while Fig. 1d shows the profile of the global mean turbulence kinetic energy at $\gamma = 60\%$. The global mean turbulence kinetic energy is given by

$$u'^2 = k = \gamma \tilde{k}_t + \frac{1}{2}\gamma(1-\gamma)\left[(\overline{u}_t - u_l)^2 + (\overline{v}_t - v_l)^2\right].$$

Acknowledgement

The research reported here was granted under contract 9.0001.91 by the Belgian National Science Foundation (N.F.W.O.) and under contract IUAP/17 as part of the Belgian National Programme on Interuniversity Poles of Attraction, initiated by the Belgian State, Prime Minister's Office, Science Policy Programming.

References

[1] Z. Yang, T.H. Shih: ICOMP-92-08 (1992)

[2] Savill A.M.: Numerical simulation of Unsteady Flows and Transition to Turbulence, Pironneau O. et al. (eds.) Cambridge University Press (1992)

[3] S. Dhawan, R. Narasimha: J. Fluid Mech. **3** 418 (1958)

[4] R. Mayle: J. of Turbomachinery **113** 509 (1991)

[5] J.P. Gostelow, G.J. Walker: J. of Turbomachinery **113** 617 (1991)

[6] J. Hourmouziadis, AGARD Lecture Series **167** (1989)

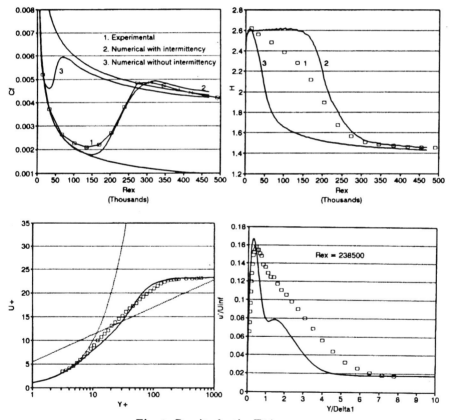

Fig. 1. Results for the T3A case

Numerical aspects of flow stability in the Veldman boundary layers

V. Theofilis, D. Dijkstra and P. J. Zandbergen

Department of Applied Mathematics,
University of Twente, P.O. Box 217,
NL-7500 AE Enschede, The Netherlands

Abstract : We present a numerical method to investigate the stability of non-unique solutions of the Falkner-Skan equation

1 Introduction

The Falkner-Skan boundary layer has been the subject of extensive study in the past and is relatively well understood. The governing equation and boundary conditions are (Rosenhead, 1963).

$$f''' + ff'' + \beta\left(1 - f'^2\right) = 0, \ f(0) = f'(0) = 0, \ f'(\infty) = 1. \tag{1}$$

The Hartree parameter β follows from the pressure gradient. With $\beta = 0$ the Blasius boundary layer is recovered. Properties of (1) have been discussed by Stewartson (1954), and Brown and Stewartson (1965). The restriction $0 < f'(\eta) < 1$ gives rise to a unique solution: the classical Falkner-Skan profile. Without this restriction non-unique solutions to the problem are possible, depending on the value of β (Craven and Peletier (1972), Veldman and Van de Vooren (1980), Oskam and Veldman (1982), Botta *et al.* (1986))

The question remains as to whether the additional numerical solutions to the Falkner-Skan equation represent physically realisable boundary layers. A similar question may be raised about the additional non-unique solutions of swirling flows discussed by Zandbergen and Dijkstra (1987).

One way of addressing this question is attempting to observe, experimentally, instability waves in the boundary layers for which additional solutions are known to exist, at Reynolds numbers lower than the linear critical value corresponding to the classical solutions. Significantly, in such a study one has to focus on the (narrow) window of frequencies that linear theory predicts to be unstable. Furthermore, in a linear stability study the spatial model is the preferred framework since its results are amenable to direct comparisons with experiment.

In the present paper we restrict ourselves in scope and foçus on a particular class of the additional solutions to the Falkner-Skan problem which pertain to $\beta > 1$. These additional solutions will be termed Veldman boundary layers. As such, the boundary layers correspond to accelerating flow, which is known to be more stable than its zero-pressure gradient counterpart. However, the existence of inflection points in the additional solutions, suggests that inviscid mechanisms may render the corresponding flows unstable at Reynolds numbers well below the established critical Reynolds numbers pertaining to the classical solutions.

We are currently engaged in obtaining stability results for the Falkner-Skan profiles presented in Table 2. These correspond to the first additional solutions, an example of which is presented in Fig. 1.

2 Numerical Methods

2.1 Base Flow

A number of options exist to calculate the base field. Veldman and co-workers obtained numerical solutions using multiple shooting. Alternatively, straightforward Newton iteration upon values obtained by (single) shooting is a possibility. Spectral collocation solution of the Falkner-Skan equations (Street $et\ al.$, 1984) is a third option. The latter procedure is preferable if a spectral method is also to be used for the stability calculations since it eliminates interpolation errors between the grids on which the base flow is obtained and the solution to the stability problem is sought.

Both single shooting and collocation have been utilised and results for the Blasius flow obtained by the two schemes are presented in Table 1. The typical behaviour of a spectral calculation is demonstrated in these results. The cost for the inversion of the dense spectral matrices has been found to be well offset by the memory requirements with the second-order scheme. Accuracy being the prime consideration in a stability calculation, we adhere to the spectral collocation base flow results for the subsequent stability calculations.

3 Spectral Collocation for the Spatial Stability Problem

To a first approximation, the hydrodynamic stability of incompressible boundary layers is governed by the classical Orr-Sommerfeld equation,

$$\frac{i}{R}\left(\phi^{IV} - 2\alpha^2\phi'' + \alpha^4\phi\right) + (\alpha U - \omega)\left(\phi'' - \alpha^2\phi\right) - \alpha U''\phi = 0. \quad (2)$$

A form $\phi(y)e^{i(\alpha x - \omega t)}$ has been assumed for the stream function ϕ of the disturbance developing around the base flow profile U, calculated in §2.1, with x and y denoting respectively the streamwise and normal coordinates, R a suitably defined Reynolds number and t time. The physical significance attached to the quantities α and ω depends upon whether one chooses to address the temporal or the spatial stability problem. In the spatial framework presently considered, the frequency of the wave is the (real) ω, while $\alpha = \alpha_r + i\alpha_i$ is taken to be complex. The (spatial) growth rate is given by the imaginary part $-\alpha_i$ of α. It is immediately apparent that the numerical complication emerges in the spatial approach, that (2) now constitutes a nonlinear eigenvalue problem. In both the temporal and the spatial approaches, one assumes that the perturbations vanish at the endpoints of the (one-dimensional) integration domain. In order to obtain spatial results, we follow Theofilis (1994b) re-casting (2) as

$$\mathbf{A}_i = \sum_{i=1}^{4} \alpha^i \mathbf{B}_i, \qquad (3)$$

namely an eigenvalue problem in which the eigenvalue α appears nonlinearly. A solution method for the stability problem, alternative but related to that presented by Bridges and Morris (1984), may be built by expanding the unknown perturbation eigenfunctions in terms of Chebyshev polynomials (Boyd, 1989; Canuto et al., 1987). A collocation approach is being utilised to derive discrete expressions for the derivative operator D with the metrics associated with the mapping functions incorporated into D (Theofilis, 1994a). Defining

$$\mathbf{X} = \left[\phi, \alpha\phi, \alpha^2\phi, \alpha^3\phi\right]^T, \qquad (4)$$

we convert (3), as depicted in Fig. 2, into a linear eigenvalue problem (Theofilis, 1994b) that may be treated using the standard QZ algorithm (Wilkinson, 1968). An essential characteristic of our solution approach is its global character. The advantage of a global over a local approach is that one may tackle a wholly new flow problem not depending on previous knowledge of eigenvalue estimates, as the case is in a local search. The disadvantage of the global approach is that it generates large dense matrices but the resulting computing effort in the incompressible regime is acceptable, amounting to a few Cray YM-P seconds.

4 Results and Conclusion

The numerical method and the code have been tested by means of stability calculations for Hagen-Poiseuille flow and the Blasius boundary layer. Good agreement with standard literature results for these flows has been obtained with 64 mesh points in the spectral Chebyshev approach.

In the present study we focus on the first additional solution which possesses a minimal number of inflection points(Table 2 and Fig. 1).

Typical results obtained are presented in Figs. 3 and 4. The global nature of the solution approach permits capturing both the viscous modes, a typical example of which is that of Fig. 3 as well as modes resulting from relaxing the boundary condition of zero perturbations at the far-field and replacing it with one of boundedness. It is to be noted that both modes correspond to growing waves at a Reynolds number of $R = 1000$; the linear critical Reynolds number for the standard Falkner-Skan solution at $\beta = 1.5$ is several tens of thousands, depending on nondimensionalisation. Results currently to be obtained involve different Reynolds numbers at the same Hartree parameter as well as at the β values presented in Table 2. Specifically, the critical Reynolds number as a function of β is sought after in the near future.

At present we may conclude that the critical Reynolds number for the Veldman boundary layers is far smaller than the critical value for their classical Falkner-Skan counterparts

Table 1. Boundary Layer Solution Quality: Blasius Flow.

2nd order RK3		Spectral Collocation	
N	τ_w	N	τ_w
11	0.49601341292	8	0.47280243216
51	0.46961004037	16	0.46984048718
101	0.46960190272	24	0.46959845843
501	0.46960000804	32	0.46959999078
1001	0.46959999089	48	0.46959998836
5001	0.46959998838		

Table 2. The Veldman boundary layers considered

β	τ_w	β	τ_w
1.25	1.29107	2.75	1.85665
1.50	1.39365	3.00	1.93731
1.75	1.49508	3.25	2.01485
2.00	1.59210	3.50	2.08958
2.25	1.68395	3.75	2.16176
2.50	1.77250	4.00	2.23164

References

[1] Botta, E.F.F., Hut, F.J. and Veldman, A.E.P., (1986). *J. Engg. Math.*, **20**, p.81.

[2] Boyd, J.P., (1989). *Chebyshev and Fourier Spectral Methods*, Lecture Notes in Engineering **49**, Springer.

[3] Brown, S.N., and Stewartson, K., (1965). *J. Fluid Mech.*, **23**, p.673.

[4] Bridges, T.J., and Morris, P.J. (1984). *J. Comp. Phys.*, **55**, p.437.

[5] Canuto, C., Hussaini, M.Y., Quarteroni, A. and Zang, T.A., (1987). *'Spectral Methods in Fluid Dynamics'*, Springer Verlag, Berlin.

[6] Craven, A.H., and Peletier, L.A. (1972). Mathematika **19** p. 135.

[7] Oskam, B., and Veldman, A.E.P., (1982). *J. Engg. Math.*, **16**, p.295.

[8] Rosenhead, L., (1963). *'Laminar Boundary Layers'*, Oxford University Press.

[9] Stewartson, K., (1954). *Proc. Camb. Phil. Soc.*, **50**, p.454.

[10] Streett, C.L., Zang, T.A., and Hussaini, M.Y., (1984). *AIAA Pap. No. 84-0170*.

[11] Theofilis, V., (1994a). *J. Engg. Math.*, to appear in Vol. **28**.

[12] Theofilis, V., (1994b). *Theor. Comput. Fluid Dyn.*, (in press)

[13] Veldman, A.E.P., and van de Vooren, A.I., (1980). *J. Math. Anal. Appl.*, **25**, p.102.

[14] Wilkinson, J.H., (1965). *The Algebraic Eigenvalue Problem*, Clarendon.

[15] Zandbergen, P.J., and Dijkstra, D., (1987). *Ann. Rev. Fluid Mech.*, **19**, p.465.

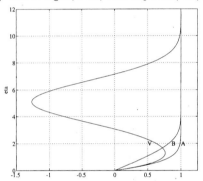

Fig. 1. A Falkner-Skan-Veldman boundary layer (V) at $\beta = 1.5$. Superimposed for comparison the Blasius profile (B) and the standard classical solution (A).

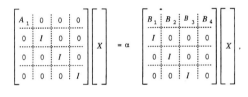

Fig. 2. Schematic of problem setup.

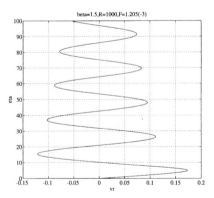

Fig. 3. Unstable mode, $R = 10^3$, $F = \omega/R = 9.5 \times 10^{-6}$ for $\beta = 1.5$.

Fig. 4. An oscillatory mode-solution of the OS equation, $F = \omega/R$.

Efficient Implementation of Turbulence Modeling for CFD

S. Venkateswaran

Propulsion Engineering Research Center
Department of Mechanical Engineering
The Pennsylvania State University

1 Introduction

The computation of turbulence flowfields is frequently fraught with difficulty. CFD codes that perform satisfactorily for inviscid and laminar flowfields may suffer serious performance degradation in the solution of turbulent problems. This degradation may manifest itself as a slowdown in convergence, lack of robustness, or both. In the present paper, we assess some of the controlling issues and identify possible methods for mitigating them.

To provide a point of reference, we focus on two-equation turbulence models and on the $k - \epsilon$ model in particular [1]. Besides the $k - \epsilon$ model, several other two equation models have been postulated such as the $k - \omega$ model, the $k - l$ model, the $k - \omega^2$ model, the $q - \omega$ model among others [2]. The issues discussed in this paper will apply to all these classes of models and may eventually be extended to include more complete full Reynolds stress closures as well.

The computational issues are discussed using contemporary time-marching methods as the basic numerical framework. These methods rely on the physics of the transient to achieve convergence. Turbulent flowfields are typically governed by a wide range of time scales. Time-derivative preconditioning [3,4] may be used to optimize these time scales and can be instrumental in accelerating convergence. Other computational issues discussed here include the degree of coupling between the fluid dynamics and turbulence equations, the effect of source terms, choice of dependent variables for the turbulence equations, sensitivity to grid stretching and high aspect ratio grids and the effects of artificial dissipation. With regard to the last issue, we contrast the use of central-differenced schemes with upwind-based methods. We note that our main emphasis is on obtaining efficient, accurate numerical solutions to a given turbulence model as opposed to improving turbulence models.

For the computational studies in the paper, we use standard test cases for which experimental data are available and which are representative of realistic internal flowfields. These include turbulent boundary layers, flow over a backward-facing step [5] and co-axial jet mixing[6]. Our results show that efficient turbulent solutions may be obtained through a combination of preconditioning, grid aspect ratio convergence control and proper linearization of the turbulence source terms. Sensitivity to initial condition specification remains an issue; however, the use of first-order upwind schemes to initialize the flowfield appears viable for most problems.

2 Theoretical Development

The preconditioned Navier-Stokes equations are given by:

$$\Gamma \frac{\partial Q_v}{\partial t} + \frac{\partial E}{\partial x} + \frac{\partial F}{\partial y} = H + V.T. \tag{1}$$

where $Q_v = (P_t, u, v, T, k, \epsilon)^T$ and the other terms are defined in Ref. 4. The preconditioning matrix Γ ensures that the system remains well-conditioned at all flow Mach numbers and Reynolds numbers, thus providing uniform convergence under a wide range of conditions.

The degree of coupling between the fluid dynamics and the turbulence equations determines whether the equations should be solved fully coupled or sequentially (i.e., weakly coupled). The primary source of coupling between the equations is the nonlinear dependence of the turbulent viscosity μ_t on k and ϵ and on the mean strain rate. The turbulent viscosity μ_t appears in the turbulence source terms as well as in the diffusion terms of the transport equations. The linearization of the turbulence source terms is extremely important and is discussed later. The diffusion co-efficient, on the other hand, may be lagged with little penalty in terms of convergence or stability. Thus, while the turbulence equations are strongly coupled to the fluids equations, the fluids equations are only weakly coupled to the turbulence. Thus, the most efficient solution method appears to be one that sequentially iterates between the Navier-Stokes and the $k - \epsilon$ equations.

Linearized analysis of the effect of source terms on stability and convergence suggest that positive sources be treated explicitly while negative sources (or sinks) be treated implicitly. For the $k - \epsilon$ equations, precise separation into sources and sinks involves the diagonalization of the source Jacobian and is rather complicated. An approximate splitting may be accomplished simply by treating the production terms explicitly and the dissipation terms implicitly. This procedure appears to work adequately for all the cases that we have tested. It is worthwhile noting that precise splitting of sources and sinks is readily accomplished for the $k - \omega$ equations, which suggests that these are a better choice of dependent variables. However, comparisons made between using the $k - \epsilon$ and $k - \omega$ variables have indicated similar behavior in practice.

Turbulent calculations usually involve regions where the grid is strongly stretched, particularly when near-wall turbulence effects are resolved. In such regions, the local grid aspect ratios can be extremely high, which in turn creates convergence difficulties for most CFD algorithms. These problems may be effectively mitigated for the ADI algorithm by proper selection of the local time step and the preconditioning matrix [4]. With the improved algorithm, substantial savings in computer time are obtained for turbulent calculations as evident in Fig. 1 for a flat plate turbulent boundary layer.

A further difficulty inherent in turbulent calculations is the presence of odd-even splitting or oscillatory solutions in the vicinity of steep gradients. These often lead to negative values of k and ϵ with disastrous consequences. Judicious addition of second-order dissipation through the use of a switches or flux limiters [7] may sometimes be necessary. Upwind schemes which possess inherent dissipative properties appear to be more robust in this regard. In fact, for complicated problems, it is often easier to initialize a flowfield with a first-order upwind scheme before carrying out a higher order accurate computation.

3 Results

The convergence rate for the Driver-Seegmiller backstep test case [5] using a third-order upwind-biased ADI scheme is shown in Fig. 2. The convergence rate is seen to be reasonably good for this more difficult case with the residuals going down about five orders of magnitude in 2000 iterations. The converged velocity contours, shown in Fig. 3, indicate a recirculation length, $x/h = 5.7$, which is in agreement with other published computational results (experimental value is 6.3). Figure 4 shows velocity profiles plotted against experimental data at several axial stations. In contrast, second-order central differencing does not yield a converged solution and the velocity contours indicate that the solution is unsteady. (Interestingly, experimental observations indicate large scale unsteadiness in the shear layer.) A small amount of second-order dissipation, however, does render the solution steady and good convergence is observed. Apparently, the third-order upwind scheme inherently possesses sufficient dissipation obviating any need to stabilize the computations.

As a further test case, we choose the experiments of Johnson and Bennett [6], which involves two co-flowing jets in a confined sudden expansion. Figure 5 shows the converged velocity contours for this case while Fig. 6 shows comparisons of velocity profiles with experimental data. Overall agreement is quite good except at the centerline in the near-injector region. Solution convergence for this case (not shown) is again quite good.

Acknowledgements

This work was supported by Pratt and Whitney, WPB, FL and by NASA/MSFC under Contract NAS8-38861. The author would also like to acknowledge the contributions of Prof. Charles Merkle and Mr. Manish Deshpande.

References

[1] Jones, W. P. and Launder, B. E., Intl. J. of Heat and Mass Transfer, Vol.15, pp. 301 - 314.

[2] Wilcox, D. C., Turbulence Modeling for CFD, 1993.

[3] Choi, Y.-H. and Merkle, C. L., J. of Comp. Phys., 105, pp. 207-223.

[4] Buelow, P., Venkateswaran, S. and Merkle, C., AIAA 93-3367, Orlando, FL.

[5] Driver, D. and Seegmiller, H., AIAA Journal, Vol.23, No.2, pp. 163 - 171.

[6] Johnson, B. and Bennett, J., Proc. ASME Fluids Engg. Conf., Boulder, CO.

[7] Jameson, A., AIAA 93-3359, Orlando, FL.

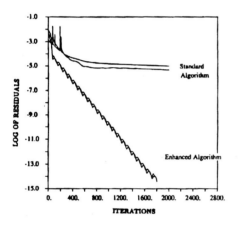

Fig. 1 Convergence of flat plate turbulent boundary layer with and without high aspect ratio enhancement.

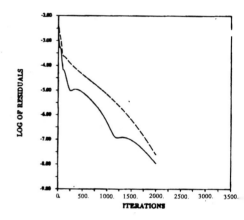

Fig. 2 Convergence of turbulent backstep computation using upwind-biased ADI.

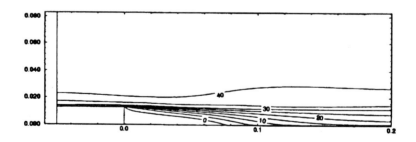

Fig. 3 Velocity contours of turbulent backstep computation.

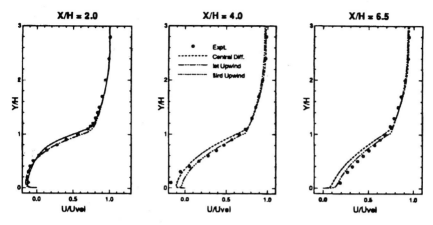

Fig. 4 Velocity profiles for turbulent backstep.

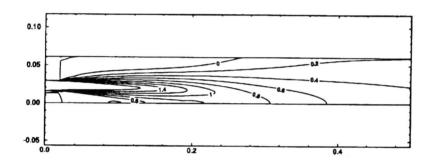

Fig. 5 Computed velocity contours for co-axial jets.

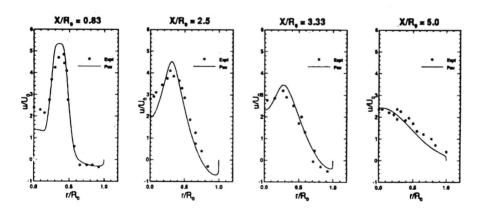

Fig. 6 Velocity profiles for turbulent co-axial jets.

Transition to Chaos in Impinging Three-Dimensional Axial and Radial Jets

M. Winkelsträter, H. Laschefski and N.K. Mitra

Institut für Thermo- und Fluiddynamik, Ruhr-Universität Bochum,
44780 Bochum, Germany

Abstract : Flow fields of arrays of rectangular axial and radial jets impinging on a flat surface have been computed from the numerical solution of 3D Navier-Stokes and energy equations. Results give Reynolds number ranges for steady, periodic and aperiodic flows. Period doubling has been observed for both axial and radial jets. Furthermore periodic window of chaos has been observed for radial jets.

1 Introduction

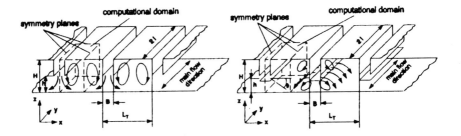

Fig. 1a. Computational domain for axial jets

Fig. 1b. Computational domain for radial jets

Rectangular axial or radial (knife) arrays of jets (see fig.1) impinging on a product surface are used for heat or mass transfer in industry. Often these jets are turbulent. However, in some practical situations (eg. electronic cooling) they are laminar at the discharge. Before or after impinging these laminar jets may go through transition and become turbulent. The transition for a given geometry depends on the Reynolds number. In a previous paper we have investigated the impinging jet flow from a round tube [1]. If the jet impingement is axial (in-line), the transition takes place at a higher Re than the radial jet for which the jet discharges from the side of the feed tube. In the present work we investigate the transition problem of impinging rectangular three-dimensional axial and "radial" (knife) jets.

2 Method of Solution

We consider the flow from a bank of jets and select an element from the jet bank as our computational domain (see fig. 2a and 2b for axial and radial jets). The flow is described by nonsteady three-dimensional continuity and Navier-Stokes equations for an incompressible fluid with constant properties.

These equations have been solved by a finite volume technique with SIMPLE-C pressure correction. The discretization is spatially second order accurate. For the sake of the reduction of computational time and memory we assumed symmetry conditions on side boundaries. The computations have been performed on 71400 and 530376 grids.

3 Results and Discussion

For the axial jet we obtained steady flow for $Re \leq 335$, $Re = u_{av} \cdot 2 B/\nu$ where B is the feed tube width, and u_{av} is the average velocity at the feed tube exit.

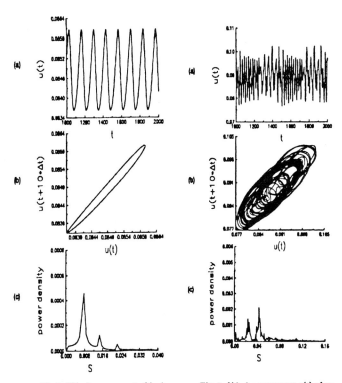

Fig. 2. Velocity component u(t) of axial jet for Re=338, (a) Time series; (b) Time phase diagram; (c) Fourier-Analyse showing power density against Strouhal number S

Fig. 3. Velocity component u(t) of axial jet for Re=414, (a) Time series; (b) Time phase diagram; (c) Fourier-Analyse showing power density against Strouhal number S

Fig. 2 shows the time series for the axial velocity u at a select point in the middle of the flow field, phase diagram with time delay reconstruction and the Fourier analysis of the time series for $Re = 338$. Notice the dominant frequency f shown by the Strouhal number $S = f \cdot 2\, B/u_{av} = 0.008$. The other Strouhal numbers are 0.015 and 0.022. A transition from periodic to aperiodic flow occurs at a Re between 412 and 414. We have also computed for $Re = 420$ and 435. The flow at these Re are qualitatively similar to that at $Re = 414$.

Fig. 4 shows Strouhal number against Reynolds number for the axial jet. Notice that for $Re \geq 416$ no dominant frequency exists. S was calculated with highest frequency.

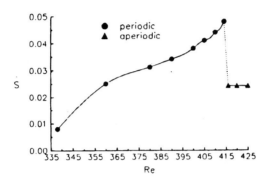

Fig. 4. Dominant Strouhal number vs Reynolds number for axial jet

Fig. 5 shows corresponding flow character for radial (knife) jets for $Re = 210$. We have also computed for $Re = 212, 215, 220, 230, 240, 245, 250, 270$ and 275, respectively. We notice periodic flows at $Re = 210, 212$ and aperiodic flow at $Re = 215$, periodic flow at $Re = 220$, aperiodic at $Re = 230$, periodic at $Re = 240, 245$ and aperiodic at $Re = 250, 270$ and 275. At $Re = 275$ no dominant frequency appears, see Fig. 6. The flow is turbulent. We have observed one period doubling. The solutions for both periodic and chaotic flows are strongly dependent on initial conditions. For the axial jet no periodic windows of chaos have been observed.

Table 1 shows the character of the flow for the different Re.

Re	S	Radial jet flow character	
210	0.047	periodic	
212	0.058	periodic	period doubling
215	0.029	aperiodic	period doubling
220	0.028	periodic	
230	0.031	aperiodic	
240	0.028	periodic	
245	0.029	aperiodic	
250	0.035	aperiodic	
270	0.039	aperiodic	
275	—	turbulent	

Acknowledgement :

This work has been supported by the Deutsche Forschungsgemeinschaft through the FG: Wirbel und Wärmeübertragung.

References

[1] Laschefski,H., Braess,D., Haneke,H., Mitra,N.K. (1994) "Numerical Investigation of Radial Jet Reattachment Flows", Int. J. for Numerical Methods in Fluids **18**, No. 7, pp. 629 - 646.

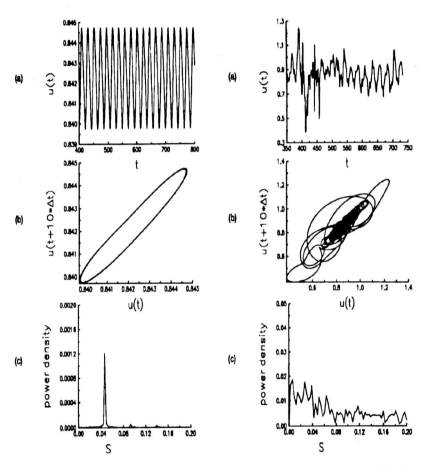

Fig. 5. Velocity component u(t) of radial jet for Re=210, (a) Time series; (b) Time phase diagram; (c) Fourier-Analyse showing power density against Strouhal number S

Fig. 6. Velocity component u(t) of radial jet for Re=275, (a) Time series; (b) Time phase diagram; (c) Fourier-Analyse showing power density against Strouhal number S

Transonic Inviscid/Turbulent Airfoil Flow Simulations Using a Pressure Based Method with High Order Schemes

Gang Zhou, Lars Davidson and Erik Olsson

Thermo and Fluid Dynamics, Chalmers University of Technology,
S-412 96 Gothenburg, Sweden

This paper presents computations of transonic aerodynamic flow simulations using a pressure-based Euler/Navier-Stokes solver. In this work emphasis is focused on the implementation of higher-order schemes such as QUICK, LUDS and MUSCL. A new scheme CHARM is proposed for convection approximation. Inviscid flow simulations are carried out for the airfoil NACA 0012. The CHARM scheme gives better resolution for the present inviscid case. The turbulent flow computations are carried out for the airfoil RAE 2822. Good results were obtained using QUICK scheme for mean motion equation combined with the MUSCL scheme for k and ϵ equations. No unphysical oscillations were observed. The results also show that the second-order and third-order schemes yielded a comparable accuracy compared with the experimental data.

1 Introduction

Traditionally most transonic aerodynamic flow simulations are carried out by using density based methods (or time-marching methods as often called) in which density is used as a primary variable in the continuity equation, while pressure is extracted from the equation of state. Recently there are a few successful computations[1, 2, 3, 4] obtained from the pressure-based method which use pressure as the dependent variable for transonic airfoil flow, which is the most challenging case for the demonstration of how well a pressure-based method can resolve physical discontinuities such as external shocks. The previous works[1, 3, 4] have shown that an advanced pressure-based method has a resolution comparable to, or better than, the traditional time-marching methods. This paper presents computations of both inviscid and turbulent transonic aerodynamic flow using the proposed pressure-based method in Refs. ([1, 3, 4]). Higher-order schemes such as QUICK[5], LUDS, MUSCL[6] and a newly proposed bounded third-order scheme CHARM[7] are used.

In recent years, many high resolution schemes have been proposed and investigated like unbounded QUICK[5], bounded SMART[8] and SHARP[9]. A piecewise linear blending function is used in SMART and a non-linear one in SHARP in combination with variable normalization to make them bounded. All these schemes are of third-order accuracy. To simplify the algorithm involved in third-order scheme (causing insignificant loss of accuracy), many second-order accuracy schemes have been proposed, like HLPA[10] and TVD MUSCL. The scheme HLPA (Hybrid Linear / Parabolic Approximation) uses a second order polynomial in the monotonic region of the normalized variable diagram (NVD). This method is similar to van Leer's scheme[11]. The TVD MUSCL scheme, which adopts the TVD flux limiter concept and variable normalization approach, employs a combination of a central differencing scheme (CDS) and a second linear upwind differencing scheme (LUDS). The result is a bounded second-order approximation which can be used for all convection transport variables including k and ϵ. These two schemes are one order of accuracy less than QUICK in terms of the Taylor series truncation error. However, they are unconditionally bounded, simple to implement in many applications and yield results with no substantial difference in accuracy compared to third-order scheme. Combining the respective advantages of the higher order accuracy of third-order schemes and that of boundedness, algorithmic simplicity of a second-order scheme, on the basis of the principle of *Leonard's* NVD characteristic, one can design a new generation of high-order schemes. The CHARM (Cubic-parabolic High Accuracy Resolution Method), presented in this work is a new unconditionally bounded scheme with essentially third-order accuracy.

Along the mutual influence on the development of high order schemes between pressure-based methods and time-marching methods, many efforts[1, 2, 3, 4] also focus on the advance of pressure-based method to external transonic flow simulations especially with the shocks. *Leonard*[9] has shown the similarities between variable normalization approach and the flux limiter which is normally adopted in time-marching methods. More recently, *Zhou and Davidson*[3] have shown the numerical dissipation models used in pressure-based method and in density-based methods have similar properties. The similarities between these two methods indicated that these two methods should have comparable abilities to simulate highly convective flows in computational fluid dynamics.

In this work we use a pressure-based Euler/Navier-Stokes SIMPLE solver in a finite volume frame employing a non-orthogonal collocated arrangement. An implicit artificial dissipation model is used, which consists of second- and fourth-order derivative of pressure in all equations. It was found that two-level filters should be used to adjust second-order dissipation for inviscid flow simulations. For turbulent airfoil flow, however, only one level filter needs to be used. The results have shown a good capability of predicting both internal and external inviscid transonic flows[3] including shock capturing. A $k - \epsilon$ turbulence closure with a near-wall one-equation model is used for the turbulent flow cases. The second-order accuracy scheme MUSCL is adopted for approximating turbulence transport properties. The results of the prediction of the turbulent transonic flow around airfoil RAE 2822 for *case* 6, and *case* 10 of Cook *et al.*[12] show good agreement with experimental data.

2 High accuracy approximation for convection

2.1 Linear schemes

Following *Leonard*'s[9] variable normalization consideration, the face normal velocity at the right-hand face of the contral volume (CV) at i can be nondimensionalized in upstream sense as

$$\hat{\Phi}_{i+1/2} = \frac{\Phi_{i+1/2} - \Phi_{i-1}}{\Phi_{i+1} - \Phi_{i-1}}. \tag{1}$$

A general form of second- and third-order linear (in NVD) schemes may be written as

$$\Phi_{i+1/2} = \Phi_i + \frac{1}{4}\left[(1-\kappa)(\Phi_i - \Phi_{i-1}) + (1+\kappa)(\Phi_{i+1} - \Phi_i)\right]. \tag{2}$$

The normalized form of Eq. (2) is

$$\hat{\Phi}_{i+1/2} = \hat{\Phi}_i + \frac{1}{4}\left[(1-\kappa)\hat{\Phi}_i + (1+\kappa)\left(1 - \hat{\Phi}_i\right)\right]. \tag{3}$$

Note that, the numerical parameter κ controls the order of scheme. Setting $\kappa = 1$, -1 and 0.5, one gets linear CDS, LUDS and QUICK scheme respectively. Adopting TVD flux limiters, one can get a piece-wise linear MUSCL scheme as follows

$$\hat{\Phi}_{i+1/2} = \begin{cases} \hat{\Phi}_i + \frac{1}{2}\min\left\{\hat{\Phi}_i, \left(1 - \hat{\Phi}_i\right)\right\}, & \frac{\hat{\Phi}_i}{1-\hat{\Phi}_i} > 0, \\ \hat{\Phi}_i, & \frac{\hat{\Phi}_i}{1-\hat{\Phi}_i} \leq 0. \end{cases}$$

This scheme combines the second linear upwind scheme and the central differencing schemes and can be written as follows,

$$\hat{\Phi}_{i+1/2} = \begin{cases} \frac{3}{2}\hat{\Phi}_i, & \hat{\Phi}_i \in [0, 0.5], \\ \frac{1}{2}\left(1 + \hat{\Phi}_i\right), & \hat{\Phi}_i \in (0.5, 1], \\ \hat{\Phi}_i, & \hat{\Phi}_i \ni [0, 1]. \end{cases}$$

Note that all schemes pass the critical point $Q(0.5, 0.75)$ (see Fig. 1) through which any scheme is at least of second order accuracy. The TVD MUSCL scheme passes the point and then follows the lower border of the second-order region. The scheme HLPA passes the point, but with a non-linear second polynomial function.

2.2 CHARM scheme

As an additional necessary and sufficient condition, as pointed out by *Leonard*[9], any approximation function passing Q with a slope 0.75 is of third-order accuracy. The scheme QUICK meets the condition, and it reads

$$\hat{\Phi}_{i+1/2} = \frac{3}{4}\hat{\Phi}_i + \frac{3}{8}. \tag{4}$$

The scheme SHARP[8] uses an exponential upwinding function which is tangent to the QUICK line at point Q. Obviously it is third-order accurate. Nevertheless, the exponential upwinding is more expensive than a polynomial. Our object is to replace the exponential upwinding function with a third-order polynomial which is tangent to the QUICK scheme at the point Q point. A <u>C</u>ubic-parabolic <u>H</u>igh <u>A</u>ccuracy <u>R</u>esolution <u>M</u>ethod, CHARM, is designed as follows

$$\hat{\Phi}_{i+1/2} = \begin{cases} \hat{\Phi}_i^3 - 2.5\hat{\Phi}_i^2 + 2.5\hat{\Phi}_i, & \hat{\Phi}_i \in [0,1], \\ \hat{\Phi}_i, & \hat{\Phi}_i \ni [0,1]. \end{cases}$$

Figure 1 shows the characteristic of CHARM. In terms of non-normalized variables, the cubic-parabolic part of the above equation has the form

$$\Phi_{i+1/2} = \Phi_i + \kappa \left(\Phi_i - \Phi_{i-1} \right) \left(\hat{\Phi}_i^2 - 2.5\hat{\Phi}_i + 1.5 \right) \tag{5}$$

with the parameter κ

$$\kappa = \begin{cases} 1, & |\hat{\Phi}_i - 1.5| \leq 0.5, \\ 0, & \hat{\Phi}_i \ni [0,1]. \end{cases}$$

It can be shown that the convective stability is satisfied. A numerical negative feedback mechanism which reduces the convective influx, $\mathbf{F_i}$, when Φ_i increases, or *vice versa*, ensures the convergency. Assuming one-dimension flow, for example, with a constant convecting velocity ($U = U_0$), we have, for $\hat{\Phi} \in [0,1]$, that

$$\frac{\partial \mathbf{F_i}}{\partial \Phi_i} = U_0 \frac{\partial \left(\Phi_{i-1/2} - \Phi_{i+1/2} \right)}{\partial \Phi_i} = U_0 \left(-2\hat{\Phi}_{i-1}^3 + 2.5\hat{\Phi}_{i-1}^2 - 3\hat{\Phi}_i^2 + 2.5\hat{\Phi}_i - 2.5 \right) < 0. \tag{6}$$

It is obvious that Eq. (5) represents a first-order upwind formulation with a third-order correction, which is implemented as explicit *sources/sink* terms in an iterative sequence leading to a steady-state solution.

3 Numerical Procedure

The SIMPLE procedure is used to achieve steady solution for both inviscid and turbulent flow cases. All the govering equations can be cast into a standard transport equation for a general dependent variable Φ, in Cartesian coordinates, as

$$\frac{\partial}{\partial t}(\Phi) + \frac{\partial}{\partial x_i}(U_i \Phi) = \frac{\partial}{\partial x_i}\left[\Gamma_\Phi \frac{\partial}{\partial x_i}\left(\frac{\Phi}{\rho}\right) \right] + S^\Phi \tag{7}$$

where Φ can be the Cartesian mass flux components, pressure correction, ρk or $\rho\epsilon$, Γ_Φ is the effective diffusivity, and S_Φ denotes source per unit volume for the dependent variables Φ. In the inviscid case $\Gamma_\Phi = 0$. Integrating Eq. (7) over all the control volumes, by using the Gaussian theorem, we get the following discretized equation

$$(\Phi - \Phi^0)\frac{\delta v}{\Delta t} + \sum_m (\mathbf{F} \bullet \mathbf{A}) - \sum_m (\Gamma_\Phi \nabla \Phi \bullet \mathbf{A}) = S_\Phi \delta v \tag{8}$$

where $\mathbf{F}$ is the flux tensor, δv is the volume of the cell, m refer to each face $\mathbf{A}$ of the control volume, and Φ^0 is value at previous time level. The discretized equation can be cast into standard form (See Ref. 1 and 3)

$$a_P \Phi_P = \sum a_{nb} \Phi_{nb} + S_C^\Phi, \quad a_P = \sum a_{nb} - S_P^\Phi. \tag{9}$$

The implicit numerical dissipation model[1, 3, 4] is a combination of Rhie-Chow type interpolation for mass flux and a retarded density. It is used to extract the damping mechanism from the second- and fourth-order differencing formulation, which is expressed in pressure for all transport equations. A standard $k - \epsilon$ turbulent closure with a near-wall one-equation model[13] is used for the turbulent flow cases. The calculation starts from uniform flow condition. Boundary conditions everywhere use the same strategy for the Euler and Navier-Stokes equations, except at solid walls where the slip condition is used for Euler and the no-slip conditions for Navier-Stokes. The values of k and ϵ were extrapolated at outlets.

4 Results and Concluding Remarks

The computed results for the inviscid flow around airfoil NACA 0012 are compared with Pulliam and Barton's[14] in Fig 2. The results of the prediction of the turbulent flow around airfoil RAE 2822 for *case* 6, and *case* 10 are shown in Figs 3 and 4. From these results we can conclude that

* An advanced pressure-based method can predict both inviscid and turbulent transonic aerodynamic flows with a satisfactory accuracy comparable to, or better than, time-marching method.

* A new proposed third-order upwinding scheme CHARM (Cubic-parabolic High Accuracy Resolution Method) give better resolution for the present inviscid case than the TVD MUSCL scheme (see Fig.2). The linear (in sense of NVD) QUICK gives an unreasonable solution for this case.

* For the turbulent flow cases, both the third-order scheme CHARM and the second-order schemes HLPA and LUDS yield solutions comparable to QUICK.

References

[1] Zhou, G. Davidson, L. and Olsson E., AIAA paper 94-2345, June, 1994.

[2] Lai, Y. G., So, R.M.C. and Przekwas, A. J., AIAA paper 93-2902, July, 1993.

[3] Zhou, G. and Davidson, L., Submitted to *Journal of Computational Fluid Dynamics*, 1994.

[4] Zhou, G., Davidson, L. and Olsson E., *Report.*, Thermo and Fluid Dynamics, CTH, 1994.

[5] Leonard, B. P., *Comp. Meth. Appl. Mech. Eng.*, Vol. 19, pp. 59, 1979.

[6] Lien, F-S and Leschziner, M. A., *Proc. 5th Int. IAHR Symp. on Refind Flow Modeling and Turbulence Measurements*, Paris, Sept. 1993.

[7] Zhou, G. and Davidson, L., *Report.*, Thermo and Fluid Dynamics, CTH, 1994.

[8] Gaskell, P. H. and Lau, A. K. C., *J. Numer. Methods Fluids*, Vol. 8, pp. 617, 1988.

[9] Leonard, B. P., *Int. J. Numer. Methods Fluids*, Vol. 8, pp. 1291 1988.

[10] Zhu J., *Comp. Meth. Appl. Mech. Eng.*, Vol. 98, pp. 345, 1992.

[11] van Leer., B., *J. Comput. Pyys.*, Vol. 32, pp. 101, 1979.

[12] Cook, P. H., McDonald, M. A., and Firmin, M. C. P., *AGARD Report*, No.138., 1979.

[13] Chen, H. C. and Patel, V. C., AIAA paper 87-1300, June, 1987.

[14] Pulliam, T. H. and Barton, J. T., AIAA paper 85-0018, January, 1985.

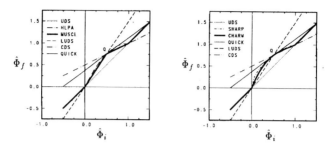

Figure 1. Characteristic of MUSCL (left) and CHARM (right)

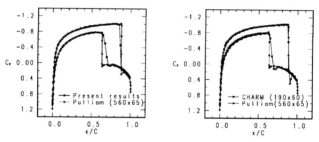

Figure 2. Euler solution for NACA 0012, $M_\infty = 0.85, \alpha = 1.00$, Pressure distribution, from MUSCL (left) and CHARM (right)

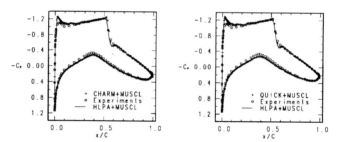

Figure 3. Turbulent N-S solution for RAE 2822, case 6.

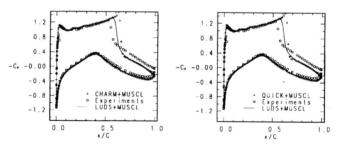

Figure 4. Turbulent N-S solution for RAE 2822, case 10.

On Oscillations Due to Shock Capturing in Unsteady Flows

Mohit Arora and Philip L. Roe

W. M. Keck Foundation Laboratory for Computational Fluid Dynamics,
The University of Michigan, Ann Arbor, Michigan, USA

1 Difficulties With Moving Shocks

Shock capturing algorithms for nonlinear systems are afflicted with a continuously generated long-wavelength error along the characteristic not belonging to the shock family (Fig. 1). This is not adequately damped by practical numerical schemes. Initially thought to be a problem only for slowly moving strong shocks (Woodward and Colella 1984), it was found by Roberts (1990) to pose problems for weaker 'slow' shocks as well, where a 'slow' shock is defined as one across which the associated characteristic changes sign. He also noticed that the Godunov and Roe schemes performed worse than the Osher scheme - this he attributed to a sharper steady shock representation by these schemes. In order to better understand this phenomenon, and to devise ways to overcome this difficulty, we have studied several systems of conservation laws of the form

$$\mathbf{w}_t + \mathbf{f}_x = 0 \quad , \quad \mathbf{w}(t=0) = \mathbf{w}_0 \quad .$$

1.1 The p-System and Orbit Plots

Here, $\mathbf{w} = (v, u)^T$, $\mathbf{f} = (-u, p)^T$, and let $p = v^{-2}$. In the sense of Roberts (1990), this system only allows 'fast' shocks since we always have $\lambda_1 = -\lambda_2$, and hence the eigenvalue never changes sign across a shock. However, for initial data resulting in a single front-shock, the back-characteristic variable C^- has a large, persistent, long-wavelength error. The other variable, C^+, has an error as well (albeit small). Further, neither of the Osher scheme variants (O, P) improve upon the Roe scheme (Fig. 1). These findings do not support previous conjectures.

In the course of the above study, we found that an efficient way to display all the relevant information in one picture is to make iterative maps in phase space, using characteristic variables (C^-, C^+) as coordinates. The state of each cell j, at a given time n, represents a point in (C^-, C^+) phase space, say P_j^n. If we plot P_j^n for all j and n on the same plot, we get the time history of all cells in this phase space (Fig. 2). The picture that emerges is reminiscent of nonlinear dynamics. The initial points (states) seem to fall randomly, but subsequent points are all attracted to the same orbit, which can be easily observed if we color-code the points by time. This is surprising, considering the intricacy of the orbit (Fig. 3).

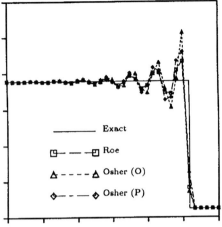 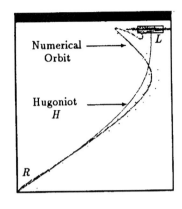

Fig. 1. Back-characteristic variable C^- vs x for a single front-shock (p–system, 1st order).

Fig. 2. A typical orbit plot for the Roe scheme (1st order) and the Hugoniot (characteristic coordinates C^+ vs C^-).

1.2 The Euler Equations

Tests on the isothermal and full Euler equations showed that both 'slow' and 'fast' shocks generated the post-shock error, although it was more pronounced for 'slow' shocks. Osher's scheme performed better here than Roe's, by generating smaller amplitude oscillations. For the full Euler equations, we found that several shock-tube problems, which appeared to be free of this phenomenon, were only so to plotting accuracy - zooming in revealed the now-familiar oscillations.

2 Defining the Problem

The observations in Sect. 1 show that post-shock oscillations are not merely a 'slow' or 'fast' shock effect. We feel that they are influenced chiefly by the degree of nonlinearity involved, of which a rather subjective measure is the disparity between the Hugoniot curve (H) joining the left (L) and right (R) states, compared with the characteristic curves (Lin 1991). This has the correct behaviour, since for very weak jumps there is no disparity (Smoller 1983), and we recover the linear behaviour. The noise is a purely nonlinear wave interaction phenomenon.

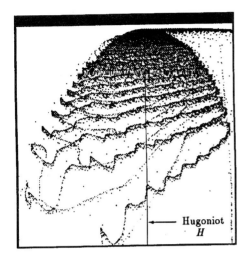
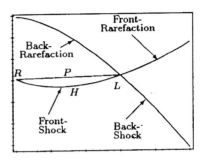

Fig. 3. Zoom of boxed region of Fig. 2 showing the intricate detail. The aspect-ratio is higher than that shown in the box.

Fig. 4. Shock and rarefaction curves for the left state (L), in conserved variable $(\rho, \rho u)$ space. P is the locus of projected states, corresponding to a right state R connected to L by a single front-shock via the Hugoniot (H) (isothermal Euler equations).

2.1 Why Do We Get Wiggles?

For the isothermal Euler equations, we considered the projection onto the grid of a shock not lying at an interface. For different shock locations within the cell, the locus of the intermediate point in the space of conserved variables $(\rho, \rho u)$ is a straight line joining states L and R (Fig. 4). However, the Hugoniot H is not straight, or even monotone in this space. Transferring to (C^-, C^+) space, we find that while the Hugoniot H is monotone, the projection P is not (Fig. 5).

Consider initial data corresponding exactly to a single front-shock. After one time-step, the newly formed intermediate state I is not on the Hugoniot but actually on the projection P. However, it was shown by Roe (1982) that stationary shocks produced by the Godunov or Roe schemes have intermediate points that do lie on the Hugoniot. Slow shocks produced by these schemes therefore have some conflicting loyalties: should I follow H or P?

If it were possible to make I follow H, then a single intermediate point would be retained, because, by the arguments in Roe (1982) no waves are created that leave the shock. However, we have not succeeded in devising such a scheme.

If I is not on H, a non-vanishing left-going wave *will* be generated by the Riemann problem (L, I). So we are led to consider schemes that do not take information exclusively from solutions of the Riemann problems. In fact, it is quite simple to devise one that follows P (retaining a single intermediate point).

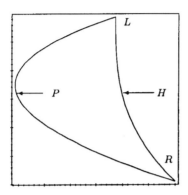
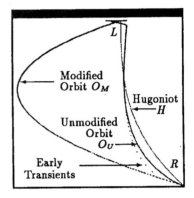

Fig. 5. Relevant part of Fig. 4, but in characteristic coordinates (C^-, C^+).

Fig. 6. Orbits for the Roe scheme and the Hugoniot H (isothermal Euler eqns).

3   A Solution Strategy and Numerical Results

Having uncovered the mechanism giving rise to post-shock oscillations, we now devise a "fix". The procedure, briefly described here, will hopefully serve as a starting point in our design of more elegant solution algorithms.

Our method resembles sub-cell resolution (Harten, 1989), but is motivated very differently. Rather than higher-order reconstruction, we seek accurate shock location and hence correct fluxes, to prevent the generation of spurious waves.

The underlying scheme is a flux-limited, explicit, Roe-averaged Lax-Wendroff scheme that typically produces shocks that are 2-3 cells wide. This is modified as follows: detect large pressure jumps over (say) 6 cells and let $L \equiv$ cell 1, $R \equiv$ cell 6. Given $\mathbf{w}_L$ and p_R, $\mathbf{w}_R^*$ is the state (ρ_R^*, u_R^*, p_R) connected to $\mathbf{w}_L$ by a shock. If $\mathbf{w}_R$ lies within some prescribed tolerance of $\mathbf{w}_R^*$, then we "find" a shock and use conservation to determine its location ($M \equiv$ shock cell), where M must lie between L and R. The shock speed is easily obtained from the jump conditions across L, R^*. The shock, thus reconstructed, gives the correct states and fluxes; hence, the limiter is switched off here. Another key point is to correctly modify the flux on the exit face if the shock leaves cell M in the current update. If a shock is not found, we simply proceed with the underlying scheme. After modification, those shocks that are detected will have only one intermediate state, as well as avoiding the spurious oscillations. The path of the shock for the modified scheme is displayed (Fig. 6) and can be seen to follow P faithfully (compare with Fig. 5).

In Fig. 7, we present a typical 'slow' front-shock for the Euler equations, where we plot the exact solution, and the solution before and after using the above fix. Also shown (Fig. 8) is the complicated wave interaction problem presented by Shu and Osher (1989), to demonstrate that this algorithm, even in its present unsophisticated state, can handle complex situations remarkably well.

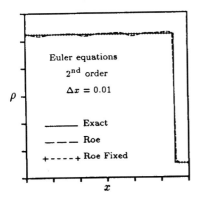
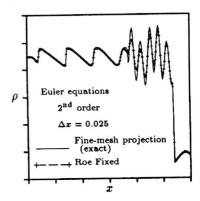

Fig. 7. Results for a 'slow' 3–shock. **Fig. 8.** Results for the Shu-Osher case.

4 Comments on Design Principles for Nonlinear Systems

Figure 5 showed that the characteristic variables are not monotone through the captured shock. Hence the Total Variation (TV) based on any of characteristic, conserved or primitive variables may legitimately increase.

A design criterion often applied to FVS schemes is to ensure that the eigenvalues of split fluxes be non-negative for $\vec{F}^+$ and non-positive for $\vec{F}^-$. This ensures smooth shock transitions in the scalar case (Jennings, 1974), but we have found that FVS schemes meeting this condition still generate post-shock noise in nonlinear systems of equations. We end by remarking that there still appears to be no rigorous design criterion that guarantees all the desirable properties of a captured shock.

References

[1] Harten, A. (1989): "ENO Schemes with Subcell Resolution", J. Comput. Phys., **83**, pp. 148–184.

[2] Jennings, G. (1974): "Discrete Shocks", Comm. Pure Appl. Math., **27**, pp. 25–37.

[3] Lin, H-C. (1991): "Dissipation Additions to Flux-Difference Splitting", 10th AIAA Computational Fluid Dynamics Conference (AIAA CP-91-1544, in Proceedings).

[4] Roberts, T. W. (1990): "The Behaviour of Flux Difference Splitting Schemes Near Slowly Moving Shock Waves", J. Comput. Phys., **90**, pp. 141–160.

[5] Roe, P. L. (1982): "Fluctuations and Signals, a Framework for Numerical Evolution Problems", in Numerical Methods in Fluid Dynamics, ed. K. W. Morton and M. J. Baines, Academic Press, pp. 219–257.

[6] Smoller, J. (1983): "Shock Waves and Reaction-Diffusion Equations", Springer-Verlag.

[7] Shu, C-W., Osher, S. (1989): "Efficient Implementation of Essentially Non-Oscillatory Shock-Capturing Schemes, II", J. Comput. Phys., **83**, pp. 32–78.

[8] Woodward, P. R., Colella, P. (1984): "The Numerical Simulation of Two-Dimensional Fluid Flow with Strong Shocks", J. Comput. Phys., **54**, pp. 115–173.

An Example of Hopf Bifurcation and Strange Attractor in Chebyshev Spectral Approximations to Burgers Equation

H. Dang-Vu[1] and C. Delcarte[2]

[1] Institut de Mathématiques pures et appliquées
Université Pierre et Marie Curie, 75252-Paris. France
[2] LIMSI-Université de Paris-Sud, 91405-Orsay. France

1 Introduction

The purpose of this work is to examine the behavior of Chebyshev collocation solutions for a spectral approximation of Burgers equation when the viscosity ν is varied. This study is partly analytic and partly numerical. We generalize the perturbation technique of Lindsted-Poincaré to study the discrete solutions and use a numerical procedure to advance these solutions to values of ν close to a value ν_c where Hopf bifurcation occurs. Under appropriate circumstances, the methods of the present model problem can be applied to study the Hopf bifurcation of other non-linear systems.

2 Burgers Equation

We are interested in this work in Burgers equation :

$$\frac{\partial u(x,t)}{\partial t} = -u(x,t)\frac{\partial u(x,t)}{\partial x} + \nu\frac{\partial^2 u(x,t)}{\partial x^2} + f(x) \qquad (1)$$

($-1 \leq x \leq 1$, $t \geq 0$, $\nu > 0$), with homogeneous Dirichlet boundary conditions $u(-1,t) = 0$, $u(1,t) = 0$ and initial condition : $u(x,0) = u_0(x)$, $f(x)$ is the forcing term. A steady-state solution $u^*(x)$ is obtained from equation (1) by setting the time derivative to zero.

A Chebyshev collocation scheme was considered for this problem in [1] : for $x_k = \cos(k\pi/N)$, $k = 0, 1, ..., N$, the spectral solution $u_k(t) = u(x_k, t)$ is defined through the collocation equations

$$\frac{du_i(t)}{dt} = H_i(u, \nu) \quad i = 1, \ldots, N-1 \qquad (2)$$

where $H_i(u, \nu)$ is given by :

$$H_i(u, \nu) = -u_i \sum_{j=1}^{N-1} D_{ij}^{(1)} u_j + \nu \sum_{j=1}^{N-1} D_{ij}^{(2)} u_j + f_i \qquad (3)$$

where $1 \leq i \leq N-1$, $f_i = f(x_i)$ and [3]:

$$D_{ij}^{(1)} = \frac{\bar{c}_i}{\bar{c}_j} \frac{(-1)^{i+j}}{x_i - x_j} \quad 1 \leq i,j \leq N-1, \quad i \neq j$$

$$D_{ii}^{(1)} = -\frac{x_i}{2(1-x_i^2)} \quad 1 \leq i \leq N-1,$$

$$D_{ij}^{(2)} = \frac{(-1)^{i+j+1}}{\bar{c}_j} \left(\frac{2 - x_i x_j - x_i^2}{(1-x_i^2)(x_i-x_j)^2} \right) \quad 1 \leq i \neq j \leq N-1,$$

$$D_{ii}^{(2)} = -\frac{(N^2-1)(1-x_i^2)+3}{3(1-x_i^2)^2} \quad 1 \leq i \leq N-1$$

($\bar{c}_i = 1 + \delta_{i0} + \delta_{iN}$). We now perform a change of variables: $v_i = u_i - u_i^*$, $i = 1, \ldots, N-1$ where $u_i^* = u^*(x_i)$ is the stationary point of (2). Then (2) becomes

$$\frac{dv_i(t)}{dt} = G_i(v, \nu) \quad i = 1, \ldots, N-1 \tag{4}$$

with:

$$G_i(v, \nu) = \sum_{j=1}^{N-1} (\nu D_{ij}^{(2)} v_j - D_{ij}^{(1)}(v_i u_j^* + u_i^* v_j)) - v_i \sum_{j=1}^{N-1} D_{ij}^{(1)} v_j \tag{5}$$

The Jacobian matrix of the system (4) is:

$$J_{ij}(v, \nu) = \frac{\partial G_i}{\partial v_j} = \nu D_{ij}^{(2)} - \delta_{ij} \sum_{k=1}^{N-1} D_{ik}^{(1)}(v_k + u_k^*) - (v_i + u_i^*) D_{ij}^{(1)}$$

3 Perturbation Analysis

In order to state our results more precisely, we quickly present the Hopf theorem. Suppose that a pair of complex conjugate eigenvalues $\lambda(\nu) = \alpha(\nu) \pm i\omega(\nu)$ of $J(0, \nu)$ cross the imaginary axis at $\nu = \nu_c$ with $\alpha(\nu_c) = 0$, $\omega_0 = \omega(\nu_c) \neq 0$ and $\frac{d\alpha(\nu_c)}{d\nu} \neq 0$, (transversality condition), while the remaining eigenvalues of $J(0, \nu)$ have strictly negative real parts. Then the system (4) admits a family of periodic solutions of period $\approx 2\pi/\omega_0$.

The computation of periodic solutions for small parameters was initiated by Lindsted and Poincaré. We will generalize this technique for the system (4). The idea is to develop the solution as a series in powers of a small parameter ϵ:

$$v_k = \epsilon v_{k1} + \epsilon^2 v_{k2} + \cdots \quad k = 1, 2, \ldots, N-1 \tag{6}$$

or in vector notation: $v = \epsilon v_1 + \epsilon^2 v_2 + \cdots$ where $v = (v_k)$, $v_1 = (v_{k1})$, $v_2 = (v_{k2})$. Since we are expanding the solution in a neighborhood of ν_c it is natural to set [2] $\epsilon^2 = \nu_c - \nu$ (Another possibility is that $\epsilon = \max_t \|v\|$). Since the period will change too, we also introduce a coordinate change:

$$\tau = (1 + \epsilon \omega_1 + \epsilon^2 \omega_2 + \cdots) t \tag{7}$$

Inserting (6) and (7) into (4) and equating coefficients of equal powers of ϵ we obtain the following sequence of equations:

$$\frac{dv_1}{d\tau} - Av_1 = 0 \tag{8}$$

$$\frac{dv_2}{d\tau} - Av_2 = g_1(v_1) \tag{9}$$

$$\frac{dv_3}{d\tau} - Av_3 = g_2(v_1, v_2) \quad \ldots \tag{10}$$

where $A = J(0, \nu_c)$ and the vectors $g_1 = (g_{i1})$, $g_2 = (g_{i2})$ are defined by:

$$g_{i1} = -\omega_1 \frac{dv_{i1}}{d\tau} - v_{i1} \sum_{j=1}^{N-1} D_{ij}^{(1)} v_{j1}$$

$$g_{i2} = -(\omega_1 \frac{dv_{i2}}{d\tau} + \omega_2 \frac{dv_{i1}}{d\tau}) - \sum_{j=1}^{N-1} D_{ij}^{(2)} v_{j1} - \sum_{j=1}^{N-1} (v_{i1} v_{j2} + v_{i2} v_{j1}) D_{ij}^{(1)}$$

The crucial idea is that we are looking for periodic solutions of (8), (9) and (10). We write the periodic solution of (8) in the form:

$$v_1 = c_1 e^{i\omega_0 \tau} + \bar{c}_1 e^{-i\omega_0 \tau} \tag{11}$$

where $c_1 = (c_{i1})$, $\bar{c}_1 = (\bar{c}_{i1})$ are the eigenvectors of A corresponding to the eigenvalues $i\omega_0$ and $-i\omega_0$ respectively ($\bar{c}_1$ denotes the complex conjugate of c_1).

Substituting (11) into (9) leads to:

$$\frac{dv_2}{d\tau} - Av_2 = -i\omega_0 \omega_1 (c_1 e^{i\omega_0 \tau} - \bar{c}_1 e^{-i\omega_0 \tau}) - b_0 - \bar{b}_0 - b_2 e^{2i\omega_0 \tau} - \bar{b}_2 e^{-2i\omega_0 \tau} \tag{12}$$

where $b_0 = (b_{i0})$, $b_2 = (b_{i2})$ are given by:

$$b_{i0} = c_{i1} \sum_{j=1}^{N-1} D_{ij}^{(1)} \bar{c}_{j1}, \quad b_{i2} = c_{i1} \sum_{j=1}^{N-1} D_{ij}^{(1)} c_{j1}$$

Thus we must set $\omega_1 = 0$ or v_2 will contain the secular term. Then disregarding the solution of the homogenous equation, we write the solution of (12) as:

$$v_2 = c_0 + \bar{c}_0 + c_2 e^{2i\omega_0 \tau} + \bar{c}_2 e^{-2i\omega_0 \tau} \tag{13}$$

where c_0, c_2 satisfy

$$Ac_0 = b_0, \quad (A - 2i\omega_0)c_2 = b_2 \tag{14}$$

Substituting (11) and (13) into (10) and recalling that $\omega_1 = 0$ we obtain:

$$\frac{dv_3}{d\tau} - Av_3 = -b_1 e^{i\omega_0 \tau} - \bar{b}_1 e^{-i\omega_0 \tau} - b_3 e^{3i\omega_0 \tau} - \bar{b}_3 e^{-3i\omega_0 \tau} \tag{15}$$

where $b_1 = (b_{i1})$, $b_3 = (b_{i3})$ are given by

$$b_{i1} = i\omega_0\omega_2 c_{i1} + \sum_{j=1}^{N-1} D_{ij}^{(2)} c_{j1}$$

$$+ \sum_{j=1}^{N-1} D_{ij}^{(1)}[c_{i1}(c_{j0} + \bar{c}_{j0}) + c_{j1}(c_{i0} + \bar{c}_{i0}) + \bar{c}_{i1}c_{j2} + \bar{c}_{j1}c_{i2}]$$

$$b_{i3} = \sum_{j=1}^{N-1} D_{ij}^{(1)}(c_{i1}c_{j2} + c_{j1}c_{i2})$$

The system (15) has no periodic solution unless the vector b_1 satisfies

$$(b_1, w) = \sum_{j=1}^{N-1} b_{j1}w_j = 0 \tag{16}$$

for all vectors $w = (w_j)$ satisfying the adjoint equation $(A^T + i\omega_0 I)w = 0$. Therefore, the periodic solution of (4) for ϵ small is given by

$$v = \epsilon(c_1 e^{i\omega_0\tau} + \bar{c}_1 e^{-i\omega_0\tau}) + \epsilon^2(c_0 + \bar{c}_0 + c_2 e^{2i\omega_0\tau} + \bar{c}_2 e^{-2i\omega_0\tau}) + O(\epsilon^3) \tag{17}$$

and is of period $2\pi/\omega_0(1 + \epsilon^2\omega_2)$ where ω_2 is determined by (16).

4 Numerical Results

Here we let $N = 16$. We require $u(x,t)$ to tend to $u^*(x) = \sin \pi x$ as t tends to infinity, so that $f(x) = \pi(\cos \pi x + \nu\pi)\sin \pi x$. The analysis of the eigenvalues of the matrix A shows a pair of complex eigenvalues which cross the imaginary axis at $\nu_c = 0.007638902877$ with $\omega_0 = 1.49629$ (Table I). The other eigenvalues of A have strictly negative real parts. Figure 1a compares the perturbation approximation (dotted lines) with the numerical solution (solid lines) for motions close to the Hopf bifurcation point ν_c. The perturbation approximation shows good agreement with the numerical solution for $\epsilon < 10^{-2}$. However its accuracy deteriorates rapidly for large values of ϵ.

TABLE I Eigenvalues of $J(0,\nu)$

λ_i	$\nu = 0.007638902877$	$\nu = 0.0075050579046$	$\nu = 0.007237040303$
λ_1	$0.00000 + 1.49629i$	$0.01478 + 1.49701i$	$0.04391 + 1.49849i$
λ_2	$0.00000 - 1.49629i$	$0.01478 - 1.49701i$	$0.04391 - 1.49849i$
λ_3	$-0.02175 + 5.21570i$	$0.00000 + 5.21782i$	$0.04329 + 5.22228i$
λ_4	$-0.02175 - 5.21570i$	$0.00000 - 5.21782i$	$0.04329 - 5.22228i$
λ_5	$-0.10109 + 10.75551i$	$-0.06749 + 10.75537i$	$0.00000 + 10.75518i$
λ_6	$-0.10109 - 10.75551i$	$-0.06749 - 10.75537i$	$0.00000 - 10.75518i$
λ_7	$-1.25624 + 1.09474i$	$-1.24780 + 1.09638i$	$-1.23140 + 1.09974i$
λ_8	$-1.25624 - 1.09474i$	$-1.24780 - 1.09638i$	$-1.23140 - 1.09974i$
λ_9	$-1.38737 + 4.65131i$	$-1.36766 + 4.65807i$	$-1.32854 + 4.67194i$
λ_{10}	$-1.38737 - 4.65131i$	$-1.36766 - 4.65807i$	$-1.32854 - 4.67194i$
λ_{11}	$-1.54007 + 10.22996i$	$-1.50639 + 10.23470i$	$-1.43870 + 10.24432i$
λ_{12}	$-1.54007 - 10.22996i$	$-1.50639 - 10.23470i$	$-1.43870 - 10.24432i$
λ_{13}	$-5.03218 + 0.00000i$	$-4.99901 + 0.00000i$	$-4.93261 + 0.00000i$
λ_{14}	$-16.79769 - 0.00000i$	$-16.36458 - 0.00000i$	$-15.49623 - 0.00000i$
λ_{15}	$-17.29347 + 0.00000i$	$-16.85415 + 0.00000i$	$-15.97323 + 0.00000i$

As ν decreases, at $\nu = 0.007505...$ a second pair of eigenvalues crosses the imaginary axis and the limit cycle bifurcates into a 2-torus. At $\nu = 0.007237...$ a third pair of eigenvalues crosses the imaginary axis with bifurcation from a 2-torus to a 3-torus and from a 3-torus to a strange attractor. It appears clear that the process represents a typical Ruelle-Takens route to chaos [4]. Finally, by further decreasing ν we find at $\nu = 0.0059$ a second limit cycle (Figure 1e) which corresponds to a spurious solution. This phenomenon is sometimes termed ghost bifurcation.

Figure 1. Projection of the motion onto the u_1, u_6 plane.

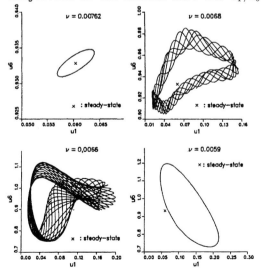

References

[1] H. Dang-Vu and C. Delcarte, *Chebyshev Spectral Solution of the Burgers Equation with High Wavenumbers*, in *Proceedings of the Second International Colloquium on Numerical Analysis*, (Plovdiv University-Bulgaria, 13-17 August, 1993) ed. by D. Bainov and V. Covachev, VSP International Science Publishers, Zeist (The Netherlands), 63-72(1994).

[2] B. J. Matkowsky, *Bull. Amer. Math. Soc.*, **76**, 620-625 (1970).

[3] R. Peyret, *Introduction to Spectral Methods*, von Karman Institute Lecture Series 1986-04, Rhode-Saint Genese, Belgium, 1986.

[4] D. Ruelle and F. Takens, *Comm. Math. Phys.*, **20**, 167-192 (1971).

Simulation of the unsteady separated flow around a finite rectangular plate using a free-divergence constraint velocity-vorticity formulation

Jean Fontaine and Loc Ta Phuoc

LIMSI/CNRS, B.P 133, 91403 Orsay cedex, France

1 Introduction

The velocity-vorticity formulation $\underline{u}, \underline{\omega}$ which is the 3D alternative version of the classical 2D stream function-vorticity formulation, is derived from the original velocity-pressure [1]. There are some advantages of this formulation, particularly for the external flow calculations. One should mention:

1. The treatment of boundary conditions is easier because the pressure term is avoided.

2. For the viscous flows around an object, where the vorticity is important, this formulation seems closer to the physical reality.

3. For incompressible flows, the form of the equations are not changed because of the non-inertial effects only enter into the solution of the problem through the discretized initial and boundary conditions [2].

But it is not easy to use it for 3D numerical simulations, because in this formulation, six differential equations have to be solved *a priori*. Moreover, the fields must be maintained solenoidal at all times by the numerical scheme. These difficulties are partially overcome using a staggered mesh. It has been demonstrated recently [3] that only four differential equations must be computed in the velocity-vorticity formulation.

2 Governing Equation

Let Ω be a bounded domain of $\mathcal{R}^3$ and Γ its boundary. The vorticity transport equation is derived from primitive formulation (**P**) of the Navier-Stokes equation:

$$(\mathbf{P}) \quad \begin{aligned} \frac{\partial \underline{u}}{\partial t} + (\underline{u}.\nabla)\underline{u} &= -\nabla p + \frac{1}{\text{Re}}\nabla^2 \underline{u} \quad &&\text{in } \Omega \quad &&(1)\\ \nabla . \underline{u} &= 0 \quad &&\text{in } \Omega \quad &&(2)\\ \underline{u} &= \mathbf{f} \quad &&\text{on } \Gamma \quad &&(3)\\ \underline{u}(x, t=0) &= \underline{u}_0(x) \quad &&x \in \Omega \quad &&(4) \end{aligned}$$

Here Re is the Reynolds number UL/v, where U is the velocity of the flow far from the obstacle, L is a representative length of this one and v the kinematic viscosity.

Applying the curl operator and using the free divergence constraint of the fields, two new formulations (**P'** and **P''**) derived from the primitive formulation (**P**), can be defined using the velocity and the vorticity fields:

$$(\mathbf{P'}) \qquad \frac{\partial \underline{\omega}}{\partial t} + \nabla \times (\underline{\omega} \times \underline{u}) = \frac{1}{Re} \nabla^2 \underline{\omega} \qquad \text{in } \Omega \qquad (5)$$

$$(\mathbf{P''}) \qquad \frac{\partial \underline{\omega}}{\partial t} + \nabla \times (\underline{\omega} \times \underline{u}) = -\frac{1}{Re} \nabla \times (\nabla \times \underline{\omega}) \qquad \text{in } \Omega \qquad (6)$$

and for both formulations:
$$\nabla^2 \mathbf{u} = -\nabla \times \omega \qquad \text{in } \Omega \qquad (7)$$

$$\nabla . \underline{\omega}_0 = 0, \underline{\omega} = \underline{\omega}_0, \mathbf{u} = \mathbf{u}_0 \text{ and } \underline{\omega}_0 = \nabla \times \mathbf{u}_0 \qquad \text{in } \Omega \qquad (8)$$

$$\nabla . \omega = 0 \text{ and } \underline{\omega} = \nabla \times \underline{u} \qquad \text{on } \Gamma \qquad (9)$$

$$\underline{u} = \underline{u}\mathbf{b} \qquad \text{on } \Gamma \qquad (10)$$

Eq.(6) has been chosen to compute the flow past the plate because it has an interesting property : Considering each component of the vorticity, the spatial operators of the transport equation are two-dimensional ones. Then, the computation of the discretized problem near the object is easier to manage accurately. On the other hand, the diffusion of the vorticity from the edges of the plate using eq.(5) requires a complex treatment in order to ensure the free-divergence constraint of the fields [4].

The equivalence between these $V - \omega$ formulations and the primitive one has been demonstrated by O. DAUBE et al [5] using strong formulations assuming that the boundary Γ is regular. In this case, the equivalence is ensured with the following conditions : $\omega \in H^1(\Omega)^3$ and $\mathbf{u} \in \mathbf{H}^2(\Omega)^3$. But, the purpose of this work is to study the flow around a rectangular plate and this obstacle cannot be considered as a regular boundary of Ω. In our case, with weaker conditions on these fields, and using a variational formalism, V. RUAS [6] proved the equivalence between the (**P**), (**P'**) and (**P''**) formulations.

3 Numerical Method

The vorticity and the velocity fields are determined according to a decoupled way. The 3D vorticity transport equation is solved by means of a fully coupled algorithm. The three components of this vector are computed together. Then, the time step used for the simulation is quite large ($\Delta t = 0.02$). But this approach requires thirty-nine coefficients in order to build the discrete transport operator. The velocity is considered with a second order time extrapolation. The time is discretized with an implicit Euler-Backward scheme when the spatial operators are discretized with second order finite differences in order to avoid numerical viscosity in the scheme.

$$\frac{3\underline{\omega}^{n+1} - 4\underline{\omega}^n + \underline{\omega}^{n-1}}{2\Delta t} + \nabla \times (\underline{\omega}^{n+1} \times \underline{u}^*) = -\frac{1}{Re} \nabla \times (\nabla \times \underline{\omega}^{n+1}) \qquad (11)$$

$\mathbf{u}^*$ is a predicted value of $\mathbf{u}^{n+1}$. The method is applied for low Reynolds number. In this way, eq.(11) is rapidly solved with a Gauss-Seidel algorithm.

The computation of the three components of the velocity field is decoupled. When three calculations of differential equations are required *a priori* to determine the velocity field, the method presented needs to solve only one 3D elliptic equation in order to define one of the components:

$$\frac{\partial^2 u_x}{\partial x^2} + \frac{\partial^2 u_x}{\partial y^2} + \frac{\partial^2 u_x}{\partial z^2} = \frac{\partial \omega_y}{\partial z} - \frac{\partial \omega_z}{\partial y} \qquad (12)$$

The algorithm for solution of this equation is an iterative method. Gauss-Seidel relaxations are used with a multigrid method. The second component u_z can be resolved *directly* using the relation derived from the definition of the vorticity :

$$\omega_y = \frac{\partial u_x}{\partial z} - \frac{\partial u_z}{\partial x} \qquad (13)$$

This 2D-equation is solved by means of a Gauss-Seidel algorithm and a multigrid method. Finally the last component u_y is resolved directly using the free-divergence constraint of the velocity :

$$\frac{\partial u_y}{\partial y} = -\frac{\partial u_x}{\partial x} - \frac{\partial u_z}{\partial z} \qquad (14)$$

But, whereas the discretized approach of the Navier-Stokes equations is equivalent to the physical continuous problem far from the plate, because of the presence of an object in the flow, the solution of the velocity cannot be decoupled from the vorticity field near the edges of the plate (see condition (9)). An influence matrix technique is introduced to assume the divergence-free velocity field. The size of this matrix is small because it depends only on the length of the edges of the plate. This matrix is computed at the begining. Then, the Reynolds number and the boundary conditions can be changed with the same influence operator. With this new algorithm, the velocity and the vorticity fields are maintained solenoidal at each time step.

4 Results

Unsteady separated flow around a spanwise finite plate has been studied. The development and the interaction between the spanwise vortex and the streamwise vortex are analysed for several obstacles. The evolution of the flow structure is given and compared with experimental visualizations [7], showing a good agreement (1).

On the plot (2), the temporal evolution of the length of the recirculation zone past different plates is represented. We represent on the figure (3) the evolution of the recirculation length versus the Reynolds number past a square plate. Then, one can notice the birth of hydrodynamic instabilities for $Re \geq 150$. An example of the evolution of vorticity sheets is given in figure (4).

Acknowledgements

We should like to thank the Director des Recherches, Etudes et Techniques for its financial support and the Institut du Devèloppement et des Ressources en Informatique Scientifique where most computations were performed.

References

1. Gatski T.B., Grosch C.E. & Rose M.E. A numerical study of the two dimensional Navier-Stokes equations in vorticity velocity computations variables *J. Comput. Phys.* **48**, $N°$ 1 1 (1982).

2. Speziale C.G., On the advantages of the velocity vorticity formulation of the equations of fluid dynamics. *J. Comput. Phys.* **73**, 476 (1987).

3. Thibeau S. Ecoulement visqueux dans une turbomachine. *Thèse de Doctorat de l'Uni. Paris 6.*

4. Fontaine J. Etude des interactions tourbillonnairesautour d'une aile. *Thèse de Doctorat de l'Uni. Paris 6. (1994).*

5. Daube O., Guermond J.L. & Sellier A. Sur la formulation vitesse-tourbillon des équations de Navier-Stokes en écoulement incompressible. *C.R.Acad.Sci. Paris 313, série II, 377-382 (1991).*

6. Ruas V. Une formulation vitesse-tourbillon es équations de Navier-Stokes incompressibles tridimensionnelles. *C.R.Acad.Sci. Paris 313, série I, 639-644 (1991).*

7. Polidori G. Etude par visualisation de sillages tridimensionnels - Application à un profil d'aile rectangulaire *Thèse de doctorat de l'université de Poitiers (1994).*

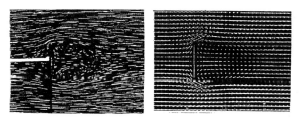

SQUARE PLATE; Re = 50; T = 5 Experiments from G. Branger.

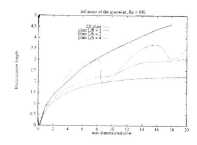

Figure 2:
Influence of the Aspect ratio of the plate.

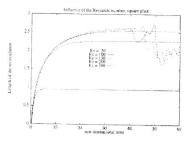

Figure 3:
Influence of the Reynold's Number

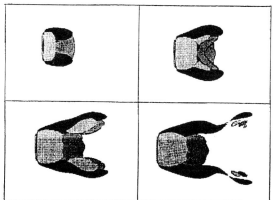

EVOLUTION OF VORTICITY SHEETS. T=2.5; 5.0; 7.5; 10.0
ASPECT RATIO : 2; Re = 100.

Time-Space Scale Effects in Numerically Computing Flowfields and a New Approach to Numerical Flow Simulation

Z. Gao[1] and F.G. Zhuang[2]

[1]Institute of Mechanics, Chinese Academy of Sciences
Beijing 100080, China
[2]Science and Technology Committee of China Aerospace Corporation
Beijing 100830, China

Abstract: Numerical solutions of the large scale (LS) average equations and small scale (SS) interacting equations describing fluid flow presented by the authors are conducted for two-dimensional channel flow with multiscale structure. A pseudospectral method is used. In the range of Reynolds number from 10^2 to 3×10^3, some important effects of nonlinear interaction between the large scale and small scale eddies are shown. Comparison of the numerical solutions of LS- and SS-equations with those of NS equations is also given.

1 Introduction

In normal cases of computing numerically flowfields of scientific and engineering interest, one needs to solve the Navier-Stokes(NS) equations in its laminar flow region and to assume the transitional region from laminar to turbulent flow and to solve the Reynolds equations or the Large Eddy Simulation(LES) equations with the aid of turbulence models in its turbulent flow region[1]. It is obviously difficult to make a guess at transitional region in complex three-dimensional flows and vortex flows. On the other hand, direct numerical simulations of the NS equations(DNS), i.e., computing numerically NS equations at the smallest time-space scale ($\Delta t_N, \Delta_N$) level, can predict the evolution of all significant scales of the flow without distinguishing between laminar-transition-turbulent flows as well as using any turbulence models[2]. The above-stated puzzled situation arises because a complex flow containing partly turbulent flow region contains a broad time-space scale range and there exist nonlinar interactions between large and small eddies with relatively large time-space scales $(\Delta t_n, \Delta_{i,n})$, where

$$\Delta t_N \ll \Delta t_n \ll \frac{L}{U}, \quad \Delta_N \ll \Delta_{i,n} \ll L \quad (i=1,2,3)$$

U and L are the characteristic flow speed and length, respectively. However, NS equations do not contain terms describing interaction between relatively large scale eddies. Our qualitative and quantitative knowledge for complex flows depends on time-space scales Δt_n and $\Delta_{i,n}(i=1,2,3)$ selected in measuring or computing the flows. Therefore, from the theoretical viewpoint, it would be valuable to develop new equations governing fluid flow and corresponding numerical methods. The new equations can be used at the smallest and relatively large time-space scales and can describe nonlinear interaction between large and small scale eddies and would be simultaneously suitable to laminar-transition-turbulent flows[3].

2 Basic Equations

The basic equations presented by the authors are composed of the large scale(LS) average equations governing the large scale average motion of the flow and the small scale(SS) interacting equations governing the small scale (or edding) motion. For incompressible flows, the LS-and SS-equations are respectively

$$\frac{\partial \bar{u}_i}{\partial x_i} = 0 \qquad (2.1a)$$

$$\frac{\partial \bar{u}_i}{\partial t} + \frac{\partial}{\partial x_j}(\overline{\bar{u}_i \cdot \bar{u}_j}) = -\frac{1}{\rho}\frac{\partial \bar{p}}{\partial x_i} + \nu \frac{\partial^2 \bar{u}_i}{\partial x_j \partial x_j} - \frac{\partial}{\partial x_j}(\bar{\lambda}_{ij}) \qquad (2.1b)$$

$$\frac{\partial}{\partial x_i}(u_i - \bar{u}_i) = 0 \qquad (2.2a)$$

$$\frac{\partial}{\partial t}(u_i - \bar{u}_i) + \frac{\partial}{\partial x_j}[(u_i - \bar{u}_i)(u_j - \bar{u}_j)] = -\frac{1}{\rho}\frac{\partial(p - \bar{p})}{\partial x_i} + \nu\frac{\partial^2(u_i - \bar{u}_i)}{\partial x_j \partial x_j}$$

$$- \frac{\partial}{\partial x_j}(\bar{u}_i \bar{u}_j - \overline{\bar{u}_i \bar{u}_j}) - \frac{\partial}{\partial x_j}[\bar{u}_i(u_j - \bar{u}_j) + \bar{u}_j(u_i - \bar{u}_i)] + \frac{\partial \bar{\lambda}_{ij}}{\partial x_j} \qquad (2.2b)$$

$$(i = 1, 2, 3)$$

where

$$(\bar{u}_i, \bar{p}, \overline{\bar{u}_i \bar{u}_j}, \bar{\lambda}_{ij}) =$$

$$\frac{1}{\Delta t_n \prod_k^3 \Delta_{k,n}} \int_t^{t+\Delta t_n} \iint \int_{x_k - \frac{1}{2}\Delta_{k,n}}^{x_k + \frac{1}{2}\Delta_{k,n}} \{u_i, p, \bar{u}_i \cdot \bar{u}_j, [(u_i - \bar{u}_i)(u_j - \bar{u}_j) + \bar{u}_i(u_j - \bar{u}_j)$$

$$+ \bar{u}_j(u_i - \bar{u}_i)]\} d\xi_k d\tau \qquad (2.3)$$

$u_i(i = 1, 2, 3)$ and p are the flow speed components and the pressure, respectively. Δt_n and $\Delta_{i,n}(i = 1, 2, 3)$ are the time and space scales selected in computing the flow; we call them cutoff scales. The second term of the right hand of the equations (2.1b) is the molecular viscous term. $\frac{\partial}{\partial x_j}(\bar{\lambda}_{ij})$ are called the eddy viscous terms, which express the nonlinear interaction between the eddies with scales smaller and larger than the cutoff scales. The fourth term of the right hand of the equations (2.2b) is the remote action term of the eddies with scales larger than the cutoff scales on the small scale eddies.

The main properties and functions of the LS-and SS-equations are as follows. (1) When the cutoff time-space scales are small enough, or they are much smaller than the intrinsic physical scales of the flows, such as the cases of low Reynolds number laminar flows, the LS-equations reduce to the Navier-Stokes equations, while the SS-equations can be cast away because they give zero solutions $(u_i - \bar{u}_i) \cong 0 (i = 1, 2, 3)$ and $(p - \bar{p}) \cong 0$. (2) For turbulent flows, if only the cutoff time scale is small enough, the LS-equations are equivalent to the Large Eddy Simulation (LES) equations; on the other hand, if only the cutoff space scales are small enough, and let $\overline{\bar{u}_i \cdot \bar{u}_j} = \bar{u}_i \cdot \bar{u}_j$, the LS-equations are equivalent to the Reynolds equations; under the above two circumstances, the SS-equations determine the small scale motion of the flow and also close the LS-equations. (3) The LS-and SS-equations open a new way to compute large and small scale eddies as well as the nonlinear interaction between different scale eddies of the flow, which may be turbulent or transitional or laminar, such as the laminar bifurcation phenomenon caused by multiple-vortex secondary flow. (4) Using LS- and SS-equations we can compute in a unified manner laminar-transition-turbulent parts of the complex flow without supposing transitional regions and distinguishing between laminar and turbulent flows and using any turbulence models. Hence the solution of LS- and SS-equations links organically direct numerical simulation at the smallest scale level with the normal numerical computation at relatively large scale level. Especially, the solving of the SS-equations can be confined within certain local regions of the flowfield and lighten greatly the burden on computer. In a sense, the LS- and SS-equations open a relatively realistic way to compute high Reynolds number complex flows.

3 Numerical Method

We use a pseudospectral method to solve simultaneously LS- and SS-equations for two-dimensional channel flow with multiscale structure. Because only two first-order partial derivative terms are added into the SS-equations (2.2) compared with the LS-equations (2.1), the SS-equations (2.2) and LS-equations (2.1) can be solved using the same algorithm. Two discretizations, Fourier collocation and Chebyshev collocation, are selected in the streamwise and normal directions, respectively. We adopt periodic boundary conditions and uniform grids in the streamwise direction and the Chebyshev collocation (grids) distribution and the boundary condition of impermeable and no-slip in the normal direction, respectively. The finest and the coarsest grids are 128×128 and 8×8, respectively. The time-stepping is a second-order semi-implicit scheme with the time-step chosen so small that the spatial errors predominate.

4 Analysis and Discussion

The two-dimensional disturbances are added to initial flow.

$$u = u_0 = 1 - y^n + \varepsilon \frac{\pi}{k_1} \cos k_1 x \sin[(2k_2 + 1)\pi y]$$

$$v = v_0 = -\varepsilon \frac{\sin k_1 x}{(2k_2 + 1)} \{1 + \cos[(2k_2 + 1)\pi y]\}$$

where $n = 2$ or 8, the amplitude parmeter ε of the initial disturbance is about 10^{-4} to 0.1 and the "wave number" k is about 1 to 80. The initial velocities satisfy the continuity equation. The evolution of all significant scales and the important influences of the eddying motion on the average variables $\bar{u}_i (i = 1, 2, 3)$ and $\bar{p}$ are given by solving simultaneously the LS- and SS-equations and compared with the corresponding numerical results of NS equations. Some main results are as follows. In all the cases of relatively low Re, the eddying motion decays with the time (see Fig.1 and 6). In the initial and intermediate stages the eddy viscous force, the nonlinear interaction between the large scale and small scale eddies, is larger than or comparable to the molecular viscous force (see Fig.2 and 3), Fig.4 and 5 show that in the middle stage of the evolution $(u - \bar{u})$ and $(v - \bar{v})$ become larger than their initial values; this is because the small scale eddies are excited by the large scale average motion. The differences between the numberical solutions of LS- and SS-equations and NS equations are remarkable (also see Table 1). In the later stage the eddying motion disappears and the solution of LS-equations becomes the same as that of NS equations. In the cases of relatively large Re, the small scale eddies do not roughly decay (see Fig.6), and the numerical solutions of both the LS- and SS-equations and NS equations are found to be sensitive to the initial disturbances and grid refinement. The critical Reynolds number at which the disturbances no longer decay is predicted numerically. The above statements agree with the theoretical inference for the mathematical properties of LS- and SS-equations and also with other study results such as[4,5].

References

[1] D.A.Anderson, et al "Computational Fluid Mechanics and Heat Transfer"

[2] (Hemisphere/McGraw-Hill, New York, 1984)

[3] W.C.Reynolds "Whither Turbulence" Cornell University (1989)

[4] Z.Gao "A New Conservation Equations of Fluid Dynamics" IMCAS STR-92026, 1992.

[5] R.D.Moser, P.Moin, A.Leonerd, Jour.Comput.Phys. Vol.52, P524~544, 1983.

[6] L.Lleiser, Schumann, Proc. of the 3rd GAMM-Conf. on Num. Meth. in fluid Mech. P165~173, 1980.

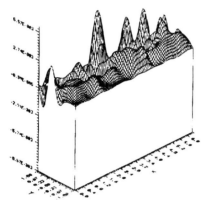

FIG.1 Evolution of average velocity

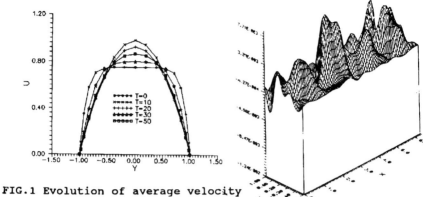

FIG.2 Sum of eddy viscous terms in x momentum eq. grid(16*17), Re=300, T=50

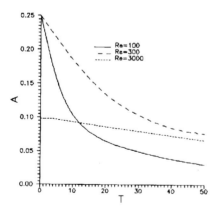

FIG.3 Sum of eddy viscous terms in y momentum eq. grid(16*17), Re=300, T=50

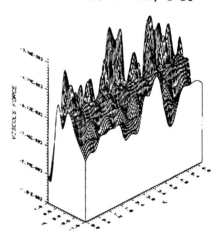

FIG.4 Sum of viscous terms in x momentum eq. grid(16*17), Re=300, T=50

FIG.6 Evolution of small scale amplitude

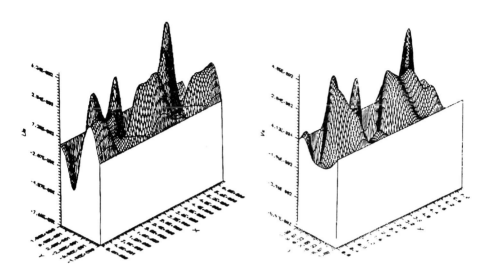

FIG.5   Small scale velocity
         grid(64*65), Re=300, T=50

Y	U	
	GZ. EQ.	NS EQ.
-0.100000E+01	-0.167711E-15	-0.955116E-16
-0.980785E+00	0.355285E-01	0.364834E-01
-0.923880E+00	0.137050E+00	0.140180E+00
-0.831470E+00	0.290095E+00	0.294611E+00
-0.707107E+00	0.469581E+00	0.474946E+00
-0.555570E+00	0.646310E+00	0.652123E+00
-0.382683E+00	0.793752E+00	0.798562E+00
-0.195090E+00	0.888307E+00	0.893743E+00
0.349148E-14	0.922699E+00	0.926544E+00
0.195090E+00	0.893403E+00	0.893745E+00
0.382683E+00	0.798655E+00	0.798566E+00
0.555570E+00	0.655533E+00	0.652127E+00
0.707107E+00	0.489344E+00	0.474950E+00
0.831470E+00	0.310566E+00	0.294613E+00
0.923880E+00	0.147432E+00	0.140181E+00
0.980785E+00	0.379175E-01	0.364837E-01
0.100000E+01	-0.182892E-15	-0.850801E-16

Table 1. Distribution of velocity U at X=0.335103E+01
         grid(16*17), Re=300, T=50

An Evaluation of a Multi-Dimensional FCT Algorithm with Some Higher-Order Upwind Schemes

Yuguo Li and Murray Rudman

Division of Building, Construction and Engineering,
CSIRO, P.O. BOX 56, Highett 3190, Australia

Abstract : Small-scale ripples appearing in solutions of two test problems with the multi-dimensional flux-corrected transport (FCT) algorithm of Zalesak (Z-FCT) [1] are investigated. An alternative derivation of the Z-FCT flux limiter is presented directly from a multi-dimensional discrete equation. Three second-order schemes are tested.

1 Introduction

Z-FCT has proved to be a very effective method for eliminating numerical oscillations in higher-order schemes [2,3] and has been widely used [e.g. 4,5]. Zalesak presented his well-known multi-dimensional FCT algorithm by directly extending his 1D formulation. An alternative derivation will be proposed in this paper, which gives further understanding of the scheme. Zalesak pointed out that Z-FCT is not strictly identical to that of monotonicity enforcement for a multi-dimensional problem. Two numerical examples will be examined here. The higher-order schemes investigated are the second-order upwind scheme (SOU), the scheme of quadratic upstream interpolation (for convective kinematics) (QUICK) and the central difference scheme (CD), which are widely used in predictions of many practical incompressible flows.

2 An alternative derivation of Z-FCT algorithm

Consider a 2D transport equation of a scalar, ϕ,

$$\frac{\partial \phi}{\partial t} + u\frac{\partial \phi}{\partial x} + v\frac{\partial \phi}{\partial y} = \Gamma_\phi(\frac{\partial^2 \phi}{\partial x^2} + \frac{\partial^2 \phi}{\partial y^2}), \qquad (1)$$

where u and v are the velocity components in x - and y - directions; t is the time and Γ_ϕ is the diffusion coefficient. Integrating eq. (1) locally over a control volume with four faces, east (e), west (w), north (n) and south (s) and four neighbouring grid points (E, W, N and S), we obtain

$$\phi_P^{n+1} = \phi_P^n - \frac{1}{\Delta S_P}(F_e - F_w + F_n - F_s), \qquad (2)$$

where ΔS_P is a 2D area element centered on grid point P, i.e. (i,j). The functional dependence of F on u, v and ϕ defines the integration scheme. The "transported and diffused" solution ϕ_P^{td} can be obtained by using F^L, the transportive flux given by a low-order monotonic scheme. Antidiffusive fluxes are defined as $A = F^H - F^L$, where F^H is the transportive flux given by a higher-order scheme. Given a correction factor C ($0 \leq C \leq 1$) to A, the solution at a new time-step $n+1$ can be computed as

$$\phi_P^{n+1} = \phi_P^{td} - \frac{1}{\Delta S_P}(C_e A_e - C_w A_w + C_n A_n - C_s A_s). \tag{3}$$

The correction factor C is determined by using the bounding principle, $\phi_P^{\min} \leq \phi_P^{n+1} \leq \phi_P^{\max}$. The methods for determining $\phi_P^{\max}$ and $\phi_P^{\min}$ were discussed in detail in [1], and satisfy

$$(C_e^- A_e - C_w^- A_w + C_n^- A_n - C_s^- A_s)_P \leq \Delta S_P(\phi_P^{td} - \phi_P^{\min}) \equiv Q_P^-;$$
$$(C_w^+ A_w - C_e^+ A_e + C_s^+ A_s - C_n^+ A_n)_P \leq \Delta S_P(\phi_P^{\max} - \phi_P^{td}) \equiv Q_P^+. \tag{4}$$

In eq. (4), C^- and C^+ are the correction factors determined from $\phi_P^{\max}$ and $\phi_P^{\min}$ respectively. Note that $Q_P^- \geq 0$ and $Q_P^+ \geq 0$. The system of equations specified in (4) cannot be solved easily and can be simplified by assuming a single C_P^+ and C_P^- for each cell P. This simplification is called the isotropic-correction assumption. Then eq. (4) becomes

$$B_P^- C_P^- \leq Q_P^-; \qquad B_P^+ C_P^+ \leq Q_P^+, \tag{5}$$

where B_P^+ and B_P^- are the net antidiffusive fluxes into and away from the grid point P respectively, that is

$$B_P^+ \equiv A_w - A_e + A_s - A_n, \qquad B_P^- \equiv A_e - A_w + A_n - A_s. \tag{6}$$

We now use the methodology of Zalesak and define the sum of all inward/outward antidiffusive fluxes at the grid point P as

$$P_P^+ \equiv \max(0, A_w) - \min(0, A_e) + \max(0, A_s) - \min(0, A_n) \geq B_P^+;$$
$$P_P^- \equiv \max(0, A_e) - \min(0, A_w) + \max(0, A_n) - \min(0, A_s) \geq B_P^-. \tag{7}$$

Therefore P_P^+ and P_P^- give maximum bounds on the values of B_P^+ and B_P^- respectively. Using P_P^+ and P_P^- in place of B_P^+ and B_P^- will guarantee that eq. (5) is satisfied provided

$$C_P^+ = \begin{cases} 0 & \text{if } P_P^+ = 0 \\ \min(1, \frac{Q_P^+}{P_P^+}) & \text{if } P_P^+ > 0 \end{cases}; \qquad C_P^- = \begin{cases} 0 & \text{if } P_P^- = 0 \\ \min(1, \frac{Q_P^-}{P_P^-}) & \text{if } P_P^- > 0. \end{cases} \tag{8}$$

It is easy to show that the final correction factors can be determined as

$$(C_e)_P = \begin{cases} \min(C_E^+, C_P^-) & \text{if } A_e \geq 0 \\ \min(C_P^+, C_E^-) & \text{if } A_e \leq 0 \end{cases}; \qquad (C_n)_P = \begin{cases} \min(C_N^+, C_P^-) & \text{if } A_n \geq 0 \\ \min(C_P^+, C_N^-) & \text{if } A_n \leq 0. \end{cases} \tag{9}$$

with similar expressions for C_w and C_s.

This gives exactly the same results as in [1]. An interesting point is that the isotropic-correction assumption is necessary to easily define the sum of all inward or outward antidiffusive fluxes (eq. (7)) in the new derivation, but gives the same result as Zalesak.

3 Two test problems and results

The equations are solved by using a finite volume method. CD is applied to the diffusion terms, and different schemes as described in the introduction are used for the convection terms. Since an explicit forward time difference scheme is used, small Courant numbers are used to minimize the numerical error arising from the temporal formulation. The first-order upwind scheme (FOU) is used as the low-order monotonic scheme for Z-FCT.

The well-known Smith-Hutton problem [5] is first tested. Γ_ϕ in eq. (1) is 1.0×10^{-6}. A grid of 40 × 20 is used. The steady-state solution is obtained by a time-marching procedure with time step Δt of 0.001. The solution obtained with QUICK on a 160 × 80 grid is used as a comparison basis. The predicted scalar field profile at the outlet for the Smith-Hutton problem is shown in Fig. 1a for the different schemes. QUICK+FCT gives the best results. As expected, the FOU scheme has the highest numerical diffusion, and oscillations arise for SOU and QUICK. CD becomes very oscillating and is unable to give a steady-state solution to this problem. However, CD+FCT produces a reasonable steady-state solution, Fig. 1b. Small-scale oscillations are observed with CD+FCT, although the solution is bounded.

Two-dimensional Burgers equations are then solved. In eq. (1), ϕ represents two velocity components u or v, and Γ_ϕ the viscosity ν ($= 1.0 \times 10^{-6}$). The computational domain is $0 \leq x \leq 5.0$ and $0 \leq y \leq 0.5$. The initial conditions and boundary conditions are defined in [7]. The grid used is 40 × 20. The time step Δt is 0.001. The maximum Courant numbers are $C_x = 0.08$ and $C_y = 0.04$. Fig. 2 shows the velocity profiles after 400 iterations for the 2D Burgers equations. Again a reasonable converged result has been obtained with CD+FCT. FCT is able to completely suppress oscillations in the u-component, but not those in the v-component when used with QUICK. These results, together with other tests done by the authors, indicate that Z-FCT is not able to completely suppress oscillations of a velocity component which has discontinuities or large gradients in a direction normal to that component's direction. However, the magnitudes of these oscillations are bounded and they are much smaller than those generated by the original higher-order schemes, e.g. QUICK. The "stronger restriction" method [1] is also tested, which does not perform well for the problems considered. However, as shown above incorporation of FCT into QUICK, SOU and CD greatly improves the solution accuracy by suppressing all the large-scale oscillations and some small-scale oscillations.

4 Conclusions

It is shown that an isotropic-correction assumption is necessary in defining the sum of all antidiffusive fluxes used in the Z-FCT algorithm. For some test schemes, Z-FCT is shown to be unable to completely suppress the oscillation of a component which has a discontinuity in other directions. However, the remaining oscillations are small and bounded.

References

[1] Zalesak, S. T. *J. Comp. Phys.*, 31:335-362, 1979.

[2] Sharif, M. A. R. and Busnaina, A. A. *J. Fluid Eng.*, 115:33-40, 1993.

[3] Tamamidis, P. and Assanis, D.N. *Int. J. Numer. Methods Fluids*,16:931-948, 1993.

[4] Löhner, R., Morgan, K., Vahdati, M., Boris, J. P. and Book, D. L. *Communications in Applied Numerical Methods,* 4:717-729, 1988.

[5] Rompteaux, A. and Estivalezes, J. L. *Lecture Notes in Physics,* 13th International Conference on Numerical Methods in Fluids Dynamics, 140-144, 1992.

[6] Smith, R. M. and Hutton, A. G.*Numerical Heat Transfer*, 5:439-461, 1982.

[7] Li, Y. and Rudman, M. "Assessment of higher-order upwind schemes incorporating FCT for linear and nonlinear convection-dominated problems." Submitted to *Numerical Heat Transfer, Part A* 1993.

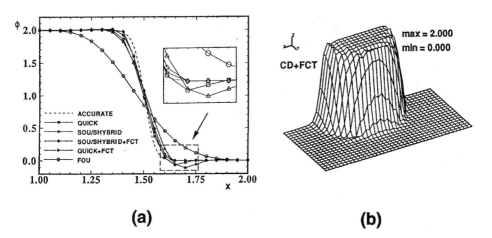

Figure 1: The Smith-Hutton problem: (a) predicted profiles of the scalar at the outlet; (b) pespective of the predicted scalar fields by CD+FCT.

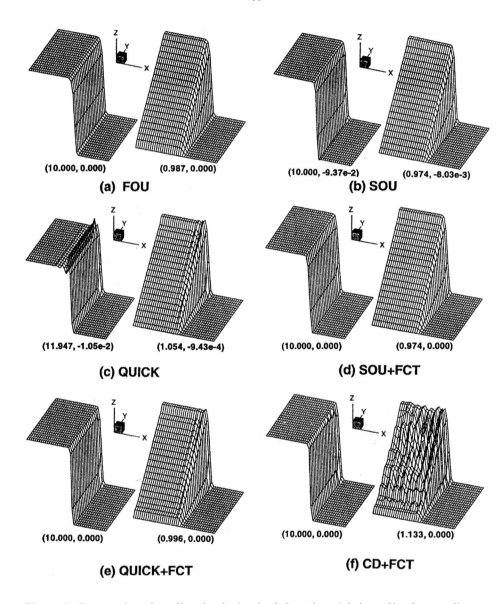

Figure 2: Perspective of predicted velocity (u, left and v, right) profiles for non-linear Burgers equations by six numerical schemes after 400 iterations, where $(x1, x2)$ are the maximum and minimum values of u and v, respectively.

Numerical Study of Unsteady Compressible Flows by a Shock Fitting Technique

Francesco Nasuti and Marcello Onofri

Dept. Meccanica e Aeronautica,
Università di Roma 'La Sapienza', Italy

1 Introduction

The numerical simulation of the transient behaviour of unsteady compressible flows requires techniques very accurate for the treatment of the shock. A possible approach is based on the use of the equations written in nonconservative form and on a handling of the shock by a fitting technique, which provides very precise predictions of location and propagation velocity of the shocks.

The present paper reports on the numerical aspects of a procedure for fitting and tracking flow discontinuities, and on the results of test cases for both steady and unstedy flows performed in order to validate the technique.

The shock fitting procedure here presented has been carried out in the spirit of the technique proposed by Moretti in [1]. According to it, the fitting of the shock is performed in three phases: i) detection, usually performed by looking for the convergence of the characteristic lines, ii) reconstruction of the shock shape, performed on geometrical basis, by connecting the shock points fitted, and iii) prediction of the flow values behind the shock, performed on the basis of the shock shape and intensity. In the common version of this technique, the shock points are fitted only on the side of the computational mesh directed according to the main flow direction. Many applications of this technique were successfully carried out for inviscid supersonic flows, as well as to analyze the interaction of shock and boundary layer [2,3]. Nevertheless some difficulties were noted, in computing flowfields where significant shock interactions occur, or when oblique or weak shocks must be fitted.

In this paper an improved version of the technique is presented. In particular, following Moretti's suggestion which had a preliminary application in [1], shock points are detected and tracked on both sides, x and y, of the computational mesh. Since a larger number of shock points is fitted compared with the standard approach, this scheme yields several improvements in the shock treatment: *i*) the shock shape reconstruction is more correct, since it is based on a larger number of shock points, located in shorter distance; *ii*) the variables behind the shocks are defined more precisely in oblique shocks, since they can be computed from neighboring computational points; *iii*) triple shock points can be easily handled by interpolation of the values between the two shock branches.

Moreover the contact discontinuities which can be detected from the beginning of the calculation are here explicitly fitted, and tracked. This avoids the smearing of the entropy jump, and may result in a more correct prediction of the shock propagation velocity when the shock and contact discontinuity are close to each other.

2 Results

The first test case performed concerns a supersonic steady inviscid flow on a ramp and aims at assessing the numerical prediction of the shock slope with an analytical solution. The comparison shows an excellent agreement (Fig.1), with errors in the location of the shock points lower than 0.05%.

The second test concerns the implosion of a cylindrical shock. The converging shock is initialized like in shock tube problems: a cylindrical diaphragm separating two uniform regions of quiescent gases, with higher pressure in the external region, is removed at t=0. Both the shock and the contact discontinuity are fitted in order to achieve an accurate solution. Since the goal was to validate the technique for any shock slope with respect to the grid lines, this 1-D problem was solved on a 2-D Cartesian grid, where obviously the angles between fitted discontinuities and grid lines have different values. As reported in Fig.2, which shows the location of shock and contact discontinuity points for three different times, the cylindrical symmetry is well maintained during the calculation. Also the prediction of the shock strength, which increases as the discontinuities approach the singular point, is in very good agreement with the semi-analytical solution proposed by Chisnell (Fig.3). The accuracy can be appreciated also from Fig.4, by comparing the pressure behaviour, reported as a function of the distance from the singularity.

The third test performed is the self-similar problem of the diffraction of a shock around an edge. This test allows assessment of the prediction of the transient for complex flow structures with experimental data and by checking the self-similarity of the solution. A shock followed by a uniform supersonic flow moves from left along a straight wall; as it reaches the wall edge it is diffracted with a 180° angle. The numerical solution of this problem requires a specific treatment when the fitted shock moves around the edge, since a minimum number of shock points is needed to have sufficient resolution to describe the 180° rotating shock. Thus, as the shock passes the edge an analytical solution proposed by Whitham is imposed locally as shock shape for a few steps, until at least ten shock points can be fitted around the edge zone. Afterwards, the calculation is continued with the general integration scheme. The solutions at different times, normalized with the distance covered by the shock along the abscissa, are reported in Fig. 5: the shock shape maintains its similarity and shows good agreement with the experimental data by Skews (filled squares). It can be seen that the location of the initial point of the second shock approaches the experimental data, indicated by a filled circle, as the resolution (i.e. time) increases. The complex flow structure, featuring contact discontinuity, slipstream and vortex, is well reproduced by the numerical solution, as it appears from the isocontour lines for pressure, entropy and Mach number (Figs.6-8).

Finally the technique has been adopted to simulate a shock tube problem in a nozzle geometry. At t=0 a diaphragm is removed, which separates two uniform regions of quiescent gases with a pressure $p_c = 100$ atm, $T_c = 3520$ K on the chamber (left) side, and with a pressure $p_a = 0.1$ atm, $T_a = 300$ K on the ambient side. The flowfield is initialized with the 1-D solution, featuring a shock and a contact discontinuity entering the nozzle, followed by a supersonic flow which holds as constant inflow. Figs. 9-12 show the solution obtained at different time steps. In Fig. 9 the original structure is still evident, with the shock (denoted by crosses) followed by a supersonic flow, and then by the contact discontinuity (denoted by circles) followed by a slower supersonic flow. Both the discontinuities are diffracted at the wall. The Mach reflection, originated when the shock passed through the concave part of the wall, propagates toward the axis. In particular the triple point is close to the axis in Fig. 10 and then reflected in Fig. 11. The contact discontinuity slows down in the vicinity of the wall (Fig. 9) due to the lower gas velocity and then interacts with the horizontal branch of the shock (Fig. 10). A new shock appears in Fig. 10: it is the recompression shock following the strong expansion in the throat region. In Fig. 11 both recompression and reflected shock are interacting with the contact surface. Fig. 12 shows the solution long time after. The incident shock as well as the contact surface are now out of the computational field. Behind them the underexpanded character of the flow yielded a sudden flow expansion at the nozzle exit with the formation of the jet plume, where the recompression shock is moving out of the flowfield.

References

1. G. Moretti, Efficient Calculations of 2-D Compressile Flows, in Advances in Computer Methods for PDE, vol. VI, IMACS, 1987.

2. G. Moretti, F. Marconi, M. Onofri,'Shock Boundary Layer Interactions Computed by a Shock Fitting Technique', XIII Int. Conf. on Numerical Methods in Fluid Dynamics, Rome, 1992.

3. M. Onofri, F. Nasuti, '2D Hypersonic Viscous Flow on a Ramp by a Shock - Fitting Technique', Workshop on Hypersonic Flows for Reentry Problems, INRIA, 1993.

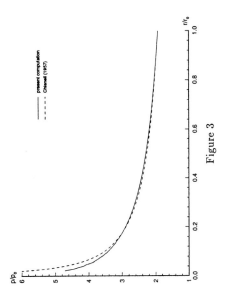

Figure 1

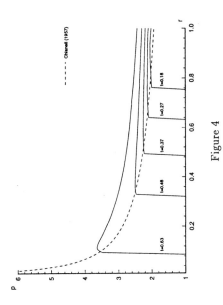

Figure 3

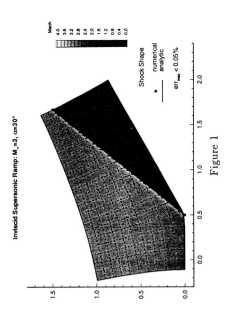

Figure 2

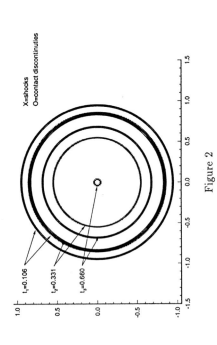

Figure 4

411

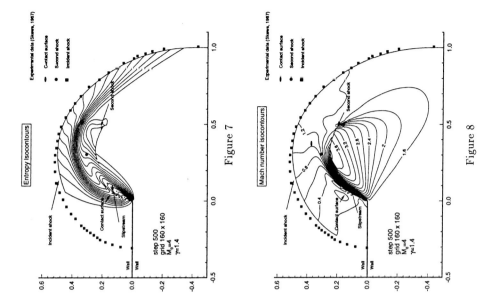

Figure 7

Figure 8

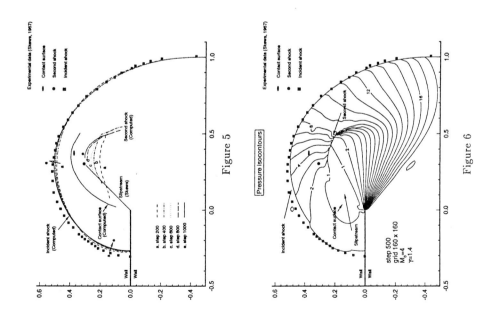

Figure 5

Figure 6

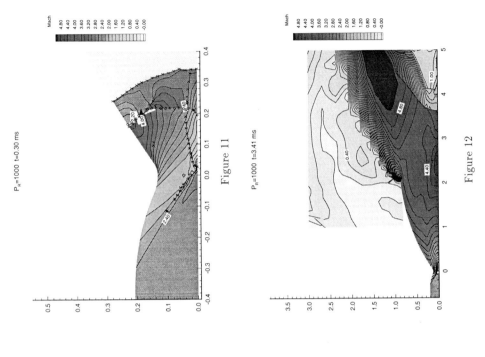

Figure 9

Figure 10

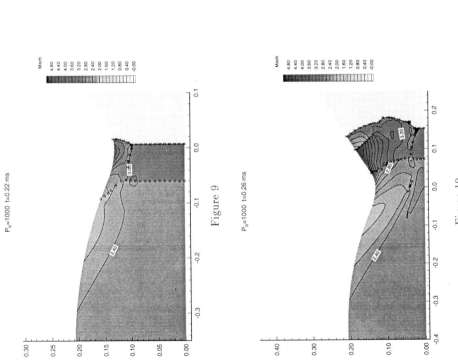

Figure 11

Figure 12

Implementation of a Nonequilibrium Flow Solver on a Massively Parallel Computer

Jani Macari Pallis and Jean-Jacques Chattot

University of California, Davis, USA

Abstract : The application of a loosely coupled fluids and chemistry solver on a massively parallel computer is investigated. The uncoupling of the fluid and chemistry equations allows coarse and medium grain parallelization to be exploited in the numerical calculation of this system of equations. Specifically, the Upwind Parabolized Navier-Stokes Solver (UPS) of Lawrence et al. with nonequilibrium chemistry enhancements due to Buelow et al. is distributed and executed over multiple computer processors concurrently. Hypersonic flows over a sharp nosed cone with 7-species/7-reactions and 11-species/17-reactions air models are considered as test cases. A speedup which approaches $1/n$ where n is the number of species is obtained within the chemistry solver on the Intel iPSC/860.

1 Introduction

Hypersonic aerothermodynamics and its application to the design of an expanding variety of air and space vehicles remains an area of enormous interest. Yet, the determination of the complex fluid and chemical interactions by existing hypersonic experimental facilities is a formidable task. Consequently, an increased interest in the development of computational methods to simulate these high temperature and high Mach number chemically reacting flows has resulted.

As the requirement and desire to study more complex chemistry models has arisen (i.e., an increasing number of reactants and reactions) alternative numerical methods are sought to solve hypersonic chemically reacting flows. Among the computational solution methods, Parabolized Navier-Stokes equation solvers are recognized as efficient and accurate computational tools for the solution of supersonic and hypersonic flowfields. The present study extends the Upwind Parabolized Navier-Stokes Solver (UPS) of Lawrence et al. 1989 to operate on a 32 node Intel iPSC/860 parallel computer. The implementation of a parallel version of UPS and the numerical results utilizing this massively parallel machine are presented.

2 Governing Equations

UPS utilizes the Parabolized Navier-Stokes equations to model the fluids. The PNS equations are derived from the time-dependent compressible Navier-Stokes equations by neglecting the unsteady terms and the viscous terms which involve derivatives in the streamwise direction (with respect to ξ). In generalized coordinates the governing equations are:

$$\frac{\partial E}{\partial \xi} + \frac{\partial F}{\partial \eta} + \frac{\partial G}{\partial \zeta} = 0. \tag{1}$$

The terms E, F and G contain both the inviscid and relevant viscous flux vectors.

To address the chemical nonequilibrium aspects of the flow, species continuity equations must also be solved and are expressed as:

$$\frac{\partial \rho_s}{\partial t} + \nabla \cdot (\rho_s V_s) = \dot{\omega}_s, \quad \text{where} \quad s = 1, 2, ...n \quad \text{and} \quad V_s = V + U_s \tag{2}$$

In these equations ρ_s is the species density, V is the fluid velocity, U_s is the species diffusion velocity, $\dot{\omega}_s$ is the production rate of the species and s is the species index.

Species mass diffusion effects are accounted for in the energy equation by redefining the heat flux terms to include the diffusion components as given by:

$$q = -\kappa \nabla T + \rho \sum_{s=1}^{n} c_s U_s h_s \tag{3}$$

where c_s and h_s are the species mass fraction and enthalpy, respectively and κ is the coefficient of thermal conductivity.

Two chemistry models are utilized. The first is the clean air model with seven species and seven reactions; the model is used by Buelow et al. 1991 and Bhutta et al. 1990. The seven species are O_2, O, N, NO, NO^+, N_2 and electrons (e^-). The second chemistry model is that of the 11 species - 17 reaction model used by Palmer 1990 which is based on the model of Park 1990. In addition to the species and first six reactions of the 7 species - 7 reaction clean air model, the species N_2^+, O_2^+, N^+ and O^+ are included. Table 1 contains the reactions for both chemistry models where M_1, M_2, M_3 are catalytic third bodies.

3 Sequential Numerical Implementation

UPS is an implicit, upwind, space marching finite-volume based code able to treat perfect, equilibrium and nonequilibrium air cases and has successfully been used to solve both supersonic and hypersonic flows. Within UPS, the chemistry and fluids are solved in a loosely coupled fashion; that is, the fluids and chemistry are solved separately and coupled in an approximate manner. Thus, as the size of the chemistry model increases, the effects to the fluid model are minimal.

As the fluids and chemistry equations are weakly coupled, the solution procedure is advanced in three basic steps. First, assuming that the chemistry is frozen, the fluid equations are marched from the n to $n + 1$ station; values for density, velocity and internal energy are obtained. Next, using these values of density and velocity the species continuity equations are solved to obtain the species mass fractions at $n + 1$. Lastly, with the species mass fractions, fluid density and internal energy at $n + 1$ now available, new values of the temperature, pressure and other thermodynamic properties at the $n+1$ marching station are obtained. This separation of fluids and chemistry in UPS presents a "natural" division within the code which is applicable to coarse grain parallelization.

4 Parallel Implementation

The inclusion of chemical reactions and species equations presents a computationally formidable problem; such calculations are inherently computationally intensive. One approach to increase the computational throughput is to exploit alternative computational methods such as parallel processing. In such cases the original sequential problem is split or decomposed into smaller subproblems. These subproblems are then distributed among several (or many) processors and executed concurrently.

To determine an appropriate strategy for the development of the parallel version of UPS, a result oriented analysis was conducted to determine the amount of parallelism available in UPS. A result oriented analysis ascertains the "results" of the code for the various possible scenarios. Knowing the results, each component of the result, how it is generated and data dependencies are determined. This highlights the critical path of processes necessary to obtain a result and also depicts what processes can occur simultaneously. The study determined that parallelism within UPS could be obtained through domain decomposition, code reorganization, concurrent processing and parallel algorithms.

Attention was also given to those approaches which would support the basic directives of the study: to focus on methods of parallelization which could utilize the natural division between the loosely coupled fluids and chemistry, to facilitate the expansion of chemistry models to include more species and reactions; and to minimize the development of new code or a significant rewriting of the UPS code,(the sequential version of UPS contains approximately 16,000 lines of Fortran statements). Architecturally, both a 32 node Intel iPSC/860 parallel computer and a variety of workstations supporting message passing systems were available for the parallel implementation at the University of California, Davis.

The parallel version of the UPS code was implemented on the Intel. Thus at the highest and coarsest level, a manager-worker strategy is engaged to coordinate the solution of the fluids and chemistry; there are specialized nodes which handle these processes. In particular, calculations for each chemical reaction are executed concurrently on different nodes. Each node then globally reports results to generate specific weights. At a finer granularity, appropriate parallel algorithms were established. For example, the UPS code utilizes a block tridiagonal solver, for which there are several accepted parallel algorithms in addition to parallel algorithms in the areas of domain decomposition.

5 Results

Since the various chemical reactions are executed on different nodes, for this segment of the code the time it takes to calculate the specific weights is the time it takes to calculate the "largest" reaction (the reaction with the most catalytic third bodies) plus a small communication cost ($\leq 5\ \%$) versus the time it would take to calculate all the reactions sequentially. Thus in the chemistry section of the code it does not take any longer to execute one reaction than it does to execute 7 or 17 reactions.

Sample results from UPS are presented in Figs. 1-3. Figure 1 depicts the speedup utilizing the Intel as the chemistry model is expanded.

A sharp nosed cone at 0-deg angle of attack and Mach 25.3 was used as a test case. Figures 2 and 3 depict profiles of the mass fraction of atomic oxygen and temperature (respectively) from both the sequential and parallel versions of the UPS code.

The parallelization of the UPS code will allow improved computer throughput and the possibility to study more complex geometries as well as increased numbers of chemical species and reaction equations for the study of nonequilibrium hypersonic flowfields. The techniques applied to this application will extend to the parallelization of other multidisciplinary codes.

Table 1. Chemistry Model Reactions

$O_2 + M_1 \rightleftharpoons 2O + M_1$	$N + N \rightleftharpoons N_2^+ + e^-$
$N_2 + M_2 \rightleftharpoons 2N + M_2$	$O + O \rightleftharpoons O_2^+ + e^-$
$NO + M_3 \rightleftharpoons N + O + M_3$	$N + e^- \rightleftharpoons N^+ + e^- + e^-$
$NO + O \rightleftharpoons O_2 + N$	$O + e^- \rightleftharpoons O^+ + e^- + e^-$
$N_2 + O \rightleftharpoons NO + N$	$O_2^+ + O \rightleftharpoons O_2 + O^+$
$N + O \rightleftharpoons NO^+ + e^-$	$N_2 + N^+ \rightleftharpoons N_2^+ + N$
$N_2 + N \rightleftharpoons 2N + N$	$N_2 + O^+ \rightleftharpoons N_2^+ + O$
	$NO^+ + O \rightleftharpoons NO + O^+$
	$NO^+ + N \rightleftharpoons N^+ + NO$
	$NO^+ + O_2 \rightleftharpoons NO + O_2^+$
	$NO^+ + O \rightleftharpoons O_2 + N^+$

References

[1] Bhutta, B.A., Song, D.J. and Lewis, C.H.(1990): "Nonequilibrium Viscous Hypersonic Flows Over Ablating Teflon Surfaces," Journal of Spacecraft and Rockets, Vol.27, pp.205-215.

[2] Buelow, P.E., Tannehill, J.C., Ievalts, J.O. and Lawrence, S.L.(1991): "Three-Dimensional, Upwind, Parabolized Navier-Stokes Code for Chemically Reacting Flows," Journal of Thermophys., and Heat Transfer, Vol.5, pp.274-283.

[3] Lawrence, S.L., Tannehill, J.C., Chaussee,D.S.(1989): "Upwind Algorithm for the Parabolized Navier-Stokes Equations," AIAA Journal, Vol.27, pp.1175-1183.

[4] Palmer, G.(1990): "Explicit Thermochemical Nonequilibrium Algorithm Applied to Computational 3-D Aeroassisted Flight Expt. Flowfield," Journal of Spacecraft & Rockets, Vol.27, pp.545-553.

[5] Park, C.(1990): "Nonequilibrium Hypersonic Aerothermodynamics" Wiley, New York.

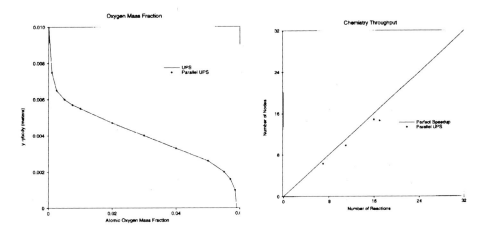

Fig. 1 Chemistry Throughput Fig. 2 O Mass Fraction

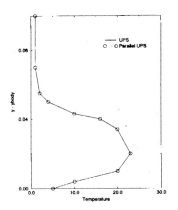

Fig. 3 Temperature Profile

Lagrangian prediction of particulate two-phase flows at high particle loadings

Bjarne B. Paulsen, Morten D. Holst, Thomas J. Condra, Jan Rusaas

Institute of Energy Technology, Aalborg University,
DK-9220 Aalborg, Denmark

Abstract: The Lagrange method for numerical prediction of particulate two-phase flows is discussed and reviewed. The numerical models used in calculating the particle trajectories, wall collisions and turbulence based movement, are presented. A venturi test case with high particle loading is treated and numerical and experimental results are compared. A circulating fluidised bed pre–separator is analysed and flow–particle coupled results at particle loadings up to 10 kg particles per kg air are presented. Convergence difficulties are discussed and a source smoothing method is introduced. Numerical and experimental results are compared and the reasons for the differences treated.

1 Introduction

Two-phase particulate flows arise in industrial applications such as gas cleaning cyclones and circulating fluidised bed particle separators. Prediction of the performance of these devices has normally been based on very approximate analyses or on experimental work with modelling factors used to scale–up separation geometries. Numerical simulation is a method of current interest for predicting particulate flows.

In the case of high particle mass loadings[1], and Stokes numbers between 0.01 and 100, the coupling between gas and particle phases becomes two–way[2]. In this kind of particle laden flows the dispersed phase calculations can be performed by employing a Lagrangian or Eulerian approach. There are a few reported gas-particle flow numerical simulations, where the two–way phase coupling has been considered (Karema and Karvinen 1992). Examples of particle laden flows with the use of the Lagrangian approach are Mostafa and Mongia (1988) with loadings from 0.2 to 0.66, and Adeniji-Fashola and Chen (1990) with loadings of 1.0 to 3.2. Durst et al.(1984) had loadings of 0.5 to 2.0, but were unable to obtain converged solutions at higher loadings.

In the test cases treated, poly–dispersed particle size distribution, particle dispersion and particle wall collisions are considered. The Lagrangian approach is found

---
[1] In this paper this is defined as the particle/gas mass flow ratio, exceeding 0.1 kg/kg
[2] The particles affect the gas-phase and vice versa

preferable due to its superior ability to handle the above mentioned problems. Problems arising from a diverging solution due to the back–coupling of high source terms to the Eulerian gas-phase calculations are described and solved in the following.

2 The numerical models implemented

The gas flow analysis was carried out using the Navier-Stokes finite volume flow solver CFDS-FLOW3D (Version 2.3) from Harwell Laboratory, England. The calculation of the particle trajectories is based on an analytical integration of the particle's momentum equation where aerodynamic drag and gravity are the forces considered. An optimum time step is calculated using an adaptive regulating method. The two-way coupling between the particle- and gas phase is modelled using the PSI-Cell model (Crowe et al. 1977). Particle wall collisions are modelled with an empirical model for non-elastic collisions (Grant and Tabakoff 1975). The influence of turbulence on the particle movement is modelled through a eddy-lifetime model (Astrup 1988).

3 Venturi test case

A simple venturi is chosen as an initial test case. In Fig. 1. the numerical results are compared to experimental results from the literature (Lee and Crowe 1981). As can be seen, the numerical results show the same linear relationship between the particle mass loading and the normalized pressure drop over the venturi as seen in the experimental results. During the numerical simulations with high particle mass loading it was necessary to introduce a considerable under–relaxation of the source terms, down to a factor of 0.1 at a load of 5.0 kg/kg, before a converged solution of the two-phase problem was achieved. Under-relaxing the source terms generally meant a considerable increase in the computing time, and this can be a major problem, especially for more complex 3D geometries, as will be presented in the following section.

4 Pre-separator test case

A pre-separator in conjunction with a circulating fluidized bed combustion reactor, on which a series of experiments at room conditions were carried out, was used as a complex 3-D geometry test case (Fig. 2.). The aim of the pre-separator is to separate the majority of the elutriated particles from the reactor from the combustion gas. Operating conditions were; inlet gas velocities at 3-6 m/s and particle mass loadings at 7.6 to 10 kg/kg. It was found that approximately 12500 cells ($23 \times 27 \times 20$) was sufficient to solve the flow. Typically 10000 trajectories were calculated and 20–50 particle time steps were taken per cell.

The above mentioned convergence problems with high loads were seen more clearly with this more complex geometry compared to the simpler venturi. Even with a load as low as 1.0 kg/kg it was very difficult to achieve a converged solution, even with significant under-relaxation. The use of the eddy-lifetime turbulence dispersion model for the particles improved convergence due to the spreading effect of turbulence. The difference in the source terms in neighbouring cells is hereby smoothed and this is a major factor in improving convergence. Even with the use of the eddy-lifetime model and under-relaxing the source terms it was not possible to obtain solutions for higher loadings.

The wall impact model used seems to exaggerate the actual loss of momentum, and because of the large number of wall impacts in certain regions, the error accumulates and causes very high source term peaks and resulting convergence problems. This problem is solved by setting a limit on the number of wall-impacts allowed in each near–wall cell. If the number is exceeded the particle is simply considered to be separated, which is physically realistic.

By this approach it was now possible to obtain solutions for loadings at 5 kg/kg. But it was still not possible to obtain solutions at 10 kg/kg. These difficulties arose because the flow solver had problems with the source terms which were both large and unevenly distributed, Fig. 3. A simple method for smoothing the source terms was introduced. The method is a Lagrangian interpolation routine, where the source terms in each cell are weighted with those of 26 surrounding cells. A more sophisticated smoothing algorithm such as the error function, was considered but the Lagrangian method proved to be very efficient, and it was now possible to obtain solutions at loadings of 10 kg/kg. Approximately 40 two-phase flow iterations were necessary. As can be seen the source terms are now more evenly distributed in the whole domain and especially at the near wall regions (see Fig. 4.).

The primary results of separation efficiency of the pre-separator are seen in Table 1. There is quantitative agreement between numerical and experimental results, but qualitatively there is not the same tendency of decreasing separation efficiency for higher loading for the numerical results, as was seen in the experiment. The reason is probably the lack of particle-particle interaction (swarm-effect) modelling.

5 Conclusion

Although the simulation of particle laden flows at high loadings generally is considered to be a problem, the developed solution strategies, have to some degree overcome the problems. This shows that it is possible to achieve a converged solution even in complex three- dimensional geometries with high particle loadings provided the source terms are handled with care.

References

1. Adeniji-Fashola, A., Chen, C.P. (1990): "Modeling of confined turbulent fluid-particle flows using Eulerian and Lagrangian schemes", in Int. J. Heat Mass Transfer, Vol. 33, No. 4, pp. 691–701

2. Astrup, P.(1988): "Development of a computer based model for stationary turbulent 3-D gas-particle flow", (Risoe National Laboratory, Denmark) (Report Risoe-M-2759 in Danish)

3. Crowe, C.T., Sharma, M.P. Stock, D.E. (1977): "The particle-source-in cell (PSI-cell) model for gas-droplet flows", in Journal of Fluids Engineering, Vol. 99, pp. 325–332

4. Di Giacinto, M., Sabetta, F., Piva, R. (1982): "Two-way coupling effects in dilute gas-particle Flows", in Journal of Fluids Engineering, Vol. 104, pp. 304–312

5. Durst, F., Milojevic, D., Schönung, B. (1984): "Eulerian and Lagrangian predictions of particulate two-phase flows: a numerical study", in Appl. Math. Modelling, Vol. 8, pp. 101–115

6. Grant, G., Tabakoff, W. (1975): "Erosion prediction in turbomachinery resulting from environmental solid particles", in J. Aircraft, Vol. 12, No. 5, pp. 471–478

7. Karema, H., Karvinen, R. (1992): "The mixing of passive scalar properties in dilute turbulent gas-particle flows", in 6th. Workshop on Two-Phase Flow Predictions, Erlangen, pp. 166-173

8. Lee, J., Crowe, C.T. (1981): "Gas-particle flow in a venturi: a numerical study", in Conference on gas borne particles, St. Catherines College Oxford (Organised by The Institution of Mechanical Engineers), pp.129–133

9. Mostafa, A.A., Mongia, H.C. (1988): "On the interaction of particles and turbulent fluid flow" in Int. J. Heat Mass Transfer, Vol. 31, No. 10, pp 2063–2075

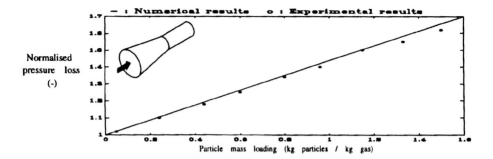

Fig. 1. Numerical- and experimental results (Lee and Crowe 1981) for venturi test case. Normalised pressure drop for varying particle mass loading.

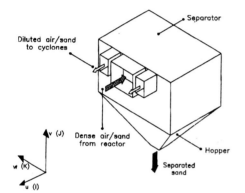

Gas velocity (m/s)	Loading (kg/kg)	Expt. sep. eff. (%)	Num. sep. eff. (%)
3.0	5.0	—	99.94
3.0	7.6	99.7	99.84
6.0	5.0	—	99.87
6.0	10.0	97.2	99.97

Table 1: Numerical and experimental results for separation efficiency for different test conditions

Fig. 2. The pre-separator modelled.

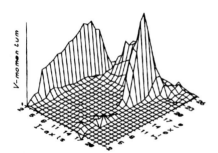

Fig 3: Unsmoothed V-momentum source terms in a vertical plane along pre-separator outlet flow axis.

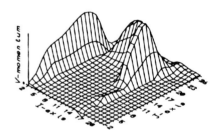

Fig 4: Smoothed V-momentum source terms in a vertical plane along pre-separator outlet flow axis.

A Numerical Method for Incompressible Viscous Flow Using Pseudocompressibility Approach on Regular Grids

Jong-Youb Sa

Yeungnam Univ., Dept. of Mech. Eng.
Kyongsan, 712-749, Republic of Korea

1 Introduction

The grid system used by the finite difference methods for incompressible viscous flow may be classified into two: regular (nonstaggered) and staggered grids. MAC[1] and SIMPLE[2] methods are based on the staggered grid, unless the checkerboard pressure field is caused. However, some recent methods such as SIMPLE[3] and INS[4] use the regular grid for fast computation of three-dimensional problems, and employ some kinds of internal smoothing such as PWIM[5] and AD4[6] to eliminate the pressure oscillation.

The present study shows that the pressure oscillation results from the inaccurate boundary condition, and examines the influence of Neumann and momentum boundary conditions on the pressure oscillation. The present study proposes the new boundary condition, PF (Pressure Filtration) condition, which can give the smooth pressure field without any interior smoothing. The new interior smoothing, PF (Pressure Filtration) smoothing, is also proposed in the present study and compared to other smoothing, PWIM and AD4.

2 Numerical methods

For the pseudocompressibility method in general curvilinear coordinates, the continuity, moment, and energy equations can be written as follows

$$\frac{\partial D}{\partial t} + \frac{\partial}{\partial \xi}(E - E_v) + \frac{\partial}{\partial \eta}(F - F_v) + \frac{\partial}{\partial \zeta}(G - G_v) = S \tag{1}$$

where

$$D = 1/J\{p, u, v, w, T\}^T, \quad S = -C_B T/J\{0, \lambda_1, \lambda_2, \lambda_3, 0\}^T,$$
$$E = 1/J\{\beta U, \ uU + \xi_x p, \ vU + \xi_y p, \ wU + \xi_z p, \ TU\}^T$$
$$F = 1/J\{\beta V, \ uV + \eta_x p, \ vV + \eta_y p, \ wV + \eta_z p, \ TV\}^T$$
$$G = 1/J\{\beta W, \ uW + \zeta_x p, \ vW + \zeta_y p, \ wW + \zeta_z p, \ TW\}^T$$
$$E_v = 1/J\{0, C_M \nabla \xi \cdot R(u), \ C_M \nabla \xi \cdot R(v), \ C_M \nabla \xi \cdot R(w), \ C_E \nabla \xi \cdot R(T)\}^T$$
$$F_v = 1/J\{0, C_M \nabla \eta \cdot R(u), \ C_M \nabla \eta \cdot R(v), \ C_M \nabla \eta \cdot R(w), \ C_E \nabla \eta \cdot R(T)\}^T$$

$$G_v = 1/J\{0, C_M \nabla \zeta \cdot R(u), C_M \nabla \zeta \cdot R(v), C_M \nabla \zeta \cdot R(w), C_E \nabla \zeta \cdot R(T)\}^T$$
$$R(\ ) = \nabla \xi \frac{\partial}{\partial \xi} + \nabla \eta \frac{\partial}{\partial \eta} + \nabla \zeta \frac{\partial}{\partial \zeta},$$
$$U = \xi_x u + \xi_y v + \xi_z w, \qquad \eta_x u + \eta_y v + \eta_z w, \qquad \zeta_x u + \zeta_y v + \zeta_z w.$$

The above coefficients are defined as for forced/mixed convection,

$$C_M = \frac{1}{Re}, \qquad C_B = \frac{G_r}{Re^2}, \qquad C_E = \frac{1}{RePr}$$

and for natural convection,

$$C_M = Pr, \qquad C_B = RaPr, \qquad C_E = 1$$

The governing equation can be written in a factored form as follows by linearizing the flux vectors with respect to the previous time step,

$$[I + \Delta\tau J \delta_\xi (A - \gamma_1)] \ [I + \Delta\tau J \delta_\eta (B - \gamma_2)] \ [I + \Delta\tau J \delta_\zeta (C - \gamma_3)] \ (D^{n+1} - D^n)$$
$$= -\Delta\tau J \left[\delta_\xi (E - E_\nu) + \delta_\eta (F - F_\nu) + \delta_\zeta (G - G_\nu) - S^n \right] \qquad (2)$$

3 Pressure Coupling on regular grid

The 2Δ-wave is assumed on oscillatory pressure field, of which the amplitude,ϵ, of oscillatory wave should be determined. The nth-order polynomials is applied to the corrected pressures, $\pm\epsilon$, at $n+2$ neighbering points, where the number of unknowns is $n+2$ ($n+1$ coefficients of polinomials and the amplitude ϵ). Now, the pressure can be corrected for the smoothness by using the amplitude, ϵ.

$$p_0^{new} = p_0^{old} + w_f \epsilon \qquad (: w_f + 0.1 \sim 0.8) \qquad (3)$$

The above equation of pressure coupling can be applied to both the boundary and the interior grid points.

$$p_0 + \epsilon = as_0^2 + bs_0 + c = c \quad (: s_0 = 0), \qquad p_1 - \epsilon = as_1^2 + bs_1 + c,$$

$$p_2 + \epsilon = as_2^2 + bs_2 + c \qquad\qquad p_3 - \epsilon = as_3^2 + bs_3 + c.$$

The new PF boundary condition of pressure is designed by using the second-order polynomials at four points near a boundary as shown in Fig. 1.

$$p_{-2} + \epsilon = as_{-2}^3 + bs_{-2}^2 + cs_{-2} + d, \qquad p_{-1} - \epsilon = as_{-1}^3 + bs_{-1}^2 + cs_{-1} +$$

$$p_0 + \epsilon = as_0^3 + bs_0^2 + cs_0 + d = d \quad (: s_0 = 0),$$

$$p_1 - \epsilon = as_1^3 + bs_1^2 + cs_1 + d, \qquad p_2 + \epsilon = as_2^3 + bs_2^2 + cs_2 + d.$$

4 Results and Discussions

4.1 Driven Cavity Flow

Fig. 3 shows a comparison of pressure field according to each boundary condition in absence of any kind of interior coupling at $Re=100$ and 1000. The Neumann boundary condition shows the worst result in which the checkerboard pressure is observed more clearly in the region with low gradient of pressure. The momentum boundary condition improves a little, but its result is still not satisfactory. The best result is available from the PF boundary condition. The pressure gradient near a boundary is automatically computed with good accuracy. Fig.4 reveals that the machine accuracy of velocity divergence-free condition can be achieved in absence of any interior smoothing.

4.2 Uniform Flow around a Circular Cylinder

The test of this case is intended to show that there exist some cases in which boundary smoothing alone is not enough for elimination of pressure oscillation without any interior coupling. When the O-type grid is used with even number of grid point in circumferential direction, the PF boundary cannot do anything in that direction. Fig.5(a) and (b) show that both Neumann and PF boundary conditions cannot suppress the pressure wiggle. However, the PF boundary condition gives better result than others.

When the PF boundary condition is used, the comparison among interior smoothings is given in Fig.5(c), (d) and (e). The comparison of accuracy is shown in detail in Table 1 and 2. In this test case, all the three interior smoothings seem to work well.

5 Conclusions

The accurate boundary condition can remove or suppress the unrealistic pressure oscillation, but cannot change the characteristic of the governing equations and the solution method. If the central differencing is used to approximate the convection term of governing equations, the pressure oscillation can never be avoided due to the intrinsic characteristic of the numerical method for the incompressible viscous flow. The accurate boundary condition merely prevents the pressure wiggle from being generated near a boundary by the inaccurate boundary condition of pressure. The present study reveals that the PF boundary condition of pressure can suppress successfully the checkerboard pressure field without adding any interior smoothing. There exists some case in which the present PF boundary condition is not enough to suppress the pressure oscillation: the O-type grid problems with even number of grid points in circumferential direction. However, even in those cases, the present PF boundary condition can improve the accuracy of solution by using smaller amount of interior smoothing than other boundary conditions.

References

1. F.H. Harlow and J.E. Welch, "Numerical Calculation of Time-Dependent Viscous Incompressible Flow of Fluid with Free Surface", Phys. Fluid, Vol.8, p.2182, 1965.

2. S.V. Patankar and D.B. Spalding, "A Calculation Procedure for Heat, Mass, and Momentum Transfer in Three-Dimensional Parabolic Flows", Int. J. Heat Mass Transfer, Vol.15, p. 1787, 1972.

3. S. Acharya and F.H. Moukalled, "Improvements to Incompressible Flow Calculation on a Nonstaggered Curvilinear Grid", Numer. Heat Transfer, Part B, Vol.15, p.131, 1989.

4. D. Kwak, J.L.C. Chang, S.P. Shanks, and S. Chakravarthy, "A Three-Dimensional Incompressible Navier-Stokes Flow Solver Using Primitive Variables", AIAA J., Vol.24, no.3, p.390, Mar. 1986

5. C.M. Rhie and W.L. Chow, "Numerical Study of the Turbulent Flow past an Airfoil with Trailing Edge Separation", AIAA J., Vol.21, p.1525, 1983.

6. T.H. Pulliam, "Artificial Dissipation Models for the Euler Equations", AIAA J., Vol.24, p.1931, Dec. 1986.

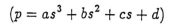

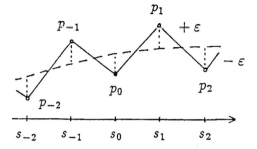

Fig. 2 2Δ-wave of pressure oscillation within a domain.

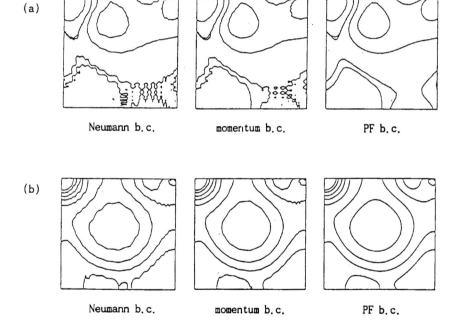

Fig. 3 A driven cavity flow without any interior smoothing at Re = 100(a) and 1000(b).

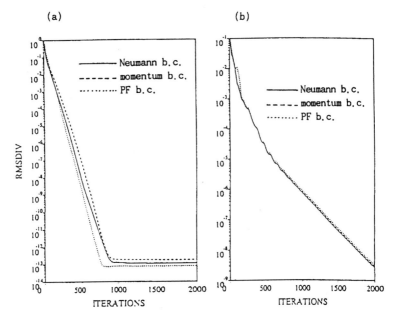

Fig. 4 Convergence history for a driven cavity flow at Re = 100(a) and 1000(b).

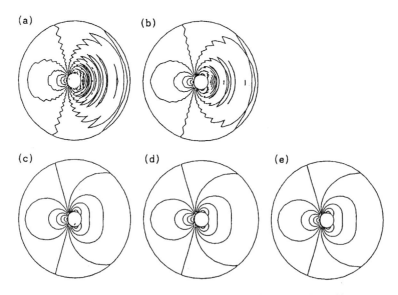

Fig. 5 A uniform flow around a circular cylinder at Re = 20:
    (a) Neumann b.c.,   no smoothing,
    (b) PF b.c.,   no smoothing,
    (c) PF b.c.,   PWIM,
    (d) PF b.c.,   AD4,
    (e) PF b.c.,   PF smoothing.

Tetrahedral Spectral Elements for CFD

S.J. Sherwin and G.E. Karniadakis

Center for Fluid Mechanics, Division of Applied Mathematics,
Brown University, U.S.A.

Abstract : In this paper we describe the foundation of a new hierarchical basis suitable for spectral and $h-p$ type finite elements in complex, three-dimensional domains. It is based on Jacobi polynomials of mixed weights and allows for a variable order in each element, which is a crucial property for efficient adaptive discretizations on unstructured meshes. In addition, it has the desirable property of an $O(N^2)$ spectrum scaling for the convective operator as in tensorial spectral methods; this is important in explicit time-integration of the Navier-Stokes equations. The general Galerkin formulation for elliptic equations and a high-order splitting scheme used to solve the two-dimensional Navier-Stokes equations has been documented in [2]. Here we elaborate on the three-dimensional basis and demonstrate its properties through different numerical examples. The new algorithms have been implemented in the general code $\mathcal{N}, \epsilon k T \alpha r$, which represents the new generation of spectral element methods for unstructured meshes.

1 Multi-Domain Tetrahedral Basis

We define a polynomial basis, denoted by $g^{1-3k}_{lmn}(r,s)$, so that we can approximate the function $f(x,y,z)$ by a C^o continuous expansion over 'k' sub-domains of the form:

$$f(x,y,z) = \sum_k \sum_l \sum_m \sum_n \overset{-k}{f}_{lmn} \overset{1-3k}{g}_{lmn}(r(x,y,z), s(x,y,z), t(x,y,z)).$$

Here $\overset{-k}{f}_{lmn}$ is the expansion coefficient corresponding to the expansion polynomial $\overset{1-3k}{g}_{lmn}$ in the kth sub-domain; (x,y,z) are the spatial co-ordinates of the function and (r,s,t) are the local co-ordinates within any given tetrahedron. We define a standard tetrahedron as the space

$$T^3 = \{-1 \leq r,s,t; r+s+t \leq -1\}$$

and introduce the new co-ordinate system:

$$a = 2\frac{(1+r)}{(-s-t)} - 1, \quad b = 2\frac{(1+s)}{(1-t)} - 1, \quad c = t.$$

The constant planes represented by a, b, c can be seen in figure Fig.1. We note the degeneracy of the co-ordinate system in the T^3 space. Planes of constant a remain planes as this co-ordinate varies from $a = -1$ to $a = 1$, and are dependent on all of the basis co-ordinates r, s and t. However, planes of constant b degenerate to a line as this co-ordinate varies from $b = -1$ to $b = 1$ although these planes only depend on the basis co-ordinates s and t. Finally, planes of constant c degenerate to a point as this co-ordinate varies from $c = -1$ to $c = 1$ and these planes only depend on the basis co-ordinate t. This transformation maps the standard triangle to the bi-unit square in two dimensions and as mapping the standard tetrahedron to the bi-unit cube in three dimensions. Although this mapping produces singularities along one edge and at the uppermost vertex, the co-ordinates have a bounded value between -1 and 1. This is analogous to polar co-ordinates where at $r = 0$, θ has values between $0° \leq \theta \leq 360°$.

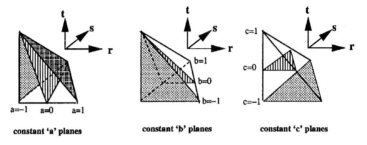

Fig. 1. Constant planes of the co-ordinates a, b and c on the standard tetrahedron.

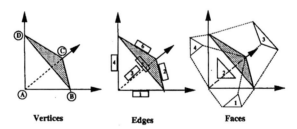

Fig. 2. Tetrahedron notation

Having defined this co-ordinate we can describe the expansion basis. Ideally, we would like to have an orthogonal expansion and Dubiner [1] has suggested such a basis; however it cannot easily be extended to form contiguous domains (see [1,2]). Nevertheless, a modified basis can be constructed which maintains a high degree of orthogonality. Using the notation given in Fig.2 and noting that the two-dimensional basis is given by $c = -1$ we have:

- Interior modes ($2 \leq l,\ 1 \leq m,n,\ l < L,\ l+m < M,\ l+m+n < N$)

$$_{1-3}g^{\text{interior}}_{lmn} = \left(\frac{1+a}{2}\right)\left(\frac{1-a}{2}\right) P^{1,1}_{l-2}(a) \left(\frac{1+b}{2}\right)\left(\frac{1-b}{2}\right)^l P^{2l-1,1}_{m-1}(b)$$
$$\left(\frac{1+c}{2}\right)\left(\frac{1-c}{2}\right)^{l+m} P^{2l+2m-1,1}_{n-1}(c)$$

- Face modes ($2 \leq l,\ 1 \leq m,n,\ l < L,\ l+m < M,\ l+m+n < N$)

$$_{1-3}g^{\text{face}-1}_{lm0} = \left(\tfrac{1+a}{2}\right)\left(\tfrac{1-a}{2}\right) P^{1,1}_{l-2}(a) \left(\tfrac{1+b}{2}\right)\left(\tfrac{1-b}{2}\right)^l P^{2l-1,1}_{m-1}(b) \left(\tfrac{1-c}{2}\right)^{l+m}$$

$$_{1-3}g^{\text{face}-2}_{l0n} = \left(\tfrac{1+a}{2}\right)\left(\tfrac{1-a}{2}\right) P^{1,1}_{l-2}(a) \left(\tfrac{1-b}{2}\right)^l \left(\tfrac{1+c}{2}\right)\left(\tfrac{1-c}{2}\right)^l P^{2l-1,1}_{n-1}(c)$$

$$_{1-3}g^{\text{face}-3}_{1mn} = \left(\tfrac{1+a}{2}\right)\left(\tfrac{1+b}{2}\right)\left(\tfrac{1-b}{2}\right) P^{1,1}_{m-1}(b) \left(\tfrac{1+c}{2}\right)\left(\tfrac{1-c}{2}\right)^{m+1} P^{2m+1,1}_{n-1}(c)$$

$$_{1-3}g^{\text{face}-4}_{1mn} = \left(\tfrac{1-a}{2}\right)\left(\tfrac{1+b}{2}\right)\left(\tfrac{1-b}{2}\right) P^{1,1}_{m-1}(b) \left(\tfrac{1+c}{2}\right)\left(\tfrac{1-c}{2}\right)^{m+1} P^{2m+1,1}_{n-1}(c)$$

- Edge modes ($2 \leq l,\ 1 \leq m,n,\ l < L,\ l+m < M,\ l+m+n < N$)

$$_{1-3}g^{\text{edge}-1}_{l00} = \left(\tfrac{1+a}{2}\right)\left(\tfrac{1-a}{2}\right) P^{1,1}_{l-2}(a) \left(\tfrac{1-b}{2}\right)^l \left(\tfrac{1-c}{2}\right)^l$$

$$_{1-3}g^{\text{edge}-2}_{1m0} = \left(\tfrac{1+a}{2}\right)\left(\tfrac{1+b}{2}\right)\left(\tfrac{1-b}{2}\right) P^{1,1}_{m-1}(b) \left(\tfrac{1-c}{2}\right)^{m+1}$$

$$_{1-3}g^{\text{edge}-3}_{1m0} = \left(\tfrac{1-a}{2}\right)\left(\tfrac{1+b}{2}\right)\left(\tfrac{1-b}{2}\right) P^{1,1}_{m-1}(b) \left(\tfrac{1-c}{2}\right)^{m+1}$$

$$_{1-3}g^{\text{edge}-4}_{10n} = \left(\tfrac{1-a}{2}\right)\left(\tfrac{1-b}{2}\right)\left(\tfrac{1+c}{2}\right)\left(\tfrac{1-c}{2}\right) P^{1,1}_{n-1}(c)$$

$$_{1-3}g^{\text{edge}-5}_{10n} = \left(\tfrac{1+a}{2}\right)\left(\tfrac{1-b}{2}\right)\left(\tfrac{1+c}{2}\right)\left(\tfrac{1-c}{2}\right) P^{1,1}_{n-1}(c)$$

$$_{1-3}g^{\text{edge}-6}_{01n} = \left(\tfrac{1+b}{2}\right)\left(\tfrac{1+c}{2}\right)\left(\tfrac{1-c}{2}\right) P^{1,1}_{n-1}(c)$$

- Vertex modes

$$_{1-3}g^{\text{vert}-A}_{100} = \left(\tfrac{1-a}{2}\right)\left(\tfrac{1-b}{2}\right)\left(\tfrac{1-c}{2}\right)$$

$$_{1-3}g^{\text{vert}-B}_{100} = \left(\tfrac{1+a}{2}\right)\left(\tfrac{1-b}{2}\right)\left(\tfrac{1-c}{2}\right)$$

$$_{1-3}g^{\text{vert}-C}_{010} = \left(\tfrac{1+b}{2}\right)\left(\tfrac{1-c}{2}\right)$$

$$_{1-3}g^{\text{vert}-D}_{001} = \left(\tfrac{1+c}{2}\right)$$

Where the indices lmn in g_{lmn} refer to the order of the principal polynomial in r, s and t respectively and L,M,N define the order of the expansion.

As can be seen, the basis is split into four types of models *interior, faces, edges* and *vertices*. Here we have used $P^{\alpha,\beta}(z)$ to represent the Jacobi polynomial which is orthogonal with respect to the weight function $(1-z)^\alpha(1+z)^\beta$. The interior modes are zero at all boundaries whilst the face modes have magnitude in the interior and on one face only. The edge modes have non-zero contributions along one edge and are zero at all vertices and other edges. The vertex modes vary from a unit value at one vertex to zero at all other vertices. These modes are identical to linear finite element modes. Every mode is a polynomial in (a,b,c) space as well as (r,s,t) space.

As a final point, we note that when $l = m+1 = n+1$ the edge modes have the same shape. Provided that the singular vertices on a face are aligned, the face modes also have the same shape. This allows the basis to be combined to form a C^0 expansion by matching the expansion coefficients of these modes. However, expansion co-efficients of odd order modes may need negating in order to ensure the C^0 continuity. In order to connect multiple domains without any redundant modes we require that $L = M = N$ and for an L order expansion there are $L(L+1)/2$ modes in two dimensions and $L(L+1)(L+2)/6$ modes in three dimensions.

2   Results

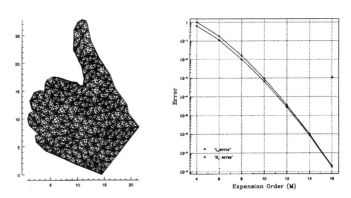

Fig. 3. Convergence to the Helmholtz problem with solution $u(x,y) = sin(\pi x)cos(\pi y)$. Spectral convergence is obtained independently of the complexity of the discretisation.

The implementation and analysis of the basis in two-dimensions has been described in [2]. The first result is one which involves a variety of triangular elements of different aspect ratios and orientations. The solution domain is shown in Fig.3. We have solved an elliptic Helmholtz problem which is the principal equation used in the solution of the incompressible Navier-Stokes equation using a high-order splitting scheme [3]. Here the Helmholtz constant was taken as $\mu = 1$ and the exact solution was $u(x,y)$

$= sin(\pi x)cos(\pi y)$. Also shown in Fig.3 are the L_∞, L_2 errors plotted with respect to expansion order. Spectral convergence is observed as indicated by the asymptotic linearity of the curves on this lin-log plot.

The second result, shown in Fig.4, demonstrates the complicated discretisation possible using tetrahedral element. Here we are considering the projection of a function which follows the shape of the domain of the form:

$$u(x,y,z) = sin(\pi(cos(0.1\pi z)x + sin(0.1\pi z)y))$$

$$sin(\pi(cos(0.1\pi z)y - sin(0.1\pi z)x))$$

The second plot in Fig.4 indicates the orientation required to allow a C^0 expansion base since singular vertices much meet singular vertices. The triangle indicates the base of each tetrahedral and points to the singular vertex whilst the circle indicates the most singular vertex of each tetrahedral. As can be seen all singular vertices do indeed connect to other singular vertices. This constraint at first seems very restrictive but it can be shown that any conforming tetrahedral discretisation can be connected in this fashion. Again spectral convergence is observed once the function is resolved.

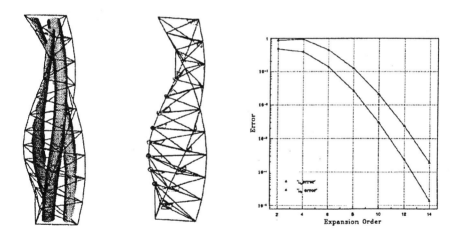

Fig. 4. Projection in a complicated tetrahedral domain. The connectivity of the mesh is indicated by the triangles and circles in the second plot. Spectral convergence in the L_∞ and L_2 norms is demonstrated in the third plot.

The final result is the solution to the Wannier-Stokes flow. We have chosen this flow since there is an exact solution in this curvi-linear complicated domain. The solution was solved for a variety of expansion orders and spectral convergence was observed in the L_∞ and H_1 norms.

We would like to acknowledge the assistance of T. Warburton on the construction of three-dimensional tetrahedral meshes. This work was supported by ONR and AFOSR.

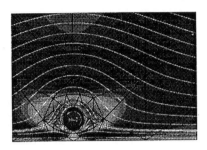
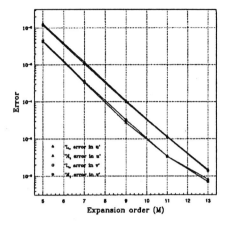

Fig. 5. Steady state solution of the Wannier-Stokes flow using an expansion order of $M = 11$ (left). Plots of the L_∞ and H_1 errors as a function of expansion order for the u and v velocity of the solution to the Wannier-Stokes flow (right).

References

1. M. Dubiner. Spectral methods on triangles and other domains. *J. Sci. Comp.*, **6** 345 (1991).

2. S.J. Sherwin and G.E. Karniadakis, A triangular spectral element method; applications to the incompressible Navier-Stokes equations. Submitted to *Comp. Meth. Appl. Mech. Eng.*

3. G.E. Karniadakis M. Israeli and S.A. Orszag. High-order splitting methods for the incompressible Navier-Stokes equations. *J. Comp. Phys.*, **97** 414 (1991).

Computation of Unsteady 3D Turbulent Flow and Torque Transmission in Fluid Couplings

L. Bai, N.K. Mitra and M. Fiebig

Institut für Thermo- und Fluiddynamik,
Ruhr-Universität Bochum, Universitätsstr.
150, D-44780 Bochum, Germany

Abstract : The turbulent flow in a fluid coupling has been calculated from the numerical solution of the three dimensional Reynolds averaged nonsteady Navier-Stokes equations in body-fitted rotating coordinate system. Turbulence has been taken care of by standard $\kappa - \varepsilon$ model. The results show complex vortex structure of the recirculating flow. The computed torque transport compares well with the available experimental results.

1 Introduction

The flow in a fluid coupling is one of the most complex problem encountered in engineering fluid mechanics. A coupling consists of two elements: a pump impeller with radial blades, and a turbine impeller, generally with different number of blades. The casing is partially filled with a fluid, usually an oil of low viscosity. The fluid transmits torque. Fig. 1 shows diagramatically a fluid coupling. As the pump begins to rotate, the fluid within its impeller moves towards the periphery and discharges into the turbine at the outer radius. Within the turbine impeller, it flows radially inward towards the centre and discharges back into the pump impeller at the inner radius. For the flow to exist, the head produced by the pump must be greater than that produced by the turbine. This is only possible if the rotational speed of the pump is greater than that of the turbine. Thus, for torque transmission there must be a speed difference, i.e. the slip. The flow in reality is two-phase, three-dimensional, unsteady, turbulent and contains separated zones. The complexity of the flow field depends on the fluid volume and the geometry and dynamical parameters (filling ratio, angular velocity of the impeller and speed differential). Some numerical studies of the laminar flow field in fluid couplings have been made by the senior authors [1]. Corresponding studies for the turbulent flow have never been reported.

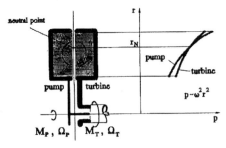

Fig. 1. Schematic of a fluid coupling with qualitative recirculating flow structure and pressure distribution; M_P, M_T: torque of pump and turbine; Ω_P, Ω_T:, angular velocity of pump and turbine; r: radius; p: pressure; r_N: neutral point radius

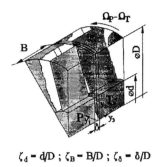

$\zeta_d = d/D$; $\zeta_B = B/D$; $\zeta_\delta = \delta/D$

Fig. 2. Computational domain; P - pump, T - turbine

The purpose of this work is to predict turbulent flow structure and the torque transmission in fluid couplings.

2 Numerical Method

The coupling is assumed to be completely filled with an incompressible viscous fluid. The flow in the coupling is described by 3D unsteady Reynolds averaged Navier-Stokes equations in conjunction with the eddy viscosity concept and $\kappa - \varepsilon$ equations.

The equations are solved in a generalized co-ordinate system that rotates with a constant angular velocity. The analysis of the flow field is limited to one pitch by assuming equal numbers of blades for the pump and the turbine impeller, Fig. 2 shows the computational domain. The basic equations are solved by a finite-volume scheme. The flow domain is subdivided into a finite number of control volumes (CV). All dependent variables are defined in the centre of the CV. The velocity and pressure fields are calculated with the SIMPLEC algorithm [3]. Because of the non-staggered variable arrangement the velocities at the CV faces are calculated from the adjacent CV-centred quantities. In order to avoid oscillations a special interpolation for the CV face mass fluxes [2] is employed. The wall functions given by Launder and Spalding [4] are employed to prescribe the boundary conditions along the walls. The discretized equations are solved by the strongly implicit procedure (SIP) of Stone [5].

3 Results and Discussion

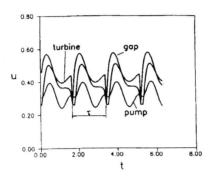

Fig. 3. Time series of u-velocity (circumferential) at three points in the pump, turbine impeller and in the gap between the coupling halves

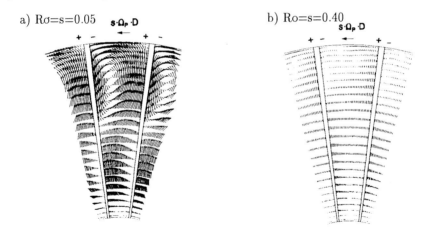

a) Ro=s=0.05

b) Ro=s=0.40

Fig. 4a Mean relative velocity fields in axial cross-section of a pump impeller for Ek=$1.5 \cdot 10^{-7}$ and Ro=0.05

Fig. 4b Mean relative velocity fields in axial cross-section of a pump impeller for Ek=$1.5 \cdot 10^{-7}$ and Ro=0.4; arrow: direction of absolute rotation

Results have been obtained for the following geometry (cf. Fig. 2): $\zeta_d = d/D = 0.326$, $\zeta_B = B/D = 0.281$, $\zeta_\delta = \delta/D = 0.01$; number of the blades $Z = 24$.

The number of grids typically used is 35x28x56. The computation time for one calculation on a Convex C120 computer is 320 CPU hours. The calculations are performed with the following flow parameters: Ekman number: $Ek = \nu/D^2\Omega_P = 1.5 \cdot 10^{-7}$ Rossby number: $Ro = u_o/\Omega_P D = 0.05, 0.15, 0.40, 0.80, 0.95$, where ν is the dynamic viscosity of the fluid, u_o is a reference velocity, defined as $u_o = (\Omega_P - \Omega_T) \cdot D$. In this sense the slip $s = 1 - \Omega_T/\Omega_P$ is equal to Ro. The Ekman number and Rossby number indicate the rotational state. Due to the periodicity of relative rotation of the turbine blades the flow is periodic in time. Fig. 3 shows a periodic history of the

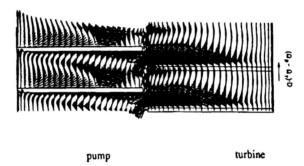

Fig. 5. Instantaneous relative velocity field in circumferential cross-section at neutral zone $r_N/D=0.370$ for $Ek=1.5\cdot10^{-7}$, $Ro=0.05$; arrow: direction of the relative rotation of the turbine

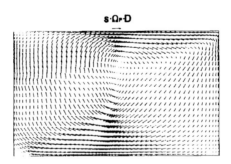

Fig. 6. Mean relative velocity field at a meridional plane for $Ek=1.5 \cdot 10^{-7}$, $Ro=0.05$

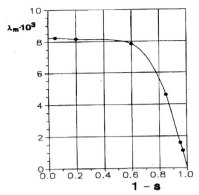

Fig. 7. Calculated operating characteristics of a coupling for Ek=$1.5 \cdot 10^{-7}$

velocity component u (in the circumferential direction) at three points respectively in the turbine, pump impeller and in the gap between the coupling halves. The flow structures are shown in Fig. 4 - Fig. 6. Fig. 4 shows velocity fields in axial (blade-to blade) cross section. The velocity vectors are displayed relative to the angular velocity of the pump.

For $Ro = s = 0.05$ (Fig. 4a) the main flow is directed radially outward and a secondary eddy (passage vortex), which hinders the torque transmission, has been established. Compared to the laminar flow [1], the eddy is longer due to the stronger Coriolis force. For $Ro = s = 0.4$ (Fig. 4b) the above-mentioned passage vortex almost disappears, and the transmitted torque rises considerably. Fig. 5 shows instantaneous velocity vectors relative to the rotating pump system in the circumferential cross sections at the neutral zone ($r = r_N$, cf. Fig. 1), a secondary eddy in the pump has also been seen. The velocity field in the meridional plane is displayed in Fig. 5. Due to the rotation and the slip between the pump and turbine one distinct circular flow is formed, which plays the dominant role in the torque transmission. By applying Euler's turbine equation and a moment balance the torque transmission can been determined with the simulated velocity and pressure fields. The dimensionless torque transmission λ vs slip s curve is shown in Fig. 7, which indicates the operating characteristics of a fluid coupling.

Acknowledgement :

This work has been supported by Deutsche Forschungsgemeinschaft through SFB278.

References

[1] Kost,A., Mitra,N.K., Fiebig,M. (1994): "Numerical simulation of 3D periodic flow in fluid coupling", in Acta Mechanica (in press).

[2] Kost,A., Bai,L., Mitra,N.K., Fiebig,M. (1992): "Calculation procedure for unsteady incompressible 3D flows in arbitrarily shaped domains", in Proc. 9th GAMM Conference on Numerical Methods in Fluid Dynamics, Notes on Numerical Fluid Mechanics **35**, Braunschweig: Vieweg, pp. 269-278.

[3] Van Doormaal,J.P., Raithby,G.D. (1984): "Enhancement of the SIMPLE method for predicting incompressible fluid flows ", in Num. Heat Transfer **7**, pp. 147 - 163.

[4] Launder,B.E., Spalding,D.B. (1974): "The numerical computation of turbulent flow", Computer Methods in Applied Mech. and Eng. **3**, pp. 269-289.

[5] Stone,H.L. (1968): "Iterative solution of implicit approximations of multidimensional partial differential equations", SIAM J. Num. Anal. **5**, pp. 530-558.

Flow investigation near the point of surface temperature and catalytic properties discontinuity

Vladimir V. Bogolepov, Igor I. Lipatov, Lev A. Sokolov

Central Aerohydrodynamics Institute (TsAGI),
Zhukovsky-3, 140160 Moscow Region, Russia
Telex: 412573 ZAKAT SU, Fax: (095) 271.00.19

Supersonic flow of a viscous gas about a flat plate at large but subcritical Reynolds numbers Re_∞ is considered. It is assumed that the gas is a binary mixture of atoms and diatomic molecules and that the temperature of the surface is not higher than the level at which molecules begin to dissociate at the local pressure values. The flow in the vicinity of the point of abrupt temperature change and/or abrupt catalytic properties of the surface change is investigated.

The solution of a such problem is needed for heat shield of space planes calculation [1], for catalytic calorimeters testing [2]. The important thing here is also the upstream propagation of disturbunces from the point of discontinuity. This phenomenon can not be described by means of a boundary layer theory [2] or viscous shock layer theory [3].

The solution of Navier-Stokes equations together with the mass conservation equation of atoms is obtained if Reynolds number is very high $Re_\infty = \epsilon^{-2} \gg 1$.

It is supposed that discontinuity in boundary conditions leads to the dimensionless nearwall temperature and mass concentration of atoms changes in the leading order: $\Delta T \sim T \sim 1$ and $\Delta c \sim c \sim 1$. This may cause leading order change of the nearwall layer thickness, where the gas density also changes in the leading order: $\Delta \rho \sim \rho \sim 1$. The thickness change and streamlines inclination are similar to those occuring in the flow near local surface distortion. The investigation of disturbed flow near surface distortions was attempted in [5] on the basis of matched asymptotic expansions method [4].

It is anticipated that in common case Prandtl and Schmidt numbers are of the order of unity $Pr \sim Sc \sim 1$ in the nearwall layer. This means that nearwall layer thickness has approximately the same value as temperature and diffusive ones. Taking into account that longitudinal velocity changes linearly with the distance from the surface $u \sim y/\epsilon$, it is possible to obtain the estimate for nearwall layer thickness in relation to its length $\Delta x \le 1$

$$\Delta y \sim \epsilon \Delta x^{\frac{1}{3}} \le \epsilon \sim \delta \tag{1}$$

where δ is undisturbed boundary layer thickness near the point of discontinuity. The change in the nearwall layer thickness (1) causes the change of the full boundary layer thickness and the displacement of its external boundary [6]. The latter induces in external supersonic flow the pressure disturbance

$$\Delta p \sim \Delta y / \Delta x \tag{2}$$

Estimates (1) and (2) allow to determine disturbed flow structure near the point of discontinuity. It may be shown using asymptotical analysis that known triple-deck structure [6] is realized here. It is worth to say that characteristic length of all disturbed layers are of order of $\Delta x \sim \epsilon^{\frac{3}{4}} \gg \delta$. On this length disturbances may propagate upstream from the point of discontinuity. The upstream disturbances propagation possibility is connected with existence of the subsonic flow (region 3) near the wall in the boundary layer. In the main part of the boundary layer (region 2), having thickness $\Delta y_2 \sim \epsilon$, streamlines are display on the nearwall layer thickness $\Delta y_3 \sim \epsilon^{\frac{5}{4}} \ll \delta$. In weekly disturbed external supersonic flow (region 1) with thickness $\Delta y_1 \sim \Delta x \sim \epsilon^{\frac{3}{4}} \gg \delta$. boundary layer thickness changes induce disturbance of pressure $\Delta p \sim \Delta y_3 / \Delta x \sim \epsilon^{\frac{1}{2}} \ll 1$. It is obvious that in the disturbed flow region induced pressure gradient takes large values $\partial p / \partial x \sim \epsilon^{-\frac{1}{4}} \ll 1$. Therefore here in the mass conservation equations the term describing the rate of formation of an individual component of the gas mixture may be dropped. Under such conditions the flow may be considered as chemically "frozen" with possible recombination reactions on the surface. It is obvious also that barodiffusion here is insignificant and as usual influence of thermodiffusion may be neglected.

After usual asymptotic analysis it was obtained that the nearwall flow in the first approximation is described by boundary layer equations for compressible gas, and induced pressure gradient is obtained from the interaction condition with external supersonic flow

$$p(x) = -\beta^{-1} dD/dx, \beta = (M^2 - 1)^{\frac{1}{2}} \tag{3}$$

where D is a displacement thickness of the nearwall layer with nonlinear changes, M is a sound velocity for "frozen" conditions in inviscid flow.

For calculation simplicity linear dependence of viscosity coefficient on temperature is supposed. It is also supposed that specific heat capacity for constant pressure, Prandtl and Schmidt numbers are constant. It is anticipated that specific heat capacities for atomic and molecular components of gas mixture are approximately equal. Using Dorodnitsin-Lees variables boundary-value problem for investigated flow near the point of discontinuity may be formulated as follows

$$\psi''' = -\ddot{d}(1 + c_1)T_1 + \psi'\dot{\psi}' - \dot{\psi}\psi'$$

$$c_i'' = Sc(\psi'\dot{c}_i - \dot{\psi}c_i')$$

$$T_i'' = Pr(\psi'\dot{T}_i - \dot{\psi}T_i'), i = 1, 2$$

$$\psi(x, 0) = \psi'(x, 0) = 0, c_2'(x, 0) = 1, T_2(x, 0) = 0$$

$$c_1'(x, 0) = Fc_1(x, 0)(1 + c_1(x, 0))^{-1}, T_1(x, 0) = (1 + \alpha)(1 + c_0)^{-1}(x \geq 0)$$

$$\psi'(x,\infty) = \int_0^\infty T_1(1+c_1)d\eta + d, c_1(x,\infty) \to c_0, T_1(x,\infty) \to (1+c_0)^{-1}$$

$$c_2'(x,\infty) \to 1, T_2'(x,\infty) \to 1$$

$$\psi(-\infty,\eta) \to \frac{\eta^2}{2}, c_2(-\infty,\eta) \to \eta, T_2(-\infty,\eta) \to \eta, d(-\infty) \to 0$$

$$c_1(x,\eta) = c_0, T_1(x,\eta) = (1+c_0)^{-1}, (x<0)$$

$$F = KSc(1+\alpha)^{-1}A^{-\frac{3}{4}}(M^2-1)^{-\frac{1}{8}}(M^2 T_0 \gamma)^{\frac{-5}{4}}(1+c_0)^{-\frac{1}{4}} \tag{4}$$

Here $\dot{(\ )} = \frac{\partial}{\partial x}$, $(\ )' = \frac{\partial}{\partial \eta}$, and ψ is a stream function, d is transformed thickness of the nearwall layer $(p = -\dot{d})$, indices $i = 1,2$ are referred to the first and second terms of asymptotic expansions for mass atoms concentration and temperature, F is a local Damkoeler number, parameter $(\alpha + 1)$ characterizing the temperature jump value, K, A, c_0 and T_0 are values of catalyticity coefficient k/ϵ, skin friction, mass atoms concentration and temperature on the surface of the plate in undisturbed boundary layer upstream of the discontinuity point, γ is the specific heats ratio for "frozen" chemical composition of gas mixture in external flow.

Effect of interaction between the nearwall flow and external flow (3) adds to the parabolic boundary problem (4) weak ellipticity property [6]. It is shown that the longitudinal momentum equation contains second derivative of displacement thickness of the nearwall layer with respect to the longitudinal coordinate. Such a property makes the solution of boundary-value problem nonunique and for the solution uniqueness it is needed to satisfy some additional conditions imposed downstream

$$\dot{d}(x \to \infty) \to 0 \tag{5}$$

Small length of the disturbed region (in comparison with the distance between the plate leading edge and point of discontinuity) is the reason that the atoms concentration up to the discontinuity point remains constant. This function may alter in the leading order if catalyticity coefficient at $x = 0$ changes from the value $k \sim \epsilon$ up to the value $k \sim \epsilon^{\frac{3}{4}}$.

In the flow region analyzed pressure and skin friction do not change in the leading order. Heat flux may undergo large changes

$$q = \epsilon^{\frac{3}{4}} q_1 + \epsilon q_2 + \ldots \tag{6}$$

Up to the discontinuity point c_1 and T_1 remain constant and value of function q is determined by the second item (6) (by values c_2' and T_2'). Downstream of discontinuity point the value of function q will be determined basically by the first item (6) (by values c_1' and T_1').

For boundary-value problem (4) numerical calculation marching method was used and for each value $x = const$ value of induced pressure gradient was selected (using additional interaction condition). For initial disturbance determination asymptotical solution (4) for $x \to -\infty$ was used. This solution takes the following form [6]

$$d \sim \exp(0.49(x - x_0))$$

where x_0 is an arbitrary constant.

Upstream of the discontinuity point boundary-value problem (4) solution is continuous. For abrupt change in boundary condition at $x > 0$ for solution accuracy keeping usually numerical grid remeshing is used. Net grid parameters usually are selected in such situations by empirical way. Matched asymptotic expansions method [4] allows to construct analytical solution immediately downstream of the discontinuity point and to formulate conditions for grid creation. For calculation additional region is introduced for which external boundary conditions are determined from the boundary-value solution (4) near the wall for $x \to -0$. Asymptotical analysis shows that the flow in this additional region is described by the boundary layer equations for compressible gas. The initial solution for this region satisfies the so called "compensation" condition

$$\psi \to \frac{\eta^2}{2} + p, (\eta \to \infty) \tag{7}$$

In [7-8] it was supposed and elaborated numerical method of direct numerical integration in both regions for $x \geq 0$. To provide analytical presentation of solution for $x > 0$ new coordinates are introduced in the additional region

$$\xi = x^{\frac{1}{3}}, n = \eta/x^{\frac{1}{3}} \tag{8}$$

and corresponding substitution of stream function are done. Then calculations in two subregions are performed simultaneously. The boundary between two subregions is determined by the following correlation

$$\eta_e = n_e \xi, n_e = N \Delta n \tag{9}$$

where e index is attributed to the values on the boundary between two subregions, N is a number of intervals along n axis. Values of $N, \Delta \eta$ and $\Delta \xi$ are selected thus for each ξ value boundary between subregions precisely corresponds to the meshes of wall layer

$$\Delta \eta = N \Delta n \Delta \xi$$

System of equations keeps its form for all meshes lying above the boundary (9) and takes the modified form (8) for meshes lying under that boundary. Immediately on boundary condition of corresponding functions equality is used. System of algebraic equations is solved then by the method of band matrix conversion, from the surface to the external boundary. For $\xi = 1$ transition to original variables is realized and calculations are made without additional region distinguishing. For each set of calculations boundary condition (5) is satisfied by means of an initial value selection.

Numerical calculations were performed for next values of determining parameters

$$\left. \begin{array}{rcl} F &=& \frac{1}{\sqrt{2}}, \quad \alpha = 0 \\ F &\to& \infty, \quad \alpha = 0 \\ F &=& 0, \quad \alpha = 1 \end{array} \right\} \quad Pr = 0.7, Sc = 0.5, c_0 = 1$$

For constant surface temperature (cases 1 and 2) atoms concentration diminishing near catalytic surface causes gas density rise and streamlines displacement to the surface – therefore expansion flow is realized. Surface temperature increasing in the case

3 causes the gas density decreasing and streamlines displacement from the surface, therefore compression flow is realized. Figure 1 shows the distribution of induced pressure p. On the figure 2 skin friction τ distributions are presented via longitudinal coordinate. For expansion flows (cases 1 and 2) skin friction before point of discontinuity increases. It was found also that for compression flow skin friction before the point of discontinuity decreases. For the case 1 τ distribution is continuous, for another two cases is discontinuous. It may be seen that the maximum possible catalytic activity increase leads to the growth in the skin friction upstream of the point of discontinuity by a factor of about 1.5. The temperature increase by a factor of 2 decreases τ to 0.4. It was obtained [9] that temperature ratio 3.87 leads to the local incipient separation of boundary layer.

In figures 3 and 4 distributions of normal gradients of temperature T_i' and mass atoms concentration c_i' on the surface ($i = 1$ solid lines, $i = 2$ dashed lines) are presented. It is seen that a sudden increase in the catalytic activity on the surface in cases 1 and 2 only insignificantly alters T_2' value distributions. At the same time surface temperature variations in the case 3 leads to a very large change of heat transfer coefficient. On the contrary, changes of catalycity degree in the cases 1 and 2 cause a very large variations of mass atoms concentration normal gradients. The surface temperature increase only insignificantly alters this characteristic in the case 3.

Acknowledgements:

This work was sponsored by the Russian Fundamental Research Foundation (grant numbers 93 − 01 − 16399, 94 − 01 − 01967, 94 − 01 − 01992).

References

[1] D.A. Stewart, J.V. Rakich and M.J. Lanfranco. Catalytic surface effects experiment on Space Shuttle. AIAA Pap., No. 1143, New York (1981).

[2] P.M. Chung, S.W. Liu and H. Mirels. Effect of discontinuity of surface catalycity on boundary layer flow of dissociated gas. Int. J. Heat Mass Transfer, v. 6, No. 3 (1963).

[3] E.A. Gershbein, V.Yu. Kazakov and V.S. Shchelin. Hypersonic viscous shock layer on a surface with a sudden change in catalytic activity. Teplofiz. Vys. Temp., v. 23, No. 5 (1985).

[4] M. Van Dyke. Perturbation Methods in Fluid Mechanics. Academic Press, New York (1964).

[5] V.V. Bogolepov and V.Ya. Neiland. Investigation of local disturbances of supersonic flows, in: Aeromechanika, Nauka, Moscow (1976).

[6] V.Ya. Neiland. Asymptotic problems of the theory of viscous supersonic flows, Tr. TsAGI [in Russian], No. 1529 (1974).

[7] P.G. Daniels. Numerical and asymptotic solutions for the supersonic flow near the trailing edge of a flat plate, Q. J. Mech. Appl. Math., v. 27, pt. 2 (1974).

[8] I.I. Lipatov and V.Ya. Neiland. Effect of a sudden change in the motion of a plate surface on flow in the laminar boundary layer in a supersonic flow, Uch. Zap. TsAGI [in Russian], v. 13, No. 5 (1982).

[9] L.A. Sokolov. Toward an asymptotic theory of plane flows of a laminar boundary layer with a temperature discontinuity on the body, Tr. TsAGI [in Russian], No. 1650 (1975).

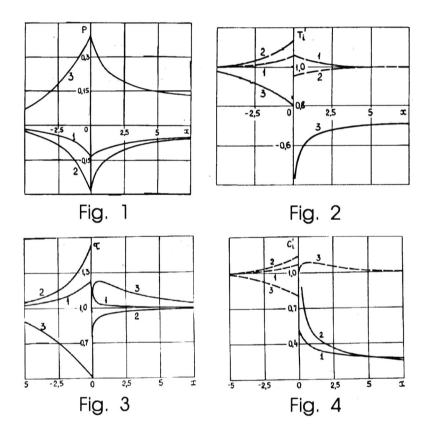

Fig. 1 Fig. 2

Fig. 3 Fig. 4

Computation of Air and Fuel Droplet Flows in S.I. Engine Manifolds

P.Y.P. Chen, M. Behnia and B.E. Milton

Schools of Mechanical & Manufacturing Engineering
The University of New South Wales
Sydney 2052 NSW

Abstract : The flow of air and fuel mixture similar to that in the induction systems of an SI engine has been studied. A standard, general purpose single phase CFD code was used with some modifications to account for momentum, mass and heat transfers between the air stream and fuel droplets. Based on a Lagrangian approach, the movement from an assumed spray distribution model of fuel droplets were individually tracked. Models using different assumed coupling between the two phases have been investigated. Under typical engine operating conditions, it was found that a fully coupled treatment for momentum is not required. For estimation of fuel evaporation, a computing time saving one-dimensional treatment was found to be adequate. Fuel film formation on various surfaces of a duct and butterfly valve simulating an engine manifold is reported.

1 Introduction

In a spark ignition (SI) engine, the precise control of the air and fuel mixture in the induction system is important in obtaining good engine performance. Under normal operating conditions, a spray of fuel is injected into the air stream either at a throttle body or now more commonly, in the inlet ports. The air-borne droplets, while being transported along the induction system, may also break-up, re-combine and evaporate. Some droplets will deposit onto duct walls to form a film, which will in turn interact further with the gas stream flow through evaporation, re-entrainment and momentum exchange.

In previous work in the current study program [1,2], it has been shown that for the base air flow without fuel, single phase CFD codes can be used to compute the velocity field with good accuracy even in geometrically complex regions such as those around the butterfly flow control valve when it is in the near closed position. It has also been demonstrated [3] that a single phase CFD code can be modified to include droplet flows if only the momentum exchange between the air stream and droplets is considered and the droplets remain intact. This approach could be extended to include any other phenomena associated with droplet flows provided that an adequate model is available. In the present research, numerical simulations have been carried out to include heat and mass transfers in addition to momentum transfer for a number of flow situations. Such computations allow studies to be undertaken of the many important phenomena such as fuel film formation on a duct wall and the effects on it of droplet sizes, velocities and injection spray characteristics.

2 Simulation of Droplet Flow

The most important addition to the single phase CFD code is the development of a tracking routine for the droplets. Based on a Lagrangian approach, when a model is considered for a single projectile without particle-particle interaction the equations of motion, based simply on bouyancy, pressure gradient and drag forces are:

$$\frac{d\tilde{V}_d}{dt} = \left(\frac{\rho_g}{\rho_d} - 1\right) g\tilde{k} - \frac{\nabla p}{\rho_d} + C_d \frac{\rho_g}{\rho_d} \frac{\left(\tilde{V}_g - \tilde{V}_d\right)\left|\tilde{V}_g - \tilde{V}_d\right|}{D_d} \quad (1) \quad \text{and} \quad \frac{d\tilde{x}}{dt} = \tilde{V}_d \quad (2)$$

Here the drag coefficient C_d is assumed to be a known function of Reynolds number, $\tilde{V}$ is velocity, ρ density, p pressure, D droplet diameter and subscripts d and g represent droplet and gas phases, respectively. Unknown factors such as added mass due to entrainment of either air vapor or neighbouring particles in a dense spray region have been for the present, ignored.

It is assumed that as the droplet spray is injected into the air stream, spray ligaments break up into droplets which consist of spherical particles of varying sizes and speeds. Each representative fuel droplet of a given size and speed is then tracked separately to produce an ensemble of flow histories from which the droplet flow field may be constructed according to the assumed size and speed compositions of the spray. This is a basic model but has the potential for modification to include further phenomena at a later stage.

3 Treatment of Momentum Coupling

If the air flow is assumed to only affect the momentum of the fuel droplets and not vice versa, the treatment is termed 'one-way' coupled, and the original CFD code can be used without modification to compute the gas stream velocity V_g in (1). However, if momentum exchange (or indeed, mass exchange) occurs as significant quantities between the phases in both directions, the so called 'two-way' coupled treatment must be used to provide an accurate gas stream assessment. An appropriate source term then needs to be added to the momentum equation governing the air flow. The source strength may be determined by the 'Particle-Source-In-Cell' (PSI-CELL) technique [4]. Thus, for a particle computational cell traversed by $\dot{n}$ number of droplets per unit time, the momentum source for the gas flow is:

$$\tilde{S}^M = \dot{n}\, m_d \left[\tilde{V}_{\text{out}} - \tilde{V}_{\text{in}}\right] \quad (3)$$

where m_d is the mass of each droplet, $\tilde{V}_{in}$ and $\tilde{V}_{out}$ refer respectively to the droplet velocities entering and leaving the computational cell. Since $\tilde{S}^M$ affects $\tilde{V}_g$ directly through the gas phase momentum equation, and also $\tilde{V}_d$ indirectly through (1)-(2), an iterative procedure is needed. For steady flows, a convenient convergence criteria is such that the source values for all the computational cells do not vary significantly between successive iterations.

4 Treatment of Heat and Mass Transfers

Since in the present application of the technique, the fuel components are quite volatile, it is necessary to consider evaporation while they are being transported through the induction system. If a one-way treatment is chosen and only the evaporation from the air-borne droplet is considered, the rate of evaporation for the jth component of a fuel droplet of diameter D_d is given by:

$$\dot{m}_{e(j)} = \pi D_d^2 \beta_{d(j)} \left(p_{s(j)} - p_{v(j)} \right) \tag{4}$$

where β_d is the mass transfer coefficient, p_s and p_v are the saturation vapour pressure and partial vapour pressure of fuel component j. The rate of temperature change for the droplet is given by:

$$\frac{dT_d}{dt} = \frac{\pi D_d^2 h_d (T_d - T_g) + \sum_j H_{fg(j)} \frac{dm_{d(j)}}{dt}}{\sum_j m_{d(j)} C_{p(j)}} \tag{5}$$

where h_d is the heat transfer coefficient, T_d the droplet temperature, T_g the airstream temperature, $H_{fg(j)}$ the latent heat of vaporisation of fuel component j, $C_{p(j)}$ the specific heat of component j. The heat and mass transfer coefficients used here follow the correlations given by Ranz and Marshall [5]. If a two-way treatment is to be employed, problems could arise on the method of computing p_v in (4) and T_g in (5), while the droplets are being tracked. As the gas phase is now a multi-component mixture, mass and heat transport effects for the evaporated fuel components in the gas phase must be considered. This is not easy if a single phase CFD mode is used. Based on an overall balance, an approach using constant equilibrium values may be used. For mass transfer, if air is entering without containing any fuel vapour and leaving with a constant equilibrium mole fraction for say the jth fuel component, $n(j)$, then:

$$n(j) = \frac{\sum_k \Delta \dot{m}_e(j)/M(j)}{\dot{n}_g + \sum_j \sum_k \Delta \dot{m}_e(j)/M(j)} \tag{6}$$

where $\sum_k \Delta \dot{m}_e(j)$ is the overall evaporation rate, $M(j)$ the molecular mass and $\dot{n}_g$ the molar air flow rate. The summation over k items implies that droplets of all sizes and initial conditions at the injection point are to be considered. Using $n(j)$, the partial pressure $p_v(j)$ for the equilibrium mixture can then be calculated. However, since $\sum_k \Delta \dot{m}_e(j)$ is related to $p_v(j)$ through (4), an iterative approach is again required until convergence is obtained.

The extensive computation associated with the above iterative approach may be eliminated if a pseudo one-dimensional treatment [6] is used, where p_v is adjusted continuously in the mean flow direction while the droplets are being tracked. Other variables are assumed to change also only in the direction of the droplet flow. For flows involving re-circulation, such as in the region behind the butterfly control valve, the one-dimensional treatment has some obvious shortcomings. However, in many cases, the errors so introduced are minimal and can be over-looked.

5 Numerical Procedures and Results

The CFD code used here is the single phase version of the multi-dimensional flow code, FLOW3D. It solves the full Navier-Stokes energy and continuity equations for either compressible or incompressible flow, using a finite volume approach together with body-fitted co-ordinates. While the code allows choices of turbulence models, that used here is the standard $k - \epsilon$ model. The code allows user-defined sources to be added to the appropriate governing equations, whenever required.

If a coupled flow is assumed, the source strength for each coupled variable needs to be computed from the droplet flow histories at each iterative cycle. The implementation of the tracking routines is contained in a specially developed computer package called TRACK[7].

A geometry typical of the problem considered here is shown in Fig. 1. It is a two-dimensional simulation of a SI engine manifold with $D = 0.06$ m. This simulates an experimental set-up rather that the induction system of an actual engine and the spray is injected downwards across the duct to avoid disturbance to the air flow by a central injector. The fuel spray used is based on a typical fuel spray described by Ingebo [8] and discretised by Chen et al [3].

It is also assumed from tests carried out in the present research program that spray has a maximum angle of $\pm 25°$ from the normal. To discretise the spray, an angular step of $2°$ and a velocity step of 1 m/s were used.

Physical properties for air provided by FLOW3D. For the fuel all computations are based on components of a typical Australian fuel analysed and used previously in experiments and computations [6]. Assuming that an SI engine is operating under stoichiometric conditions, the maximum mass loading ratio of the fuel droplets to air is 1 to 15. With such a lean mixture, it has been shown [9] that the base air flow field is unlikely to vary by more than 1% if full momentum coupling is replaced by one-way coupling. For evaporation studies, it is found that, although the fully coupled approach based on the equilibrium assumption requires only 2 or 3 iterations to reach convergence, the actual computer time usage for a given run is 3 times longer than that used by the one-dimensional approach. The results obtained, however, are given within 1% of each other. It was therefore decided to use the latter approach with no momentum coupling for all the case studies here.

For a straight duct with a butterfly control valve, a typical velocity field for the base air flow is shown in Fig.2. For the fuel film distribution on different duct surfaces, results without and with evaporation are shown in fig. 3, where the differences between the two models are clearly shown. Fig.4 shows the same results for a duct with a bend.

6 Conclusions

In a situation representing air and fuel droplet flows in an experimental situation of an SI engine manifold, due to the fact that fuel is injected at an angle to the air flow, most of the droplets end up on the duct walls where they form a fuel film. The remaining small portion (about 20%) of air-brone droplets has a residence time of only a fraction of a second. This, together with the fact that the mass loading ratio is quite small, results in the possible use of the one-dimensional evaporation model.

For the purpose of studying the behaviour of fuel droplets, the use of a single phase CFD code together with the appropriate tracking routines has been found to be feasible. The momentum, mass and heat transfer accounting seems to be adequate even when some simplified models such as a one-way coupled treatment for momentum and a one-dimensional treatment for evaporation are used.

References

1. Behnia, M., Jones, I.P. and Milton, B.E.: "Simulation of Flow Pass on Engine Intake Manifold Throttle Valve", Proc. 4th Asian Congress of Fluid Mechanics, Hong Kong, pp.F105-F108, 1989.

2. Kim, C.H., Behnia, M., Milton, B.e. and Chen, P.Y.P., "Computation of the Flow Field around a Butterfly Valve", Trans. Korea Soc. Auto. Eng. Vol. 14, pp.47-53, 1991.

3. Chen, P.Y.P., Milton, B.E. and Behnia, M., "A Numerical Examination of Fuel Film Formulation in Manifolds", Transport Phenomena in Heat and Mass Transfer, Elsevier, Vol. 2, pp.1167-1178, 1992.

4. Crowe, C.T., Sharma, M.P. and Stock, D.E., "The Particle-Source-In-Cell (PSI-CELL) Model for Gas-Droplet Flows", Trans. ASME, J. Fluid Emgr., Vol.99, pp.325-332, 1977.

5. Ranz, W.E. and Marshall, W.R., "Evaporation from Drops", Chem. eng. Prog., Vol.48, pp.141-146, 1952.

6. Milton, B.E. and Behnia, M., "A Numerical Study of the Interchanging Vapour, Droplet and Film Flows in IC Engine Manifolds", Heat and Mass Transfer in Engines, Hemisphere Publications, pp.245-258, 1989.

7. Blakeley, M.R., "Computer Program for the Numerical Prediction of Fuel Particle Tracks in Duct and Bends", UNSW Report 1993/FMT/1, School of Mechanical & manufacturing Engineering, The University of New South Wales, 1993.

8. Ingebo, R.D., "Vaporisation Rates and Drag Coefficients for Iso-octane Spray in a Turbulent Air Stream", NASA Tech. Note 3265, Washington DC, 1954.

9. Kim, C.H., Chen, P.Y.P., Behnia, M. and Milton, B.E., "Effect of Injected Liquid Fuel Droplets on the Airflow Field in SI Engine Manifolds", Computational Mechanics - from Concepts to Computations, Balkema Publications, Vol. 1. pp.561-566, 1993.

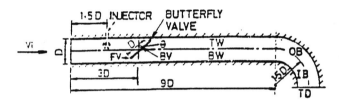

Fig. 1. Geometry of a combined bend and valve

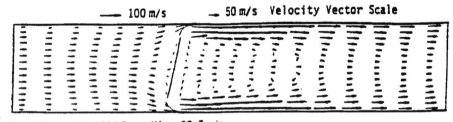

TA = 10°, ΔP = 10 kPa, Vi = 32.7 m/s

Fig. 2. FLOW3D base air flow computation

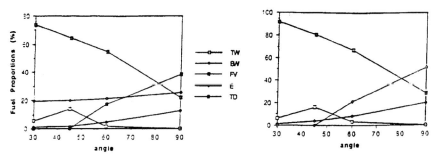

Fig. 3. Fuel film formation of different surfaces (straight duct)

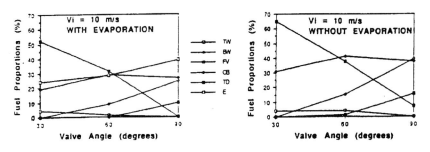

Fig. 4. Fuel film formation of different surface (duct with bend)

TW	:	TOP WALL	E	:	EVAPORATED
BW	:	BOTTOM WALL	TD	:	THROUGH DUCT
FV	:	FRONT OF VALVE	Vi	:	INLET AIR VELOCITY

Validation of TVD-Schemes for Unsteady Turbomachinery Flow

K. Engel, F. Eulitz[1], M. Faden, S. Pokorny[2]

[1]DLR, Institute for Propulsion Technology,
D-51147 Cologne, Germany
[2]DLR, Department for High Performance Computing,
D-51147 Cologne, Germany

Abstract : Well documented explicit TVD schemes are assessed in terms of time accuracy, propagation properties and computational work load with emphasis on their application to unsteady turbomachinery flow computations. Non-reflecting boundary conditions are in use for the flow entry and exit boundaries. As inviscid test cases serve the flow through a flat plate cascade excited by oscillating back pressure and the unsteady flow past an oscillating NACA-0012 profile. The shock induced turbulent boundary layer separation on an arc profile in a channel is studied qualitatively.

1 Introduction

For the computation of unsteady compressible turbomachinery flow, an interactive flow simulation system is in development. Presently, the two-dimensional Reynolds-averaged, unsteady Navier-Stokes equations for a compressible perfect gas are solved. Turbulence effects are accounted for by means of either the algebraic Baldwin-Lomax model (1978) or the one-equation transport model according to Spalart and Allmaras (1992). Implementation of the latter is still in progress. Quasi three-dimensional effects are included through the specification of a streamtube thickness in the third dimension. For an accurate capturing of the compressible flow physics, various time explicit total-variation-diminishing schemes (TVD), have been implemented. To cope with the computational work load and the data flood, the code has been implemented on a massively parallel computer (Pokorny 1991). Parallelization of the code is achieved through domain decomposition. The full scalability of the system (Faden 1993) allows the hardware size to be matched up with the flow problem size with undiminished parallel efficiency. The current state of the flow solver and an application to the unsteady transonic compressor stage interaction is documented in a recent work by Engel et al. (1994).

2 Numerical Methods Investigated

The behaviour of the following well documented flux-difference splitting TVD method is tested, namely

(i) the Runge-Kutta (four step), MUSCL [1] scheme with upwind and upwind biased extrapolation according to Yee (1989),

(ii) the Fractional Step, non-MUSCL, fully upwind scheme by Harten (1984) and Yee (1989),

(iii) the Predictor-Corrector, non-MUSCL, central scheme according to Roe, Davis (1985) and Yee (1989).

For all schemes, the unknown flow values at the cell interfaces are determined using Roe's approximate Rieman solver (Roe 1981). According to the authors of the above original works, the schemes are of second order accuracy in space and time. Two formulations of non-reflecting boundary conditions are used for the treatment of the inflow and outflow boundaries (Engel et al. 1992). One is based on the nonlinear approach originally developed by Hedstrom (1979) for the one-dimensional case and later extended by Thompson (1986) for the multiple dimensions. The other is based on the linear approach by Giles (1988) and generalized by the authors for curvilinear coordinates (Engel et al. 1994).

3 Validation

The order of the following three test cases represents a step by step increase of flow complexity. Quantitative comparisons are presented for the first two test cases. The study of the final test case is done in a more qualitative manner.

3.1 Flat Plate Cascade

The first test case is the inviscid flow through a flat plate cascade excited by an unsteady small-amplitude perturbation of the uniform, steady flow. For sufficiently small amplitudes of the perturbations, nonlinear effects are neglegible and a comparison of the numerical results to a linear, inviscid theory as described in Smith (1979) is justified. Similar to thin airfoil theory, the linear analysis leads to an integral equation which is solved computationally according to Whitehead (1987). The linear analysis has been previously applied by Giles (1991) for the validation of his unsteady flow solver. Here, the unsteady response of the cascade flow to an oscillating back pressure is considered. The flat plate cascade has a pitch to chord ratio of 1/2 and a stagger angle of 30 degrees.The back pressure is varied sinusoidally with a reduced frequency of 1.068. The mean flow condition of the cascade is given by a Mach number of 0.7. The mentioned parameters were taken from Giles (1991) to allow for an additional code-code comparison.

Convergence of the numerical solution to the periodic state was generally obtained after computation of two to four perturbation periods. A representative residual history plot, as obtained by Predictor-Corrector time stepping, is given in figure 1.

---

[1] MUSCL stands for monotonic upstream schemes for conservation laws.

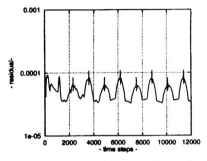

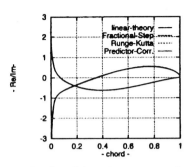

Fig. 1. Residual history for excited flat plate cascade flow

Fig. 2. Real and imaginary part of complex pressure jump on flat plate

After calculating the first Fourier harmonic of the pressure jump across the flat plate, the numerical results can be compared to the linear results. Figure 2 shows the real and imaginary parts of the complex pressure jump on the flat plate as a result of the numerical methods and linear theory. The plot can be directly compared to Giles' result (1991). The corresponding amplitude and phase distributions of the complex pressure jump are compared in figures 3, 4.

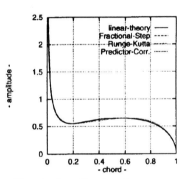

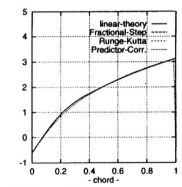

Fig. 3. Amplitude of pressure jump on flat plate

Fig. 4. Phase of pressure jump on flat plate

The curves of linear theory, the Fractional-Step, and Runge-Kutta method practically overlap. There is not an as perfect agreement for the Predictor-Corrector method. Also, the phase distribution has a slightly different shape. This means that the pressure disturbance, imposed at the exit boundary, propagates upstream with a slightly modified velocity. A possible explanation for the error may be the spatial unsymmetry of the predictor and corrector step. In contrast, each Runge-Kutta step shows full spatial symmetry.

The computations took about twenty minutes each on a SUN work station for a time period. Relative to the CPU time consumed by the Runge-Kutta scheme for the computation of a time period, the Fractional-Step scheme took about 90% and the Predictor-Corrector scheme 65%.

3.2 Oscillating Profile

Next, the flow past a NACA-0012 profile oscillating about the quarter-chord point is considered to evaluate the numerical schemes in the nonlinear regime. Depending on the incidence, shock waves develop and vanish in turn on either side of the airfoil. The numerical results are compared to measurements by Landon (1977) for a freestream Machnumber of 0.75, a frequency of 62.5 Hertz and an angle of attack range from -2.49 to $+2.52$ degrees. The unsteady simulation is started with a steady-state solution for the profile at rest (at the mean angle of attack of 0.016 degrees) and careful acceleration of the rotation speed. Due to the rather low reduced frequency value of 0.08 the numerical solution becomes time-periodic after the first to third period. Comparison to the experimental data is provided in terms of the pressure coefficient C_p. In figures 5 and 6, numerical and experimental chordwise C_p-distributions are shown for two selected angles of attack (α). The experimental data are provided for orientation.

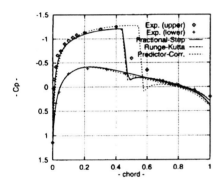
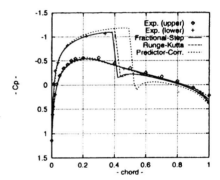

Fig. 5. C_p-distribution for $\alpha = 2.01$ deg on oscillating NACA-0012 airfoil

Fig. 6. C_p-distribution for $\alpha = -0.54$ deg on oscillating NACA-0012 airfoil

Inviscid, numerical results should differ from the viscous reality. As expected, the second order TVD schemes resolve the shock within three mesh cells (Yee 1989). Approximately, the well defined computed shock location lies in a midway position of the measured pressure ramps. Away from the shock, there is good agreement of computations and experiment. In figure 5, however, the experimental data indicate substantial separation of the boundary layer downstream of the shock which cannot be captured by an inviscid computation.

3.3 Shock Induced Flow Separation

In the final test case, the TVD-schemes which incorporate inviscid physics (Roe 1986) are applied for the calculation of the transonic high-Reynoldnumber flow past a double circular arc profile in a channel. For this case, the full Reynolds-averaged Navier-Stokes equations are solved using the one-equation turbulence model of Spalart and Allmaras (1992).

In a first step, it is of interest to check the behaviour of TVD-schemes in the strongly viscous flow regions, e.g. in separated flow zones, boundary layers, and wakes. Kroll (1994) reported principal deficiencies of TVD schemes in high-Reynoldnumber applications. Similarly, the authors observed spurious oscillations in iso-pressure lines near the solid boundary. The difficulty is mainly due to the TVD-limiting in wall-normal direction. One way to suppress the spurious oscillations for the MUSCL scheme is to use the limited pressure gradient for the MUSCL-extrapolation of all variables (von Lavante 1990).

In a steady computation on a 5-block H-/O-grid topology (with 310*62 nodes) the Mach number at the entry boundary is 0.7, the non-dimensional back pressure is 0.67 and the Reynoldsnumber is 500,000. In figure 7, the computed iso-density lines of the flow field as a result of the modified Runge-Kutta time stepping is shown. Figure 8 shows a detail of the velocity vectors in the region of the boundary layer separation.

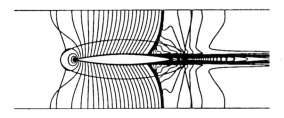

Fig. 7. Local Mach number distribution of transonic flow past dca profile in a channel

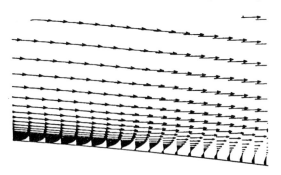

Fig. 8. Turbulent flow separation on dca profile in a channel

The advantage of the present case is that the converged steady-state solution can be used as an initial solution for the simulation of self-excited shock oscillation. According to Yamamota and Tanida (1989), the unsteadiness arises for a certain pessure range which is set at the channel exit. The unsteady simulation will be presented in a future work.

4 Conclusion

Different explicit TVD schemes were tested and validated with emphasis on time accuracy by comparison against theoretical and experimental data. Very good agreement was found for the linear, inviscid test cases. A small error has been found for the Predictor-Corrector scheme. For the nonlinear test case of the oscillating NACA-0012 airfoil, the error is even more pronounced. The Runge-Kutta and Fractional-Step method produced results of equal accuracy. Of the three schemes tested, the Predictor-Corrector method was the most economic in computation time, the Runge-Kutta method was the most expensive. Further, it has been shown that by a simple modification the MUSCL scheme can be successfully applied to complex turbulent flows. The combination of the TVD schemes with a rather new one-equation turbulence model is promising and will be further pursued for unsteady simulations.

References

[1] Baldwin, B., Lomax, H. (1978): "Thin Layer Approximation and Algebraic Model for Separated Turbulent Flow", AIAA paper 78-0257

[2] Engel, K., Faden, M., Pokorny, S. (1992): "Implementation of Non-Reflecting Boundary Conditions into an Unsteady Flow Simulation System", 3rd Int. Conf. Num. Methods Fluid Dynamics

[3] Engel, K., Eulitz, F., Faden, M., Pokorny, S. (1994): "Numerical Investigation of the Rotor-Stator Interaction in a Transonic Compressor Stage", AIAA paper 94-2834

[4] Faden, M., Pokorny, S., Engel, K.: (1993): "Unsteady Flow Simulation on a Parallel Computer", 11th AIAA Comp. Fluid Dynamics Conf., Orlando, U.S.A.

[5] Giles, M., B.: (1988): "Non-Reflecting Boundary Conditions for the Euler Equations", CFDL-TR-88-1

[6] Giles, M., B.: (1991): "Validation of a Numerical Method for Unsteady Flow Calculations", ASME Paper No. 91-GT-271

[7] Harten, A. (1984): "On a Class of High-Resolution Total-Variation-Stable Finite-Difference Schemes", in SIAM J. Num. Anal., Vol.21, pp. 1–13

[8] Kroll, N., Radespiel, R., Rossow, C.-C. (1994):"Accurate and Efficient Flow Solver for 3D Applications on Structured Meshes", in VKI-Lecture Notes in Comp. Fluid Dynamics

[9] Landon, R., H. (1977): "Oscillatory and Transient Pitching", in AGARD Report No. 702, *Compendium of Unsteady Aerodynamics Measurements*

[10] von Lavante, E., Warda, H., A. (1990): "Simple Explicit Upwind Schemes for Solving Compressible Flows", Proc. of 8th GAMM-Conf., ed. by P. Wesseling

[11] Pokorny, S., Faden, M., Engel, K. (1991): "An Integrated Flow Simulation System on a Parallel Computer. Part I: The Basic Idea, Part II: The Flow Solver", 7th Int. Conf. Numerical Methods in Laminar and Turbulent Flow, Stanford, U.S.A.

[12] Roe, P., L. (1981): "Approximative Rieman Solvers, Parameter Vectors and Difference Schemes", J. Comp. Physics, Vol. 43, pp. 357–372

[13] Roe, P., L. (1986): "Characteristic-based Schemes for the Euler Equations", *Annual Review of Fluid Mechanics*, pp. 337, Annual Reviews Inc.

[14] Smith, S., N. (1979): "Discrete Frequency Sound Generation in Axial Flow Turbomachines", Report CUED/A-Turbo/TR 29, University of Cambridge

[15] Spalart, P., R., Allmaras, S., R., (1992): "A One-Equation Turbulence Model for Aerodynamic Flows", AIAA-92-0439

[16] Whitehead, D., S. (1987): "Classic Two-Dimensional Methods in Aeroelasticity in Axial Flow Turbomachines",ed. by M. Platzer and F. O. Carta, AGARD-AG-298, Vol. 1

[17] Yee, H., C. (1989): "A Class of High-Resolution Explicit and Implicit Shock-Capturing Methods", in VKI-Lecture Notes in Comp. Fluid Dynamics

Numerical Simulation of Steady and Unsteady Flows Through Plane Cascades

J. Fořt, M. Huněk, K. Kozel, J. Lain, M. Šejna, M. Vavřincová

Dep. of Technical Mathematics, Fac. of Mechanical Eng.,
CTU Prague, Czech Republic[1]

Abstract : This paper of a few co-authors presents some works of the group of the Department of Technical Mathematics, Faculty of Mechanical Eng., TU Prague, which deals with numerical methods in fluid dynamics. We present numerical methods for a solution of different physical and mathematical models of flow through plane cascades. We use the Mac Cormack's scheme, Ron - Ho - Ni's scheme and Runge - Kutta schemes on H - type structured grid and upwind schemes on an unstructured triangular grid. This methods are used for simulation of steady or unsteady inviscid flow and for simulation of viscous laminar flow. We deal with comparison of different methods mutually and with experimental data and with comparison of different physical and mathematical models of flow used for numerical simulation.

1 Mathematical Models

The basic mathematical model is the 2D system of Euler or laminar Navier - Stokes equations in the cartesian coordinates

$$\mathbf{W}_t + \mathbf{F}_x + \mathbf{G}_y = \frac{1}{Re}(\mathbf{R}_x + \mathbf{S}_y), \tag{1}$$

where $\mathbf{W} = (\rho, \rho u, \rho v, e)^T$, $\mathbf{F} = (\rho u, \rho u^2 + p, \rho uv, (e+p)u)^T$, $\mathbf{G} = (\rho v, \rho uv, \rho v^2 + p, (e+p)v)^T$, $\mathbf{R} = (0, \tau_{xx}, \tau_{xy}, u\tau_{xx} + v\tau_{xy} + kT_x)^T$, $\mathbf{S} = (0, \tau_{xy}, \tau_{yy}, u\tau_{xy} + v\tau_{yy} + kT_y)^T$, $p = (\gamma - 1)(e - \rho(u^2 + v^2)/2)$, $\tau_{xx} = \mu(4/3 u_x - 2/3 v_y)$, $\tau_{yy} = \mu(4/3 v_y - 2/3 u_x)$, $\tau_{xy} = \mu(u_y + v_x)$. The right hand side of (1) is equal to zero for the Euler equations. The AVDR (Axial Velocity Density Ratio) influence is simulated by modified system of Euler equations with the continuity equation in form

$$(h\rho u)_x + (h\rho v)_y = 0 \tag{2}$$

where $h = h(x,y)$ is the suitable distribution function of AVDR in a blade channel. The following model of inviscid flow through cascade lying on an arbitrary rotationally symmetric surface (S_1) of variable thickness b has form

$$\mathbf{W}_t + \frac{1}{r}(r\mathbf{F})_m + \frac{1}{r}\mathbf{G}_\varphi = h(W) \tag{3}$$

---
[1] The research was supported by the grants 0461/93, 0455/93, 2127/93 of the Grant Agency of the Czech Republic.

where m is the shape of S_1 surface in the meridian plane, φ the cylindrical coordinate, r the distance from the axis. The nonzero terms on right hand side of (3) describe the influence of the variable radius r, the variable thickness b, the rotation of blade with constant angular velocity ω - see [5] for more details.

We prescribe the non-permeability (Euler equations) or the non-slip conditions (Navier - Stokes equations) on the walls, the periodicity conditions, the uniform (in space variables) inlet conditions (three or four components of $\mathbf{W}$) and one outlet condition (the value of the pressure).

2 Numerical Methods

The Lax - Wendroff type schemes are based on the numerical approximation of the integral form of (1). The values of $\mathbf{W}^{n+1}$ in time step $(n+1)$ are computed as

$$\mathbf{W}_{i,j}^{n+1} = \mathbf{W}_{i,j}^n - \frac{1}{\mu_{i,j}} \oint_{\partial D_{i,j}} (\tilde{\mathbf{F}} - \frac{1}{Re}\tilde{\mathbf{R}})dy - (\tilde{\mathbf{G}} - \frac{1}{Re}\tilde{\mathbf{S}})dx, \tag{4}$$

where $\mu_{i,j}$ is the area od $D_{i,j}$, $\tilde{\mathbf{F}}$, $\tilde{\mathbf{G}}$ are the numerical approximations of fluxes $\mathbf{F}$ and $\mathbf{G}$ respectively and $\tilde{\mathbf{R}}$, $\tilde{\mathbf{S}}$ the numerical approximations of viscous terms $\mathbf{R}$, $\mathbf{S}$. Two schemes of this type are used. The method $\mathbf{M_1}$ is based on the cell centered finite volume discretisation and Mac Cormack explicit predictor corrector scheme (older form or TVD form):

1. predictor

$$\mathbf{W}_{i,j}^{n+1/2} = \mathbf{W}_{i,j}^n - \frac{\Delta t}{\mu_{i,j}} \left(\sum_{k=1}^{4} \mathbf{F}_k^n \Delta y_k - \mathbf{G}_k^n \Delta x_k \right) \tag{5}$$

$\Delta y_k = y_{k+1} - y_k, \; y_5 = y_1 \quad \Delta x_k = x_{k+1} - x_k, \; x_5 = x_1 \; k = 1, \cdots, 4$
$\mathbf{F}_1 = \mathbf{F}_{i+1,j}, \mathbf{F}_2 = \mathbf{F}_{i,j+1}, \mathbf{F}_3 = \mathbf{F}_4 = \mathbf{F}_{i,j}, \qquad$ and the similar is true for $\mathbf{G}$.

2. corrector

$$\widetilde{\mathbf{W}}_{i,j}^{n+1} = \frac{1}{2} \left(\mathbf{W}_{i,j}^n + \mathbf{W}_{i,j}^{n+1/2} - \frac{1}{2}\frac{\Delta t}{\mu_{i,j}} \left(\sum_{k=1}^{4} \mathbf{F}_k^n \Delta y_k - \mathbf{G}_k^n \Delta x_k \right) \right) \tag{6}$$

$\mathbf{F}_1 = \mathbf{F}_2 = \mathbf{F}_{i,j}, \mathbf{F}_3 = \mathbf{F}_{i-1,j}, \mathbf{F}_4 = \mathbf{F}_{i,j-1}, \qquad$ and the similar is true for $\mathbf{G}$.

3. implementation of artificial viscosity terms

$$\mathbf{W}_{i,j}^{n+1} = \widetilde{\mathbf{W}}_{i,j}^{n+1} + D\mathbf{W}_{i,j}^n \quad D\mathbf{W}_{i,j}^n = D_x \mathbf{W}_{i,j}^n + D_y \mathbf{W}_{i,j}^n \tag{7}$$

$D_x W_{ij}^n$ is considered in older form [3] or TVD form [9]. We denote by $\mu_{i,j}$ the measure of finite volume $D_{i,j}$, with $\mathbf{W}_{i,j}$ in the center, and by index k the vertices of $D_{i,j}$ so, that $k = 1$ denotes SE corner, and boundary ∂D is assumed positively oriented. This method is used for the steady inviscid computations of flow through turbine and compressor cascades (together with multi-grid FAS algorithm), for the solution of the model (2), for computations of unsteady flow through cascades and next for the solutions of viscous steady flow (1). The method M_2 is based on the one stage Ron - Ho - Ni's scheme and cell vertex finite volume concept. This method is developed only for the inviscid computation.

1. **computation of fluxes in finite volume** (i, j)

$$DW_{i,j} = (\Delta t)/(2\mu_{i,j}) \sum_{k=1}^{4} ((F_k^n + F_{k+1}^n)\Delta y_k - (G_k^n + G_{k+1}^n)\Delta x_k) \quad (8)$$

2. **computation of elementary increments in vertices of finite volume** (i, j)

$$\begin{aligned}
\delta W_1 &= -DW_{i,j} - f + g + D_{L_1}, & \delta W_2 &= -DW_{i,j} - f - g + D_{L_2}, \\
\delta W_3 &= -DW_{i,j} + f - g + D_{L_3}, & \delta W_4 &= -DW_{i,j} + f + g + D_{L_4},
\end{aligned} \quad (9)$$

where f and g are the redistribution terms of second order and D_L the artificial viscosity terms.

3. **computation of new values** $W_{i,j}^{n+1}$ **in vertex** (i, j)

$$W_{i,j}^{n+1} = W_{i,j}^n + (\mu_{i,j} + \mu_{i-1,j} + \mu_{i-1,j-1} + \mu_{i,j-1})^{-1}$$
$$(\mu_{i,j}(\delta W_4)_{i,j} + \mu_{i-1,j}(\delta W_1)_{i-1,j} + \mu_{i-1,j-1}(\delta W_2)_{i-1,j-1} + \mu_{i,j-1}(\delta W_3)_{i,j-1}) \quad (10)$$

We can solve the steady and unsteady flow through plane cascades, the the flow on S_1 surface (3).

The methods M_3 uses the Runge - Kutta type schemes, when the following numerical approximation (1)

$$(\frac{\partial W}{\partial t})_{i,j} = \widetilde{Rez}(W) = -\frac{1}{\mu_{i,j}} \oint_{\partial D_{i,j}} (\tilde{F} - \frac{1}{Re}\tilde{R})dy - (\tilde{G} - \frac{1}{Re}\tilde{S})dx. \quad (11)$$

is solved by

$$\begin{aligned}
W_{i,j}^{(0)} &= W_{i,j}^n \\
W_{i,j}^{(k)} &= W_{i,j}^{(0)} - \alpha_k \widetilde{Rez}(W_{i,j}^{(k-1)}), \quad k = 0, \cdots, m \\
W_{i,j}^{n+1} &= W_{i,j}^{(m)}.
\end{aligned} \quad (12)$$

We use (12) with the different values of m and α_k and with some acceleration techniques, like local time stepping, HRA (Hierarchical Residual Averaging) algorithms, for simulation of steady inviscid and viscous flow through plane cascades or for the solution of model (2). The all up to now described methods uses the structured H - type grid and artificial damping terms of Jameson's type. The method M_4 is developed for unstructured triangular adaptive grid. The implicit and explicit TVD upwind schemes of first and second order are used (see e.g. [5] [6]). It can be used for inviscid steady plane flow.

3 Some Numerical Results

The results of steady inviscid computation of flow through a turbine cascade achieved by M_1 and M_2 are presented on Fig.1. The iso-Mach lines in radial cascade for various outlet and fixed incomplete inlet conditions (model (3), method M_2) are on Fig.2. The unsteady flow through turbine cascade with periodical outlet condition $p_2 = p_2(t)$ are presented on Fig.3 (method M_2 - upper, M_1 - lower series of figures). The results of inviscid and viscous flow simulation (method M_3) are compared on Fig.4 with experimental data for DCA 8% cascade. This is our usual test case, we can compare the results of different models (potential eq., small perturbation potential eq., Euler equations, viscous flow) and experimental data. The example of unstructured grid, numerical results achieved by M_4 and experimental data are on the last Fig.5. We can observe the region of local compression near the suction side of the profile on Fig.5b. We did not obtain this agreement with experimental data (Fig.5c) on structured H type grid.

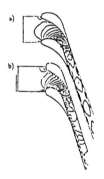

Fig.1 Turbine cascade. Iso-Mach lines
a) method M_1, $M_2 = 1.09, p_2 = 0.51 p_1$
b) method M_2, $M_2 = 1.16, p_2 = 0.48 p_1$

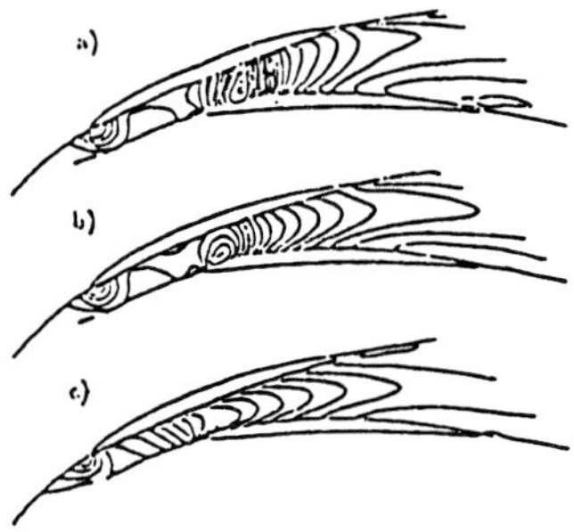

Fig.2 Radial cascade. Iso-Mach lines
Method M_2
$p_2 = 0.85 p_0, p_2 = 0.92 p_0, p_2 = 97 p_0$

References

[1] Hulek T., Huněk M., Kozel K.: Proceedings of the FECFD conf. Brussel, 1992, Ed. Ch. Hirch, J. Périaux, W. Kordulla, pp. 61 - 68

[2] Fořt J., Kozel K., Vavřincová: Proceedings of the IMACS'91 conf., Dublin 1991, Ed. R. Vichnevetsky, J., H. Miller, pp. 489 - 490

[3] Fořt J., Kozel K., Vavřincová: Proceedings of the 5th ISCFD conf., Sendai 1993, Ed. H. Daiguji

[4] Kozel K. (Ed.): Report K 201 TU Prague, No 201-3-91-111, Prague, 91

[5] Kozel K. (Ed.): Report K 201 TU Prague, No 201-3-92-114, Prague, 92

[6] Hwang C. J., Liu J. L.: AIAA J., Vol 29, No 10, 1991

[7] Svanson R. C., Turkel E. : ICASE Report No 84-62, 1985

[8] Arnone A., Svanson R. C. : ICASE Report No 88-32, 1988

[9] Fürst J., Kozel K.: GAMM Jahres Tagung, Braumschweig, 1994

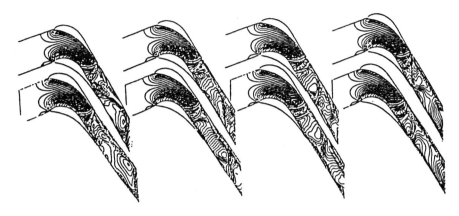

Fig. 3. Turbine cascade, unsteady flow, $p_2 = (0.6 + 0.15\sin(3t))p_1$ Methods M_2 (upper series), M_1 (lower series)

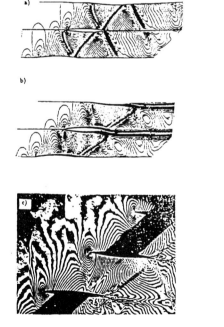
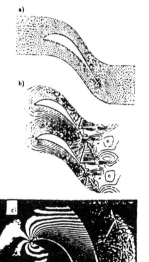

Fig.4 DCA 8% cascade, method M_3
a) inviscid flow $M_2 = 1.15$
b) viscous flow $M_2 = 1.15$, $Re = 10000$
c) experimental data $M_2 = 1.15$

Fig.5 Turbine cascade, method M_4
a) used grid
b) Iso-Mach lines $M_2 = 1.2$
c) experimental data $M_2 = 1.2$

Computation of unsteady viscous flows past oscillating airfoils using the CPI method

Guilmineau E. and Queutey P.

CFD Group, LMF, URA 1217 CNRS
ECN, 1, rue de la Noë, 44072 NANTES, France

Abstract : Numerical solution of the incompressible two-dimensional Navier-Stokes equations, with the help of the CPI discretization, are presented for different airfoils. The strongly conservative equations are discretized with a finite volume method. The method uses a system of numerically generated curvilinear coordinates and retains the pressure and the cartesian velocity components as dependent variables on a non-staggered grid. Two flows around an airfoil are computed and compared to experimental results. First, the starting flow past a NACA 0012 airfoil oscillating at large incidences is investigated. Secondly, the turbulent flow past an AS 240 airfoil at a fixed incidence is studied.

1 The Equations

The unsteady incompressible Navier-Stokes equations can be written in the following strong-conservation form:

$$\nabla \cdot \vec{U} = 0; \quad \frac{\partial \vec{U}}{\partial t} + \nabla \cdot \vec{F} = 0 \qquad (1, 2)$$

where the cartesian components F_{kj} of the momentum flux $\vec{F}$ are given by:

$$F_{kj} = U_j U_k + p\delta_{jk} - \frac{1}{Re}\frac{\partial U_k}{\partial x_j} + \overline{u_j u_k} \qquad (3)$$

They involve the cartesian velocity components U_k, the pressure p, the Reynolds stress tensor components $\overline{u_j u_k}$ and the Reynolds number $Re = U_\infty L/\nu$ where L is the airfoil chord, U_∞ is the freestream velocity and ν is the kinematic viscosity of fluid.

The resulting turbulent closure problem is solved by means of an algebraic viscosity model as follows:

$$\overline{\vec{u}\,\vec{u}} = \frac{2}{3} k \vec{1} - \nu_t \left(\nabla \vec{U} + \nabla^T \vec{U} \right) \qquad (4)$$

The effective Reynolds number R_{eff} is defined by $R_{eff} = 1/[\nu_t + Re^{-1}]$.

In order to allow the study of the flow field past complex geometrices, the curvilinear body-fitted coordinates (ξ^1, ξ^2 ξ^3) are introduced but the dependent variables are the cartesian velocity components and the scalar pressure field.

This partial transformation of equations (1), (2) and (3) yields the following equations:

$$\frac{1}{J}\frac{\partial(b^i_j U_j)}{\partial \xi^i} = 0; \qquad \frac{\partial U_k}{\partial t} + \frac{1}{J}\frac{\partial(b^i_j F_{kj})}{\partial \xi^i} = 0 \qquad (5,6)$$

$$F_{kj} = U_j U_k + p\delta_{jk} - \frac{a^m_j}{R_{eff}}\frac{\partial U_k}{\partial \xi^m} - v_t a^n_k \frac{\partial U_j}{\partial \xi^n} \qquad (7)$$

where the jacobian J of the transformation from the curvilinear coordinates $\{\xi^i\}$ to the cartesian coordinates $\{x_i\}$ measures the "physical" volume and can be expressed by $J\delta^j_i = \vec{a}_i \cdot \vec{b}^j$. The area vectors $\vec{b}_i = \vec{a}_j \times \vec{a}_k$ (i, j, k in cyclic order) measure the oriented area of a small surface of unit sides along ξ^j and ξ^k on a $\xi^i =$ const. surface in the computational space. The moduli of the covariant vector $\vec{a}_i = \partial \vec{R}/\partial \xi^i$ measures the sides of a curvilinear hexadron in the physical space.

2 The Numerics

2.1 Grid layout and discrete equations

The so-called colocated cell-centered grid layout is used : the cartesian velocity components and the pressure share the same location at the center of the control volume (figure 1).

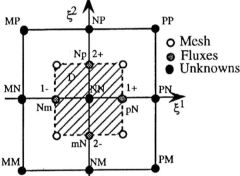

Fig. 1. Schematic sketch of notations

In the following, $U_k(NN)$ will be the unknown k-th cartesian velocity component at point NN.

The discrete divergence of the flux Φ over the control volume D is simply :

$$(\Delta_i \Phi^i) = \Phi^1(pN) - \Phi^1(mN) + \Phi^2(Np) - \Phi^2(Nm) \qquad (8)$$

so that the discrete continuity equation results from $\Phi = \vec{b}^i \vec{U}$.

The time-derivative is discretized using a second-order accurate backward Euler method involving the time levels $t^0 \equiv t - \Delta t$ and $t^{00} \equiv t - 2\Delta t$, besides the actual time level t. We then have :

$$\frac{\partial \Phi}{\partial t} \approx e_1 \Phi + e_0 \Phi^0 + e_{00} \Phi^{00} \text{ with } \Phi = \phi(t); \; \Phi^0 = \phi(t^0) \; \Phi^{00} = \phi(t^{00}) \qquad (9)$$

Using (8) and (9) yields the following motion equations :

$$J^{-1}(NN)\Delta_i(b_j^i U_j)(NN) = 0 \qquad (10)$$

$$e_1 U_k(NN) + \left[e_0 U_k^0(NN) + e_{00} U_k^{00}(NN)\right] + J^{-1}(NN)\Delta_i(b_j^i F_{kj})(NN) = 0 \quad (11)$$

In the discrete divergence at point NN in (11), the linearized momentum flux $\vec{b}^i \; \vec{F}_k$ is defined at interfaces pN, mN, Np, Nm as indicated in (8). For instance :

$$\left[\vec{b}^i \; \vec{K}_k\right](pN) = \left[(\vec{b}^i \; \vec{U}^*) U_k + pb_k^i - \frac{\vec{b}^i \; \vec{a}^m}{R_{eff}} \frac{\partial U_k}{\partial \xi^m} - v_t \, a_k^n \frac{\partial U_j}{\partial \xi^n}\right](pN) \qquad (12)$$

where $\vec{U}^*$ is a prediction of the velocity field at the actual time.

2.2 The reconstruction problem

It appears that besides unknown nodal values of the cartesian components, expression (12) involve the values $U_k(pN)$ which are also unknown, but at points which are not nodal points. This introduces the so-called reconstruction problem : fluxes, like $U_k(f)$ - if f is the interface pN, mN, Np or Nm - which are not defined at nodal points must be expressed in terms of nodals unknowns. We use the Consistent Physical Interpolation (CPI) [1], [2] which determines $U_k(f)$ from the solution of the convective form of the momentum equations at point f. This interpolation involves the set of neighbours $NB(f)$ of influencing nodes. For the control volume D, the set of active neighbours are :

$$NB(pN) = NP, NN, NM, PP, PN, PM \quad NB(mN) = MP, MN, MM, NP, NN, NM$$

$$NB(Np) = PN, NN, MN, PP, NP, MP \quad NB(Nm) = PM, NM, MM, PN, NN, MN$$

Upon substitution of closures written at pN, mN, Np and Nm into the discrete momentum equation, we obtain the discrete scheme for the momentum equation where the velocity and the pressure unknowns are located at NN and at the eight nodal neighbours of the set $NB(NN) = \{MM, MN, MP, NM, NP, PM, PN, PP\}$.

Upon substitution of the same closures into the continuity equation, we obtain a discrete scheme for a nine-point pressure equation.

2.3 Pressure-Velocity coupling

The algorithm which yields a coupled of the momentum equation and the continuity equation is enforced by the PISO algorithm [3].

3 Results

The instantaneous incidence $\alpha(t)$ is given by $\alpha(t) = 30° - 15°* \cos(2\pi f t)$. The laminar flow at a Reynolds number Re=3000 around a NACA 0012 airfoil is computed with f=0.1 on a 180*80 grid with a time step $\Delta t = 0.0025$ until $t = 2.5$ and then with a time step $\Delta t = 0.001$. Some comparisons with computations and visualisations [4] are possible. The agreement with data appears better than [4] where a streamfunction-vorticity formulation was used, especially at time instants $t = 1.5$ and 2.5 (figure 2). Figure 3 illustrates the time evolution of the unsteady wake past the airfoil oscillating at the half-chord. After the detachment of starting vortex (existing until t - 1.5), the initial separation bubble develops into a large-scale leading edge vortex which spreads all over the upper surface ($t < 3.0$). This vortex detaches slowly from the airfoil ($t < 4.5$) while the experimental value is $t < 4.0$.

3.1 AS 240

The turbulent flow around an AS 240 airfoil at a Reynolds $Re = 2.10^6$ and at an incidence of 8° degrees is computed using the Baldwin-Lomax model. Figure 4 shows the comparison between the computed wall pressure coefficient and data [5], [6]. The measured pressure coefficient presents a small plateau of pressure at the trailing edge. The calculated drag coefficient is $C_D = 0.0286$ while the experimental value is $C_D = 0.0113$. The same discrepancy is found also in [7] with a mixing length model. The calculated lift coefficient is $C_L = 1.402$ while the experimental value is $C_L = 1.375$. The use of standard turbulence models, as Baldwin-Lomax, $k - \epsilon$, predicts a lift coefficient which does not slow a significant decrease as the angle is increase above stall anges. Some improvements in this respect seems possible with the ASM model [8].

Acknowledgements :

The authors gratefully acknowledge the attribution of cpu time on C98 by the Scientific Commitee of IDRIS and on Cray YMP of PACA Region by the GdR Mecanique des Fluides Numèriques.

References

1 Deng G.B., Piquet J., Queutey P & Visonneau M., Gamm Workshop. Notes on Numerical Fluid Mechanics, Vol. 36, pp. 34-45, Deville M., Lê T.H. & Morchoisne Y. Eds, 1992.

2 Deng G.B., Guilmeneau E., Piquet J., Queutey P. & Visonneau M., International Journal for Numerical Methods in Fluids, (to be published).

3 Issa R.I., Gosman A.D. & Watkins A.P., Journal of Computational Physics, Vol. 62, $n°$ 1, pp. 66-82, 1986.

4 Ohmi K., Coutanceau M., Daube O. & Ta Phuoc Loc, Journal of Fluid Mechanics, Vol.225, pp. 607-603, 1991.

5 Gleyzes C., Report ONERA/CERT 57/5004.22, Toulouse, 1988.

6 Gleyzes C., Report ONERA/CERT 57/5004.22, Toulouse, 1989.

7 Chevalier G., PhD Thesis, ENSAE, 1994.

8 Davidson L. & Rizzi A., Journal of Spacecraft and Rockets, Vol. 29, pp. 794-800, 1992.

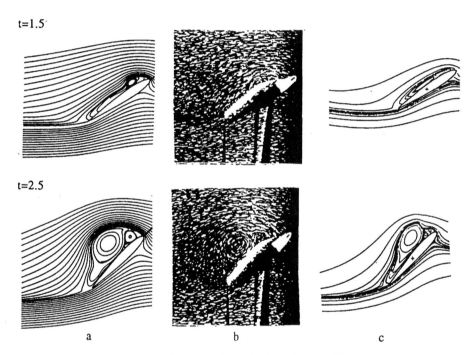

Figure 2. Comparison of calculations and [4]
(a : calculation, b experiment [4], c : calculation [4])

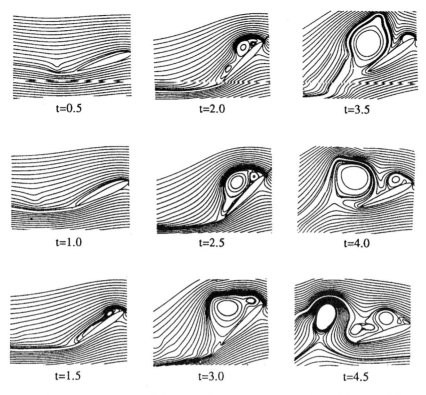

Figure 3. Streamlines of the unsteady vortex wake past a NACA 0012 airfoil

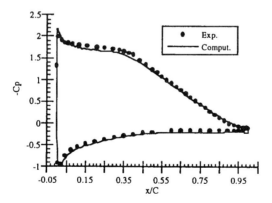

Figure 4. Pressure coefficient

Investigation of 3-D Shock Focusing Effects on a Massively Parallel Computer

Hartmann Chr. and Roesner K. G.

Institut für Mechanik, Technische Hochschule
Darmstadt, 64289 Darmstadt, Germany

1 Introduction

The subject of the present paper is the evolution and interaction of shock waves leading to shock focusing effects within a gas-filled cubical cavity with rigid walls. The three-dimensional, time-dependent flow is invoked by the following initial condition: In the center of the cubical cavity a high pressure region of spherical shape is released at the beginning of the calculation, which leads immediately to a shock wave formation. The aims of this investigation are, on the one hand, the examination of the physical character of the shock focusing. We want to demonstrate the symmetry conserving character of the invariant numerical scheme. In this case no preferential direction due to the orientation of the grid is engendered in the numerical solution during the calculation. On the other hand, the possibility of a strong CPU-time reduction by the use of a massively parallel computer (type SIMD) is demonstrated.

2 Basic Equations

The simulation is based on the strong conservation form of the three-dimensional, time-dependent Euler equation and on the equation of state for an ideal gas.

$$\frac{\partial \vec{W}}{\partial t} + \frac{\partial \vec{F}}{\partial x} + \frac{\partial \vec{G}}{\partial y} + \frac{\partial \vec{K}}{\partial z} = 0 \qquad (1)$$

with $\vec{W} = \begin{pmatrix} \rho \\ \rho u \\ \rho v \\ \rho w \\ E \end{pmatrix}$ $\begin{cases} \rho = & \text{density} \\ u = & \text{velocity in x-direction} \\ v = & \text{velocity in y-direction} \\ w = & \text{velocity in z-direction} \\ E = & \text{total energy per unit volume} \end{cases}$

$$\vec{F} = \begin{pmatrix} \rho u \\ \rho u^2 + (\gamma - 1)\left[E - \frac{1}{2}\rho(u^2 + v^2 + w^2)\right] \\ \rho uv \\ \rho uw \\ \gamma uE - (\gamma - 1)u\frac{1}{2}\rho(u^2 + v^2 + w^2) \end{pmatrix}$$

$$\vec{G} = \begin{pmatrix} \rho v \\ \rho v u \\ \rho v^2 + (\gamma - 1)\left[E - \frac{1}{2}\rho(u^2 + v^2 + w^2)\right] \\ \rho v w \\ \gamma v E - (\gamma - 1)v\frac{1}{2}\rho(u^2 + v^2 + w^2) \end{pmatrix}$$

$$\vec{K} = \begin{pmatrix} \rho w \\ \rho w u \\ \rho w v \\ \rho w^2 + (\gamma - 1)\left[E - \frac{1}{2}\rho(u^2 + v^2 + w^2)\right] \\ \gamma w E - (\gamma - 1)w\frac{1}{2}\rho(u^2 + v^2 + w^2) \end{pmatrix}$$

3 Numerical Method

The numerical method consists of an invariant three-dimensional explicit difference scheme. It is based on a two-dimensional version found by Rusanov [1], [2] on the base of Lie-group theory of partial differential equations. The Lie-group theory is applied to both the partial differential equation and the difference equation in the first order differential approximation. If the group generators of the partial differential equation and those of the corresponding first differential approximation of the difference equation are coming out to be the same, the scheme is called invariant or of symmetry conserving character. The scheme is of first-order-accuracy in time and space. Here, the finite-difference-equation is given only for one dimension:

$$\frac{W_j^{n+1} - W_j^n}{\tau} + \frac{F_{j+1}^n - F_{j-1}^n}{2h_1} + \cdots = \frac{\tau}{2h_1}\left[(\Omega_{11})_{j+\frac{1}{2}}^n \frac{W_{j+1}^n - W_j^n}{h_1} - (\Omega_{11})_{j-\frac{1}{2}}^n \frac{W_j^n - W_{j-1}^n}{h_1}\right] + \cdots \quad (2)$$

$\Omega_{ml} = A_m \cdot A_l + C_{ml}, \quad (m, l = 1, 2, 3), \quad C_{ml} =$ artificial viscosity matrices

$$A_1 = \frac{d\vec{F}}{d\vec{W}}, \quad A_2 = \frac{d\vec{G}}{d\vec{W}}, \quad A_3 = \frac{d\vec{K}}{d\vec{W}}$$

This method has been object of several investigations. In the past, it could be shown by calculations on coarse meshes that symmetry is conserved over a great number of time-steps [3], [4], [5], [6]. To obtain a better knowledge about the discontinuities in the flow field, a certain interest to refine the mesh in the whole cubical cavity has arisen in order to obtain a higher resolution of the shock.

4 Results

As results of the examination we state that symmetrical behaviour of the numerical scheme can be observed on refined meshes over a long time of calculation. Since the difference scheme is of first order approximation, the shock is smeared over a certain number of grid points. It could be observed that the smearing does not exceed four grid points over a long time of calculation. Both, Fig. 1 and Fig. 2 give a distribution of a physical value along a center axis of the cube parallel to the faces. Furthermore, the initial distribution is given in the diagrams. The computation has been performed on a 64 × 64 × 64 mesh.

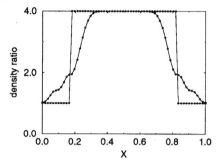

Fig. 1. Density at T=0.0824 dimensionless time, shock fully developed

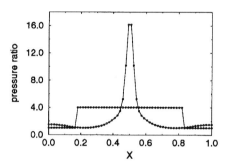

Fig. 2. Focusing effect at T=0.6239 dimensionless time

5 Parallelisation

As the computation on fine meshes over a long time became to expensive even on a powerful serial computer, the code now is implemented on a massively parallel computer, the MasPar 1216, containing 16384 processors with each 16 kbytes local memory. This MP1216 is a computer of SIMD-type (Single-Instruction-Multiple-Data-Stream) which means that all processors are managed by a single instruction processor. In this special case, where the geometry of the problem is similar to the architecture of the computer and where the explicit character of the numerical scheme is appropriate for the machine type, a strong reduction of computation time can be achieved. In comparison to the IBM RS6000/350 with 64 Mbytes core memory the execution time can be reduced to 1.4% for the computation on the above mentioned mesh size. In comparison to the DEC AXP the reduction is approximately of the same magnitude. Severe problems had to be overcome during the implementation due to the small local memory on each processor of the parallel machine. This can only be compensated by a modified mapping of the grid points on the processor array. Of course, a higher amount of communication is necessary compared to the standard mapping.

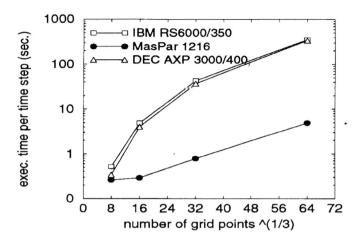

Fig. 3. CPU-time comparison for different types of computers

6 Conclusion and Perspectives

The symmetry conserving behaviour of the numerical scheme can be observed over more than 500 time-steps although it is of first order accuracy. A strong focusing of pressure and energy is achieved. A factor 4.05 between the maximum pressure and the initial high pressure has been obtained. The computation time could be reduced to 1.4 % of that of a powerful workstation by parallelisation. A further project will be the increase of accuracy and the modelling of the viscous terms of the Navier-Stokes-Equation.

References

[1] Rusanov,V.V, USSR, 'The calculation of interaction of nonstationary shock waves with obstacles', Zh. Vychisl. Mat. Mat. Fiz 1 (2) (1961), 267-279

[2] Shokin,Yu.I., 'The Method of Differential Approximation', Springer-Verlag, Berlin Heidelberg New York Tokyo, 1983

[3] Srivastava R.C., Roesner K.G. and Leutloff D., Astrophys. Space Science 135 (1987), 399-407

[4] Leutloff D., Srivastava R.C. and Roesner K.G., 'Numerical Solution of a Converging Shock Problem', Computers and Fluids 16 (1988), 175-182

[5] Srivastava R.C., Roesner K.G., Leutloff D. and Kuwahara K., Proc. Int. Symp. Computational Fluid Dynamics, Davis, CA (1991), 1101-1106

[6] Leutloff D., Srivastava R.C. and Roesner K.G., 'Numerical Investigation of Three-Dimensional Shock Focusing Effects by an Invariant Difference Scheme', Fluid Dynamics Research 10 (1992), 469-482

The mechanism of entrainment in circular jets

Joseph Mathew and Amit J. Basu

National Aerospace Laboratories, Bangalore 560 017, India

Abstract : The evolution to turbulence of a thin circular mixing-layer (jet-like flow) was computed using a Fourier-spectral, 4th-order predictor-corrector scheme. Data for Reynolds number (based on initial jet-diameter and velocity difference across shear layer) of 1600 on a 64^3 grid and 2400 on a 128^3 grid were used to study the fluxes associated with entrainment to the turbulent region. The flux that leads to a growth in the size of the region is but slightly larger than that which leads to its reduction. The data suggest that engulfment is therefore an incomplete picture of turbulent entrainment: A fluid packet is drawn to the turbulent region by the vorticity, entering and leaving it several times as it acquires vorticity by viscous diffusion. The entraining flows are strong and clearly associated with vortex rings at early times, but following ring-breakdown, clear organization on the largest scales is not found. Plots of the vorticity and velocity fields do suggest organization on intermediate scales (thickness of vortical structures).

1 Introduction

The flow near the nozzle of circular jets is known to be dominated by the dynamics of large-scale vortex rings that grow, sometimes pair and later break down. Further downstream, the presence of large-scale structure is not as unambiguous as in mixing layers, but axisymmetric, helical and double-helical forms have been found in laboratory jet data. Dimotakis *et al* (1983) and Dahm & Dimotakis (1985) suggested that, even far downstream, these structures dominate the dynamics, while Tso & Hussain (1989) observed that the spatial development of jets can be viewed as being due to differential, radial advection of these structures. In this paper, we present an examination of our direct simulation data in terms of the associated entraining fluxes with a view to identifying the importance of underlying structure in turbulent circular jets.

The evolution of uniform, circular, thin shear-layers (jet-like flow) subject to small perturbations were computed using Fourier-spectral discretization in space and fourth-order predictor-corrector integration in time; de-aliasing according to the two-thirds rule has been used (see Basu *et al* 1992 for details). The flow is incompressible, nominally aligned with the z-axis and is periodic with respect to all three Cartesian coordinates for all time. The computation is part of our program of direct simulation studies of free shear flows using the parallel computer, 'Flosolver', developed in-house; the parallelization has been discussed elsewhere (Basu 1994). Two cases are used for analysis below. Case I: The Reynolds number based on initial jet-diameter and velocity difference across the shear layer is 1600; the grid is 64x64x64 on a cube of side 4 (the jet has an initial diameter of unity). Case II: The Reynolds number is 2400 and the grid is 128^3.

Unlike in mixing layers, flow visualization of laboratory jets is not sufficient encouragement to view their dynamics as being closely associated with structures of the order of the jet-width. Yet it does seem plausible that a most significant role is played by structures of moderate size, whose signature must appear in measures of the associated entrainment. Entrainment of ambient fluid by the jet leads to an increase in the region occupied by fluid that is rotational (vorticity exceeding some threshold); in turbulent flow, this is also a region of fluctuating vorticity. The mechanism of entrainment is no longer viscous diffusion alone: the jet draws towards itself and ingests irrotational fluid which then acquires vorticity by diffusion. This process has been studied here by calculating entrainment fluxes across jet boundaries.

The turbulent region was defined as the region containing significant vorticity fluctuations. At any time t, by averaging over streamwise gridpoints, distributions of mean and fluctuating (r.m.s.) vorticity in the transverse plane (Oxy) were obtained. Then, we defined the turbulent region to be bounded by cylindrical surfaces whose generators were parallel to the streamwise axis (Oz) and which intersected the transverse plane where the fluctuation level was (say) 10% of the maximum fluctuation level. Initially, this region forms a hollow cylinder as the fluctuations in the potential core are low. As this cylindrical boundary is based on streamwise averaging, the flux across it at different places can be rotational (vorticity exceeding 10% of the streamwise mean vorticity) or irrotational.

Bearing in mind the definition of the turbulent region in terms of fluctuating vorticity, we note that the turbulent region tends to grow when there is a flux of irrotational fluid into it, or, a flux of rotational fluid out of it. Similarly, this region tends to shrink following an influx of rotational fluid or an efflux of irrotational fluid. Accordingly, in the flux computations, the four kinds of fluxes (rotational or irrotational, influx and efflux) were summed separately.

2 Results

The data (*e.g.* contours of the vorticity field) over the dimensionless time interval $0 \leq t \leq 40$ show the initial roll-up of the circular shear-layer into four vortex rings per period, pairing, azimuthal instability, growth of streamwise structures and the relatively rapid transition to a disordered state. The computed solutions thus exhibit features observed in laboratory (spatially developing) jets.

Figure 1 shows the growth of the of the region wherein significant vorticity fluctuations are present, scaled with the initial jet volume. Initially, the growth rate is slow while vortex-ring dynamics are dominant; later, the growth is faster, uniform and linear. In the lower Reynolds number flow (case I), a transition stage, which shows the most rapid growth, connects these two stages. This transition stage occurs following the breakdown of the rings and the disappearance of the potential core, during which period, streamwise structures are dominant. In case II, this breakdown occurs sooner and more quickly.

Figure 2 shows the variation of the four kinds of fluxes across the jet boundary: rotational, irrotational, that entering the jet (positive) and that leaving (negative). The larger fluxes at early times are in part due to the larger surface area, roughly twice when the potential core is large and in part due to the stronger induction of the vortex rings. The most important feature, however, is that the fluxes are all much larger than the differences between them, though it is the differences that account for the growth of the jet. Note that after transition ($t \approx 15$), the net irrotational flux is slightly positive and the net rotational flux is slightly negative. Figure 3 shows the fluxes that lead to growth (sum of expelled rotational fluid and ingested irrotational fluid) and shrinkage as discussed above. The underlying mechanism is nearly symmetrical; the small difference results in the net growth.

Engulfment is therefore an incomplete picture of turbulent entrainment: A fluid packet is drawn to the turbulent region by vortical structures, which are like a collection of strings, and travels around these structures due to induction, being ingested and expelled several times as it acquires vorticity by viscous diffusion. Figure 4 is suggestive of this process. A cross-section of the jet at $t = 24$ (case II) is shown. Streamwise vorticity is positive over the dark shaded patches, negative over the lighter patches. Velocity vector components in this plane have been superposed. Organization of the flow on intermediate scales (thickness of the vortical structures) and the entraining flow away from the jet can be seen.

Figure 5 is a grey-scale plot of the variation of the total influx (rotational plus irrotational) over streamwise stations and time; the darkest areas correspond to the largest fluxes. The nearly straight bands at small times correspond to the passage of the vortex rings. Initially there are four thin rings which undergo pairing, becoming intense and causing large fluxes. Not surprisingly, the sharpest variations occur in the vicinity of the rings. The transition is sharp and changes the character of the entrainment. Strong organization on the scale of jet-width is no longer apparent.

Acknowledgments :

We thank our colleagues, especially Dr. U. N. Sinha of the Flosolver unit which provided computing support. We also thank Prof. R. Narasimha with whom we had valuable discussions of the ideas presented here. We thank CMMACS for some post-processing support.

References

[1] Basu, A. J., Narasimha, R., Sinha, U. N. (1992): "Direct simulation of the initial evolution of a turbulent axisymmetric wake", *Current Science*, **63**, No. 12, 734–740

[2] Basu, A. J. (1994): "A parallel algorithm for spectral solution of the three-dimensional Navier-Stokes equations" *Parallel Computing*. In press.

[3] Dimotakis, P. E., Miake-Lye, R. C., Papantoniou (1983): "Structure and dynamics of round turbulent jets", *Phys. Fluids*, **26**(11) 3185–3192

[4] Dahm, W., Dimotakis, P. E. (1985): "Measurements of entrainment and mixing in turbulent jets", *AIAA 23rd Aerospace sciences meeting*, AIAA-85-0056

[5] Tso, J., Hussain, F. (1989): "Organized motions in a fully developed turbulent axisymmetric jet", *J. Fluid Mech.*, **203**, 425–448.

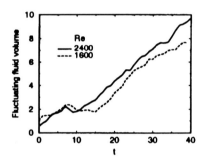

Fig. 1. Volume of fluid with significant vorticity fluctuations

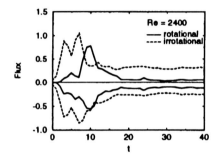

Fig. 2. Variation of fluxes across jet boundary. ——: rotational; - - - -: irrotational; positive: into fluctuating region.

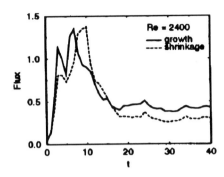
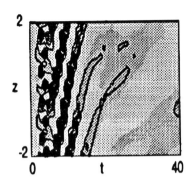

Fig. 3. Variation of fluxes leading to growth (—) and shrinkage (- - - -) of jet

Fig. 4. Distribution of flux into jet (increases with grey-scale intensity).

Fig. 5. Entrainment by structures of intermediate size. Cross-section of jet at $t = 24$, $z = 0$. Light areas: streamwise vorticity negative; dark areas: streamwise vorticity positive; arrows: velocity vector component in the plane.

Numerical Computation of Critical Flow over a Weir

George Mejak

University of Ljubljana, Department of Mathematics and Mechanics, Slovenia

The aim of the paper is to present some advances in numerical computation of critical flow over a weir. As is recognized this is rather difficult problem as not only position of the free surface is unknown in advance but also a value of the critical discharge has to be found as a part of the solution. Traditionally numerical computation of such problems consists of two iteration levels, interior where for a given discharge coefficient a corresponding free boundary value problem is solved and the exterior where a correct value of the discharge is sought. In this paper a new approach is proposed where both levels are combined into one iteration.

The problem is modelled within potential theory. To allow generalization to three dimensional geometry formulation of the problem is in terms of the velocity potential function φ. With reference to Fig. 1 we denote by $\Gamma_{\text{in}} = \text{GA}$ ($\Gamma_{\text{out}} = \text{EF}$) a part of the boundary parallel to y-axis where the fluid enters (exits) the computational domain $\Omega \subset \mathbb{R}^2$, by $\Gamma_{\text{free}} = \text{FG}$ the free boundary the position of which is unknown beforehand and by $\Gamma_{\text{bed}} = \text{ABCDE}$ a part of the boundary which refers to the flat bed $\text{AB} \cup \text{DE}$ and the weir BCD. It is assumed that the outlet Γ_{out} is the equipotential line and that along Γ_{in} x-component of the inlet velocity is uniform. The problem is put into dimensionless form using the stagnation level H_0 and x-component of the inlet velocity V_0 as the representative length and velocity. Discharge Q is put into dimensionless form by $Q = Q_{\text{D}}/\sqrt{gH_0^3}$ where Q_{D} is the actual flow rate per unit width and g is acceleration due to gravity acting in the direction of negative y-axis. Along the free boundary the Bernoulli equation $\frac{1}{2}|\nabla\varphi|^2 + y/\text{Fr} = 1/\text{Fr}$ applies. Here $\text{Fr} = V_0^2/gH_0$ is the Froude number. Assumption that the flow has uniform inlet x-velocity V_0 and that Γ_{in} is parallel to y-axis implies that $\text{Fr} = Q^2/S_2^2$ where S_2 is the depth of the flow at the inlet. Recalling that in the computation of critical flow over a weir the upstream depth S_2 is assumed to coincide with the subcritical depth of the uniform flow over a flat bed. i.e. S_2 is defined as the largest root of the equation

$$2y^3 - 2y^2 + Q^2 = 0 \tag{1}$$

we have arrived to the following mathematical formulation of the problem: Find φ, position of the free boundary Γ_{free} and a value of the critical discharge Q such that

$$\nabla^2 \varphi = 0 \quad \text{in } \Omega, \tag{2}$$
$$\partial \varphi / \partial \vec{n} = 0 \quad \text{on } \Gamma_{\text{free}} \cup \Gamma_{\text{bed}}, \tag{3}$$
$$\partial \varphi / \partial \vec{n} = -1 \quad \text{on } \Gamma_{\text{in}}, \tag{4}$$
$$\varphi = 0 \quad \text{on } \Gamma_{\text{out}}, \tag{5}$$
$$|\nabla \varphi|^2 - (1-y)/(1-S_2(Q)) = 0 \quad \text{on } \Gamma_{\text{free}}, \tag{6}$$
$$y_G - S_2(Q) = 0. \tag{7}$$

Here notation $S_2(Q)$ denotes that S_2 depends upon Q by the functional relation (1) while y_G denotes y-coordinate of the inlet point G. Note that for a given domain Ω equations (2)–(5) constitute a flow problem for φ.

The problem is put into the discrete level of approximation by introducing the finite element discretization to the flow problem and collocation method to condition (6). In particular, (6) is imposed to hold at the finite element nodes along the free boundary. Since (6) in this way involves values of gradient $\nabla \varphi$ at the nodal points along Γ_{free}, Hermite finite element approximation of the flow problem is proposed. Denoting by $\mathbf{u} \in \mathbb{R}^r$ degrees of freedom corresponding to the finite element discretization of φ and by $\mathbf{y} \in \mathbb{R}^s$ a vector of y-coordinates of the free nodal points the problem (2)–(7) is discretized into

$$\mathbf{f}(\mathbf{u}, \mathbf{y}, Q) = \begin{bmatrix} \mathbf{K}(\mathbf{y})\mathbf{u} - \mathbf{b}(\mathbf{y}) \\ \mathbf{c}(\mathbf{u}, \mathbf{y}, Q) \\ d(\mathbf{y}, Q) \end{bmatrix} = \begin{bmatrix} \mathbf{0} \\ \mathbf{0} \\ 0 \end{bmatrix} \tag{8}$$

where $\mathbf{K}(\mathbf{y})$ is the element stiffness matrix which depends upon $\mathbf{y}$, $\mathbf{b}(\mathbf{y})$ is the load vector which due to the variable position of point G depends upon y coordinate of G, $\mathbf{c}(\mathbf{u}, \mathbf{y}, Q) \in \mathbb{R}^s$ is a vector with components

$$c_i(\mathbf{u}, \mathbf{y}, Q) = |\nabla \varphi_i|^2 - \frac{1 - y_i}{1 - S_2(Q)} \tag{9}$$

and $d(\mathbf{y}, Q) = y_G - S_2(Q)$. Noting that the first set of equations in (8) is linear, $\mathbf{u}$ is eliminated and this results in

$$\mathbf{g}(\mathbf{y}, Q) = \begin{bmatrix} \mathbf{c}(\mathbf{K}^{-1}(\mathbf{y})\mathbf{b}(\mathbf{y}), Q) \\ d(\mathbf{y}, Q) \end{bmatrix} = \begin{bmatrix} \mathbf{0} \\ 0 \end{bmatrix}. \tag{10}$$

This non-linear system of $s+1$ equations is solved by the Newton-Raphson method. Since computation of the Jacobian requires quite a lot of computation, Broyden's method is proposed after first few Newton-Raphson iteration steps.

Differentiation of the element stiffness matrix requires detailed specification how displacements of the free boundary affect the finite element mesh. There is somewhat conflicting situation; namely, to reduce computational cost of computing $\partial \mathbf{K}(\mathbf{y})/\partial y_i$ respond of the mesh to a single free node displacement should be local and on the other hand to prevent creation of too distorted or even overlapping elements remeshing of Ω should not be restricted only to the neighbouring elements of Γ_{free}. This dilemma is solved by giving priority to the requirement that the method is robust and thus the global remeshing procedure is preferred. However, to reduce the computational cost the Jacobian is actually computed as if only elements along Γ_{free} were variable. Next we observe that at the solution of (1)-(7) the Jacobian is analytically singular and thus the theory does not guarantees the superlinear convergence rate of Broyden's method. However, in agreement with common practice of solving nonlinear equations with simple singularities, the exhibited convergence rate of the method is of acceptable linear order which typically requires about $15 - 20$ iterations to reduce the infinity norm of g to the order of 10^{-7} from the initial error of order unity. The validity of the approach is studied by means of a comparison with results in the literature (Aitchison 1980), (O'Caroll et al. 1984) as well as with comparison of computed results for different mesh gradings. The former is given in Table 1 for three different cases. Geometry of the weir is chosen as in Fig. 1 and the lengths $U = \overline{AB}$, $D = \overline{DE}$ and H_0 relative to the height p of the weir are given in the first half of Table 1.

Table 1. Comparison of the computed critical discharge. Columns H_0/p, U/p and D/p define dimensions of the computational domains for cases A, B and C. In the column O'Carroll and Toro acronyms FEM and RK refer to results computed by the finite element and Ritz-Kantorovich method, respectively. Mesh parameters in the present study are, case A 2295 elements and 103 free nodes along Γ_{free}, case B 2444 elements and 105 free nodes, and case C 2340 elements and 105 free nodes.

Case	H_0/p	U/p	D/p	Aitchison	O'Carroll and Toro		Present study
					FEM	RK	
A	7.90	7	7	0.5110	0.5101	0.5102	0.51001
B	5.69	4	4	0.4798	0.4796	0.4791	0.47916
C	4.63	2	4	0.4517	0.4493	0.4477	0.44898

Investigating convergence radius of the method with respect to the initial value of Q it was found that the method converges to two different values of Q, denoted by Q^- and Q^+, $Q^- < Q^+$. From the numerical point of view both solutions are equivalent. However, analysing Γ_{free} we found that Q^- results in formation of a wave upstream of the weir while Q^+ gives the free boundary which monotonically decreases from the inlet to the outlet and therefore, according to the standard criterion that the free boundary of the critical flow has no waves, Q^+ was chosen for a correct value of critical discharge. Nature of the existence of two convergent values of Q is revealed by the mesh refinements, see Table 2. As it appears, presence of two convergent values of Q is property of the discrete problem. With mesh refinements difference between Q^- and Q^+ diminishes, height of the crest decreases and both free boundaries corresponding to Q^- and Q^+ converge to the same position.

Table 2. Influence of the mesh parameters upon the critical flow discharge. n_x is a number of x-layers, n_y of y-layers, N_e is a number of elements, s is a number of free nodes and Crest is the height of the crest. Mesh for case A_5 is shown in Fig. 1.

Case	n_x	n_y	N_e	s	Q^+	Q^-	Crest
A_1	11	4	143	23	0.51085	0.50850	0.01501
A_2	21	8	441	43	0.51022	0.50956	0.00420
A_3	30	12	870	61	0.51011	0.50959	0.00264
A_4	42	16	1554	85	0.51004	0.50971	0.00139
A_5	51	20	2295	103	0.51001	0.50976	0.00091

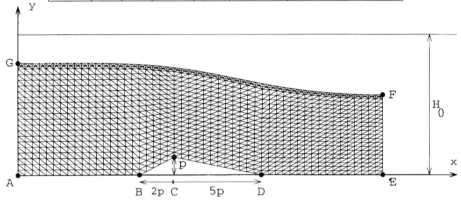

Fig. 1. Computational domain for flow over a weir. Finite element mesh consists of 2295 elements and has 103 free nodes along Γ_{free}.

References

[1] Aitchison, J.M. (1980): "A finite element solution for critical flow over a weir", in Proc. 3rd International Conference on Finite Elements in Flow Problems, ed. D.H. Norrie (University of Calgary, Calgary) Vol. 1, pp. 52–59

[2] O'Carrol, M.J., Toro, E.F. (1984): "Numerical computations of critical flow over a weir", Int. j. numer. methods fluids, **4**, pp. 499–509

Numerical Simulation of Natural Convection in a Cavity of an Ellipsoid of Revolution, using a Spectral Finite Difference Method

Yoshihiro Mochimaru

Dept. of Control & Systems Engineering,
Tokyo Institute of Technology, Tokyo 152, Japan

Abstract : Analyzed is a natural convection followed by a steady-state natural convection of liquid metals filled in a lower half part of an ellipsoid of revolution. The free surface of the liquid metal is assumed to be facing to a gas atmosphere with a uniform constant temperature sufficiently higher than a room temperature. A Fourier spectral finite difference scheme is applied to the equations of vorticity transport and energy in a boundary fitted coordinate system, using a non-uniform grid spacing. It is found that the strength of circulating motion of liquid metal layers due to natural convection can be well controlled by the variation of profiles of heat transfer through the wall of the ellipsoid.

1 Introduction

Natural convection in liquid metal layers is focussed in settling of inclusions[1]. Here, analyzed is a natural transient convection followed by a steady-state natural convection of liquid metals filled in a lower half part of an ellipsoid of revolution, the symmetrical axis of which is assumed to be vertical. The velocity and thermal field treated is limited to an axially symmetric one with respect to the symmetrical axis and is assumed to be laminar.

2 Analysis

2.1 Basic Equations

It is assumed that the free surface of the liquid metal is facing to a gas atmosphere with a uniform constant temperature sufficiently higher than a room temperature, and that all the liquid metal layers are initially at rest and at the same temperature as the above gas atmosphere, being subject to such a small heat transfer (loss) through walls of the ellipsoid to the ambient air of a room temperature afterward that the maximum temperature drop in the liquid metal bath is small compared with the temperature difference between the gas and the ambient air, so that the local heat transfer rate along the wall is regarded to be (almost) invariant with time. Hereafter, length, velocity, and time are assumed to be made dimensionless with respect to a

(one half of the diameter of the circle formed by the free surface and the ellipsoid), $(\nu/a)\sqrt{Gr}$, and $a^2/(\nu\sqrt{Gr})$ respectively, where ν is a kinematic viscosity, Gr being a Grashof number based on the length a and the temperature difference between the gas temperature and the room temperature. Dimensionless temperature T is defined as

$$T = \frac{\text{local liquid metal temperature} - \text{gas temperature}}{\text{gas temperature} - \text{room temperature}}.$$

Using a cylindrical polar coordinate system (the z-axis being coincident with the symmetric axis of the ellipsoid, the r-axis being perpendicular to the z-axis, θ being the azimuth angle), the vorticity vector can be given through the only one non-zero θ-component, $-\zeta$. In addition, to match large variation of velocity near the surface, introduced is a boundary fitted coordinate system (α, β) [$\alpha = \alpha_0$ corresponds to the wall of the ellipsoid] specified by the transformation of the type $r + iz = f(\alpha + i\beta)$ in the $r-z$ plane, where f is a complex function, either one of $(1/\cosh\alpha_0) \cdot \cosh$ or $(1/\sinh\alpha_0) \cdot \sinh$ according to an oblate spheroid ($\alpha_0 = \tanh^{-1}(b/a)$) or a prolate spheroid ($\alpha_0 = \tanh^{-1}(a/b)$) respectively [a semisphere case is not treated here], and b is the depth of the liquid metal bath. The range of coordinates is $0 \leq \alpha \leq \alpha_0, -\pi/2 \leq \beta \leq 0$. Thus the equation of vorticity transport is expressed with the aid of an axisymmetric stream function ψ (dimensionless quantity based on $\nu a\sqrt{Gr}$) [t : dimensionless time] as

$$K\frac{\partial}{\partial t}\left(\frac{\zeta}{r}\right) + r\frac{\partial(\zeta/r, \psi/r^2)}{\partial(\alpha, \beta)} + \frac{2\psi}{r^2}\frac{\partial(\zeta/r, r)}{\partial(\alpha, \beta)}$$

$$= \frac{1}{\sqrt{Gr}}\left(\frac{\partial^2}{\partial\alpha^2} + \frac{\partial^2}{\partial\beta^2}\right)\left(\frac{\zeta}{r}\right) + \left(\frac{1}{r}\frac{\partial r}{\partial\alpha}\frac{\partial}{\partial\alpha} + \frac{1}{r}\frac{\partial r}{\partial\beta}\frac{\partial}{\partial\beta}\right)\left\{\frac{3}{\sqrt{Gr}}\left(\frac{\zeta}{r}\right) + T\right\}, \quad (1)$$

$$K \equiv \frac{\partial(r,z)}{\partial(\alpha,\beta)} = \left(\frac{\partial r}{\partial\alpha}\right)^2 + \left(\frac{\partial r}{\partial\beta}\right)^2, \quad (2)$$

$$K\left(\frac{\zeta}{r}\right) = -\left(\frac{\partial^2}{\partial\alpha^2} + \frac{\partial^2}{\partial\beta^2}\right)\left(\frac{\psi}{r^2}\right) - 3\left(\frac{1}{r}\frac{\partial r}{\partial\alpha}\frac{\partial}{\partial\alpha} + \frac{1}{r}\frac{\partial r}{\partial\beta}\frac{\partial}{\partial\beta}\right)\left(\frac{\psi}{r^2}\right). \quad (3)$$

The energy equation under a Boussinesq approximation can be expressed as

$$K\frac{\partial T}{\partial t} + r\frac{\partial(T, \psi/r^2)}{\partial(\alpha, \beta)} + \frac{2\psi}{r^2}\frac{\partial(T, r)}{\partial(\alpha, \beta)} = \frac{1}{Pr\sqrt{Gr}}\left\{\left(\frac{\partial^2}{\partial\alpha^2} + \frac{\partial^2}{\partial\beta^2}\right)T \right.$$
$$\left. + \left(\frac{1}{r}\frac{\partial r}{\partial\alpha}\frac{\partial}{\partial\alpha} + \frac{1}{r}\frac{\partial r}{\partial\beta}\frac{\partial}{\partial\beta}\right)T\right\}, \quad (4)$$

where Pr denotes a Prandtl number. The r- and z- components of velocity, i.e. U, W, are given by $(\partial\psi/\partial z)/r$ and $-(\partial\psi/\partial r)/r$ respectively, and $\zeta \equiv (\partial W/\partial r) - (\partial U/\partial z)$.

2.2 Boundary Conditions

As boundary conditions for velocity fields, no slip flow on the surface of the ellipsoid and no shear on the free surface (the effect of the gas atmosphere can be neglected) are used, whereas $T = 0$ on the free surface (from the definition) and the dimensionless heat flux q on the surface on the ellipsoid (based on a thermal conductivity of the liquid metal and the length a) are specified. In case of $b > a$ introduced is an auxiliary condition that $\psi/r^2, \zeta/r$, and T are smooth at $\alpha = 0$ (a part of the y-axis).

2.3 Numerical Analysis

By expanding $\zeta/r, \psi/r^2$, and T into the following form of Fourier series of β:

$$\begin{bmatrix} \zeta/r \\ \psi/r^2 \\ T \end{bmatrix} = \sum_{n=1}^{\infty} \begin{bmatrix} \zeta_n \\ \psi_n \\ T_n \end{bmatrix} \sin(2n-1)\beta \ , \tag{5}$$

a Fourier spectral finite difference scheme[2,3] is applied to Eqs.(1),(3), and (4) (including boundary conditions), which can be decomposed, using addition formulae of trigonometric functions etc., into Fourier components. In case of $a > b$, boundary conditions on the free surface ($\alpha = 0$ or $\beta = 0$) are reduced to $\zeta_n(0) = \psi_n(0) = T_n(0) = 0$ for any positive integer n, whereas in case of $a < b$ boundary conditions on the free surface ($\beta = 0$) are identically satisfied; instead the auxiliary conditions lead to $(\partial/\partial\alpha)\zeta_n(0) = (\partial/\partial\alpha)\psi_n(0) = (\partial/\partial\alpha)T_n(0) = 0$ in case of $a < b$. On the wall of the ellipsoid, no slip condition and the heat flux profile q lead to

$$\psi_n(\alpha_0) = 0 \ , \tag{6}$$

$$\frac{\partial}{\partial\alpha}\psi_n(\alpha_0) = 0 \ , \tag{7}$$

$$\frac{\partial}{\partial\alpha}T_n(\alpha_0) = -\frac{4}{\pi}\int_0^{\pi/2} q \sqrt{(K)\alpha = \alpha_0} \sin(2n-1)\beta \ d\beta \tag{8}$$

for any positive integer n. Especially, vanishing of the first order derivative (Eq.(7)) can be coupled with Eq.(3) in discretization at $\alpha = \alpha_0$.

2.4 Discretization

The set of decomposed equations (including boundary conditions) can be discretized in time and space by a finite difference approximation, using a non-uniform grid spacing in α, where the n-th coordinate α_n (counted from the wall) is specified, using a suitable positive constant γ, by

$$\alpha_n = \alpha_0 - h\left\{\frac{\sinh(n-1)\gamma}{\sinh\gamma} + 1\right\} \ . \tag{9}$$

2.5 Numerical Integration Scheme

By specifying, as an initial field, uniform temperature and stationary velocity fields, such discretized equations can be integrated semi-implicitly with respect to time to get a steady-state solution. A circulation time of a liquid metal particle passing through a given point at a steady-state, which is defined as a time required to circulate along a closed contour of streamlines, can be given by

$$\text{Circulation time}$$

$$= \int K\sqrt{(d\alpha)^2 + (d\beta)^2} \bigg/ \sqrt{\left(2\frac{\partial r}{\partial\alpha}\frac{\psi}{r^2} + r\frac{\partial\psi/r^2}{\partial\alpha}\right)^2 + \left(2\frac{\partial r}{\partial\beta}\frac{\psi}{r^2} + r\frac{\partial\psi/r^2}{\partial\beta}\right)^2} \ . \tag{10}$$

3 Numerical Results and Discussion

Figure 1 shows the transient character of mean Nusselt number Nu_m (based on the thermal conductivity of the liquid metal) averaged over the free surface. For small time of t, pure-heat-conduction approximation can be applied; the solution can be expressed as

$$T = \sqrt{t} \sum_{m,n} \exp\left(-\frac{\alpha_0 - \alpha}{\sqrt{t}} \gamma_m\right) A_{mn} \sin(2n-1)\beta \quad , \tag{11}$$

where γ_m's denote eigen values, and the set $A_{mn}(n = 1, 2, \cdots)$ denotes an eigen vector corresponding to γ_m. The approximate solution using 4 eigen values and corresponding first 4 components is shown in Fig.1. Figures 2 ~ 5 show steady-state streamlines for a variety of conditions in the section passing through the central axis, where heat flux distribution q along the meridian is assumed to be given in the form of $q = \sum Q_k \sin(2k-1)\beta$ and Nu denotes a mean Nusselt number averaged over the wall of the ellipsoid. Figures 6 and 7 show the characteristics of circulation time with variation of a heat flux profile. For these (and moderate) Grashof numbers, the shorter the circulation time is, the stronger the convection would be.

4 Conclusions

It is found that the strength of circulating motion of liquid metal layers due to natural convection, which is typically given by the inverse of the circulation time, can be well controlled by the variation of profiles of heat transfer through the wall of the ellipsoid.

References

1. C.Sztur, F.Balestreri, J.L.Meyer, B.Hannart, 1990, Settling of inclusions in holding furnaces: modeling and experimental results, *Light Metals*, pp.709-716.

2. Y.Mochimaru,1989,Natural and forced convection from a horizontal circular cylinder to liquid metals, *Numerical methods in laminar and turbulent flow*, Vol.6, part 2, pp.1207-1215.

3. Y.Mochimaru,1992, Numerical simulation of forced convection heat transfer from a circular cylinder under a magnetic field, *Transport phenomena in heat and mass transfer /Proc. 4th Int. Sym. Transport Phenomena in Heat & Mass Transfer*, Vol.1, pp.373-383.

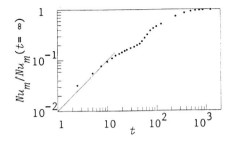

Fig.1 Transient characteristics of Nu_m at $Gr = 10^8$, $b/a = 1.5$, $Pr = 0.04$(aluminum), $Nu = 0.01$ with a uniform heat flux. ———: Eq.(11).

Fig.2 Steady-state streamlines at $Gr = 10^6$, $b/a = 0.5$, $Pr = 0.04$, $Nu = 0.01$; $Q_k = 0$ ($k \geq 2$). $\psi = 2 \times 10^{-5}$(inside), 10^{-5}, 0.5×10^{-5}, 0.

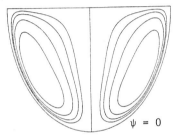

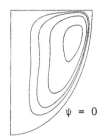

Fig.3 Steady-state streamlines at $Gr = 10^8$, $b/a = 1.5$, $Pr = 0.04$, $Nu = 0.01$; $Q_k = 0$ ($k \geq 2$). $\psi = -4 \times 10^{-5}$ (inside), -2×10^{-5}, -10^{-5}, -0.5×10^{-5}, 0.

Fig.4 Steady-state streamlines for a nearly stratified case at $Gr = 10^8, b/a = 1.5, Pr = 0.04, Nu = 0.01$ in the right half plane. $\psi = 10^{-7}, -10^{-7}, 10^{-7}$ (from the top to the bottom).

Fig.5 Steady-state streamlines with a uniform heat flux at $Gr = 10^8, b/a = 1.5, Pr = 0.04, Nu = 0.01$ in the right half plane. $\psi = -4 \times 10^{-4}$ (inside), $-2 \times 10^{-4}, -10^{-4}, -0.5 \times 10^{-4}$.

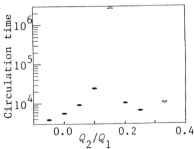
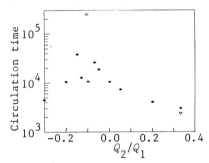

Fig.6 Characteristics of circulation time through a point near the bottom ($r = 0.3333, z = -0.4089$) at $Gr = 10^6, b/a = 0.5, Pr = 0.04, Nu = 0.01$. •: $Q_k = 0(k \geq 3)$, △: nearly stratified, ▽: with a uniform heat flux.

Fig.7 Characteristics of circulation time through a point near the bottom ($r = 0.2500, z = -1.3524$) at $Gr = 10^8, b/a = 1.5, Pr = 0.04, Nu = 0.01$. •: $Q_k = 0(k \geq 3)$, △: nearly stratified, ▽: with a uniform heat flux.

Multielement Systems with Moving Surface Boundary-Layer Control: Analysis and Validation

V.J. Modi[1], S.R. Munshi[2], G. Bandyopadhyay[3] T.Yokomizo[4]

[1,2,3] Department of Mechanical Engineering
The University of British Columbia
Vancouver, B.C., Canada V6T 1Z4
[4] Department of Mechanical Engineering,
Kanto Gakuin University
Matsuura, Kanzawa, Yokohama, Japan 236

1 Introduction

Ever since the introduction of the boundary-layer concept by Prandtl, there has been a constant challenge faced by scientists and engineers to minimize its adverse effects and control it to advantage. Over the years, boundary-layer separation has been achieved with varying degree of success by employing vortex-generators, suction, blowing, etc. A vast body of literature accumulated over years has been reviewed rather effectively by several investigators including Modi et al. [1,2] and others.

Irrespective of the method used, the main objective of a control procedure is to prevent, or at least delay, the separation of boundary-layer from the wall. A moving surface attempts to accomplish this in two ways:

(a) it retards growth of the boundary-layer by minimizing relative motion between the surface and the free stream;

(b) it injects momentum into the boundary layer.

Using rotating cylinders as momentum injecting elements, wind tunnel tests with a family of two-dimensional aerofoils have shown the Moving Surface Boundary-layer Control (MSBC) to be remarkably successful [1] in delaying the separation. This, in turn, resulted in increasing the maximum lift coefficient by more than 150% and delaying the stall to as high as $\alpha = 44°$. Similar application of the MSBC to two and three-dimensional rectangular prisms as well as tractor-trailer truck configurations [2] showed an impressive reduction in drag (24-53%) during wind tunnel studies using scale models.

With this as background, the present study focuses on:

---

[1] Professor; Fellow AIAA, ASME
[2] Graduate Fellow
[3] Visiting Research Associate; Associate Professor, Department of Aerospace Engineering, Indian Institute of Technology, Kharagpur, India
[4] Professor

(i) Numerical simulation of the MSBC as applied to both streamlined (aerofoil) as well as bluff geometry (D-section) multielement systems (Figure 1). Effects of the angle of attack and the momentum injection parameter U_C/U (U_C = cylinder surface velocity; U = freestream velocity) on the surface pressure distribution and aerodynamic coefficients are assessed.

(ii) Validation of the numerical results through extensive wind tunnel tests.

(iii) Flow visualization study to gain better appreciation of the complex flow.

2 Numerical Simulation

Simulation of fluid dynamical problems has been classically approached in two fundamentally different ways:

(a) modelling of the physical character of a phenomenon, as approached by Prandtl, through insight into the physics of the problem;

(b) simulation of the governing equations of motion as accomplished through finite element or difference schemes.

The precise numerical solution of this complex problem involving moving boundaries can be obtained by solving the general time-dependent Navier-Stokes equations incorporating a suitable turbulence model. However, for realistic values of the Reynolds number, this would demand enormous computational efforts and cost. On the otherhand, judicious modelling of the flow character can provide information of sufficient accuracy for all engineering design purposes with nominal computational tools and insignificant cost. To that end extension of the well developed panel code to multielement systems with momentum injection appeared quite attractive.

During the past three decades, the classical panel method involving distribution of surface-singularities has evolved to a sophisticated level where it can tackle complex geometrices and flow separation condition [3-5]. Maskew and Dvorak [3] modelled separated flow by 'free-vortex lines' having a known constant vorticity but initially of unknown shape. Successive iterations yield the converged wake shape. Ribaut [4] accounted for the vorticity dispersion through dissipation and diffusion leading to a finite wake. The first-order panel method employing linear varying vorticity along each panel and incorporating in the wake to model the separated flow [5] is also attractive.

In the present study of the multielement systems, represented by an aerofoil and the D-section with the momentum injecting rotating cylinder(s), each of the element is represented by a large number of panels (100-150). Each panel has a continuous distribution of linearly varying vorticity and a constant source strength. The separated flow is modelled by the 'free-vortex lines' emanating from the lower and upper separation points. The free-vortex lines are also discretized into panels. At the beginning of the free-vortex line, the vorticity strength is taken to be equal to that to the separation point. The vorticity is allowed to dissipate at a given rate along the free-vortex lines resulting in a finite wake. An iterative scheme leads to the final solution.

3 Validation thro' Wing Tunnel tests

To help validate the numerical procedure, an extensive wind tunnel test program was undertaken. Both the aerofoil and the D-section models with momentum injection elements were tested in the subsonic regime to assess the effects of system parameters such as the speed ratio U_C/U, angle of attack (α), cylinder surface roughness, etc. on the aerodynamic coefficients (C_P, C_L, C_D representing pressure, lift and drag coefficients, respectively). Here only a small sample of data, useful in discerning trends, are presented.

Figure 2(a) compares numerically obtained surface pressure distribution on a two-dimensional Joukowsky aerofoil with that measured during the wind tunnel tests for both without and with momentum injection. Note, even for such a complex multielement configuration with momentum injection the correlation is remarkably good. The momentum injection creates a large suction peak on the top surface near the leading edge and significantly delays the separation as indicated by a constant pressure in the wake ($x/c \approx 0.6$, $U_C/U = 4$). It may be pointed out that the aerofoil, in absence of momentum injection, stalled at around 10° . Thus the effect of the MSBC is to increase lift and delay stall for the streamlined body such as the Joukowski aerofoil. Figure 2(b) shows variation of the lift coefficient C_L with α, and angle of attack, as obtained through integration of the pressure distribution results obtained over a range of the angle of attack. Note, the panel method predicts the results with considerable accuracy sufficient for all practical design applications. A change in $C_{L,max}$ from 0.72 at $U_C/U = 0$ to 1.85 for $U_C/U = 4$ represents an increase of more than 150%. The corresponding delay in stall from 10° to 44° is remarkable and suggests promising V/STOL application.

The corresponding pressure distribution plots for a bluff geometry represented by the D-section are shown in Figure 3(a) for $\alpha = 0$. Again the numerical procedure predicts the trends rather accurately. Note, the effect of the momentum injection is to increase the base pressure as well as generate high pressure in the front of the D-section. As can be expected, this would result in a drag reduction. In the present case, it was 15% for $U_C/U = 4$. Corresponding reduction in the wake-width in presence of the MSBC is indicated in Figure 3(b).

4 Flow Visualization

To get better appreciation as to the physical character of the complex flow field as affected by the angle of attack and momentum injection parameters, extensive flow visualization study was undertaken. It also gave useful information about relative importance of the various system parameters and hence assisted in planning of the experiments as well as the numerical analysis. The flow visualization tests were carried out in a closed circuit water channel using polyvinyl chloride particles as tracers in the presence of slit lighting. The study confirming the trends suggested by the numerical results. Both still photographs as well as video movies were taken. Typical photographs for the Joukowsky aerofoil and the D-section are presented in Figure 4 and 5 respectively. They show, rather dramatically, effectiveness of the MSBC with the flow remaining essentially attached even at a high angle of attack.

5 Concluding Remarks

Based on the numerical, wind tunnel and flow visualization studies with a two-dimensional aerofoil and a D-section bluff geometry, it can be concluded that the momentum injection through moving surface represents an effective procedure for boundary-layer control. For a Joukowsky aerofoil, it resulted in an impressive increase in $C_{L,max}$ (1.85) and delay in stall (44°). A reduction in the drag coefficient by 15% at $\alpha = 0$ for the D-section is equally remarkable. The semipassive character of the procedure makes it quite attractive for real-life prototype application. The study of multielement streamlined and bluff geometries in the presence of MSBC, using the panel method and its substantiation through wind tunnel and flow visualization tests, has never been reported before. The current efforts are aimed at application of the concept to square and rectangular prisms, and tractor-trailer truck configurations.

Acknowledgments :

The investigation reported here was supported by the Natural Sciences and Engineering Research Council of Canada, Grant No. A-2181.

References

1. Modi, V.J., Mokhtarian, F., Fernando, M.S.U.K. and Yokomizo, T., "Moving Surface Boundary-Layer Control as Applied to Two-Dimensional Airfoils," *Journal of Aircraft, AIAA*, Vol.28, No.2, February 1991, pp.104-112.

2. Modi, V.J., Fernando, M.S.U.K. and Yokomizo, T., "Moving Surface Boundary-Layer Control: Studies with Bluff Bodies and Application," *AIAA Journal*, Vol.29, No.9, September 1991, pp.1400-1406.

3. Maskew, B. and Dvorak, F.A., "The Prediction of $C_{L_{max}}$ Using a Separated Flow Model," Journal of American Helicopter Society, Vol.23, April 1978, pp. 2-8.

4. Ribaut, M., "A Vortex Sheet Method for Calculating Separated Two-dimensional Flows," *AIAA Journal*, Vol.21, August 1983, pp. 1079-1084.

5. Mukherjea, S., and Bandyopadhyay, G., "Separated Flows about a Wedge," *The Aeronautical Journal of the Royal Aeronautical Society*, June-July 1990, pp. 196-202.

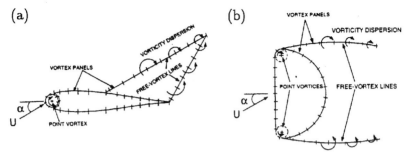

Figure 1 Schematic diagrams showing two distinctly different class of multielement systems, with the Moving Surface Boundary-layer Control (MSBC), used in the present study: (a) streamlined symmetrical Joukowski aerofoil with the leading edge replaced by a rotating element for momentum injection; (b) D-section, with twin rotating cylinders, representing a bluff system.

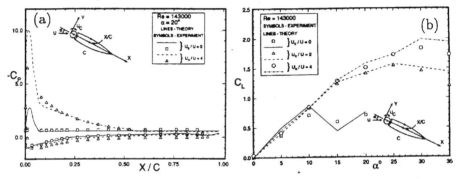

Figure 2 Comparison of the numerically obtained results with the experimental data for a Joukowski aerofoil with the MSBC: (a) pressure distribution at $\alpha = 20°$; (b) variation of the lift coefficient with angle of attack.

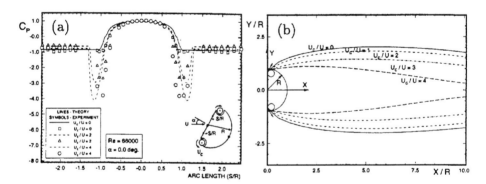

Figure 3 Typical results for the D-section at $\alpha = 0$ as affected by the MSBC: (a) comparison between numerically predicted and experimentally obtained pressure plots; (b) numerically obtained narrowing of the wake suggesting reduction in the pressure drag with the momentum injection.

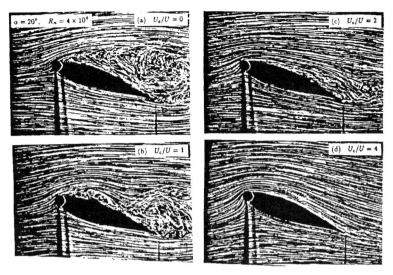

Figure 4 Representative flow visualization pictures showing, rather dramatically, effectiveness of the boundary-layer control through momentum injection. The observed results compare well with the delay in separation predicted by the panel code.

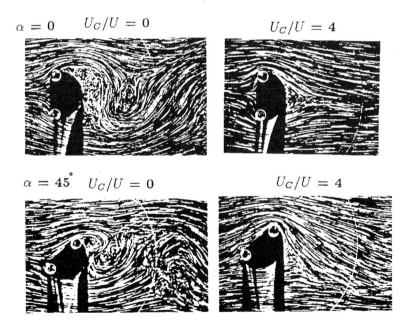

Figure 5 Flow visualization photographs showing effectiveness of the Moving Surface Boundary-layer Control (MSBC) as applied to the D-section prism. Note at $U_C/U = 4$, the flow tends to approach the potential character.

Prediction of turbulent delta wing vortex flows using an RNG $k - \epsilon$ model

N.O'Shea and C.A.J. Fletcher

CANCES, University of New South Wales,
Sydney 2052 Australia

1 Introduction

The formation of vortices over a delta wing at angle of attack is an important feature of such flows. With the demands for increased maneuverability in high performance aircraft, delta wing planforms or leading edge extensions (LEX) have become standard on modern fighter aircraft in order to promote primary vortex formation, and hence facilitate controlled flight in the post-stall flight regime. Development in computer capability over the last decade have allowed CFD methods to be employed to be model this phenomenon. While computations based on Euler equations [1,2] show that the primary vortex from sharp leading edge wings is well captured, secondary separation has only been captured when solving the laminar [3,4], and more recently the turbulent [5], Navier-Stokes equations.

The Reynolds number encountered in the real flow is very high and the flow is dominated by a separated shear layer from the leading edge that rolls up to form the primary vortex. Hence for an accurate prediction of the flow, some level of turbulence modelling is required. Generally, studies of these flows have employed the Baldwin-Lomax turbulence model [6,7] and the Johnson-King model [8] in the transonic regime. Good predictions have also been reported when using the two equation $k-\epsilon$ turbulence model [9,10]. However, all these models are empirically based and thus suffer from a lack of generality. The recently developed Renormalization Group theory (RNG) of turbulence [11] does not contain any empirically adjustable parameters and is therefore applicable to a wide range of flow situations. Recent reports in the literature [12] have indicated that the differential $k - \epsilon$ turbulence model based on the RNG theory of turbulence is effective in a 2-D situation when predicting the behaviour of flows that exhibit vortex shedding and are subjected to rapid streamline curvature. The standard $k - \epsilon$ model has tended to overpredict the turbulence and hence potentially damp out any of the finer flow details such as, for the delta wing case, secondary separation. However, inherent in the derivations of both the standard $k - \epsilon$ and the RNG $k - \epsilon$ models is the assumption of isotropic eddy viscosity. In this paper, the predictive capability of the RNG $k - \epsilon$ turbulence model is investigated and compared with results from a $k - \epsilon$ turbulence model for incompressible turbulent vortex flow over a delta wing.

2 Method

The Reynolds-averaged Navier-Stokes equations for steady flow are:

$$\frac{\partial}{\partial x_i}(\rho U_i) = 0 \tag{1}$$

$$\frac{\partial}{\partial x_i}(\rho U_i U_j + \delta_{ij}p - \tau_{ij} + \rho \overline{u_i u_j}) = 0 \tag{2}$$

where U_i is the mean velocity, u_j is the fluctuating velocity, ρ is the density, p is the pressure, τ_{ij} are the viscous stresses and $\rho \overline{u_i u_j}$ are the Reynolds stresses. The Reynolds stresses in (2) are calculated both by the standard $k - \epsilon$ turbulence model and the RNG $k-\epsilon$ turbulence model. The transport equation for the turbulent kinetic energy, k, is:

$$U_i \frac{\partial k}{\partial x_i} - \frac{\partial}{\partial x_i}\left[\alpha V_T \frac{\partial K}{\partial x_i}\right] = V_T \frac{\partial U_i}{\partial x_j}\left[\frac{\partial U_i}{\partial x_j} + \frac{\partial U_j}{\partial x_i}\right] - \epsilon \tag{3}$$

and is used in both models. The dissipation rate equation for the standard $k - \epsilon$ model is:

$$U_i\left[\frac{\partial \epsilon}{\partial x_i}\right] - \frac{\partial}{\partial x_i}\left[\alpha V_T \frac{\partial \epsilon}{\partial x_i}\right] = C_{\epsilon 1} \frac{\epsilon}{k} V_T \frac{\partial U_i}{\partial x_j}\left[\frac{\partial U_i}{\partial x_j} + \frac{\partial U_j}{\partial x_i}\right] - C_{\epsilon 2} \frac{\epsilon^2}{k} \tag{4}$$

And for the RGN $k - \epsilon$ model is:

$$U_i\left[\frac{\partial \epsilon}{\partial x_i}\right] - \frac{\partial}{\partial x_i}\left[\alpha V_T \frac{\partial \epsilon}{\partial x_i}\right] = C_{\epsilon 1} \frac{\epsilon}{k} V_T \frac{\partial U_i}{\partial x_j}\left[\frac{\partial U_i}{\partial x_j} + \frac{\partial U_j}{\partial x_i}\right] - (C_{\epsilon 2} + R)\frac{\epsilon^2}{k} \tag{5}$$

where R is a rate-of-strain term defined as:

$$R = \frac{C_\mu \eta^3 (1 - \eta/\eta_0)}{1 + \beta \eta^3}, \quad \eta = \frac{k}{\epsilon}\left[\frac{\partial U_i}{\partial x_j}\left[\frac{\partial U_i}{\partial x_j} + \frac{\partial U_j}{\partial x_i}\right]\right]^{1/2} \tag{6}$$

$$\beta = 0.015, \eta_0 = 4.38$$

The RGN theory gives values of the constants $C_\mu = 0.0845$, $C_{\epsilon 1} = 1.42$, $C_{\epsilon 2} = 1.68$, and $\alpha = 1.39$, as opposed to the standard $k - \epsilon$ model values of $C_\mu = 0.09$, $C_{\epsilon 1} = 1.40$, $C_{\epsilon 2} = 1.90$, and $\alpha = 1.0$. A single-layer wall function has been used to provide the solution and satisfy the boundary conditions close to the wing surface, with the first grid point at $y^+ \approx 50$–100.

A finite volume spatial discretisation of the governing equations is used [13]. The discretised equations are solved using a Strongly Implicit Procedure (SIP) at each iteration to obtain values for the velocity components, pressure and turbulence quantities. The velocity–pressure corrections are performed using an irrotational velocity potential correction and are coupled via a SIMPLEC algorithm [14].

3 Results & Discussion

The governing equations are solved around a 65° round leading edge delta wing, possessing a taper of 15% and an aspect ratio of 1.38. The cross–section is a symmetric modified NACA 64A005 section having a maximum thickness of 5% at 0.4c. A C-O mesh of dimension $50 \times 20 \times 18$ is generated about the wing using a transfinite interpolation technique. Fig. 1 shows a detail of the meshed delta wing within the C-O domain. Only half the wing is required given the symmetric nature of the flow.

All of the results presented are for a solution at angle of incidence equal to 20° and at a Reynolds number of 2.7×10^6. The freestream Mach No. is $M_\infty = 0.1$. Plots are shown at the $x/c = 0.9$ wing location. Fig. 2 indicates the cross–flow velocity vectors for the standard $k - \epsilon$ case. The RNG $k - \epsilon$ cross-velocity plot is very similar. The gross features of a delta wing flow are evident in this plot, showing the formation of the lee–side vortex and the flow separation from the leading edge. Figures 3 and 4 indicate the pressure distributions for both the standard $k - \epsilon$ and the RNG $k - \epsilon$ models. Again the general features of the flow are evident, with the minimum pressure occurring in the vortex core region. The solutions for pressure for both the turbulence models are very similar. The turbulent energy and turbulent dissipation were markedly reduced for the RGN $k - \epsilon$ model, thus resulting in lower values for turbulent viscosity V_t. Figures 5 and 6 indicate the variations in V_t for the two models. Despite the lower values in V_t, the computations show no evidence of separation occuring.

An investigation on the effect of 'R' in equation (5) revealed that this quantity was negligible compared to $C_{\epsilon 2}$, as evidenced in the plot shown in Figure 7. The η term in (6) is a function of velocity gradients in the flow field, and is plotted in Figure 8. Thus, in regions of large strain rate, such as in the separated shear layer that sheds off the leading edge to form the primary vortex, η should be very high. In such a situation, R can become negative, and the net '$C_{\epsilon 2+R}$' term reduces even further, resulting in even lower values for k and ϵ, and hence a further reduction in V_t.

The following factors are suggested as reasons for the unexpected behaviour of the R term and the lack of a secondary recirculating region in the flow solution:

a. Mesh Density - the separated shear layer occupies a very small area of the flow domain. In order to capture it's properties and thus improve the predictions of both the standard and $k - \epsilon$ models, a much finer grid resolution is required in the spanwise and normal directions.

b. Mesh stretching - errors are introduced into a finite volume formulation when control volumes are far from uniform or when spatial variation of the mesh occurs at too great a rate. Inspection of the grid used for this study indicate that this may be a factor.

c. Wall functions - although the RGN $k - \epsilon$ model can be used in conjunction with wall functions, its true value lies in the ability to solve for k and ϵ in low Re regions, such as the near-wall region. The results obtained here support the literature regarding viscous stress in regions of high strain rate. However further refinement of the mesh and/or treatment of the near-wall region must be addressed in order to capture the finer details of the flow topology when using the RGN $k - \epsilon$ model.

d. Turbulence Model - even though the RGN $k - \epsilon$ turbulence model overcomes of the deficiencies associated with the standard $k - \epsilon$ turbulence model, it shares the reliance on the isotropic eddy viscosity. The present problem has a marked variation in the magnitude of the Reynolds stresses [10]. Thus it is expected that the RGN formulation needs to be combined with an ASM or full Reynolds stress turbulence model to adequately predict the detailed flow behaviour for the above problem.

References

1 A. Rizzi, 'Euler Solutions of Transonic Vortex Flows around the Dillner Wing', J. Aircraft, 22, 325-328, 1985

2 A. Kumar & A. Das, 'Numerical Solutions of Flow Fields around Delta Wings using Euler Equations', Proc. International Vortex Flow Experiment on Euler Code Validation, Stockholm, pp. 175-186, 1986

3 A. Kumar, 'Flow Calculation over a Delta Wing using the Thin-Layer Navier-Stokes Equations', Proc. 3rd. Intl. Symposium on CFD, Nagoya, pp. 371-376, 1989

4 Joseph Vadyak & David M. Schuster, 'Navier-Stokes Simulation of Burst Vortex Flowfields for Fighter Aircraft at High Incidence',

5 Shreekant Agrawal, Raymond Matt Barnett & Brian Antony Robinson, 'Numerical Investigation of Vortex Breakdown on a Delta Wing', AIAA Journal, Vol. 30, No. 3, March 1992

6 C.-H. Hsu & C.H. Liu, 'Numerical Study of Vortex-Dominated Flows for Wings at High Incidence and Sideslip', J. Aircraft, Vol.29, No. 3, May-June 1992

7 J. Vadyak & D.M. Schuster, 'Navier-Stokes Simulation of Burst Vortex Flowfields for Fighter Aircraft at High Incidence', J. Aircraft, 28:638-645, 1991

8 U. Kanyak, E. Tu, M. Dindar & R. Barlas, 'Nonequilibrium Turbulence Modelling Effects on Transonic Vortex Flows About Delta Wings', AGARD-FDP Vortex Flow Aerodynamics, 1990

9 Y. Takakura, S. Ogawa & T. Ishioguro, 'Turbulence Models for 3D Transonic Viscous Flows II', Proc. 3rd. Intl. Symp. on CFD, nagoya, 680-685, 1989

10 A. Kumar & C.A.J. Fletcher, 'Prediction of Turbulent Lee-Side Vortex Flow', 13th International Conference on Numerical Methods in Fluid Dynamics, Rome 1992

11 Victor Yakhot & Steven A. Orszag, 'Renormalization Group Analysis of Turbulence'. Journal of Scientific Computing, Vol. 1, No. 1, 1986

12 Luigi Arttinelli & Victor Yakhot, 'RGN-based Turbulence Transport Approximations With Applications to Transonic Flows', AIAA 89-1950, 221-231

13 C.A.J. Fletcher, 'Gas Particle Industrial Flow Simulation using RANSTAD', Surveys in Fluid Mechanics, Vol. 18, 657-681, 1993

14 C.A.J. Fletcher, 'Computational Techniques for Fluid Dynamics', 2nd Edition, Vol. 2, Springer Verlag, 1991

15 M.V. Lowson, 'Visualisation Measurements of Vortex Flows', J. Aircraft, 28:320-327, 1991

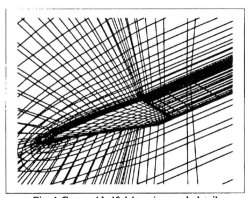

Fig. 1 Cropped half-delta wing mesh detail

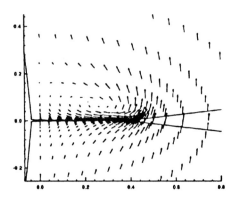

Fig. 2 Cross vector plot

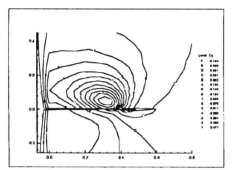

Fig. 3 Pressure Dist'n., k-ε model

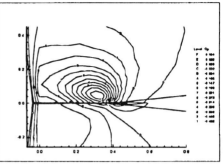

Fig. 4 Pressure Dist'n., RNG k-ε model

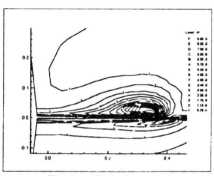

Fig. 5 v_t dist'n, k-ε model

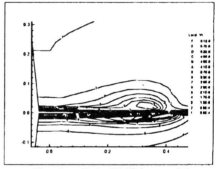

Fig. 6 v_t dist'n, RNG k-ε model

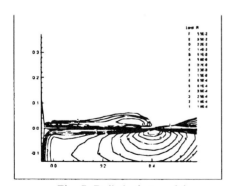

Fig. 7 R dist'n, k-ε model

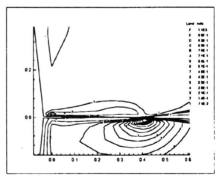

Fig. 8 η dist'n, RNG k-ε model

EVAPORATIVE FLOW IN HEAT PIPES
-An Object Oriented System Analysis-

Koichi Oshima[1] and Hanlin Chen[2]

[1]Toshiba Advanced Systems Corporation
[2] Canadian Space Agency

Abstract : Firstly, object oriented paradigm for system analysis is introduced, which is; to analyze the object into the elements, to construct the mathematical model to describe physical processes involved, and to derive the mathematical model of the total system. Simulation of the operating characteristics on a virtual reality environment is carried out. Secondly, this paradigm is applied to analyze a heat pipe system and a mathematical model of it is introduced and their operation characteristics are simulated. In particular, the phase changing flow taking place in heat pipes is discussed in detail.

1 Introduction

Heat pipes was introduced as a highly efficient heat transfer device in early 1960's, particularly for thermal control of spacecraft. This is a closed pipe provided with a wick structure on the inner pipe surface, in which a specified amount of fluid is enclosed. When one end of it is heated, the liquid contained evaporates and the vapour diffuses into the other end and condenses. The condensed liquid forms a liquid layer on the inside surface of the container and flows back through wick or groove structure to the heated end along the inside container wall. The driving force of the vapour from the heated end to the cooled end is the diffusion effect and the one of the liquid in the opposite direction is the capillary effect.

In such classical heat pipe theory, it is assumed that the vapour phase in the pipe is in thermodynamically equilibrium, and the fluid evaporated at the evaporator section is refilled without energy loss [1]. Therefore, it can predict only the maximum heat transfer rate limited by the driving force of liquid due to this capillary effect, assuming that the whole thermodynamic state in the pipe is equilibrium and stationary, which results in zero heat transfer resistance of the heat pipe. Then, an obvious question arises, that is, where goes the momentum loss due to the viscous resistance of the returning flow.

The answer is simple: the vapour in the pipe is in thermodynamically equilibrium state, in fact, at the saturation condition, which means that the vapour pressure and temperature are uniquely related. Throughout the whole length of the pipe, the vapour and the liquid contact each other forming a meniscus. The liquid phase occupies continuously whole length of the pipe and its pressure decreases due to the viscous resistance along the pipe length. It is compensated by the capillary pressure at the meniscus in order to balance the vapour and liquid pressures. The liquid which flows into the evaporator section has lower Ω pressure and necessarily lower temperature, which must be raised by heating up to the vapour temperature in order to be able to evaporate. This is the energy needed to operate the heat pipes under thermodynamically equilibrium condition. From this simple story, we may understand that the heat pipes are one of the complicated system and in order to analyze them it must be treated as a system, taking into account the interactions of the liquid and vapour flows. This is of nominal operational condition, which tends to be disturbed at excessively high heat load as well as at a low temperature close to the freezing point of the working fluid. Such limiting cases have been also analysed extensively [2,3].

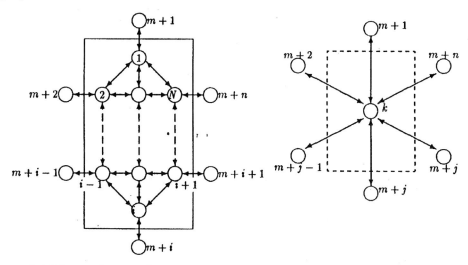

(a) Thermal network (b) Higher class network

Fig.1: Thermal nodal system

2 Object Oriented Paradigm

Owing to the recent progress in computer technology, system analysis of such complex systems as spacecraft, nuclear power plant or celestial environmental problems has become a reality, in which the mathematical modelling of the objects, the numerical analysis of the model equations and the simulation using the artificial reality environment are combined to produce an indispensable tool of system analysis. These process are based on the paradigm of object oriented analysis, in which the object system is divided into subsystems and these subsystems are divided into subsubsystems [1]. These hierarchy structures are analyzed starting from the lowest elements and are combined to construct the higher systems.

The fundamental principle of these system analysis is the conservation laws and the control volume formulation. Each element which consists of a system is assumed to have a scalar conservation quantity, fluxes of the conservation quantity to and from the other elements, and the source terms which are the conservation quantities supplied from outside of the system. As the conservation quantities, mass, momentum and energy are most commonly used. For thermal problems, temperature times heat capacity is sometimes considered [2].

In the case of continuum media, the finite volume formulation of difference form reduces to the same form of equations, this control volume formulation is applicable to discontinuous boundary, such as solid body and surrounding fluid, or two solid bodies contacting each other. In the latter case, taking temperature as the conserving quantity, conservation quantity is the volume integral of the temperature, and the flux is the heat flow through the contacting, and the radiation heat exchange with the other element and the convective heat exchange with the surrounding fluid are included. In order to determine these flux terms, numerical analysis of heat and mass flow is the most powerful tool.

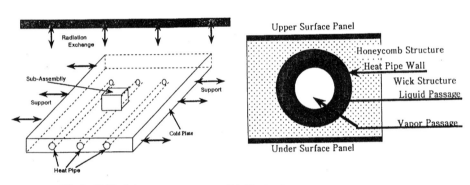

Fig.2: Cold plate   (b) Fig.3: Cross section of heat pipe

3 Finite Volume Formulation

We consider a scalar quantity which is a function of time t and space coordinate x, and a closed region Ω filled by a continuum media, which is surrounded by the boundary Γ. The conservation law of u contained in this region Ω is expressed as

$$\int_\Omega \frac{\partial u}{\partial t} dV = -\int_\Gamma \Phi \times n\, dS + \int_\Omega f\, dV \tag{1}$$

where dV is the volume element ($= dx_1\, dx_2\, dx_3$), n is the outward unit normal vector of the surface Γ. Here the quantity u is called "stock", and the first and second terms of the right hand side are called "flux" and "source", respectively. Using Gauss's divergence theorem, we have

$$\int_\Omega \left(\frac{\partial u}{\partial t} + \text{div}\,\Phi - f \right) dV = 0 \qquad (2)$$

This is the integral form of the conservation law of u, and, if u is continuous inside Ω, the integrand of (1) is also zero, which is called the differential form of the conservation law.

When we take u as the thermal energy $\rho c T$ contained in the unit volume, thermal nodal network is constructed.

4 Derivation of higher class equation

Here we construct the higher class nodal equation from the lower class ones. A subsystem, consisting of m nodal point, is considered and find out the relation so that this subsystem can be considered as a single node. In Fig.1, this relation is schematically shown. This system consists of m nodal point, out of which n nodes are exchanging heat with the outer nodes. Then, the nodes 1 to m belong to this subsystem, out of which the node 1 to n are interacting with the outer nodes, which are correspondingly numbered as from $m+1$ to $m+n$.

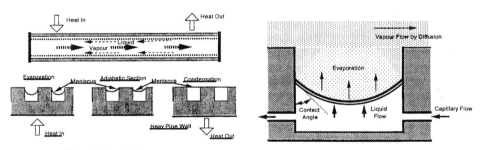

Fig.4: Physical process in heat pipe   Fig.5: Meniscus in heat pipes

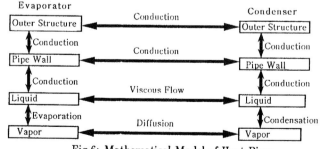

Fig.6: Mathematical Model of Heat Pipes

Then the thermal nodal equations are

$$C_i \frac{dT_i}{dt} + \sum_{j=1}^{m} D_{i,j}(T_i - T_j) = -D_{i,i+m}(T_i - T_{i+m}) + Q_i \qquad i = 1, 2, \cdots, m, \qquad (3)$$

where $D_{i,i+m}|_{i>n} = 0$. Therefore, this subsystem may be substituted by a single node k as

$$C_k = \sum_{i=1}^{m} C_i, \qquad T_k = \frac{\sum_{i=1}^{m} C_i T_i}{C_k}, \qquad Q_k = \sum_{i=1}^{m} Q_i \qquad (4)$$

Since

$$D_{k,i}(T_k - T_i) = D_{i,i+m}(T_i - T_{i+m}) \quad \text{Then} \quad D_{k,i} = \frac{T_i - T_{i+m}}{T_k - T_I} D_{i,i+m} \qquad (5)$$

Thus, higher class of nodal system is constructed.

5 Heat pipe system

An example of heat pipe system is shown in Fig.2, which is a cold plate provided with heat pipes for temperature equalization. An example of heat pipe cross section is shown in Fig.3. In order to construct the detailed heat transfer model, the heat transfer and the mass flow processes in a system of vapour, liquid and solid pipe wall have been modeled as seen in Fig.4. Figure 5 shows detail of meniscus. Figure 6 is the mathematical model of the heat pipe operation. Figure 7 schematically shows the distributions of temperature and liquid pressure in heat pipes.

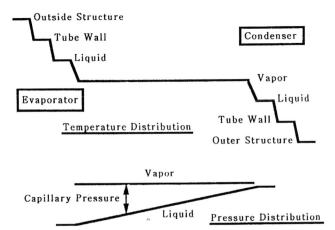

Fig.7: Physical Properties in Heat Pipes

In this study, the flow and heat transfer equations applicable to such system are solved numerically, which two-dimensionally simulate a region in the evaporating or the condensing zones. As the first step, the meniscus shape is determined based on the mass conservation relation within a capillary channel under the surface tension forces. Dynamical effect is known to be minor, and the meniscus shape depends only on the Bond number and the contact angle, which is left as a parameter to be determined later. Thus the relation between meniscus curvature and the pressure difference is found with the parameter of the contact angle [3].

Then the flow equation; two-dimensional, incompressible viscous Navier-Stokes equations is solved, taking into account the thermocapillary convection. As the boundary conditions are taken into account the evaporative heat transfer across the liquid-vapour boundary-evaporation. The finite volume formulation with 4th order in space and 2nd order in time is applied on a body fitted grid system. The buoyant convection was not taken into account. The simulated results give the flow fields due to the Marangoni effect. The heat transfer rate through the meniscus-the evaporation resistance - is found as the functions of the Marangoni number and the pressure difference.

This analysis gives quantitative estimation of the heat transfer at the evaporation zone, and the same analysis is applied to the condensing zone also. Thus a total heat transfer rate is reduced numerically without using experimental correlation.

References

1 K. Oshima et al. (1988) Heat Pipe Engineering; Asakura 1988.

2 Peter Coad, Edward Yordon (1991) Object-Oriented Analysis; Prentice Hall 1991

3 H. Chen and Y. Lee; Numerical analysis of thermocapillary convection and heat transfer in two-dimensional capillary cavity; CFD Journal, vol. 1 pp. 481-494.

Numerical Study of Viscous Compressible Flow Inside Planar and Axisymmetric Nozzles

B. Ramakrishnananda, M. Damodaran

School of Mechanical and Production Engineering,
Nanyang Technological University, Singapore

1 Introduction

The computation of internal flows is primarily motivated by the desire to improve the design and performance of components such as aircraft engine exhaust nozzles, inlets to propulsion systems, turbomachinery blade passages and so on. In this work steady, subsonic, transonic and supersonic flowfields in planar and axisymmetric nozzles are analysed numerically by modifying the numerical scheme implemented in Damodaran et al. [1] which incorporates the formulation strategies proposed by Jameson et al. [2–3] and Martinelli et al. [4].

2 Formulation

The finite volume formulation is established by employing a cartesian system of co-ordinates (x-y) for two-dimensional flow. For the axisymmetric flow, while x denotes the distance along the centerline of the flow, y denotes the radial distance from the centerline. For a two-dimensional or axisymmetric flow field analysis, the integral form of the Navier Stokes equations may be expressed as:

$$\frac{\partial}{\partial t} \iint W \, dx \, dy + \int (F \, dy - G \, dx) + \alpha \iint (H + H_v) \, dx \, dy = 0 \qquad (1)$$

where
$$W = (\rho, \rho u, \rho v, \rho E)^T$$

$$F = \left(\rho u, \rho u^2 + p + \sigma_{xx}, \rho uv + \tau_{xy}, (\rho E + p)u + u\sigma_{xx} + v\tau_{xy} - k\frac{\partial T}{\partial x}\right)^T$$

$$G = \left(\rho v, \rho uv + \tau_{yx}, \rho v^2 + p + \sigma_{yy}, (\rho E + p)v + u\tau_{yx} + v\sigma_{yy} - k\frac{\partial T}{\partial y}\right)^T$$

$$H = \frac{1}{y}(\rho v, \rho uv, \rho v^2, (\rho E + p)v)^T, \quad H_v = \frac{1}{y}(0, H_{v2}, H_{v3}, H_{v4})^T$$

where
$$H_{v2} = \tau_{xy}, \quad H_{v3} = \sigma_{yy} + \tau_{\theta\theta}, \quad H_{v4} = u\tau_{xy} + v\sigma_{yy} - k\frac{\partial T}{\partial y}$$

The components of the viscous stress terms are defined as follows:

$$\sigma_{xx} = \frac{2}{3}\mu\left(\frac{\partial u}{\partial x} + \frac{\partial v}{\partial y}\right) - 2\mu\frac{\partial u}{\partial x} + \alpha\frac{2}{3}\mu\frac{v}{y}, \quad \sigma_{yy} = \frac{2}{3}\mu\left(\frac{\partial u}{\partial x} + \frac{\partial v}{\partial y}\right) - 2\mu\frac{\partial v}{\partial y} + \alpha\frac{2}{3}\mu\frac{v}{y}$$

$$\tau_{xy} = \tau_{yx} = -\mu\left(\frac{\partial u}{\partial x} + \frac{\partial v}{\partial y}\right), \quad \tau_{\theta\theta} = -\mu\left(\frac{2}{3}\left(\frac{\partial u}{\partial x} + \frac{\partial v}{\partial y}\right) - \frac{4v}{3y}\right)$$

In the above equations p, ρ, u, v, T, E, μ and k are the pressure, density, cartesian x- and y- velocity components, temperature, specific total energy, total viscosity and coefficient of thermal conductivity. The pressure is obtained from the equation of state for a perfect gas. Nozzles with a circular arc constriction in the middle are chosen for the analysis. The Euler equations of motion are derived by setting all the viscous stress terms in the flux vectors F and G and H_v to zero in Eq. (1). The pressure and density are scaled by their inlet stagnation values. The velocities are scaled by $a_\infty/\sqrt{\gamma}$ where a_∞ is the speed of sound at the inlet and γ is the ratio of the specific heats of the gas. All lengths are scaled by the extent of the constriction in the nozzle. The coefficient of viscosity is scaled by $\mu_\infty Re/\sqrt{\gamma}M_\infty$ where Re is the Reynolds number and M_∞ is the Mach number of the flow at the nozzle inlet. A simple Sutherland's formula [5] is used to calculate the laminar viscosity coefficient. The turbulent transport coefficients are modelled by the Baldwin-Lomax turbulence model [6]. The switch $\alpha = 0$ represents two-dimensional planar flow while $\alpha = 1$ represents axisymmetric flow.

3 Numerical Scheme

Applying the integral law of Eq. (1) separately to each cell and supplementing them by numerical dissipative terms results in a system of ordinary differential equations for the average values of the dependent variables of the following form:

$$\frac{d}{dt}(hW)_{i,j} + Q_{i,j} + \alpha(H + H_v)h_{i,j} - D_{i,j} = 0$$

where $h_{i,j}$ is the cell area and $Q_{i,j}$ is the net flux out of the cell whose location in the physical space is denoted by the i-th and the j-th coordinate. This flux can be written as:

$$Q_{i,j} = \sum_{k=1}^{k=4}(F\Delta y - G\Delta x)$$

where Δx and Δy are the increments in the x and y directions along the k-th face of the cell. Green's theorem is applied to evaluate the first-order derivatives of the viscous terms appearing in the flux evaluation. On a sufficiently smooth mesh, this approximation is second-order accurate. The numerical dissipative terms $D_{i,j}$ are constructed by blending second and fourth difference terms in the flow variables and are scaled by the adaptive dissipative fluxes in the streamwise and normal directions. These equations are then advanced in time from a set of initial conditions using a suitable multi-stage time-stepping scheme till a steady state is achieved.

Uniform inlet conditions are specified throughout the domain as initial conditions. Flow tangency condition is applied at the walls for inviscid flow computations. Pressure at the wall is derived from the momentum equations. For the viscous computations, no-slip condition is applied at the nozzle walls which are assumed to be adiabatic. At the inflow boundary, characteristic formulation of boundary conditions [7] is applied. Appropriate symmetry conditions are imposed at the centerline for axisymmetric nozzles. For inviscid computations, the characteristic formulation is retained at the outflow boundary. For viscous outflow boundaries, all the variables are extrapolated from the interior if the local Mach number at the outflow boundary is supersonic. Otherwise, the exit pressure is specified and all the other variables are extrapolated from the interior. An initial estimate of this exit pressure is obtained from the corresponding inviscid quasi one-dimensional nozzle flow solution. This exit pressure is adjusted during the course of the computation so that the specified average inlet Mach number is obtained at the inflow boundary. Convergence acceleration strategies for the numerical scheme such as local time stepping, implicit residual smoothing and multi-grid strategies [3-4] have been implemented.

4 Results and Discussion

All the computations have been carried out on an algebraically generated 192x32 cell grid with clustering near the solid boundaries. About 4 to 5 orders of reduction in the r.m.s value of the density residual have been obtained for most cases at convergence. The Reynolds number for all the viscous calculations shown is about 1.6×10^6. For planar nozzle computations, a rectangular nozzle with a circular arc bump on the lower wall tested in Ref [8-10] is chosen. The bump is 4 percent thick for supersonic flows and 10 percent thick for all the other cases. The axisymmetric nozzles chosen for the analysis have the same quasi-one-dimensional inviscid flow solution as the corresponding planar nozzle. The centerline of the axisymmetric nozzle is made to coincide with the upper wall of the planar nozzle.

Figs. 1-8 show selected results obtained for inviscid and viscous flows. The planar inviscid results obtained from the present work are in good agreement with similar results already available in literature [10]. Figs. 1-4 pertain to planar nozzles while Figs. 5-8 pertain to axisymmetric nozzles. Fig. 1 and Fig. 2 compare supersonic inviscid and viscous turbulent flowfields inside a planar nozzle when the inlet Mach number is 1.6. The structure of the shock reflection and interaction patterns can be clearly seen in Fig. 1 and Fig. 2. Fig. 3 shows the inviscid flowfield when the inlet Mach number is 0.675 leading to transonic flow inside the planar nozzle. A shock is formed in the aft portion of the bump on the lower wall. The viscous transonic flow in Fig. 4 showing a separated region after the shock occurs when the inlet Mach number is around 0.66. Figs. 5-7 show the flowfield inside an axisymmetric nozzle when the inlet Mach number is such that it leads to choking. For inviscid flow, choking takes place when the inlet Mach number is 0.678 while for viscous flow this inlet Mach number is 0.663. The reduction in the choking inlet Mach number for the viscous case is attributed to the boundary layer on the nozzle walls. Fig. 7 shows the total pressure losses associated with the trailing edge shock and the boundary layer. Fig. 8 shows the flowfield results inside an axisymmetric nozzle when the flow is subsonic throughout the nozzle for an inlet Mach number of 0.5.

Acknowledgements:

The computational resources of the Centre for Graphics and Imaging Technology of Nanyang Technological University have been used for this study. Financial support for the first author is from a Graduate Research Scholarship offered by Nanyang Technological University.

References

[1] Damodaran, M. and Lee, A.K.H. (1993): "Finite Volume Calculations of Steady Viscous Compressible Flow Past Airfoils", in Journal of The Institution of Engineers, Singapore, Vol. 33, No. 5

[2] Jameson,A., Schmidt,W. and Turkel,E. (1981): "Numerical Solution of The Euler Equations by Finite Volume Methods Using Runge-Kutta Time-Stepping Schemes", AIAA Paper 81-1259

[3] Jameson,A. and Baker,T.J. (1984): "Multi-Grid Solution of The Euler Equations For Aircraft Configurations", AIAA Paper 84-0093

[4] Martinelli,L. and Jameson,A. (1988): "Validation of a Multigrid Method For The Reynolds Averaged Equations", AIAA Paper 88-0414

[5] Schlichting, H. (1979): *"Boundary Layer Theory"* (McGraw-Hill, New York)

[6] Baldwin, B.S. and Lomax, H. (1978): " Thin layer Approximation and Algebraic Model for Separated Turbulent Flows", AIAA Paper 78-257

[7] Pulliam, T.H.: (1981): "Numerical Boundary Conditions Procedures", NASA CP-2201

[8] Rhie,C.M. (1986): "A Pressure Based Navier-Stokes Solver Using The Multigrid Method", AIAA Paper 86-0207

[9] Kallinderis,J.G. and Baron,J.R. (1987): "Adaptation Methods For a New Navier-Stokes Algorithm", AIAA Paper 87-1167

[10] Jentink, T.N. (1989): " Formulation of Boundary Conditions for the Multigrid Acceleration of the Euler and Navier-Stokes Equations", M.S. Thesis, Purdue University

Fig. 1. Inviscid supersonic flow - Pressure contours

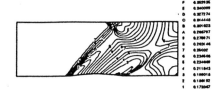

Fig. 2. Turbulent supersonic flow - Pressure contours

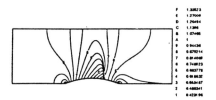

Fig. 3. Inviscid transonic flow - Mach number contours

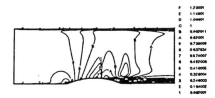

Fig. 4. Turbulent transonic flow - Mach number contours

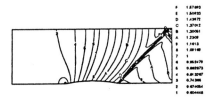

Fig. 5. Inviscid choked flow - Mach number contours

Fig. 6. Turbulent choked flow - Mach number contours

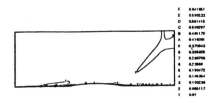

Fig. 7. Turbulent choked flow - Total pressure loss contours

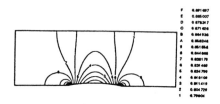

Fig. 8. Inviscid subsonic flow - Pressure contours

Numerical Simulation of an Air Flow past a Moving Body with/without Deformation

Nobuyuki Satofuka[1] and Toshio Nishitani[2]

[1] Department of Mechanical & System Engg., Kyoto Institute of Tech.,
Matsugasaki, Kyoto 606, Japan.
[2] Engg. Lab., TORAY Company Inc., Sonoyama, Otsu 520, Japan.

Abstract : Numerical solution of flowfields around a deformable body is carried out. The governing compressible equations are spatially discretised using a central difference approximation and subsequent time integration is carried out using the rational Runge-Kutta(RRK) scheme. Furthermore, higher-order far-field boundary conditions utilizing the Riemann-like variables have been applied. An overset grid technique is adopted in solving the flowfields around a deforming body in a duct flow. Firstly, numerical calculations are carried out for a flowfield past a deformable cylindrical body in a duct with parallel walls. Secondly vibrating string in an air flow in a duct is computed. From above studies, the present numerical method indicates its effectiveness in solving moving body problems with/without deformation.

1 Introduction

There are a number of unsteady fluid flow phenomena in which the surface of the body deforms in accordance with time, in both external and internal flow problems. A flowfield past a super high-rise building in rolling motion, a blood flow in a vessel with an elastic wall and propulsion mechanism of swimming motion of fish etc. are typical examples of this kind of problems. Numerical analysis of moving or deforming body in the flowfield is of great importance to practical use and application[1].

In this paper numerical results are presented for compressible flows past moving bodies with/without deformation. Firstly, unsteady flowfields around a deforming cylindrical body are investigated. In the second computed result is presented for a flowfield past a string vibrating in an air flow.

2 Governing Equations

The two-dimensional compressible Reynolds-averaged Navier-Stokes equations transformed to the general curvilinear coordinate system (ξ, η, τ) can be written in dimensionless, conservation law form as follows [2];

$$\frac{\partial \hat{\mathbf{q}}}{\partial \tau} + \frac{\partial \hat{\mathbf{E}}}{\partial \xi} + \frac{\partial \hat{\mathbf{F}}}{\partial \eta} = \frac{1}{Re}\left(\frac{\partial \hat{\mathbf{E}}_v}{\partial \xi} + \frac{\partial \hat{\mathbf{F}}_v}{\partial \eta}\right) \qquad (1)$$

where the convective flux vectors $\hat{\mathbf{q}}$, $\hat{\mathbf{E}}$, $\hat{\mathbf{F}}$, are

$$\hat{\mathbf{q}} = J^{-1} \begin{bmatrix} \rho \\ \rho u \\ \rho v \\ e \end{bmatrix}, \quad \hat{\mathbf{E}} = J^{-1} \begin{bmatrix} \rho U \\ \rho u U + \xi_x p \\ \rho v U + \xi_y p \\ (e+p)U - \xi_t p \end{bmatrix}, \quad \hat{\mathbf{F}} = J^{-1} \begin{bmatrix} \rho V \\ \rho u V + \eta_x p \\ \rho v V + \eta_y p \\ (e+p)V - \eta_t p \end{bmatrix} \tag{2}$$

$$\hat{\mathbf{E}}_v = J^{-1} [\xi_x \mathbf{E}_v + \xi_y \mathbf{F}_v]$$
$$\hat{\mathbf{F}}_v = J^{-1} [\eta_x \mathbf{E}_v + \eta_y \mathbf{F}_v] \tag{3}$$

Here ρ is the density, u and v are the velocities, p is the pressure, and e is the total energy per unit volume defined by

$$e = \frac{p}{\gamma - 1} + \frac{\rho}{2}(u^2 + v^2) \tag{4}$$

The so-called contravariant velocities considering the motion of the grids are as follows.

$$U = \xi_t + \xi_x u + \xi_y v \qquad V = \eta_t + \eta_x u + \eta_y v \tag{5}$$

The viscous flux vectors Ev, Fv are

$$E_v = \begin{bmatrix} 0 \\ \tau_{xx} \\ \tau_{xy} \\ \phi_x \end{bmatrix}, \quad F_v = \begin{bmatrix} 0 \\ \tau_{xy} \\ \tau_{yy} \\ \phi_y \end{bmatrix} \tag{6}$$

The stress terms are given as follows.

$$\tau_{xx} = \frac{\mu}{3} [4(\xi_x u_\xi + \eta_x v_\eta) - 2(\xi_y v_\xi + \eta_y v_\eta)]$$
$$\tau_{xy} = \mu (\xi_y u_\xi + \eta_y u_\eta + \xi_x v_\xi + \eta_x v_\eta)$$
$$\tau_{yy} = \frac{\mu}{3} [-2(\xi_x u_\xi + \eta_x u_\eta) + 4(\xi_y v_\xi + \eta_y v_\eta)] \tag{7}$$
$$\varphi_x = u\tau_{xx} + v\tau_{xy} + \frac{\gamma}{\gamma - 1} \frac{\mu}{Pr} (\xi_x T_\xi + \eta_x T_\eta)$$
$$\varphi_y = u\tau_{xy} + v\tau_{yy} + \frac{\gamma}{\gamma - 1} \frac{\mu}{Pr} (\xi_y T_\xi + \eta_y T_\eta)$$

Re represents the reference Reynolds number, Pr the Prandtl number, μ the viscosity coefficient, and γ the ratio of specific heats.

In general coordinate transformation, the inverse Jacobian J^{-1} is defined by

$$J^{-1} = x_\xi y_\eta - x_\eta y_\xi \tag{8}$$

and the metric coefficients are derived by use of chain rule expressions as follows:

$$\xi_x = J y_\eta$$

$$\xi_y = -J x_\eta$$

$$\eta_x = -J y_\xi \tag{9}$$

$$\eta_y = J x_\xi$$

$$\xi_t = -x_\tau \xi_x - y_\tau \xi_y$$

$$\xi_t = -x_\tau \eta_x - y_\tau \eta_y$$

Hereafter, the symbol ∧ , which denotes the vector quantity divided by Jacobian J, may be omitted for simplicity without any confusion, unless a specific note is provided. In the case of $Re \longrightarrow \infty$, Eqs.(1) reduce to the Euler equations.

3 Numerical Procedure and Boundary Conditions

In our method of lines approach[3], the Navier-Stokes equations (1) are first discretized by the conventional central finite-difference approximation as follows:

$$\tfrac{d}{dt} \mathbf{q}_{i,j} = -(\mathbf{E}_{i+1,j} - \mathbf{E}_{i-1,j})/(2\Delta\xi) - (\mathbf{F}_{i,j+1} - \mathbf{F}_{i,j-1})/(2\Delta\eta)$$

$$+ (\mathbf{E}_{vi+1/2,j} - \mathbf{E}_{vi-1/2,j})/(Re\Delta\xi) + (\mathbf{F}_{vi,j+1/2} - \mathbf{F}_{vi,j-1/2})/(Re\Delta\eta) + D_\xi + D_\eta. \tag{10}$$

where subscript i and j denote the grid indexes such as $q_{i,j} = q(i\Delta\xi, j\Delta\eta)$. In order to eliminate spurious oscillations and capture shock waves, the dissipative terms D_ξ and D_η are added in the right hand side of Eq.(10). The dissipative terms used here are based on the second and fourth differences, which was first introduced by Jameson[4]. The term can be written for ξ-direction as:

$$D_\xi = d_{i+1/2,j} - d_{i-1/2,j}, \tag{11}$$

with the dissipative flux

$$d_{i+1/2j} = \left[\epsilon^{(2)}_{i+1/2,j} \Delta_\xi (J\mathbf{q})_{i,j} - \epsilon^{(4)}_{i+1/2,j} \Delta_\xi^3 (J\mathbf{q})_{i-1,j} \right] /(J\Delta t_\xi)_{i+1/2,j} \tag{12}$$

Here Δ_ξ is forward difference operator, Δt_ξ is the certain time step specified later, and $\epsilon^{(2)}$ and $\epsilon^{(4)}$ are defined as

$$\epsilon^{(2)}_{i+1/2,j} = \omega^{(2)} \max(v_{i+1,j}, v_{i,j}), \epsilon^{(4)}_{i+1/2,j} = \max(0, \omega^{(4)} - \epsilon^{(2)}_{i+1/2,j}) \tag{13}$$

where $\omega^{(2)}$ and $\omega^{(4)}$ are adaptive coefficients and $v_{i,j}$ is obtained from the second difference of pressure.

For a time stepping scheme, the rational Runge-Kutta (RRK) scheme of Wambecq[5] is used. The system of ordinary differential equations(10) can be rewritten as:

$$\frac{d\vec{q}}{dt} = \vec{W}(\vec{q}) \tag{14}$$

Then the second order RRK scheme may be written in the following two stage form.

$$\vec{g}_1 = \Delta t \vec{W}(\vec{q}^{\,n}), \vec{g}_2 = \Delta t \vec{W}(\vec{q}^{\,n} + c_2 \vec{g}_1),$$

$$\vec{g}_3 = b_1 \vec{g}_1 + b_2 \vec{g}_2, \tag{15}$$

$$\vec{q}^{n+1} = \vec{q}^{\,n} + [2\vec{g}_1(\vec{g}_1, \vec{g}_3) - \vec{g}_3(\vec{g}_1, \vec{g}_1)]/(\vec{g}_3, \vec{g}_3),$$

where superscript n denotes the index of time steps and $(\vec{g}_i, \vec{g}_j)$ denotes the scalar product of two vector $\vec{g}_i$ and $\vec{g}_j$. An efficient second order scheme is given by the coefficients

$$b_1 = 2, b_2 = -1, c_2 = 0.5 \tag{16}$$

Higher-order far-field boundary conditions utilizing the Riemann-like variables[6] have been applied to the inflow and outflow boundary conditions. A no-slip or free-slip velocity condition is imposed on the body boundary with time dependent information. An overset grid technique[7] is adopted and the minor grid is generated at each time step by a combined use of both algebraic method and elliptic PDE method.

4 Numerical Results

4.1 Flowfield around periodically deforming body

Firstly we investigated the flowfield around periodically deforming cylinder in a duct. The inlet width of the duct is 4 and the length of the duct is 12, and uniform Mach number is 0.3 and Reynolds number is 5000. Initially the cylindrical body the diameter of which is one is set in the duct and then deforms its cross-sectional shape periodically between a circle and an ellipse.

If we take the x axis the direction of the uniform flow, the location of the surface of the body is defined as follows,

$$\frac{x_s^2}{a^2} + \frac{y_s^2}{b^2} = 1 \tag{17}$$

$$\begin{aligned} a &= 0.5 + 0.25 \sin \omega t \\ b &= 0.25/a \\ \omega &= 0.25 \end{aligned} \tag{18}$$

keeping the area of the body constant in time advance.

Fig. 1. shows the computed time history of the lift coefficient (CL), drag coefficient (CD) and a length of one axis of the ellipse(BT) change with time(a) and iso-bar lines at $t = 8(b)$.

The lift coefficient changes in time in an inconsistent manner, owing to the differences between upper and lower flowfield of the body.

4.2 Flowfield past a string vibrating in an air flow

Secondly, we investigated the flow around a string vibrating in an air flow in a duct. The motion of the string is assumed to be governed by the wave equation. Fig. 2. shows a schematic of the problem. The center line shape of the string with thickness D is described by the following equation,

$$L(x) = 0.05\sin(s\pi x) \tag{19}$$

and the positions of H1 and H2 in Fig. 2. are fixed.

Fig. 3. displays the evolution of the computed iso-bars lines, while Fig. 4. shows the computed time history of the drag coefficient (CD), the lift coefficient (CL) and y coordinate at 3/4 chord of the string(BTX).

5 Conclusions

An explicit method of lines approach combined with an overset grid technique is adopted for solving the unsteady compressible flowfields around a deformable body in the air. The computed results both of a deforming cylindrical body and of a vibrating string in a duct flow indicate that this method is effective for obtaining the unsteady flowfields around a deforming body in the air and can show the possibility of resolving more complicated problems.

References

1. Jari Hyvärinen, Integration of Computational Fluid Dynamic and Structural Mechanics Simulations, Technical Paper of the 5th Int. Symp. on Computational Fluid Dynamics - Sendai, 1993.

2. Nakamichi, J., A Verification of Unsteady Navier-Stokes Solutions Around Oscillating Airfoils, NASA-TM-88341, 1986.

3. Satofuka, N., Morinishi, K and Tokunaga, H., Notes on Numerical Fluid Mechanics, 13, 319, Vieweg, 1986.

4. Jameson, A., and Baker, T.J., "Solution of the Euler Equations for Complex Configurations", AIAA Paper 83-1929, 1983.

5. Wambecq, A., Rational Runge-Kutta Method for Solving System of Ordinary Differential Equations, Computing 20, pp333-342, 1978.

6. Verhoff, A., and Stookesberry, D., "Far-Field Computational Boundary Conditions for Internal Flow Problems", Naval Postgraduate School, NPS67-88-001CR, Monterey, CA, Spet. 1988.

7. Steger, J.L., and Benek, B.L., On the Use of Composite Grid Schemes in Computational Aerodynamics, First World Congress on Computational Mechanics, Austine, Texas, Sept. 1986.

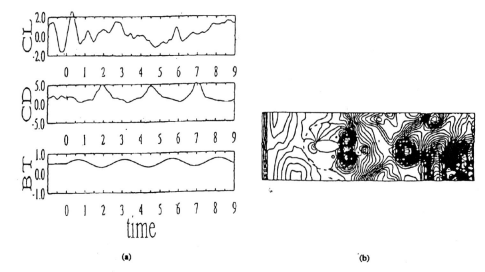

(a) (b)

Fig.1 Computed time history of the lift coefficient, the drag coefficient and the length of one axis of the ellipse through periodically deforming cylinder (a), and iso-bar lines at $t = 8$ (b).

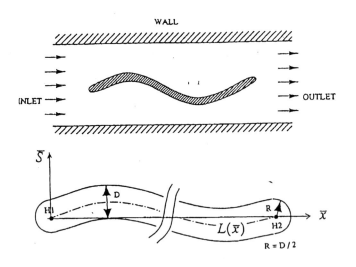

Fig.2 Schematic of the problem.

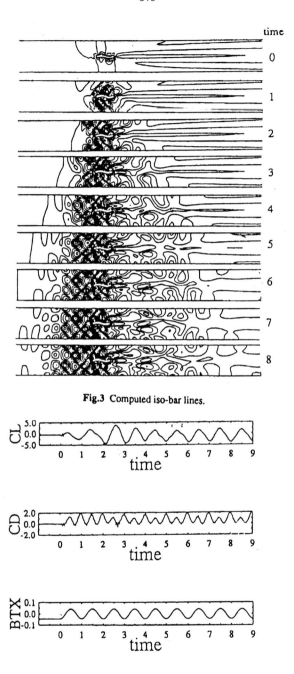

Fig.3 Computed iso-bar lines.

Fig.4 Computed time history of the lift coefficient, the drag coefficient and y coordinate at 3/4 chord of the string.

Computation of Three Dimensional Supersonic and Hypersonic Blunt Body Flows using High Resolution TVD Schemes based on Roe's Approximate Riemann Solver

S.K. Saxena and K. Ravi

Computational and Theoretical Fluid Dynamics Division
National Aerospace Laboratories, Bangalore, India

1 Introduction

Inspite of some known shortcomings, Roe's Approximate Riemann Solver is being widely used [1,2] to discretize the Euler terms in three dimensional supersonic/hypersonic application codes for solution of both Euler and Navier Stokes Equations. This is because the scheme is known to provide a sharp description of shocks and accurate results in the viscous region. The pitfalls in the Roe Solver are overcome by an appropriate choice of the Entropy Fix. This paper discusses some experiences of the authors with the application of an explicit cell centered finite volume upwind TVD (MUSCL) formulation of the Roe's Approximate Riemann Solver for the computation of 3D inviscid and viscous high speed flows past blunt body configurations of practical interest.

Local Time Stepping and Parallelization Techniques are employed to accelerate convergence in terms of effective wall clock times. The parallel machine used is the FLOSOLVER which is based on the Intel i860 chip. A 4 node version of the machine has been employed in the present work. A streamwise domain decomposition technique is employed for the parallelization. The study involved axisymmetric bodies for which a full three dimensional grid (360°) is generated and no symmetry boundary conditions are imposed.

2 Governing Equations and Numerical Scheme

The Euler Equations are employed to compute inviscid flows, while the Thin Layer Navier-Stokes Equations simulate laminar viscous flows. The non dimensional conservation law form of these equations with ideal gas assumption is cast in generalized curvilinear coordinate system [3]. The complete discretization of the Euler terms based on Roe's Approximate Riemann Solver is available in Ref. [4]. The viscous terms are central differenced in the usual way.

3 Results and Discussions

The following test cases have been considered in the present study

1. **Inviscid flows**
   (a) Supersonic flow past a hemisphere–cylinder configuration at zero angle of attack with $M_\infty = 2.94$. A grid size of (27x21x22) was used in the computation.
   (b) Hypersonic flow past a hemisphere–cone with a semi cone angle of $15°$ at an angle of attack of $15°$ with $M_\infty = 10.6$. The grid size used was the same as above.

2. **Viscous flows**
   (a) Hypersonic flow past a hemisphere–cone with a semi cone angle of $15°$ at an angle of attack of $15°$ with $Re_\infty = 1.1\text{x}10^5$, $M_\infty = 10.6$, $Pr_\infty = 0.723$, $T_o = 1111K$ and isothermal wall temperature $T_{wall} = 300K$. A grid size of (42x37x22) was used in the computation.
   (b) Hypersonic flow past a bulbous heat shield of a typical Launch Vehicle with a semi cone angle of $20°$, boattail angle of $15°$ and angle of attack of $10°$ with $Re_\infty = 2.5\text{x}10^4$, $M_\infty = 5.75$, $Pr_\infty = 0.723$, $T_{o,\infty} = 1830K$ and isothermal wall temperature $T_{wall} = 305K$. A grid size of (92x47x22) was used in the computation.

In all these computations free stream conditions were used for the starting solution.

4 Comparative study of TVD Limiters and Entropy Fixes

The formulation of six well known TVD Limiters are compared in terms of convergence in Fig. 1. It may be observed that the Roe Diffusive Limiter performs the best and the Davis Limiter the worst. This study confirms that the choice of Limiters has a significant effect on the convergence characteristics of the computation.

The performance of three Entropy Fixes, Chakravarthy [1], Harten-Yee [2] and Harten-Hyman [5], are compared in terms of convergence characteristics in Fig. 2. The best convergence history is from Harten-Yee Fix and this is the only Fix found to be reliable for Hypersonic flows.

4.1 Computed flow field

Some important flow field results are presented and discussed below. These have been obtained with Roe Diffusive TVD Limiter and Harten-Yee Entropy Fix.

The pressure contours for the Case 1(b) are presented in Fig. 3. It can be seen that the shock is captured sharply. The C_p distribution is shown in Fig. 4. The comparison with computational/experimental results is excellent [6,7].

The velocity vector plot for the viscous Case 2(a) is presented in Fig. 5 and the well developed boundary layer velocity profiles can be clearly seen. The wall heat flux distribution in the pitch plane is shown in Fig. 6 and it compares well with experiment [8].

The results of hypersonic flow computations for the Case 2(b) are presented in Fig. 7 for the pitch plane. The computational grid is shown in Fig. 7(a). The pressure contours are given in Fig. 7(b). The interaction between the expansion fan at the cone-cylinder junction and the blunt body shock can be clearly seen on the windward side. The weak expansion fan and compression shock are also visible in the boat tail region; these features are still weaker on the leeward side. A streamline plot shows a recirculation bubble on the leeward side boattail in Fig. 7(c). The separation and reattachment points can be easily identified. A similar bubble but much smaller in size is also seen on the windward side, although not shown here. The pressure distribution is presented in Fig. 7(d). It may be observed that on the leeward side boattail, the pressure is almost constant due to the large separation bubble mentioned earlier.

The Stanton number distributions are shown in Fig. 7(e) and Fig. 7(f) for the windward and leeward side respectively. A steep fall in heating rates can be observed near the cone-cylinder and cylinder-boattail junction where expansion fans occur. The computed results seem to agree resonably well with the experiment by Srinivasa [9].

5 Concluding Remarks

An explicit upwind finite volume TVD (MUSCL) formulation of the Roe scheme has been presented for the solution of three dimensional Euler and Navier-Stokes equations. It is shown to provide accurate solutions of 3D inviscid and viscous high speed flows past blunt body configurations of practical interest.

Ackowledgements

The present study has been carried out under an AR & DB project No. CF-1-121. Many useful discussions with Prof. S.M. Deshpande and Dr. S.S. Desai are acknowledged with thanks. Thanks are due to Dr. U.N. Sinha and his team at the FLOSOLVER laboratory for their initial help in the parallelization of the code.

References

[1] Chakravarthy, S.R. (1988): VKI Lecture Series 1988-05, Comp. Fluid Dynamics
[2] Yee, H.C. (1989): NASA Technical Memorandum 101088
[3] Saxena, S.K. (1993): NAL-UNI Lecture Series, NAL SP 9301
[4] Saxena, S.K. and Ravi, K. (1994): AIAA Paper No.94-1875, 12th Applied Aerodynamics Conf., Colorado Springs
[5] Harten, A. and Hyman, J.M.(1983): Journal of Comp. Phy., Vol.50, pp. 235-269
[6] Riedelbauch, S. and Müller, B. (1987): DFVLR-FB 87-32
[7] Cleary, J.W. (1965): NASA TN D-2969
[8] Cleary, J.W. (1969): NASA TN D-5450.
[9] Srinivasa, P.(1991): Ph.D. Thesis, Dept. of Aerospace Engg., IISc.,Bangalore

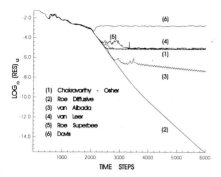

Fig. 1. Comparison of various TVD Limiters for Case 1(b).

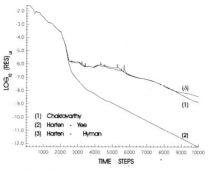

Fig. 2. Comparison of various Entropy Fixes for Case 1(a).

Fig. 3. Pressure contours in the pitch plane for Case 1(b).

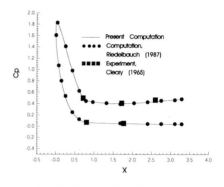

Fig. 4. Pressure distribution on the body in the pitch plane for Case 1(b).

Fig. 5. Velocity vector plot in the pitch plane for Case 2(a).

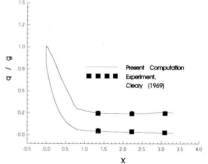

Fig. 6. Heat flux on the body in the pitch plane for Case 2(a).

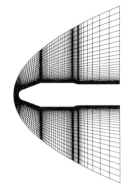
(a) Computational grid (92 x 47 x 22).

(b) Pressure contours.

(c) Streamline plot on the leeward side boattail.

(d) Pressure distribution.

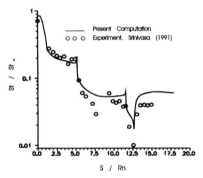

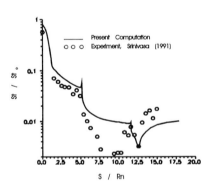

(e) Stanton number distribution (windward).

(f) Stanton number distribution (leeward).

Fig. 7. Hypersonic Viscous flow past a bulbous heat shield, Case 2(b).

3-D KFVS Euler code "BHEEMA" as aerodynamic design and analysis tool for complex configurations

S Sekar[1], M Nagarathinam[1], R Krishnamurthy[1],
P S Kulkarni[2] and S M Deshpande[2]

[1] Aerodynamics Division(CFD), Defence Research & Development Laboratory,
Kanchanbagh, Hyderabad, INDIA 500 258.
[2] CFD Laboratory, Department of Aerospace Engineering,
Indian Institute of Science, Bangalore, INDIA 560 012

Abstract : The usefulness of the 3D Euler code BHEEMA (**B**oltzmann **H**ypersonic **E**uler **E**quation solver for **M**issile **A**erodynamics) based on KFVS method has been demonstrated for a variety of design and analysis problems arising in vehicle aerodynamics.

1 Introduction

A three dimensional time marching Euler code BHEEMA (Boltzmann Hypersonic Euler Equations solver for Missile Aerodynamics) has been developed. This code is based on Kinetic Flux Vector Splitting (KFVS) method[1,2] and works on a network of finite volumes. Deshpande et al. [3] have applied BHEEMA to compute the flow around hypersonic reentry vehicle. This code has been extensively validated by computing high speed flows around many standard shapes such as cone, sphere, cone- cylinder and axisymmetric configurations. The aim of the present paper is to demonstrate the power, versatility and the robustness of BHEEMA. This is done by applying BHEEMA to (i) Bulbous nose in very low subsonic speed, (ii) Transonic flow around the fore body of generic shape, iii) Supersonic flow around wing-body-tail configuration and non circular cross section shapes, (iv) Supersonic flow through axisymmetric spike type isolated air intake, (v) Supersonic coaxial mixing of flows and (vi) Hypersonic flow past configurations consisting of hemisphere-cone- flare with cruciform control surfaces.

2 Description of "BHEEMA" and grid generation

Earlier version of the code called BHEEMA1 is a first order accurate code which has been found to give various aerodynamic parameters $(C_A, C_N, C_M and X_{cp})$ within engineering accuracy. This code was modified to BHEEMA2 by using second order accurate KFVS method with minmod operator for suppressing wiggles. BHEEMA2 also operates on a multiblock grid[4].

The grid generation for various geometries mentioned in the introduction is done by stacking. The configuration is cut into several cross sectional planes in each of which a 2D grid is generated by Thomas-Middlecoff procedure[5] with grid control parameters ϕ and ψ respectively controlling the angle of intersection of the grid lines with the body and the grid spacing[6]. Fig 1 shows the far field boundary and the body. Fig 2 shows a 2D grid in a typical cross section of cruciform configuration. Fig 3 and 4 shows the surface grids.

3 Results and discussions

BHEEMA1/BHEEMA2 have been run on all the configurations listed in the introduction. Fig.5 shows pressure level colour plot for case (i) with a free stream Mach 0.088 (30 m/s)$\alpha = 0^0$ TFV (total no. of finite volumes) being 45X21X21. This is an extremely low Mach number case primarily studied from the point of view of finding how BHEEMA performs on such cases. Preliminary computations and suitable comparison of these results with water tunnel experiments [7] show that BHEEMA performs fairly well even at these low Mach numbers. The high pressure bubbles behind the backward facing step are clearly visible in the colour plot. Results for case (ii) have been shows in Fig.6. The C_p values given by BHEEMA2 are compared with the experimental measurements given in[8]. There is an excellent agreement between the computed values of C_p and the experimental measurements. BHEEMA1 was run for case (iii) (shown in Fig. 4) and the results obtained for the axial force coefficients C_A are shown in Table 1. Here the skin friction coefficient for the configuration was obtained independently by using semiemperical formula and was added to C_A obtained by BHEEMA1. The total axial force coefficients so obtained compares very well (within at most 12%) with the wind tunnel measurements[9]. BHEEMA2 was next applied to case (iv). Fig.7 shows the pressure level colour plots. The conical shock and the normal shock after fixing back pressure are clearly visible in this plots. The code BHEEMA1 was thereafter applied to supersonic coaxial mixing with different pressure and Mach number to study the mixing characteristics of the flows. An inviscid code, strictly speaking, is not meant for simulating the mixing which in the present case takes place due to numerical diffusion. The isomach lines are shown in Fig.8 and the results of comparison with experiments[10] are shown in Fig.9. The code BHEEMA2 was next applied to a typical hypersonic vehicle consisting of hemisphere-cone-cylinder-flare with cruciform wings (shown in Fig. 3) for free stream mach numbers varying between 4 to 14 with $\alpha = 0^0$ and 2^0 [3]. The cross flow at the leading edge and the trailing edge of the control surface have been shown in Fig .10 and 11 which also show isopressure contours.

4 Conclusions

These extensive computations performed using **BHEEMA** demonstrate its versatility and robustness in computing flows with speeds varying from very low subsonic to hypersonic cases. The KFVS based BHEEMA has been found to be very useful in both analysis and design modes.

Acknowledgments :

The authors are grateful to Dr. V J Sundaram, Director, DRDL, Dr. A Sivathanu Pillai, Program Director, DRDL, Dr. B S Sarma, Director, Vehicle Dynamics, DRDL, Hyderabad for extending valuable encouragement and technical support to carry out this work. The thanks are due to Mr. P K Sinha of DRDL Hyderabad for their help in various forms.

References

[1] Deshpande S M, "Kinetic theory based new upwind methods for inviscid compressible flows", AIAA paper 86-0275,1986.

[2] Mandal J C and Deshpande S M, "Kinetic Flux Vector Splitting for Euler Equations." Computers & Fluids vol 23, No2, pp447-478, 1994.

[3] Deshpande S M, Sekar S, Nagarathinam M, Krishnamurthy R, Sinha P K, Kulkarni P S, "3-Dimensional upwind Euler solver using Kinetic Flux Vector Splitting method(KFVS)", Lecture notes in Physics, No.414 Springer Verlag, Proc. of 13th Int. Conf. on Numerical Methods in Fluid Dynamics, 1992, pp 105-109.

[4] Sekar S, Deshpande S M and Nagarathinam M, "Development of BHEEMA and its applications", Report No.5120.1040.029, Aerodynamics Division, DRDL, Hyderabad, India.

[5] Thomas P D and Middlecoff J F, "Direct control of the grid point distribution in meshes generation by elliptic equations", AIAA journal 18, p652, 1980.

[6] Krishnamurthy R, Nagarathinam M, Sekar S, and Sinha P K, "A two dimensional elliptic grid generator for a wing-body section involving grid control functions", Jl. of Aero. Soc. of India, Aug-Sep 1992, pp 205-208, Vol. 44, No.3.

[7] Kukillaya S P, "Private communication with Project Director, PJ08", 1993, DRDL, Hyderabad.

[8] Hsieh T, "Flow field study about a hemisphere-cylinder in the transonic and low supersonic Mach numbers range", AEDC-TR-75-114, 1974.

[9] Paneerselvam S, Theerthamalai P, Thakur D N and Balu G, "Preliminary aerodynamic document", Internal report No. 5120.1001.513, DRDL, Dec 1984.

[10] Narayanan A K, "An experimental study on the mixing of two coaxial high speed streams", Ph.D thesis, 1992, Indian Inst. of Tech., Madras.

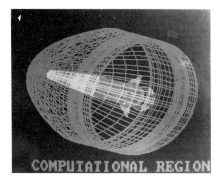

Fig. 1 computational domain

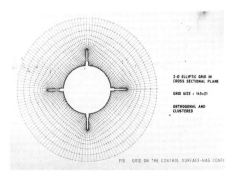

Fig. 2 An elliptic grid in a typical cross section

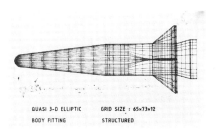

Fig. 3 surface grids on a reentry vehicle

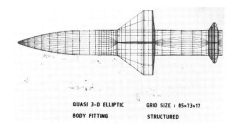

Fig. 4 surface grids on wing-body-control surface configuration

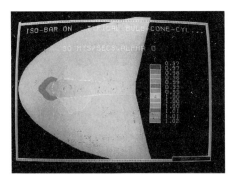

Fig. 5 pressure contour for bulbous nose (30 m/sec, $\alpha = 0^o$)

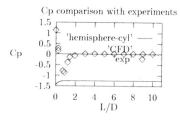

Fig. 6 C_{cp} Distribution on hemisphere-cylinder

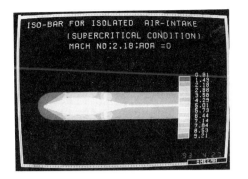

Fig. 7 Pressure contour for isolated air-intake($M_\infty=2.18$, $\alpha=0°$)

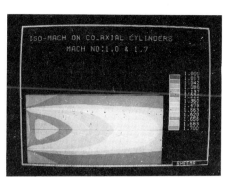

Fig. 8 Isomach contours for coaxial mixing jets ($M_1=1.0, M_2=1.7, P_1=1.162, P_2=1.0$)

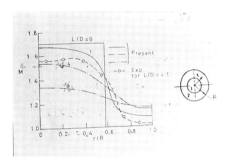

Fig. 9 Mixing charcterstics of two coaxial supersonic streams($M_1=1.7$, $p_1=1.0, p_2=1.2$)

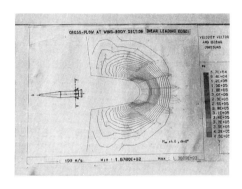

Fig. 10 Pressure contours sup vector plot at a typical cross section. edge($M_\infty=4.0$)

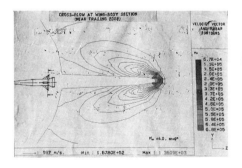

Fig. 11 Pressure contours superimposed with vector plot at a typical cross section near trailing edge ($M_\infty=4.0$)

Axial force coefficient

M_∞	2.0	3.0	4.0
CFD BHEEMA	0.21	0.159	0.135
CFD + skin friction contribution	0.28	0.209	0.173
Experimental data	0.25	0.209	0.179

Fig. 12 comparison of C_a with experimental results

Separation Analysis of the Pegasus XL From an L-1011 Aircraft

W. L. Sickles[1], M. J. Rist[1], C. H. Morgret[1], S. L. Keeling[1]
and K. N. Parthasarathy[2]

[1] Calspan Corporation / AEDC Operations
Arnold AFB, TN 37389-6001
[2] The Johns Hopkins University / Applied Physics Laboratory
Johns Hopkins Rd., Laurel, MD 20723-6099

1 Introduction

The Orbital Sciences Corporation (OSC), as part of its commercial space venture, has designed and built the Pegasus XL (Fig. 1), a longer and heavier version of the baseline winged-rocket booster, to deliver small satellites into space. The Pegasus XL will be launched from a modified Lockheed L-1011 Tristar aircraft, the next generation launch platform for both the baseline and XL versions. In carriage, the Pegasus XL is mounted symmetrically beneath the L-1011 fuselage with a clearance of approximately six inches while the Pegasus vertical fin extends into a fin-penetration box on the L-1011.

Nielsen Engineering and Research (NEAR)[1] used a panel code to predict the separation of the baseline Pegasus from a B-52 aircraft. In the absence of wind tunnel data, their predictions were compared to flight data and showed acceptable agreement[2].

Again no wind tunnel tests of the L-1011/Pegasus XL configuration are planned to determine aerodynamic separation characteristics. A purely computational approach is also taken here to determine the trajectory for the Pegasus XL. Trajectory simulations are performed for two release conditions. NEAR performed an independent Pegasus XL/L-1011 separation analysis using panel codes[3].

During captive-carriage flight tests of the Pegasus XL/L-1011, changes to the carriage configuration were incorporated to reduce buffeting. The current investigation does not include these carriage modifications. Assessment of the effects of these changes is under investigation.

2 Analytical Methods for Separation Analysis

As described below, a combination of engineering methods and computational fluid dynamic (CFD) codes is used for the separation analysis. Engineering methods are used extensively to perform aircraft/store safe-release certifications based on wind tunnel data[4]. In the current investigation, the CFD methods provide the information that is normally provided by wind tunnel data.

2.1 Engineering Methods

Store trajectories are determined by a marching process starting at carriage. At a given time, the aerodynamic loads on a store are calculated for the current position and orientation. Next, the rigid-body equations of motion are numerically integrated over a small time step to determine the new position and orientation. The process is repeated until the store is safely away from the aircraft. The aerodynamic loads acting on the store during separation are determined using the direct flow-field approach. In this approach, the aircraft flow field, without the store, is interpolated onto a specified grid beneath the aircraft. Store loads at each point in a trajectory simulation are determined analytically using engineering methods based on superposition of the store on the aircraft flow-field grid.

Calculation of store loads based on superposition does not account for mutual interference effects between the aircraft and the store. However, these effects decay rapidly and are only significant in the immediate vicinity of the aircraft[4]. To account for these effects, store loads at and near carriage are computed with CFD techniques for the combined configuration and are used to adjust the loads determined by the superposition method. A detailed description of the engineering methods and their specific application to the Pegasus XL separation simulation is given in Ref. 5.

2.2 CFD Methods

In the current investigation, the time-dependent Euler equations are time marched to a steady state using a locally-implicit algorithm[6]. Because of the complexity of the Pegasus XL and L-1011 configurations, the chimera overset-grid approach[7] is used. In the chimera approach, the configuration is segmented into regions, and a grid is generated for each region. The approach requires that the grids overlap, and establishes communication of the flow variables among the grids.

3 Results

Trajectory simulations were performed for the on-design release condition and an off-design release condition. The on-design release is at a Mach number (M_∞) of 0.79, an angle of attack (α) of four degrees, a sideslip angle (β) of zero degrees, and an altitude of 39,000 feet. The off-design release is at $\beta = -4°$. During the trajectory simulations, it was assumed that the Pegasus XL control surfaces were locked, that the attachment hooks released simultaneously, and that no ejector forces were employed.

3.1 CFD Calculations

A matrix of eight free-stream flow-field solutions was performed about the Pegasus XL over a range of attitudes that the vehicle might experience during separation for both release conditions. From these flow-field solutions, the computed loads were used to calibrate the engineering methods to determine the free-stream loads at the expected vehicle attitudes.

Solutions were also obtained about the L-1011 alone, for both release conditions, to provide the aircraft flow-variable maps. To determine carriage load increments and mutual interference decay, flow-field calculations were performed about the combined configuration at carriage and when the Pegasus XL was 0.2 and 1.0 store diameters below carriage. These calculations involved an additional eight CFD solutions.

3.2 Trajectory Simulations

The trajectories were calculated over a two-second interval for both release conditions. Figure 2 shows the temporal variation of the normal-force coefficient(C_N), pitching-moment coefficient (C_m), and axial-force coefficient (C_A) on the Pegasus XL for the on-design release. The vehicle experiences a nose-down pitching moment for the first 1.2 seconds. The effects on vehicle pitch angle relative to the aircraft ($\Delta\Theta_A$) can be seen in Fig. 3. Also shown in Fig. 3 is the axial(X_A) and vertical (Z_A) movement of the vehicle center of gravity relative to the aircraft. Corresponding NEAR predictions[3] in Fig.3 show excellent agreement except for pitch angle. The pitch discrepancy results from differences in C_m predictions between the CFD Euler calculations and the NEAR panel codes. Of critical importance is the clearance of the vertical fin that extends into the L-1011 cavity. Analysis shows that the fin clears the cavity as it separates. However, at the off-design condition, collision is detected as the rudder exits the fin-penetration cavity. The detected collision is not a major safety concern, because the maximum achievable sideslip for the L-1011 aircraft during normal flight conditions is a $|\beta| \sim 2.5°$. In Ref. 3, NEAR has determined that the Pegasus XL will safely clear the cavity for $|\beta| \leq 3.0°$. In addition, OSC has restricted the launch envelope to a $|\beta| \leq 1.0°$.

4 Conclusions

A purely computational approach was taken to predict the separation trajectory of the Pegasus XL for two release conditions. In this approach, the traditional wind tunnel data base was replaced by CFD predictions. Analysis of the trajectories shows that the Pegasus XL separates cleanly at the on-design release condition.

Acknowledgements :

The current analysis was funded by the Orbital Launch Services Project of the NASA Goddard Space Flight Center. The authors gratefully acknowledge their sponsorship.

References

[1] Mendenhall, M. R., et al. AIAA Paper No. 91-0109 (1991)

[2] Mendenhall, M. R., et al. AIAA Paper No. 93-0520 (1993)

[3] Mendenhall, M. R., et al. AIAA Paper No. 94-1910 (1994)

[4] Keen, K. S., et al. AIAA Paper No. 90-0274 (1994)

[5] Sickles, W. L., et al. AIAA Paper No. 94-3454 (1994)

[6] Reddy, K. C. and Benek, J. A. AIAA Paper No. 90-1525 (1990)

[7] Benek, J. A., et al. AEDC-TR-85-64 (AD-A167466) (1986)

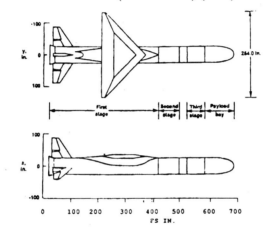

Fig. 1. Pegasus XL configuration

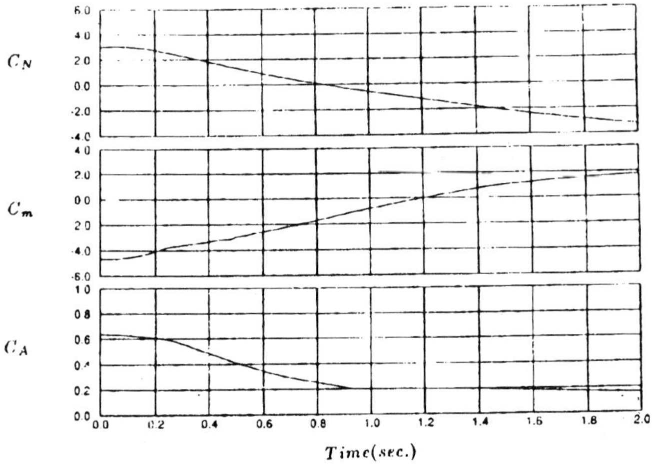

Fig. 2. Calculated Pegasus XL forces and pitching moment for on-design release condition, $M_\infty = 0.79$, $\alpha = 0°$, $\beta = 0°$, Alt = 39,000 ft.

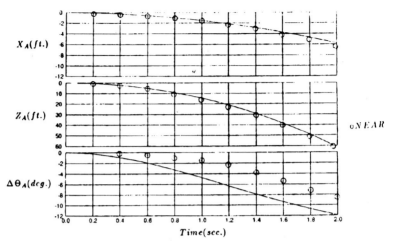

Fig. 3. Calculated Pegasus XL trajectory relative to L-1011 for on-design release condition, $M_\infty = 0.79$, $\alpha = 0°$, $\beta = 0°$, Alt = 39,000 ft.

Application of an Euler Code on a Modern Combat Aircraft Configuration

K P Singh[1], K Murali Krishna[1], S.Saha[2] and S K Mukharjea[2]

[1]ADA, Post Box 1718, Bangalore 560017, India
[2]Dept. of Aerospace Engg., IIT, Kharagpur 721302, India

Abstract : An Euler Code is applied on a modern combat aircraft configuration to study the effect of deflections of leading edge high lift devices and control surfaces on the aerodynamic coefficients. The calculations for deflection of Leading Edge Vortex CONtroller (LEVCON) and Elevons are done at Mach No. = 0.7, and angle of attack = 10^0 and for deflection of leading edge slats at Mach No. = 0.95, angle of attack = 7.5^0. It is found that the incremental change in the aerodynamic coefficients - lift, induced drag, and the pitching moment due to the deflections of LEVCON, slats and elevon are predicted reasonably accurate both quantatively and qualitatively.

1 Introduction

A modern combat aircraft is designed to perform a variety of missions. These missions require that the combat aircraft should be designed for a good manoeuverability. A tail-less delta configuration is considered a good candidate for a combat aircraft with desirable manoeuverability and acceleration. In order to improve its performance further, high lift devices, namely, leading edge slats (LES), leading edge flaps (LEF), or strakes are usually employed. One of the most effective leading edge devices are slats (Fig.1a) which are deployed to delay leading edge flow separation, thereby improving lift/drag ratio and buffet boundary. Another leading edge device called LEVCON (Leading Edge Vortex CONtroller) (Fig.1a) can be used to improve take-off performance, lift/drag ratio and landing performance. During design phase, an aircraft designer is often interested to have knowledge of incremental changes in aerodynamic coefficients due to deflection of high lift devices to arrive at optimal deflections. For this purpose, in order to save both cost and time, a CFD tool applicable to these devices, which are effective only with on-set of a vortical flow, will be an ideal choice. In this paper, an attempt is made to compute numerically the effect of two high lift leading edge devices: i.e., (i) leading edge slats and (ii) LEVCON and also one control surface deflection i.e., elevons on aerodynamic coefficients C_l, C_d and C_m (Lift, Induced drag and Pitching moment coefficient) of a combat aircraft.

2 Method

In the present study, the results are obtained using a time-marching Euler code AMES (Aircraft Multi-block Euler Solver), developed under an ADA sponsored project at IIT, Kharagpur. This code is based on Jameson's finite volume method [1]. The 3-D grids around complete aircraft are generated using an indigenously developed grid generator-GEMS (Grids using Elliptic multi-block solver). The grids for LEVCON, slats and elevons with the small deflection are obtained by deflecting the original grid. However, for the large deflections, the grids are regenerated. The geometry definitions are obtained from the CAD Package CATIA. The grids generated about different components of aircraft are shown in Fig.1b. All the computations are done on an IBM RISC 6000/560 work-station.

3 Results and Discussions

The present code is validated with the results available for RAE configuration wing-body, ONERA wing and experimental results available for a combat aircraft configuration. (The result for ONERA wings are not presented here due to lack of space). In Fig.2, computed pressure distribution for RAE wing-body at $M = 0.9$, $AOA = 0^0$ (M = free stream Mach number, AOA = Angle of attack) compares well with the experimental values [2] at all three spanwise stations over the wing. The computed results for the present combat aircraft are compared with experimental values [3] in Figs 3 and 4. In Fig 3(a), the pressure distribution over the front fuselage at station a-b at $M = 0.7$, $AOA = 20^0$, $AOSS = 5^0$ (AQSS = Angle of side slip) compares well with the experimental values [3]. The comparison of the pressure distribution on top and bottom generator is also good (Figs. 3b and 3c). As shown in Fig.4, the comparison of the global coefficients C_l and C_d at $M = 0.9$ is good except at higher AOA. C_m compares well only in the linear range. The Euler solution fails to capture pitch up phenomenon attributed to flow separation.

The change in aerodynamic coefficients due to deflections of slats, LEVCONS and elevons is shown in Fig 5-7. Fig.5 shows the comparison of the incremental change in aerodynamic coefficients due to the slat deflection with the experimental values at $M = 0.95$, with $AOA = 0^0$ and 7.5^0 (dC_l, dC_d, dC_m - difference in C_l, C_d and C_m for zero slat and full slat condition). The comparison is good both qualitatively and quantitatively. It can be seen from the figure that the at a $AOA = 0^0$ slat increases untrimmed drag significantly but reduces it substantially at $AOA = 7.5^0$. Hence, the slats are beneficial only at higher AOA.

The effect of elevon deflection is shown in Fig.6 (dC_l, dC_d, dC_m - difference in C_l, C_d, C_m for deflected elevon and non-deflected elevon). For this case also the comparison with the experimental values is good. The maximum error of about 2.5 per cent is found in the value of dC_m. Fig. 7 shows the results of dC_l, dC_d, dC_m for four LEVCON deflections i.e. $10^0, 20^0, 30^0$ down and 30^0 up at $M = 0.7$ $AOA = 10^0$ - (dC_l, dC_d and dC_m - difference in C_l, C_d, C_m for deflected LEVCON and no-LEVCON configuration). It can be seen from the figure that with the addition of LEVCON, a nose-up moment is created, which destabilises the aircraft. With deflection of LEVCON, the maximum variation in C_l is found to be marginal. However, C_d is found to increase considerably for 30^0 up deflection and decrease substantially for 30^0 down deflection. There is considerable gain in untrimmed lift/drag ratio with down deflection. Further more, no appreciable gain is obtained from 20^0 - 30^0 down deflection. These CFD findings are being experimentally verified.

From the present analysis, it is concluded that using an Euler code, the incremental changes in aerodynamic coefficients due to deflection of high lift devices/control surfaces at transonic high angle of attack can be computed accurately both qualitatively and quanitatively.

References

[1] A. Jameson, W. Schimdt, E. Turkel: AIAA Paper 81-1259 (1981)

[2] D. A. Treadgold, A. F. Jones, K. H. Wilson: AGARD-AR-138 (1979)

[3] A. Arokkiaswamy, ADA Bangalore: Personal Communication.

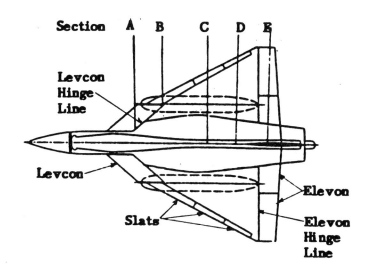

Fig. 1a A Combat Aircraft Configuration with Control Surfaces

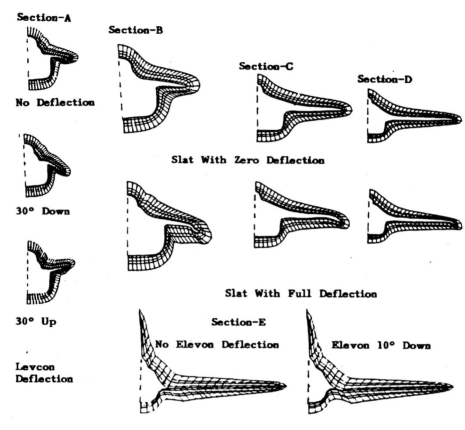

Fig. 1b. Grids about different components of a combat aircraft configuration

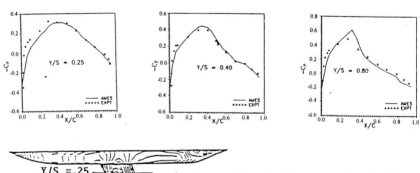

Fig. 2. Pressure Distribution on RAE-Wing-Body at $M = .9$, $AOA = 0.0$

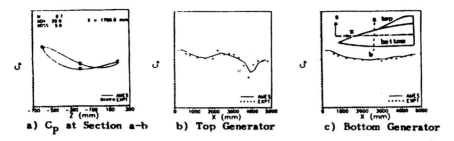

a) C_P at Section a-b   b) Top Generator   c) Bottom Generator

Fig. 3. Pressure distribution on front fuselage of a generic combat aircraft

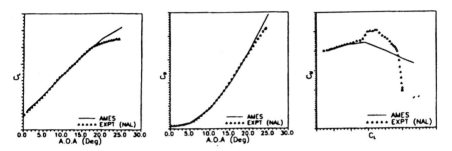

Fig. 4. Variation of aerodynamic coefficients for a generic combat aircraft at M=0.90

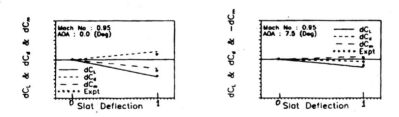

Fig. 5. Change in aerodynamic coefficients due to L.E. slat deflection

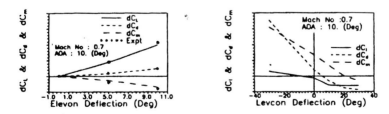

Fig. 6. Change in aerodynamic coefficients due to elevon deflection

Fig. 7. Change in aerodynamic coefficients due to levcon deflection

Study on Hypersonic Finite-Rate Chemically Reacting Flows Using Upwind Method

Song, D.J.[1], Kwon, K.D.[2], Won, S.Y.[2]

Yeungnam University, Gyongsan, 712-749, Korea

Abstract : A finite-difference method based on conservative supra characteristic method type upwind flux difference splitting has been developed to study the equilibrium/nonequilibrium chemically reacting high speed flow. For nonequilibrium air, NS-1 species equations were strongly coupled with flowfield equations through convection and species production terms. The effective gamma(γ) was used to couple chemistry with the fluid mechanics for equilibrium chemically reacting flow. Inviscid, equilibrium/nonequilibrium chemically reacting air mixture flows in a quasi-one dimensional divergent nozzle were solved to demonstrate the capability of the current method. The shock waves were captured clearly within 2 points. At low altitude flight conditions, the equilibrium/nonequilibrium air models predicted almost same temperature density, pressure behind the shock; however, at high altitudes they showed substantial differences due to nonequilibrium chemistry effect.

1 Introduction

Equilibrium/nonequilibrium chemically reacting gas flow has been a very important research area due to recent interests in hypersonic scram jets and atmospheric entry of space vehicles, such as the aero-assisted orbital transfer vehicles, space shuttle and reentry vehicles. Due to its importance many research works have been done already to study such flowfields particularly in the early 70's and most part of the 80's.

Navier-Stokes methods generally took much longer computing time and efforts when compared with boundary layer(BL) method(Blottner et al.,1971), viscous shock-layer(VSL) method(Miner and Lewis,1975) and parabolized Navier-Stokes (PNS) method (Bhutta and Lewis, 1985). Hypersonic nonequilibrium chemically reacting flow has been analyzed chiefly using BL, VSL and PNS methods based on the Blottner's(1971) type linearization of special production terms i.e., $\dot{w}_i/\rho = \dot{w}_i^0 - \dot{w}_i^1 \, C_i$. Until recently such reduced subsets of Navier-Stokes method have been used to analyze the nonequilibrium flowfields (Blottner et al., 1971, Finson et al., 1985, Miner and Lewis, 1975, Bhutta and Lewis,1985); however, with the progress in computers, lots of efforts(Grossman and Cinnella, 1990, Lee and Deiwert, 1990, Molvik and Merkle,1989, Widhopf and Wang,1988) have been put on the development of fully coupled nonequilibrium chemistry Navier-Stokes codes. Upwind flux vector splitting

---

[1] Assistant Professor
[2] Graduate Students

methods of Steger and Warming(1981), Van Leer(1982) and Roe's upwind flux difference splitting method have been used by various researchers. Among the flux difference splitting methods, the work of Lombard et al.(1983) deserves some attention in solving 3-D complex hypersonic flowfield.

In this study, the flux difference splitting algorithm of Lombard et al.(1983) has been developed to include fully coupled equilibrium/nonequilibrium chemistry. Parts of Widhopf and Wang's(1988) coupling algorithm are also used to incorporate the nonequilibrium chemistry effect more efficiently.

2 Analysis

2.1 Quasi 1-D Formulation

Nonequilibrium Chemistry Reacting Flow

The governing equations for an inviscid quasi one-dimensional flow with nonequilibrium chemistry may be written in vector conservation form as

$$\delta_t \begin{bmatrix} \rho \\ \rho u \\ E \\ \rho_i \end{bmatrix} + \nabla_x \begin{bmatrix} \rho W \\ \rho W_u + \xi_x P \\ W(E+P) \\ \rho_i W \end{bmatrix} = \begin{bmatrix} 0 \\ 0 \\ 0 \\ \dot{w}_i \end{bmatrix}, \quad i = 1, NS-1 \tag{1}$$

where $W = \xi_x u$, $p = (\gamma - 1)P$, ξ_x = local area and $P = \rho e$. The $\dot{w}_i$ is a reacting source term for production and loss of the chemical species through chemical reactions. The conservation supra characteristic method (CSCM) flux difference splitting utilizes the following properties :

$$\partial_x F = A\partial_x q = MT \wedge T^{-1}M^{-1}\partial_x q \tag{2a}$$
$$MT \wedge T^{-1}\partial_x \tilde{q} = M\vec{A}\partial_x \tilde{q} \tag{2b}$$
$$MT \wedge \partial_x \tilde{\tilde{q}} \tag{2c}$$

where $\tilde{q}$ = primitive variables, $\tilde{\tilde{q}}$ = characteristic variables, $\wedge$ is a diagonal matrix whose diagonal elements correspond to the eigenvalues (u, $u+c$, $u-c$, u, u, u, u), and variables q, $\tilde{q}$, and $\tilde{\tilde{q}}$ are related as follows :

$$\partial \tilde{q} = M^{-1}\partial q, \qquad \partial \tilde{\tilde{q}} = T^{-1}\partial \tilde{q}$$

M matrix transforms primitive variables $\tilde{q}$ into conservative variables q. The characteristic variables can be obtained from the primitive variables through the relation

$$T^{-1}(\vec{A}\Delta \tilde{q} = T^{-1}(T \wedge T^{-1}\Delta \tilde{q} = \wedge \Delta \tilde{\tilde{q}} \tag{3}$$

where T^{-1} is a matrix which transforms primitive variables into characteristic variables and which incorporates density and Mach number in logarithmic difference form.

The inviscid flus ΔF can be divided into $\Delta \hat{F}^+$ and $\Delta \hat{F}^-$ using diagonal matrix truth function $D^{\pm}$ depending upon the sign of eigenvalues. The CSCM upwind flux difference splitting method thus relates the flux difference to the conservative variables in the Roe's conservative form as follows :

$$\Delta \hat{F}^{\pm} = \overline{MT}D^{\pm}\overline{T^{-1}}\hat{M}^{-1}\Delta q = A^{\pm}\Delta q \quad (4)$$

where

$$\hat{M}^{-1} = \begin{vmatrix} 0 & \xi, \Delta & 0 & 00000 \\ -\frac{\bar{u}}{\bar{\rho}}\overline{\rho W}\Delta + \bar{\xi}_x \bar{u}^2 \bar{\rho}\Delta\frac{(\gamma-1)}{\bar{\rho}} & & & \\ -\frac{\bar{u}^2}{2}\bar{\xi}, \Delta(\gamma-1) & \frac{\overline{\rho W}}{\bar{\rho}}\Delta - \bar{u}\frac{\bar{\xi}_x}{\bar{\rho}}\bar{\rho}\Delta(\gamma-1) & \bar{\xi}_x, \Delta(\gamma-1) & 00000 \\ \overline{W}\left(\bar{u}^2 - \frac{\bar{u}^2}{2}\right)\Delta - \overline{\gamma P}\frac{\overline{W}}{\bar{\rho}}\Delta & \overline{W}\bar{u}\Delta + \bar{\gamma}\frac{\overline{P}}{\bar{\rho}}\bar{\xi}, \Delta & \overline{W} & 00000 \\ -\overline{W}\frac{\bar{\rho}_1}{\bar{\rho}} & \overline{\rho_1}\frac{\bar{\xi}_x}{\bar{\rho}} & 0 & \overline{W}0000 \\ -\overline{W}\frac{\bar{\rho}_5}{\bar{\rho}} & \overline{\rho_5}\frac{\bar{\xi}_x}{\bar{\rho}} & 0 & 0000\overline{W} \end{vmatrix}$$

and

$$D^{+} = \frac{1}{2}\left[\frac{\Lambda}{|\Lambda|} + 1\right], \quad D^{-} = \frac{1}{2}\left[1 - \frac{\Lambda \cdot}{|\Lambda|}\right]$$

2.2 Chemical System and Reaction Source Term

Seven species and seven reaction equations with catalytic third bodies efficiency matrix of Blottner et al.(1971) are chosen to describe the hypersonic nonequilibrium chemically reacting flow. The species considered in nonequilibrium chemically reacting flow are O_2, N_2, NO, O, NO^+, N and e^-. The reaction equations are as follows

$$\begin{array}{llllll}
O_2 & + & M_1 & + & 2O & + & M_1 \\
N_2 & + & M_2 & + & 2N & + & M_2 \\
N_2 & + & N & + & 2N & + & N \\
NO & + & M_3 & + & N & + & O & +M_3 \\
NO & + & O & + & O_2 & + & N \\
N_2 & + & O & + & NO & + & N \\
N & + & O & + & NO^+ & + & e^-
\end{array}$$

where M_1, M_2 and M_3 are catalytic third bodies efficiences relative to Argon.

For a set of NR reactions for the form $\sum_{i=1}^{NS} \alpha_{i,r} X_i \Leftrightarrow \sum_{i=1}^{NS} \beta_{i,r} X_i$, the production term of the $(NS-1)$ chemical species equations has the following expression.

$$\dot{w}_i = \frac{d\rho_i}{dt} = M_i \sum_{r=1}^{NS}(\beta_{i,r} - \alpha_{i,r})\left[k_{f,r}\Pi_{j=1}^{NS}\left(\frac{\rho_j}{M_j}\right)^{\alpha_{j,r}}\left(\frac{\rho - \sum_{m=1}^{NS}\rho_m}{M_{NS}}\right)^{\alpha_{NS,r}}\right.$$

$$\left.-k_{b,r}\Pi_{j=1}^{NS}\left(\frac{\rho_j}{M_j}\right)^{\beta_{j,r}}\left(\frac{\rho - \sum_{m=1}^{NS}\rho_m}{M_{NS}}\right)^{\beta_{NS,r}}\right] \quad (5)$$

where $\alpha_{i,r}$ and $\beta_{i,r}$ are stoichiometric coefficients, ρ_i = species mass density and M_j = molecular weight of species j. The X_i, k_f and k_b are the symbols of the reactant species, forward and backward reaction rate constants respectively. The source term can be linearized around the previous iteration level 'n' as follows

$$\dot{w}_i^{n+1} = \dot{w}_i^n + (\partial \dot{w}_i/\partial t)^n \Delta t \qquad (6)$$

where

$$\left(\frac{\partial \dot{w}_i}{\partial t}\right)^n = \sum_{k=1}^{NS} \left(\frac{\partial \dot{w}_i}{\partial \rho_k}\right)^n \frac{\rho_k^{n+1} - \rho_k^n}{\Delta t} + \left(\frac{\partial \dot{w}_i}{\partial \rho}\right)^n \frac{\rho^{n+1} - \rho^n}{\Delta t}$$

$$+ \left(\frac{\partial \dot{w}_i}{\partial T}\right)^n \frac{T^{n+1} - T^n}{\Delta t}$$

As discussed by Widhopf and Wang(1989), $\rho, \rho_1, \ldots, \rho_{NS-1}$ and T have been chosen because they are the natural thermodynamic variables and conform with the governing equations. Since the Jacobian $\partial T/\partial q$ may not be easily derived by using conservative variables, it would be convenient to treat this term, $\partial \dot{w}_i/\partial q$ explicitly on the right hand side.

2.3 Thermodynamic Relations

In order to close the system of equations, the set of partial differential equations is complemented by additional thermodynamic relations. The pressure, temperature and the parameter γ are related with the dependent variables of the solution. The relations are the equation of state (with Dalton's Law)

$$p = \rho \frac{R_o}{\overline{M}} T = \sum_{i=1}^{NS} \frac{\rho_i}{M_i} R_o T \qquad (7)$$

where $\rho_{NS} = \rho - \sum_{i=1}^{NS} \rho_i$ and a volumetric internal energy relation

$$P = \rho e = \sum_{i=1}^{NS} \rho_i e_i \qquad (8)$$

The specific internal energy of each species is given by curve fit relations. The frozen specific entropy of the mixture is determined by summing the individual contributions of each species from the data table as

$$h = \sum_{i=1}^{NS} C_i h_i \qquad (9)$$

where C_i is the concentration of species i and $h_i = \int_{293}^{T} Cp_i dT + \Delta h_i^F$.

Since the temperature is not a variable directly obtainable from the conservative variables, the temperature of the equilibrium gas mixture can be obtained from the equation of state of thermally perfect but calorically imperfect gas. The temperature can be computed using the Newton-Raphson iteration method around T^k as

$$T^{k+1} = T^k \frac{g(T^k) - e}{g'(T^k)} \qquad (10)$$

where

$$g(T^k) = \sum_{i=1}^{NS} C_i(h_i(T) - RT/M_i)$$
$$g'(T^k) = \sum_{i=1}^{NS} C_i(Cp_i(T) - RT/M_i)$$

e = specific internal energy : $e = (E - \rho u^2/2)/\rho$. and k indicates the iteration index. Then at a given T, take Eqs.(8) and (9) as the basis for defining the ratio of frozen mixture enthalpy to specific internal energy, $\gamma = h/e$.

2.4 Equilibrium Chemically Reacting Flow

At or near chemically equilibrium flight regime the state variables and species adjust instantaneously to local flow condition. To describe such flow conditions, the code was modified to include equilibrium chemistry effect through effective γ. A modified ratio of specific heats (enthalpy-internal energy ratio) as used by Saladino et al.(1990) was used to relate pressure with conservative variables at chemical equilibrium

$$p = (\bar{\gamma} - 1)[E - \rho u^2/2]$$

The curvefit data of Srinivasan et al.(1987) was used to obtain $\tilde{\gamma}$ at given states. After we obtained the converged equilibrium air flowfield solution, the equilibrium air species concentrations were obtained from the law of mass action applied individually to the various reactions(Vincenti and Kruger(1982)) as follows

$$\Pi_i(n_i)^{v_i} = \frac{1}{\rho \sum_i v_i} K_{c,r}(T)$$

where $K_{c,r}(T)$ is the equilibrium constant for reaction r and $n_i = c_i/M_i$, mole-mall ratio(or number of moles per unit mass of mixture). We obtain 7 species composition from 4 chemical reactions, 2 conservation of oxygen and nitrogen nuclei and the fact of zero net electron charge. Newton-Raphson method was used to obtain the species concentration. For computational purpose, the natural log;ln (n_i), was used to reduce the number of nonlinear equations.

2.5 Boundary Conditions

The general boundary conditions used in current quasi one dimensional divergent nozzle flow are supersonic inflow and subsonic outflow boundary conditions. They are defined as

Supersonic inflow (8 physical b.c.s , 0 numerical b.c.)
- constant entropy and total enthalpy

- specify velocity at the inlet
- specify the species density at the inlet
- ($C_{N2} = 0.7635$, $C_{O2} = 0.234$)

Subsonic outflow (7 numerical b.c.s , 1 physical b.c.)
- specify the exit pressure ($p = p_e$)

3 Results and Discussion

A quasi one-dimensional divergent nozzle(Fig.1) was chosen to test the new coupled upwind flux difference splitting, equilibrium/nonequilibrium chemically reacting Euler code. Hypersonic inflow conditions and exit pressure have been prescribed to get a standing strong shock of strength typical of bow shocks on the space vehicles at various altitudes and Mach numbers. We assumed the solution was converged when the root mean square errors were reduced by 5-6 orders of magnitude from the initial rms errors.

Figures 2a and 2b show the pressure distribution comparison along a divergent nozzle among current result, NACA-1135 data and Hoffmann's(1993) numerical result for checking accuracy of the perfect gas code at supersonic inlet and : inflow velocity 1153 ft/sec, pressure 1000 ft/sec, pressure 1000 $lb_f/ft2$ and subsonic exit pressure 2459.6 $lb_f/ft2$, and at same supersonic inlet and supersonic exit conditions, *i.e.*, isentropic flow expansion, respectively. As shown in Fig. 2a, the current Euler code predicted wall pressure distribution quite well. The shock was captured within 2 grid points. The current result was compared well with the data of NACA 1135 report; however, the Hoffmann's result with same number of grid points was not good. the current numerical result for isentropic flow expansion case was almost same as NACA 1135 data and Hoffmann's computational result. These results assured the accuracy and robustness of the current perfect gas upwind flux difference splitting Euler code.

3.1 Effects of Chemistry Modelling

Figures 3a and 3b show the pressure and temperature distribution comparison among different chemistry models at Mach 15 altitude 30km. Back pressure was adjusted to get a strong standing shock in the middle of the divergent nozzle. A shock was captured within two grid points in all cases. Test cases conditions are given in Table 1. The fully coupled nonequilibrium air result showed almost the same shock position as the equilibrium air result; however, the position of perfect gas shock was located upstream of equilibrium/nonequilibrium gas shock with same back pressure ratio (P_e/P_∞). Figure 3b shows a temperature distribution along the nozzle. Temperature behind the shock for perfect gas model was approximately 10200K, which was much higher than those of equilibrium/nonequilibrium air model(about 5100K). A small peak in temperature after the shock was observed in nonequilibrium air model and quickly reduces to an equilibrium air value.

Figures 4 and 5 show temperature distribution comparison among different chemistry models at Mach 15 and altitudes 50km and 70km, respectively. The differences in temperature between equilibrium air and nonequilibrium air results were significant in both 50km and 70km altitudes. Due to rarefied gas flow region at high altitudes the flow may not be in complete chemical equilibrium at once. Therefore, the flow is essentially in chemically nonequilibrium state. Using equilibrium air model in 50km, 70km altitudes ($T/T_\infty = 12.2$, 19.7) predicted substantially lower temperature and very high density behind the shock comparing with nonequilibrium air ($T/T_\infty = 21.4$, 28.6) and perfect gas models ($T/T_\infty = 44.9$, 44.9). Especially at 70km, the temperature behind the shock reduced gradually to an exit value, which was a different tendency from the lower 50km and 30km altitude cases. In these flight altitudes, it is necessary to consider the effect of equilibrium air chemistry.

3.2 Altitude Effect

Nonequilibrium air temperature distributions at various flight altitudes were compared in Fig. 6 at Mach 15 with the same exit pressure ratio (P_e/P_∞). Since the nonequilibrium air model predicted almost the same temperature as the equilibrium air model at relatively low altitudes such as 30km and 40km, the nonequilibrium air code was chosen to compare the temperature distribution at different flight conditions. A strong shock was captured at the same location at various flight altitudes and Mach 15. The 70km altitude case shows a substantially higher temperature distribution(lower level of dissociation and ionization) than the outer cases(30km, 40km, 50km), which indicated an appreciable nonequilibrium air effect in that altitude. At Mach 25, the shock location of altitude 50km case was slightly downstream of the other altitude cases. Due to high freestream temperature(270K) at 50km the temperature behind the shock was very high causing more molecular dissociation and ionization, eventually leaded to lower temperature when compared with other altitude cases. The more dissociation and ionization occurred in the flowfield, the farther the shock was located downstream of nozzle.

A similar temperature distribution at Mach 25 is shown in Fig. 7. Due to high flight Mach number, a peak temperature ratio (T/T_∞) behind the shock was substantially increased when compared with Mach 15 case(i.e., $T/T_\infty \cong 31$, at $M = 15$, $T/T_\infty \cong 52$, at $M = 25$, at same 50km). The differences in exit temperature ratio among 30, 40 and 50km at Mach 25 were much larger than the corresponding ones at Mach 15. At high Mach number cases, the altitude effect was more apparent than at low Mach number cases.

3.3 Effect on Chemical Species

Figures 8a-8c show the nonequilibrium air O_2, N_2, NO, O, NO^+ and N along the nozzle at Mach 15 and altitude 30km, 50km and 70km, respectively. O_2 species was completely dissociated below 50km at Mach 15; however, only half of it was dissociated gradually at 70km, N_2 species was slightly dissociated(about 5%) below 50km comparing with almost no dissociation at 70km. A peak of NO species was existed just behind the shock below 50km, but NO species concentration increased linearly at 70km and Mach 15.

Figure 9a-9c show nonequilibrium air species mass fraction at Mach 25, altitudes 30km, 50km and 70km, respectively. At Mach 25, O_2 species were dissociated completely at all flight altitudes and N_2 species were substantially dissociated producing N species. N_2 species concentration at the exit, were 0.35, 0.18 and 0.4 at altitudes 30km, 50km and 70km. At 50km almost 3/4 of N_2 species were dissociated. Much less NO species was produced at Mach 25 comparing with that of Mach 15.

4 Conclusion

The CSCM type characteristic based upwind flux difference splitting Euler method has been used to study the effect of the equilibrium air and the nonequilibrium air chemistry. A strong shock was captured within 2 grid points and the temperature was obtained within a few Newton-Raphson iterations for nonequilibrium air calculation.

The chemistry equations were coupled with the fluid mechanics through effective γ. The pressure distribution of the perfect gas code was exactly agreed with the theoretical value(NACA-1135) and we obtained better solution than Hoffmann did with same grid.

The perfect gas shock was located upstream in front of those of equilibrium air and nonequilibrium air. The equilibrium air shock was located nearly the same or slightly behind the nonequilibrium air shock.

The temperature behind the shock of equilibrium air and nonequilibrium air was substantially reduced from that of the perfect gas at all Mach numbers and flight altitudes. At low altitudes, the equilibrium air and nonequilibrium air models predicted almost same temperature, density behind the shock; however, at high altitudes they showed substantial differences due to significant nonequilibrium chemistry effect. The more molecular dissociation and ionization occurred in the flowfield, the farther the shock was located downstream of nozzle.

At Mach 15, O_2 species was completely dissociated except 70km altitude case. O_2 species was completely dissociated at all flight altitudes and even N_2 species was substantially dissociated at Mach 25.

The new equilibrium/nonequilibrium seven species chemistry reacting upwind flux difference splitting Euler method has been developed successfully. Further works will include the extension of current work to multi-dimensional and to other complex chemistry system.

Table 1. Test Cases Condition

Alt. (km)	Pres. (Pa)	Dens. (kg/m^3)	Temp. (K)
30	1190.45	1.83×10^{-2}	227
40	286.97	3.98×10^{-3}	251
50	79.74	1.14×10^{-3}	270
70	4.5	8.756×10^{-5}	218

References

1. Bhutta, B.A. and Lewis, C.H. 1985, An Implicit Parabolized Navier-stokes Scheme for High-Altitude Reentry flows, AIAA 85-0362

2. Blottner, F.G., Johnson, M., and Ellis, M. 1971, Chemically Reacting Viscous Flow Program for Multi-Component Gas Mixtures, Sandia Lab., SC-RR-70-754.

3. Finson, M.L. and Ameer, P.G. 1985, Non-Equilibrium Boundary-layer Code, PSI-069/TR-512, physical Sciences Inc., MA.

4. Grossmann, B. and Cinnella, P. 1990, Flux-Splitt Algorithms for Flows with Non-Equilibrium Chemistry and Vibrational Relaxation, J. of Computational Physics, 88, pp.131-168.

5. Hoffmann, K.A., 1993, Computational Fluid Dynamics For Engineers, Vol. II, Engineering Education System, Wichita, U.S.A.

6. Lee, S.H. and Deiwert, G.S. 1990, Calculation of Nonequilibrium Hydrogen-Air Reactions with Implicit Flux Vector Splitting Method, J. Spacecraft & Rockets, Vol.27, pp167-174.

7. Lombard, C.K., Bardina, J. 1983, Venkatapathy, E. and Oliger, J., Multidimensional Formulation of CSCM-An Upwind Flux Difference Eigenvector Split Method for the Compressible Navier-Stokes Equations, AIAA-83-1895.

8. Miner, E.W. and Lewis, C.H. 1975, Hypersonic Ionizing Air Viscous Shock-Layer Flows Over Nonanalytic Blunt Bodies, NASA CR-2550.

9. Molvik, G.S. and Merkle, C.L. 1989, A Set of Strongly Coupled Upwind Algorithms for Computing Flows in Chemical Nonequilibrium, AIAA 89-0199.

10. Saladino, A., Praharaj, S., Collins, F., and Seaford, C., 1990, "Upwind of PARC2D to include Real Gas Effects," AIAA paper No. 90-0552.

11. Srinivasan, S., Tannehill, J.C., and Weilmuenster, K.J. 1987, Simplified Curve Fits for the Thermodynamic Properties of Equilibrium Air, NASA RP 1181.

12. Steger, J.L. and Warming, R.F. 1981, Flux Vector Splitting of the Inviscid Gasdynamics Equations with Application to Finite Difference Methods, J. Computational Physics, Vol. 40, pp. 263-293.

13. Van Leer, B. 1982, Flux-Vector Splitting for Euler Equations, Lecture Notes in Physics, Vol. 170, pp. 507-512.

14. Vincenti, W.G. and Kruger, C.H. Jr 1982, Introduction to Physical Gas Dynamics, R.E. Krieger publishing Co, Malabar Florida.

15. Widhopf, G.F. and Wang, J.C.T. 1988, A TVD Finite-Volume Technique for Nonequilibrium Chemically Reacting Flows, AIAA-88-2711.

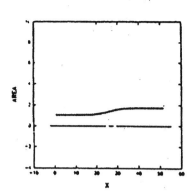

Fig. 1 The test case divergent nozzle geometry

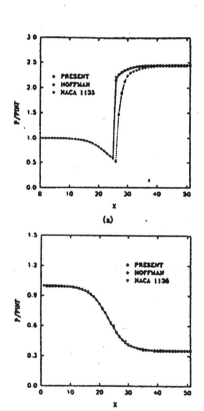

Fig. 2 Comparison of pressure distribution (a) at supersonic inflow and subsonic outflow and (b) at supersonic inflow and supersonic outflow

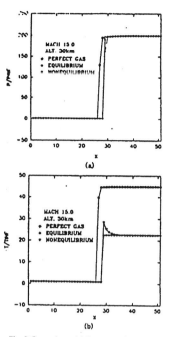

Fig. 3 Comparison of (a) pressure, (b) temperature distribution along the nozzle among different gas models at M=15 and altitude of 30km

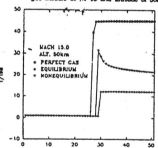

Fig. 4 Comparison of temperature distribution along the nozzle among different gas models at M=15 and altitude of 50km

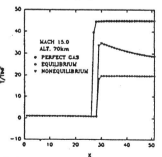

Fig. 5 Comparison of temperature distribution along the nozzle among different gas models at M=15 and altitude of 70km

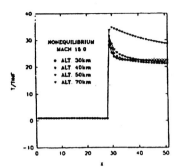

Fig. 6 Comparison of temperature distribution along the nozzle among different flight altitudes at M=15

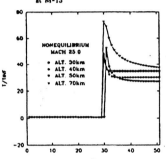

Fig. 7 Comparison of temperature distribution along the nozzle among different flight altitudes at M=25

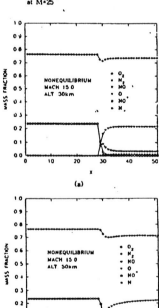

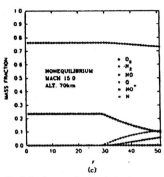

Fig. 8 Species mass fraction distribution along the nozzle at M=15 and altitudes of (a) 30km, (b) 50km and (c) 70km

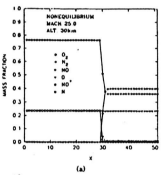

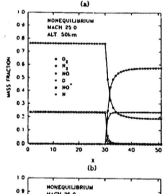

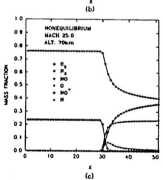

Fig. 9 Species mass fraction distribution along the nozzle at M=25 and altitudes of (a) 30km, (b) 50km and (c) 70km

Study of the Erosion of Stably Stratified Medium Heated from Below

K. R. Sreenivas, P. K. Dey, Jaywant H. Arakeri, and J. Srinivasan

Department of Mechanical Engineering
Indian Institute of Science, Bangalore-560 012, India

Abstract : We present, results from a two dimensional numerical simulation of a stably stratified layer heated from below. The aim of the simulation is to study the mechanism of erosion of a stably stratified medium by the convective velocity. We solve transport equations for momentum, heat and species using Patankar's SIMPLER method. Tracer particles are used to get flow pattern inside the mixed layer and in the gradient zone. We have observed randomly moving thermals issuing from the bottom heated surface and from the interface separating the mixed layer and the gradient zone. Stable fluid from the gradient zone gets entrained into the mixed layer with the falling thermal from the interface. This is the main mechanism of the turbulent entrainment. Results from the numerical simulation indicate the need to include the effects of the turbulent entrainment in modelling the equilibrium condition.

1 Introduction

The study of erosion of a stably stratified medium by the convective velocity is useful because of its relevance to atmospheric boundary layer, oceanography and solar ponds. In the atmosphere a stable layer formed due to the cooling of earth surface during the night gets eroded due to convective velocities from the heated earth surface during morning hours. This problem involves diffusion of heat only. In the ocean and in solar ponds, a stable stratification is due to decreasing salinity with height and convective velocities are due to wind shear and evaporative cooling near the top surface and due to the heating near the bottom surface. The dynamics of the system depends on the diffusion coefficients of the two components, heat and salt; it is a double diffusive system.

In this paper we have considered the erosion of a stably stratified gradient zone by convective velocities caused by bottom heating. Parameters which govern the erosion of the gradient zone are the thermal and solutal Rayleigh numbers (Ra_t and Ra_S), Richardson number (Ri), density stability ratio (R_ρ) and the ratio of diffusivity of salt to that of heat (τ). Various models have been proposed to predict the rate of erosion of the gradient zone. In these models the entrainment velocity, which is the rate of change of mixed layer height with time, is given as a function of the governing parameters. These models can be grouped under two categories. The first class of models, known as turbulent entrainment models, proposed by Turner (1968) and

Deardorff et al. (1980), consider only the effect of convective velocities on the erosion of gradient zone and the effect of diffusion is neglected. The models proposed by Witte (1989), Zangrando and Fernando (1991), can be called 'diffusion models', and consider the effect of diffusion on entrainment.

Observations in the laboratory and in the field indicate that there are conditions under which the interface does not move; this is known as equilibrium condition. Nielsen gives, based on the experimental and field data, an empirical correlation for the equilibrium condition (Hull et al. (1989)). Turbulent entrainment models fail to predict the observed equilibrium condition; diffusion models predict the equilibrium condition different from Nielsen's equilibrium condition. Hence, there is a need for understanding the mechanism of entrainment and behaviour of interface separating the bottom mixed layer and the gradient zone. This will help in developing a better model to predict the entrainment velocity and to explain the Nielsen's equilibrium condition. For this purpose, we have carried out two dimensional numerical simulations of the double diffusive system heated from below.

2 Numerical Solution

A schematic diagram of the system considered for the simulation with the boundary conditions is given in Figure 1. A constant heat flux is applied at the bottom wall which is impermeable to the species. Slip condition for vertical velocity is applied at the side walls. We have solved the unsteady equations governing the transport of momentum, energy and species together with the continuity equation, using Patankar's SIMPLER method (Patankar (1980)). Initially salinity and hence the density decreases uniformly with the height in the gradient zone. Simulations are carried out for different salinity gradients and bottom heat flux. A 0.05 second time step and 121×121 or 81×81 uniformly spaced grid are used. Sixty hours of CPU time is required, for the simulation of one hour physical time on IBM work station (RISC 6000).

3 Results and Discussion

Figure 2., shows the temperature variation with the time, just below the interface, in the mixed layer. This type of temperature traces have been observed in experiments and are interpreted as the 'bursting' events. Temperature contours separated by 40 seconds shown in Figure 3., indicate the presence of randomly moving thermals issuing from the interface and the bottom heated surface. A thermocouple probe fixed at some point near the interface would give a signal similar to the one indicated in Figure 2. Hot thermals from the bottom heated surface penetrate into the gradient zone to a height of order of 10^{-3} m.

We used tracer particles to study the effect of fluid motion on the interface. Simulations are started with tracer particles at different horizontal and vertical positions of the computation domain and their positions are followed with time. The movements of the tracer particles (Figure 4) indicate that the convective motions in the mixed layer induce velocities in the gradient zone. Induced velocities are mostly in the horizontal direction and the motion in the vertical direction are damped by the stable density gradient. These particles are eventually entrained into the mixed layer as the interface moves up. In Figure 5., we present the density anomaly (difference between the local density and average mixed layer density) contours. This contour clearly shows; the lighter fluid from the gradient zone geting entrained into the mixed layer, under the influence of falling thermal. This is the main mechanism of entrainment.

From the numerical study, we conclude that the convective motions in the mixed layer cause the erosion of gradient zone by entraining the stable fluid from the gradient zone. This entrainment occurs even in the absence of wave breaking phenomenon. Weak circulation in the gradient zone induced by convective motion will weaken the gradien zone. Turbulent entrainment has to be considered in addition to the effects of diffusion in modelling of erosion of the gradient zone and cannot be neglected as proposed by Zangrando and Fernando (1991).

References

[1] Deardorff, J., Willis, G., Stockton, B., (1980): "Laboratory studies of the entrainment zone of a covectively mixed layer" in J.Fluid Mech. V- 100, pp 41-64.

[2] Hull, J.R., Nielsen C.E., Golding P. (1989): *Salinity-gradient solar ponds* (CRC press, Inc.).

[3] Patankar, S.V. (1980): *Numerical heat transfer and fluid flow* (Hemisphere publishing corporation).

[4] Turner, J.S. (1968): "The behavior of a stable salinity gradient heated from below" in J.Fluid Mech. V-33. pp 183-200.

[5] Witte, J.M. (1989): *Thermal burst model of a double-diffusive 'diffusive' interface.* (Ph.D. Thesis University of Illinois at Urbana-Champaign).

[6] Zangrando, F., Fernando, H.J.S. (1991) "A predictive model for migration of double-diffusive interfaces" in J.Solar Energy V-113. pp 59-65.

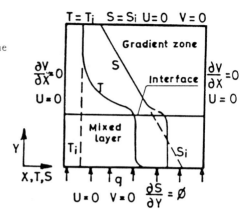

Fig. 1. A schematic diagram of the system considered for computation; T temperature, S salinity, q heat flux, subscript i denotes initial condition and U and V are horizontal and vertical velocities.

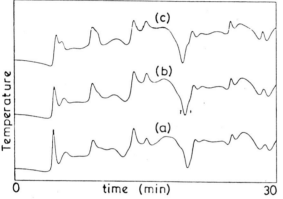

Fig. 2. Temperature variation with time, just below the interface at different horizontal positions.
(a) $X = 0.40$,
(b) $X = 0.43$ and
(c) $X = 0.45$.

Fig. 3. Evolution of temperature contours (a-d) separated by 40 sec.: contours show the formation and the movement of thermals.

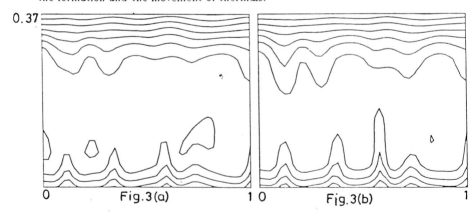

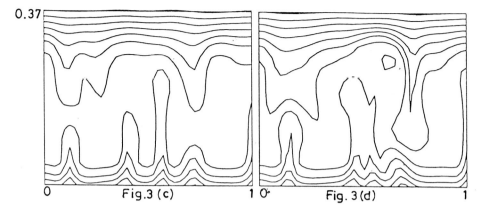

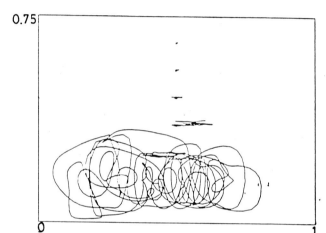

Fig. 4. Trajectories of tracer particles; located at different vertical positions. Convective motions inducing particle motion mostly in horizontal direction. Effect is reduced with the distance from the interface into the gradient zone.

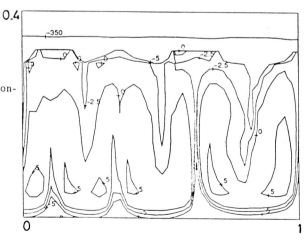

Fig. 5. Density anomaly contours at one instant of time. Negative contours, fluid of lesser density than average mixed layer density, have moved down with the falling thermals.

Advances in Euler and Navier-Stokes Methods For Helicopter Applications

G. R. Srinivasan

JAI Associates, Inc. and Sterling Software, Mail Stop 258-1
NASA Ames Research Center, Moffett Field, CA 94035-1000, USA

1 Introduction

Numerical simulation of the flowfield of a helicopter is a challenging problem. The complexity of the flow stems from several peculiar features that are unique to a helicopter. Much of the flow in the vicinity of the rotating blades is nonlinear, three-dimensional, and often unsteady. In addition, there are transonic regions near the blade tips of the advancing blades and dynamic stall pockets inboard of the retreating blades. Also, the blades shed complex vortical wakes. The presence of the vortical wake and its interaction with rotating and nonrotating parts affects the performance, blade loads, vibrations, acoustics, and safety of helicopter. Accurate prediction of the rotor wake system (its structure and trajectory) is one of the most studied problems in rotorcraft aerodynamics.

This paper outlines the growing importance of the use of the Euler and Navier-Stokes computational fluid dynamics (CFD) methods to simulate the nonlinear problems of helicopter aerodynamics and aeroacoustics. Although a complete flowfield simulation of helicopter has not been demonstrated by using CFD, impressive gains have been made in analyzing individual components in great detail. State-of-the-art numerical methods in conjunction with powerful supercomputers have yielded noticeable progress in modeling viscous-inviscid interactions, blade-vortex interactions, tip vortex simulations, wake effects, as well as radiated noise in hover and forward flight.

The present study describes a free-wake Euler and Navier-Stokes method, called TURNS (Transonic Unsteady Rotor Navier-Stokes), and its capabilities for calculating the aerodynamics and acoustics [1-5]. With this method, it is possible to use a single calculation to model both aerodynamic and acoustic flowfields. Recent modifications to this method have been used to model complex configurations by using Chimera overset grids [6-8].

2 Numerical method

The numerical method of TURNS is described in Srinivasan et al. [1]. It is an implicit numerical method that solves the Euler and thin layer Navier-Stokes equations in either blade-fixed or inertial coordinates system. The finite-difference numerical method uses a Roe upwind-biased scheme with high-order MUSCL-type limiting for the right-hand side to model shocks and capture acoustics accurately. An LU-SGS implicit operator on the left-hand side makes the numerical scheme computationally efficient as well as robust. The method is third-order accurate in space and first- or second-order accurate in time with provision for Newton sub-iterations. An algebraic eddy viscosity model is used to compute turbulent flows. The computational meshes are body conforming and typically have 100,000-950,000 grid points. Boundary conditions are applied explicitly. No-slip or tangential boundary conditions are used on solid surfaces. Characteristic boundary conditions are used at the farfield. In addition, the hovering problem requires special treatment at the periodic and farfield boundaries.

3 Results

The TURNS code has been applied to both model rotors as well as full scale modern helicopter rotors to calculate aerodynamics and acoustics in hover and advancing flight. Typical solutions are presented with comparison to experimental data. Figure 1 shows a plot of steady surface pressures of a two-bladed model rotor [9] for two radial stations at a transonic flow condition. Although the vortical wake is diffused, the blade surface pressures are well captured. The ability of the numerical method to calculate hover performance characteristics is demonstrated in Fig. 2 for realistic blades with twist and taper. The four-bladed Black Hawk UH-60A rotor and the BERP (British Experimental Rotor Program) rotor are evaluated for planform effects. Performance curves of thrust (C_T) vs torque (C_Q) and Figure of Merit are presented in Fig. 2 for two thrust conditions based on collective pitch (θ_c) of 9° and 13° and a tip Mach number M_{tip}=0.628 and a Reynolds number, based on the tip speed and equivalent chord, of 2.75 million. Also shown in this figure are the induced power (C_{Q_i}), calculated using an empirical relation [10], the ideal power, and the experimental data [11]. The results at low thrust condition show good agreement with experiments. The high thrust condition is apparently beyond the model test range. This condition produces severe shock-induced flow separation over large parts of the blades. It is interesting to note that the Figure of Merit *including viscous effects* is almost the same at the two thrust conditions, indicating that this particular measure of total hover performance seems to stay relatively uniform for these rotors over the range of θ_c used. The nearfield wake trajectory, the wake contraction and descent, of the Black Hawk rotor is shown in Fig. 3 compared to the experiments. At least over the range of one rotation of the blade, the agreement of the wake trajectory with experiments is satisfactory.

The interactional flowfield of an advancing nonlifting rotor encountering a line vortex is also calculated for parallel and oblique blade-vortex interactions using a vortex-fitting technique [12] for preserving the structure of the interacting vortex. Figure 4(a) shows the experimental configuration [13] for a parallel interaction. First, the unsteady flowfield is calculated in the absence of vortex interaction and this solution is then utilized as the starting solution for the blade-vortex interaction calculation. Figures 4(b) and 4(c) show sample instantaneous surface pressures of an advancing rotor ($M_{tip} = 0.8$ and advance ratio $\mu = 0.2$) without and with vortex (nondimensional strength $\Gamma = 0.177$) interaction compared to the experiments at a radial station of $r/R = 0.893$ and at different azimuthal locations. The agreement of surface pressures is very good.

In the TURNS code, a single calculation will capture both aerodynamics and acoustics. Because of the finite size of flowfield grid the radiated noise cannot be propagated to very long distances. Baeder [3-4], by choosing a refined grid and orienting the grid lines beyond the blade tip to follow characteristics, was successful in capturing high speed impulsive (HSI) noise to distances of three rotor radii from the blade tip. The acoustic wave forms of high speed impulsive noise calculated by Baeder for hovering and advancing blade are reproduced from Ref. [3]. Comparison of the calculation with wind tunnel [14] and flight data [15] indicate the sound intensity and wave forms are well captured by the numerical method. Recently, Strawn and Biswas [5], also using the TURNS code in conjunction with a Kirchhoff formulation, were able to successfully propagate the HSI noise over much larger distances.

Future directions point to using multi-block version of TURNS code to model complex configurations containing fixed and rotating components. Some limited progress about this using Chimera overset grids is described in Refs. [6-8].

4 Conclusions

The good agreement of the numerical results with experiments in both hover and forward flight indicates the accuracy and suitability of the TURNS code for helicopter aerodynamics and acoustics simulations.

Acknowledgements :

Most of the research described herein is supported by the U. S. Army Research Office under contracts DAAL03-88-C-0006 and DAAL03-90-C-0013 to JAI Associates, Inc.

References

[1] G. R. Srinivasan, J. D. Baeder, S. Obayashi, W. J. McCroskey: AIAA J. **30** 2371 (1992)

[2] G. R. Srinivasan, V. Raghavan, E. P. Duque, W. J. McCroskey: J. AHS **38** 3 (1993)

[3] G. R. Srinivasan, J. D. Baeder: AIAA J. **31** 959 (1993)

[4] J. D. Baeder: AHS Int. Tech. Spec. Meeting (1991)

[5] R. C. Strawn, R. Biswas: 19th Army Science Conference (1994)

[6] E. P. N. Duque, G. R. Srinivasan: Proc. 48th AHS Ann. Forum 429 (1992)

[7] G. R. Srinivasan, J. U. Ahmad: 19th Euro. Rot. Forum (1993)

[8] J. U. Ahmad, E. P. N. Duque: AIAA Paper 94-1922 (1994)

[9] F. X. Caradonna, C. Tung: NASA TM-81232 (1981)

[10] W. Johnson: Princeton U. Press (1980)

[11] P. F. Lorber, R. C. Stauter, A. J. Landgrebe: Proc. 45th AHS Ann. Forum (1989)

[12] G. R. Srinivasan, W. J. McCroskey: Vertica, 11 3 (1987)

[13] F. X. Caradonna, G. H. Laub, C. Tung: NASA TM-86005 (1984)

[14] D. A. Boxwell, Y. H. Yu, F. H. Schmitz: Vertica 3 35 (1979)

[15] F. H. Schmitz, D. A. Boxwell, W. R. Splettstoesser, K. J. Schultz: Vertica 4 395 (1984)

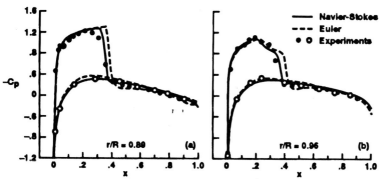

Fig. 1. Surface pressures of a hovering 2-bladed rotor; M_{tip}=0.877, collective pitch-8°, and Re=3.93x10^6.

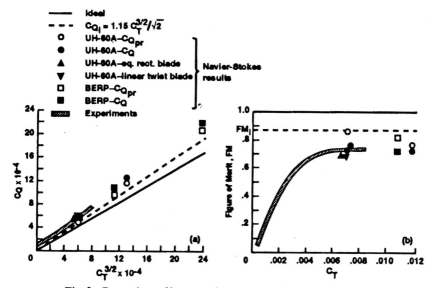

Fig. 2. Comparison of hover performance for different 4-bladed rotors.

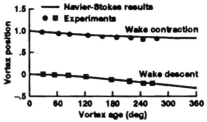

Fig. 3. Wake contraction and descent of a 4-bladed hovering Black Hawk rotor; M_{tip}=0.628, collective pitch=9°, and Re=2.75x10^6.

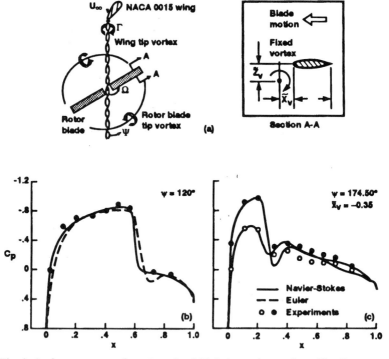

Fig. 4. Surface pressures of an advancing 2-bladed rotor interacting with a line vortex; a) schematic of experimental setup; b) without vortex interaction, and c) with vortex interaction; M_{tip}=0.8, μ=0.2, Γ=0.177, r/R=0.893, and Re=2.89x10^6.

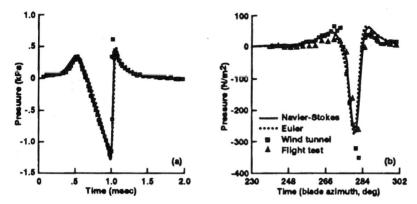

Fig. 5. Pressure-time histories of high speed impulsive noise in-plane of rotor blade; a) hovering UH-1H rotor, $M_{tip}=0.95$, $Re=1.6 \times 10^6$, and $r/R=3.09$; b) advancing OLS rotor of AH-1 helicopter, $M_{tip}=0.665$, $\mu=0.348$, $Re=2.17 \times 10^6$, and $r/R=3.44$.

Numerical Simulation of Unsteady Compressible Viscous Flow NACA 0012 Airfoil - Vortex Interaction

C. Tenaud, Ta Phuoc Loc

L.I.M.S.I. - UPR CNRS 3251
BP.133, 91403 ORSAY Cedex, France.

Abstract : The numerical simulation of the compressible flow around a NACA 0012 airfoil has been undertaken using the unsteady Navier-Stokes and total energy equations. The resolution of the governing equations is based on a Roe's approximate Riemann solver. Their space integration is performed by means of an upwind Total Variation Diminishing (TVD) scheme, developed by Harten and Yee. The time integration is performed by a linear conservative implicit form through the use of an Alternating Direction Implicit (ADI) formulation. The interaction between a vortex and a NACA 0012 airfoil has been studied. The Mach and Reynolds numbers considered are respectively equal to 0.6 and 22000. and the CFL number used is 20. The numerical results are then reviewed.

1 Introduction

The interaction of concentrated vortices with lifting surfaces occurs in many aerodynamic applications. For instance, this kind of interaction could be encountered in an helicopter rotor flow field, which is of particular interest. It is well known that in such a flow field, the interaction of trailing vortex wake with the oncoming rotor blades can induce unsteady blade loading and aerodynamic noise. Following several experimental results, the configuration where the vortex axis is parallel to the leading edge of the airfoil provides with the highest pressure variations which play an important role in the noise generation. Of course, this impulsive noise is mainly sensitive to the compressibility of the flow. Therefore, to understand the mechanisms involved in the interaction of an incomming vortex with an airfoil, the numerical simulation of the compressible flow around a NACA 0012 airfoil has been undertaken using the Navier-Stokes and total energy equations. In order to obtain a well estimated solution, the numerical method should at least be conservative, have as weak as possible numerical dissipation and the integration scheme should be sufficiently accurate in space and especially in time since long time integrations are required.

2 Numerical approach

2.1 Basic equations

The equations of the flow are the Navier Stokes and total energy equations written in conservative form in a cartesian coordinate system. The static pressure P and temperature T are calculated through the use of an equation of state, written for a perfect gas. The dynamic viscosity (μ) as well as the heat conductivity (λ) are supposed to be only functions of the static temperature. The heat conductivity is related to the viscosity through the use of a constant Prandtl number assumption $\mathcal{P} = 0.725$.

2.2 Grid and coordinate transformation

The grid is a C-grid which fits to the body. The mesh is tightened near the wall in the normal to the wall direction (η) as well as in the direction parallel to the wall (ξ) in the vicinity of the leading and trailing edge of the airfoil. To solve the governing equations, a generalized coordinate transform ($\xi = \xi(x,y)$ and $\eta = \eta(x,y)$) is used which preserves the conservative form of these equations.

2.3 Space and Time integration

The resolution of the Navier-Stokes and total energy equations has been performed by means of a finite difference method.

As regards the convective terms, their integration is based on a Roe's approximate Riemann solver [2]. The governing equations are projected along the characteristic direction ; a set of 4 characteristic equations in then obtained. Therefore, a scalar scheme could be applied on each characteristic direction in order to integrate these equations. This integration is performed by means of a shock capturing finite difference method based on an upwind Total Variation Diminishing (TVD) scheme, developed by Harten [1] and Yee [5]. Using appropriate limiter function, this scheme is second order accurate in space. To express the anti-diffusive fluxes used in this scheme (see Yee [7]), we have used the relationship proposed by Van-Leer [6] [4]. In order to evaluate the different terms at the mid-point of the grid (i+1/2, j or i, i+1/2), we employ a mass-weighted averaging technique introduced by Roe [2] between the states at the grid points.

Concerning the diffusive flux functions, a central differencing scheme is applied giving a second order accuracy in space.

The convective terms are linearized and we applied a Linear Conservative Implicit formulation [5]. Notice that, using the previous discretization, this scheme is second order accurate in both time and space ; nevertheless, it may no longer be unconditionally TVD. Important values of the time-step can however be employed, corresponding to a Courant numbers (CFL) in the range of 10 to 50.

The implicit operator is factorized using an Alternating Direction Implicit (ADI) form [5]. Therefore, in each coordinate direction, a block tridiagonal matrix is solved using a classic algorithm.

2.4 Boundary conditions

On the airfoil, a no-slip condition is applied and a wall temperature (T_w) is prescribed. At the inlet ($\eta = \eta_{max}$), we suppose that the flow is inviscid and the governing equations are projected along the characteristic directions. Therefore, if the characteristic speeds are positive we employ an upwind discretization of the derivative in the η direction ; while if the characteristic speeds are negative, we impose a non-reflecting condition (there exists no variation in time). it is important to get an appropriate treatment of the conditions at this boundary since, if there exists pressure waves, they must be convected out of the domain without reflection in the computational domain.

At the outlet ($\xi = 0.$ or $\xi = \xi_{max}$), the derivatives are supposed to be equal to zero. Let us add that, all the boundary conditions are applied in the implicit part of the algorithm.

3 Results and discussion

First, this numerical code was checked on the calculation of the flow in the vicinity of a NACA 0012, airfoil without incidence (without vortex interaction). The Mach number is equal to 0.6, the temperature is 300°K and the Reynolds number, based on the chord length, is about 22000. The CFL number is equal to 20.

The pressure coefficient (C_p) distribution along the chord of the airfoil is presented on figure (1)(solid line). We can see that the calculated distribution is perfectly symmetric since the upper and lower-side distribution coincide. The calculated distribution is compared with experiments (dots) conducted at ONERA-CERT (Toulouse-France) at higher Reynolds number ($R_e \neq 3.10^6$). A good agreement is obtained on the main part of the airfoil (Fig. 1), except close to the trailing-egde where the computation exhibits a plateau-evolution. This plateau behaviour is due to the presence of a slight recirculation region occuring at the trailing-edge of the airfoil. According to the distribution of the skin friction coefficient, the recirculation region starts at 0.65 chord length from the leading edge. We can say that the Reynolds number has no influence on the C_p distribution in the front-part of the airfoil. Nevertheless, the influence of the Reynolds number is very sensitive close to the trailing-edge. In fact, experimental studies have been shown that the higher the Reynolds number is, the more downstream the separated point is situated.

Secondly, the computation of the interaction between a concentrated viscous vortex and a NACA 0012 airfoil has been undertaken. The calculation, using a CFL number equal to 20, starts from the steady state using the previous results. At the initial state, we have abruptly superimposed on the previous solution a concentrated vortex with a dimensionless strength equal to 0.2. This vortex is initially situated at the following location : $x_v/C = -0.6$; $y_v/C = -0.25$ (where C is the chord length and $x = 0., y = 0$ refers to the leading edge of the airfoil).

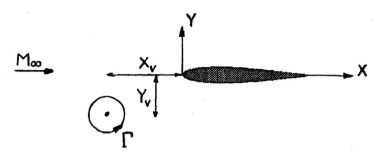

When the vortex is closed to the leading edge of the airfoil $x_v/C = 0$. (Fig. 2), the distribution of the pressure coefficient (C_p) on the upper side of the airfoil stays comparable to the undisturbed distribution (that means the distribution obtained without vortex interaction); while the C_p distribution on the lower side is greatly affected (Fig. 2). Therefore, the lift force, due to the pressure distribution on the airfoil, is positive. When the vortex is located at about the mid-chord of the airfoil ($x_v/C = 0.5$) (Fig. 2), the C_p distribution on the lower side of the airfoil has almost recovered the undisturbed distribution, except in the vicinity of the trailing edge. On the upper side, the pressure distribution seem to be greatly affected everywhere along the chord length. Compared with the previous vortex position, the sign of the lift force has changed and become negative. The change in the sign of the lift force occurs when the vortex is located at about $x/C = 0.3$ which corresponds roughly at the position of the maximum thickness of the airfoil. After $x_v/C = 1.$, corresponding to the trailing edge position, the pressure distribution tends to recover its undisturbed distribution (Fig. 2), except close to the trailing edge where the C_p distribution is affected by unsteady, extended separations (Fig. 2). That time, the lift force oscillates and tends to recover its undisturbed zero-value. At last, let us mention that when the vortex is close to the leading edge of the airfoil, alternate vortices occur in the vicinity of the trailing edge of the airfoil ; therefore a vortex wake is created. The more the interacting vortex moves downstream, the more the vortex-wake grows. The dimensionless frequency (Strouhal number based on the upstream velocity and the chord length) of the vortex shedding at the trailing-edge of the airfoil is close to $S_t = 3$. On the static pressure field, a pressure wave is visible due to the interaction between the concentrated vortex and the airfoil. This wave grows from the wall of the airfoil to the upper boundary of the computation domain. Let us mention that no reflection of the pressure wave is observed using the appropriate boundary condition (based on the characteristics variables) on the external boundary. When the concentrated vortex is situated downstream of the airfoil, pressure waves are then created from each side of the vortex wake due to the alternate vortices. The frequency of these pressure waves is highly related to the frequency of the vortex shedding. All these pressure waves might contribute to noise generation.

4 Conclusions

The numerical simulation of the compressible flow around a NACA 0012 airfoil has been undertaken using the Navier-Stokes and total energy equations. The integration of these governing equations is based on a Roe's approximate Riemann solver through the use of an upwind TVD scheme, developed by Harten [1] and Yee [5]. The time integration has been performed by a linear conservative implicit formulation through the use of an Alternating Direction Implicit (ADI) resolution. First, the calculation of the compressible flow in the vicinity of a NACA 0012 airfoil without incidence has been performed using the implicit algorithm with a CFL number equal to 20. The numerical results are compared with a good agreement with experimental data provided by ONERA-CERT. Secondly, the interaction between a concentrated vortex and a NACA 0012 airfoil has been investigated. The influences of the concentrated vortex on the distribution of C_p along the chord of the NACA 0012 airfoil have then been analysed. When the concentrated vortex is close to the leading edge of the airfoil, alternate vortices occur in the vicinity of the trailing edge of the airfoil. Therefore, due to these alternate vortices, pressure waves have been created from each side of the vortex wake. The frequency of these pressure wave is highly related to the frequency of the vortex shedding ($S_t = 3.$).

Acknowledgements :

The computations have been carried out on the Cray C98 of I.D.R.I.S. (Orsay-France) ; the authors greatly acknowledge the support of this institution.

References

1. A. Harten *High Resolution Schemes for Hyperbolic Conservation Laws.* Jal Comput. Phys. 49, 1983. pp. 357-393

2. P.L. Roe *Approximate Riemann Solvers, Parameter Vectors and Difference Schemes.* Jal Comput. Phys. 43, 1981, pp. 357-372

3. C. Tenaud - T.P. Loc *Numerical Simulation of Unsteady Compressible Viscous Flow Around NACA 0012 Airfoil.* Workshop on Num. Methods for the Navier-Stokes equations, Heideelberg, Oct. 1993

4. B. Van Leer *Towards the Ultimate Finite Difference Scheme II : Monotonicity & Conservation combined in a IInd order scheme.* Jal Comput. Phys. 14, 1974, pp. 361-370

5. H.C. Yee - A. Harten *Implicit TVD Schemes for Hyperbolic Conservation Laws in Curvilinear Coordinates.* AIAA Jal 25, No 2, 1987, pp. 266-274

6. H.C. Yee *Construction of Explicit and Implicit Symmetric TVD Schemes and Their Applications.* Jal Comput. Phys. 68, 1987, pp. 151-179

7. H.C. Yee *Numerical Experiments With a Symmetric High Resolution Shock-Capturing Scheme.* NASA TM-88325, (1986)

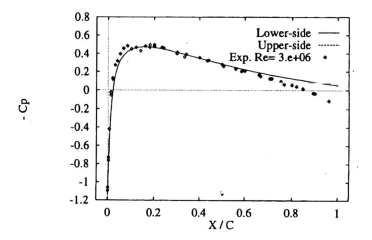

Figure 1: Distribution of the pressure coefficient along a NACA 0012 airfoil without vortex interaction ($M_\infty = 0.63$; $Re_C = 22000$.)

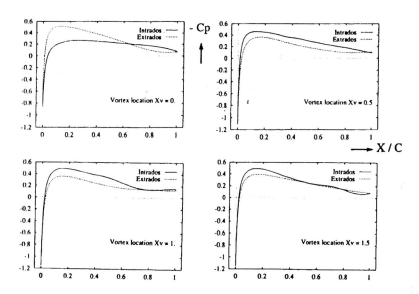

Figure 2: Distribution of the pressure coefficient along a NACA 0012 airfoil with vortex interaction ($M_\infty = 0.63$; $Re_C = 22000.$; $\Gamma = 0.2$)

A Numerical Study of Unsteady Flow Phenomena in the Driven and Nozzle Sections of Shock Tunnels

Susan Tokarcik-Polsky[1] and Jean-Luc Cambier[2]

[1]Sterling Software, [2] Thermosciences Institute
MS 230-2 NASA Ames Research Center,
Moffett Field, Ca 94035, USA

Abstract : This work examined transient flows that occur in shock tunnels. Two areas were examined. The first area was the end wall region of the driven tube. This region forms the reservoir for the nozzle section. The second area was the transient flow in the nozzle itself. These transient flows are important when determining test times and test conditions. It is very difficult to study these transient flows experimentally; therefore, computational fluid dynamics (CFD) was used.

1 Numerical Method

The computations presented in this paper were performed using a computer code developed at NASA-Ames Research Center by Dr. J.-L. Cambier. The code used a finite- volume, second-order TVD scheme to solve the Navier-Stokes equations. Multiple grids were used to handle complex geometries, and sliding grids were used to enhance the resolution of transient flow features. In the sliding grid method, a coarse grid was established as a "background" grid. This grid generally encompassed most of the domain of study. Another grid of higher resolution was then established as a subset of the background grid. This new grid was able to move within the background grid so that transient flow features could be tracked with the high resolution grid. This sliding grid method allowed grid refinement in areas of interest while avoiding unnecessarily dense grids in areas of little activity or interest. For more details on the numerical method, see Ref. [1].

2 Reflected Shock Simulation

For this portion of the study, the dynamics of a primary-shock-wave reflection at the end wall of a shock tunnel driven tube were examined. When a primary shock wave reflects off the end wall of the driven tube, a reservoir of nearly stagnant gas forms behind it. This reservoir gas is expanded through the nozzle and is used as the "test gas." When the test gas becomes contaminated with driver gas, the test is considered to be over. It has been found by previous CFD studies [1,2] that when the primary shock reflects, a vortical disturbance is generated at the axis of symmetry of the driven tube. The vortices cause the reflected shock to bulge out in the direction of propagation (up the driven tube). The bulge in the reflected shock could cause premature contamination of the test gas reservoir thus reducing the test time. It should be noted that the bulging effect does not occur in a two-dimensional set-up. It has only been observed in CFD results for axisymmetric and three-dimensional formulations.

The CFD calculations were started from the moment the primary shock reached the end wall of the driven tube and began to reflect. It was assumed that the secondary diaphragm broke ideally at this moment. All times given in this paper were measured assuming t=0 sec at this point. The initial conditions in the driven tube were as follows: air at p=50.76 atm, T=3674 K, V=2710 m/s, and an equilibrium chemical composition. The initial conditions in the nozzle were: air at $p = 1.3 \times 10^{-4}$ atm, T=298 K, V=0.0 m/s. For the numerical analysis, the flow was assumed to be inviscid and chemically frozen.

Because no experimental data exist, the formation of the vortices has only been observed in CFD results. To investigate whether the vortices were artifacts of the numerical method, several numerical experiments were conducted. First, grid effects were examined. It was found that if the grid aspect ratio was such that long (in the axial direction) and short (in the radial direction) cells existed near the axis, a jetting effect occurred along the singular axis. The jetting greatly accentuated the vortices which in turn caused the reflected shock to bulge more. As the jetting increased the reflected shock formed a cone-type shape (i.e. <) around the axis of symmetry (top half of Fig. 1). While shock interactions can produce jets, a result such as this in a CFD solution, especially when it is near a singular axis, was cause to doubt whether a cone-shaped reflected shock was a physically realistic result. This jetting effect has previously been observed by the present authors [1] and can also be seen in the results of P. Jacobs [2].

It was suggested in Ref. [1] that coarsening the grid in the radial direction would alleviate the jetting problem. Shown on the bottom half of Fig. 1 is the same case as computed on the top half; however, the grid aspect ratio was changed so that the cells near the axis were tall (radially) and thin (axially). In this case the bulge in the reflected shock formed a smooth arc instead of a sharp cone. This indicated that the jetting effect could be eliminated by a judicious choice of grid aspect ratio near the axis. It should be noted however, that the bulge in the reflected shock was never completely eliminated. A number of other numerical experiments, including further grid studies and boundary condition studies, were also conducted [1]. Again, in none of these cases was the bulging in the reflected shock eliminated. This does not, however, resolve the question of whether or not the formation of the bulge is truly a physical phenomenon. Ultimately, experimental data may be required to resolve this issue.

In order to examine the effects of geometric changes on the size and character of the bulge in the reflected shock, the geometry in the driven-tube/nozzle interface region was varied by: 1) changing the throat size, 2) changing the shape of the juncture between the two sections, and 3) changing the shape of the end wall of the driven tube.

Small, medium and large throat sizes were examined. It was found that, although the transient flows within the nozzle sections were different for each case (in particular, compression wave interactions in the subsonic portion of the nozzle were intensified by the smaller throats), the vortex structures in the driven tube were nearly identical. This suggests that the formation of the vortex system in the driven tube is not strongly dependent on throat size.

Three changes were made to the geometry at the end wall of the driven tube. These changes increased the diameter of the aperture in the end wall by 35% (1), rounded the juncture between the nozzle and the end wall (2), and tapered the end wall into the driven tube (3). The throat size was held fixed.

Increasing the diameter of the aperture initially caused a slight increase in the deformation of the reflected shock. However, by 120 micro-sec after shock reflection, the solution looked very similar to the original solution shown at the bottom of Fig. 1. Rounding the juncture between the nozzle and the end wall again caused the reflected shock to be more distorted, especially near the axis where the bulging effect was increased. Again, however, by 120 micro-sec the solution looked very similar to the original solution.

For the tapered-end-wall case, the flow field looked considerably different than those examined above. There were strong, transient compression waves in both the subsonic and supersonic portions of the nozzle that were not seen in the other cases. Also, the bulge in the reflected shock was initially larger; however, once again, by 120 micro-sec the overall structure in the reflected shock region looked very similar to the case with the original geometry. The important exception was that at $t=120$ micro-sec the reflected had traveled farther up the driven tube than for the other cases. This occurred because the tapered end wall effectively shortened the driven tube.

It is important to note that all the above calculations assumed that the flow was inviscid. Therefore, the viscous effects which may occur due to geometric changes were not examined. Viscous effects may not be insignificant, and further study is required to determine their importance to this problem.

For the bulging of the reflected shock to cause premature contamination of the test gas, it would have to persist long enough to interact with the contact discontinuity. The original case was continued until $t=400$ micro-sec, and it was found that the bulge did diminish as the reflected shock continued to propagate up the driven tube. Eventually, the reflected shock became almost planar. Therefore, the likelihood of premature mixing depends on how much time elapses before the contact discontinuity and the reflected shock interact. Note that for this study, the driven tube walls were assumed to be inviscid; therefore, the effects of shock-wave/boundary-layer interactions at the walls, which could also cause premature contamination, were not examined.

3 Transient Nozzle Flow Simulation

The nozzle flow examined computationally in this study was originally examined experimentally by Amann and Reichenbach[3]. The nozzle was two dimensional and had a rounded inlet. The test was initiated with a Mach 3.0 shock in air upstream of a nozzle with a 15 deg semi-divergence angle. The shock propagated into stagnant gas with $T=293$ K and $p=6.3$ kPa.

The numerical study was initialized with a planar, Mach=3 shock just upstream of the shock tube exit. The shock tube section was calculated with slip wall boundary conditions. The walls of the nozzle, however, were calculated with no-slip boundary conditions since viscous interactions in the nozzle produced some of the dominant features in the flow field. In all, two sub-scale, high-resolution grids were used: one to track the primary shock and contact surfaces and a second to track the secondary shock. This nozzle flow was also examined computationally by Prodromou and Hillier[4]; however, those calculations were entirely inviscid.

The general evolution of the wave system within the nozzle is shown in Figs. 2ah. The computational equivalents of Figs. 2a-h are shown by way of temperature contours in Figs. 3a-h. All features shown in Figs 2a-h, which are described below, were observed in the computational results.

The nozzle flow develops as follows. Initially the incident shock (IS) enters the nozzle and a regular reflection occurs at the nozzle walls (Fig. 2a). The regular reflection quickly breaks down, and a Mach reflection develops (Fig. 2b). A contact surface (I) is associated with each Mach reflection. The Mach reflection consists of a Mach stem (MS), the reflected shock (RS), and an incident shock. These three features meet at the triple point. The triple point eventually meets the symmetry plane, and a regular reflection of the Mach stem (which must now be considered another incident shock) occurs (Figs. 2c & 2d). Again a Mach reflection (including the associated contact surface) develops on either side of the symmetry plane (Fig. 2e). The newly formed triple point then travels toward the nozzle wall setting the stage for the formation of yet another Mach reflection (Fig. 2f). As the flow continues to develop, a secondary shock wave is formed which is facing upstream but being swept downstream (Fig. 2g). This secondary shock causes the boundary layer along the nozzle walls to separate (Fig. 2h).

A series of experimental shadowgraphs were compared to synthetic shadowgraphs generated from the computed solutions (the last frames in the series are shown in Figs. 4a and 4b). The comparison showed that initially the shock systems and the contact surfaces were captured very well by the computations. However, as the flow progressed, the resolution of the contact surfaces in the computational solutions began to degrade. Numerical and viscous dissipation tend to dissipate contact surface, and the sub-scale, high-resolution grids were required to capture the detail observed in the CFD results. Solutions computed using only the coarse grids showed that the contact surfaces were all but non-existent in the downstream portion of the nozzle.

The boundary layer separation (Fig. 4a) was captured well by the computations (Fig. 4b). However, as the secondary shock moved downstream, the CFD solutions predicted that this shock became curved (Fig. 4b) instead of remaining mostly planar as was shown in the experimental shadowgraph (Fig. 4a). The reason for this is not fully understood; however, one possible cause may be that the boundary layer upstream of the secondary shock was not modeled precisely enough causing the shock-wave/boundary-layer interaction to be different than in the experiment. There was some indication in the experimental shadowgraphs that the separated region may have been turbulent. The calculations assumed that the entire flow field was laminar; therefore, any turbulent effects were not captured.

The experimentally measured density along the nozzle axis at six different times during the starting process is shown in Fig. 5a. The density profiles from the CFD solutions at the same times are shown in Fig. 5b. The computations agree very well with the experimental results. In particular the locations of the primary and secondary shocks were captured well throughout the simulation indicating that the shock speeds were predicted accurately. Also, the features within each profile were captured well.

4 Conclusions

It was found that refining the grid in the radial direction produced a non-physical jetting behavior along the axis of the driven tube. Changing the grid aspect ratio was presented as a solution to the jetting problem. Several geometric changes were made to both the nozzle section and the driven tube to determine their effects on the flow field. In all cases the reflected shock became almost planar by $t=400$ micro-sec after shock reflection. Overall, the geometric changes did not have a significant affect on the bulge in the reflected shock. It was determined that further study was required to examine viscous effects which may occur due to geometric changes.

The transient flow in a shock tube nozzle was calculated and the results, including shock speed and overall flow structure, compared well with experimental results. The CFD solutions incorporated a grid tracking technique where transient flow features were tracked by high resolution grids. This technique greatly improved the accuracy of the CFD solutions.

References

1. Tokarcik-Polsky, S. and Cambier, J.-L., "Numerical Study of Transient Flow in Shock Tunnels," AIAA Journal, Vol. 32, No. 5, May 1994.
2. Jacobs, P.A. "Simulation of Transient Flow in a Shock Tunnel and a High Mach Number Nozzle," 4th Int. Symp. on Comp. Fluid Dynamics, Davis, Sept. 1991.
3. Amann, H.O. and Reichenbach, H. "Unsteady flow phenomena in shock tube nozzles," Recent Developments in Shock Tube Research, pp. 96-112, edited by D. Bershader & W. Griffth, 1973.
4. Prodromou, P. and Hillier, R. "Computation of unsteady nozzle flows," Shock Waves Proc., ed. K. Takayama, Sendai, Japan 1991, vol. II, pp. 1113-1118.

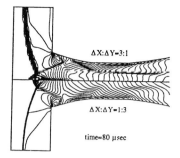

Fig. 1 Effect of grid aspect ratio on the computed reflected-shockwave.

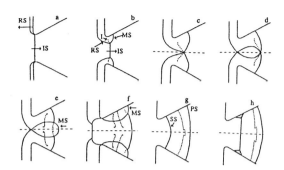

Fig. 2 (a-h) Schematic of the transient nozzle flow.

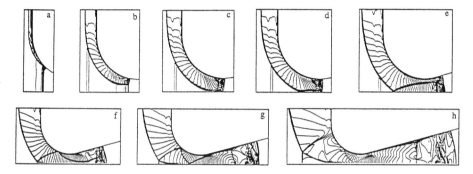

Fig. 3 (a-h) Computational solution at times corresponding to the flow schematics presented in Fig. 2 (a-h).

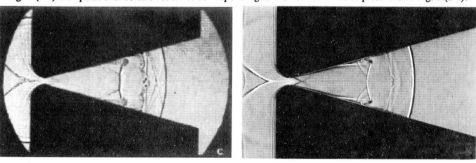

Fig. 4 Shadowgraph taken during the starting process of a reflection nozzle (left), and a shadowgraph constructed from the computational results (right).

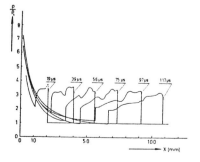

Fig. 5a Experimental density on the nozzle axis at six stages in the starting process.

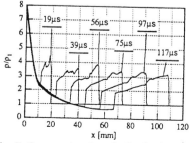

Fig. 5b Computational results for density on the nozzle axis at six stages in the starting process.

The Dynamics of Reacting Shear Layers Adjacent to a Wall

Hsin-Hua Tsuei and Charles L. Merkle

Propulsion Engineering Research Center
Department of Mechanical Engineering
The Pennsylvania State University

1 Introduction

The large scale vortices that are generated by convective instabilities in shear layers have been widely studied. As these vortices grow and begin to roll up, adjacent vortices pair and merge causing their size and spacing to increase as first demonstrated by Brown and Roshko [1]. Although these dominant structures are generated in simple shear layers with only a velocity difference between streams, the Brown-Roshko's experiments also included widely different molecular weights. More recently, other researchers using both experimental [2,3] and numerical [4,5] methods have shown that similar large structures are also generated in reacting shear layers.

The present paper stems from interest in the mixing layers that arise from film cooling rocket engine combustors[5,6,7]. Of particular interest is a gaseous hydrogen-oxygen engine that uses about two-thirds of the hydrogen for wall film cooling while the remaining one-third is mixed with the oxidizer and used for primary combustion. As the hydrogen coolant is convected along the wall from the injection plane to the throat, an unknown fraction of it mixes with the oxidizer-rich core gas and burns. This combustion improves the overall efficiency of the thruster, but reduces the effectiveness of the coolant layer. The resulting shear layer that is created between this coolant layer and the hot combustion gases in the core region includes large molecular weight differences, chemical reactions and strong heat release. Classical shear layer instability, unsteady vortex roll-up and related shear layer dynamics can therefore be expected in the mixing layer. The vortex dynamics are, however, impacted by the presence of the adjacent wall so that subsequent roll-up will be weakened. The issues of interest in the analysis are to document the role that these various mechanisms have on unsteady mixing in the shear layer and the integrity of the film of coolant adjacent to the wall.

2 Computational Model

The numerical algorithm is based on extending earlier supersonic reacting flow calculations [8,9] to the subsonic combustion problem [5,10]. The analysis uses an unsteady, three-dimensional, finite volume Navier-Stokes procedure that includes chemical nonequilibrium effects. The governing equations are solved by standard upwind, finite-volume methods. Time-accurate solutions are achieved by expressing the equations in a fully implicit fashion using three point backward differencing in time. The left hand side matrix is then solved at each physical time step by means of a pseudo time iteration. This dual time stepping procedure uses an LU approximate factorization in the pseudo time to converge the solution within each physical time step [5]. The physical time step is chosen based on the time evolution of the unsteady shear layer while the pseudo time is determined from the stability characteristics of the numerical algorithm. A hydrogen-oxygen chemical kinetics model involving nine chemical species and eighteen elementary reactions is used [8].

3 Results

To validate the accuracy of the numerical procedure, we first compute the disturbance amplification in a non-reacting mixing layer of unequal molecular weight and compare with predictions of parallel flow stability theory. Nitrogen (lower portion) and helium (top portion) are chosen to impose a molecular weight ratio of 7 and a hyperbolic tangent velocity profile with a velocity ratio of 0.25 (lower to top) is implemented to compare with the stability theory predictions. Computational results were obtained by specifying upstream boundary conditions corresponding to the velocity and species profiles plus a 1 % perturbation of the eigenfunction obtained from the stability solution at a particular frequency. These conditions were subsequently allowed to propagate through the computational domain and the growth or decay of the perturbations was observed. Contour plots of the v-component of the perturbation velocity are given in Fig. 1a for the computational results and Fig. 1b for the stability solution. Quantitative comparisons of the wavelength and the spatial growth rate of the disturbances between the numerical and analytical solutions are as follows: analytical, $c = 0.4834$, $\alpha_i = -0.0727$; computational, $c = 0.4813$, $\alpha_i = -0.0743$.

To demonstrate the dynamics of unforced hydrogen-oxygen reacting shear layers, we begin by investigating the effects of the wall location. Figure 2 shows results of calculations for four different outer wall locations. The location of the splitter plate (radius of 10.3 mm) and the width of its base (1.3 mm) remain unchanged. The flow velocities are chosen to match the engine baseline operating condition: the hydrogen fuel (top portion) is injected at a speed of 125 m/s while the oxygen core flow (lower portion) enters at 185 m/s, giving a nominal velocity ratio of 0.67 and a density ratio of 16. The plots show contours of the OH concentration, a reasonable indicator of the instantaneous diffusion flame location. In Fig. 2a, the outer wall is sufficiently far away from the shear layer (radius = 30 mm) that it has little effect on the vortex dynamics. The hydrogen-oxygen interface undergoes strong distortion and roll-up in the classical sense. The small bulge near the splitter plate represents the nucleus of the next succeeding roll-up. At an intermediate wall location (radius = 22 mm) some wall interaction is observed (Fig. 2b), but strong distortion is still present. The presence of the wall appears to cause the vortex to sweep forward more strongly than in Fig. 2a when the wall is further away. It is clear that the presence of the nearby wall affects the shape of the distorted shear layer. Figure 2c shows results for when the outer wall is brought still closer to the shear layer (radius = 17 mm). At this location the wall begins to prevent roll-up and diminish the unsteadiness. The distortion hits the outer wall before the entire roll-up process is completed, and the subsequent roll-up is forced to decay. Figure 2d shows results for the case when the outer wall is very close to the splitter plate (radius = 12.7 mm), the geometry of interest for the gas-gas combustor. A quick visual inspection shows the dominant influence of the adjacent wall. For this case, the wall is too close to the shear layer to allow the strong distortions seen in the previous two plots, and consequently the ensuing vortex roll-up does not occur. The shear layer does, however, maintain a substantial amount of unsteady oscillation.

To provide numerical perspective, additional calculations were made using two other time-accurate methods to contrast with the dual time stepping procedure described in the previous section; a first order Euler implicit method and a three-point, non-iterative method. The case in Fig. 2b is chosen as a representative geometry. Time histories of integrated water vapor mass flow rate at an axial location ($x = 33$ mm) are shown in Fig. 3 for both the iterative dual time approach and the non-iterative three-point backward methods. Although these methods are both second order accurate, their results are considerably different as Fig. 3 shows. The solution for the non-iterative method becomes periodic in time shortly after the calculations are started. Only one dominant frequency, representing the vortex shedding from the base region, is observed in this solution. However, the dual time stepping results exhibit additional frequencies in the early time calculation along with a notable shift of the dominant frequency. The results suggest that the non-iterative method is more dissipative than that of the dual time stepping methodology. This additional dissipation serves to wash out the weaker frequencies and causes the dominant frequency to shift. Comparisons of the Euler implicit and the three-point non-iterative methods exhibit only small differences.

Figure 4 shows global mixing effects for the cases of Figs. 3b and 3c for the two different outer wall locations. When the outer wall is sufficiently far away from the shear layer (Fig. 3b), the mixing and roll-up processes are more complete, and the amount of water vapor produced is higher. With the outer wall moved closer to the shear layer (Fig. 3c), the roll-up is limited. The effects of wall interaction becomes the dominant mechanism in controlling the dynamics of the shear layer, and the mixing is reduced to result in a smaller amount of water vapor formed. The impact of the outer wall on the reacting shear layer clearly not only controls the vortex dynamics but also reduces the mixing and chemical reaction effects in the shear layer.

Representative solutions of the unsteady, reactive mixing layer inside an actual combustor are given in Fig. 5. The upper half of this figure shows the spatial variation of the OH radical concentration in the combustor, while the lower half shows the temperature contours. The calculations indicate that unsteady flow exists in the mixing layer between the heavy gases in the hot core region and lighter hydrogen cooling film. Vortex roll-up is, however, minimized by the proximity of the combustor wall as noted above. A large temperature gradient is observed in the reacting layer because of the presence of a diffusion flame between the core gas and the cooling layer. The temperature first increases to a maximum (3400 K), and then decreases to the coolant film temperature (670 K). The OH radical concentration is here again used as an indicator of the diffusion flame. The OH concentration in the thin flame zone is very high but it diminishes quickly outside the flame zone, and finally decreases to zero in the hydrogen coolant layer. Computations of the thrust and specific impulse show they fluctuate by about ±3% because of the shear layer unsteadiness.

Acknowledgements

This work has been supported by NASA Grants NAG 3-1020 and NAGW-1356. Partial computational support has been provided by Penn State's Center for Academic Computing and the NAS Program at NASA Ames. The stability results are provided by Mr. Douglas Schwer.

References

[1] Brown, G. and Roshko, A., J. of Fluid Mechanics **64** 775 (1974)

[2] Hermanson, J. and Dimotakis, P. , J. of Fluid Mechanics **199** 333 (1989)

[3] Chang, C., Marek, C., Wey, C., Jones, R. and Smith, M., AIAA Paper 93-2381.

[4] Grinstein, F. and Kailasanath, K., Physics of Fluids **A4** no. 10 2207 (1992)

[5] Tsuei, H.-H. and Merkle, C., AIAA Paper 94-0553.

[6] de Groot, W. and Tsuei, H.-H., AIAA Paper 94-0220.

[7] Weiss, J. and Merkle, C., AIAA Paper 93-0237.

[8] Yu, S.-T., Tsai, Y.-L. P. and Shuen, J.-S., AIAA Paper 89-0391.

[9] Molvik, G. and Merkle, C., AIAA Paper 89-0199.

[10] Venkateswaran, S., Weiss, J., Merkle, C. and Choi, Y.-H., AIAA 92-3437.

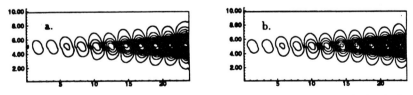

Fig. 1. Contours of the v-component perturbation velocity. a. Numerical solution, b. Analytical solution.

a. Outer wall radius = 30 mm

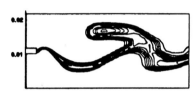

b. Outer wall radius = 22 mm

c. Outer wall radius = 17 mm

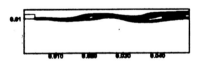

d. Outer wall radius = 12.7 mm

Fig. 2. The OH radical concentration contours for four different outer wall locations.

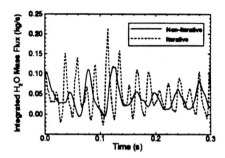

Fig. 3. Local flowfield comparisons between iterative and non-iterative second order time accurate methods.

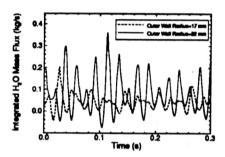

Fig. 4. Effects of outer wall location on global mixing characteristics of the shear layers.

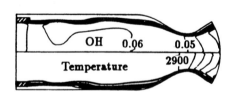

Fig. 5. Spatial variation of the OH radical concentration and temperature contours in the combustor.

Validation of Two 3-D Numerical Computation codes for the flows in an annular cascade of high turning angle Turbine Blades

Wang Wensheng, Zhang Fengxian, Xu Yanji and Chen Naixing

Institute of Engineering Thermophysics, P.O. Box 2706
Chinese Academy of Sciences, Beijing 100080, P.R.China

Abstract : This paper describes and validates two improved three-dimensional numerical methods employed for calculating the flows in an annular cascade of high turning angle turbine blades tested by the authors in the annular cascade wind tunnel of the Institute of Engineering Thermophysics. Comparisons between the predictions and measurements were made on the static pressure contours of blade pressure and suction surfaces, and the spanwise distributions of pitchwise area-averaged static pressure coefficient and flow angle in the downstream of the cascade. The agreement between the calculated results and experimental data shows good and validates the reliability and applicability of the computation codes.

1 Introduction

For predicting and reducing the cascade losses for modern highly loaded, high efficient turbomachinery is required more accurate and detailed knowledge of the complex internal flow mechanism. Because the experimental studies on three-dimensional cascade flows are expensive and difficult to obtain the whole flow features in detail, the importance of numerical techniques in improving the aerodynamic performance of turbomachinery and reducing time and costs consumed is now widely recognized. With the rapid developments of computer science and computational techniques nowadays, the numerical methods for calculating the internal flows on turbomachinery have been developed from two-dimensional to three-dimensional, from inviscid to viscous[1]. In recent years, great efforts have been made by the authors in China and abroad on CFD applied to blade design and internal flow predictions in turbomachinery, and the computational codes developed have worked quite well in many cascade flow cases[2]−[9].

In this paper, two three-dimensional numerical methods proposed by the authors are presented and validated for the flows in an annular cascade of high turning angle turbine blades. Comparisons between the computations and experiments were made on the static pressure contours of blade pressure and suction surfaces and spanwise distributions of the static pressure and flow angle downstream of the cascade. The results have shown the good agreement between both of the numerical codes and measured data. Comparatively, the pressure correction Navier-Stokes solutions are more accurate and reasonable with the measured data than the time-marching Euler solutions.

2 Numerical Methods

2.1 Time Marching Euler Algorithm

Governing Equations The three-dimensional Euler governing equations for compressible inviscid flows in an absolute Cartesian coordinate system are written in integral form as:

$$\left.\begin{array}{l} \int_\Omega \frac{\partial \rho}{\partial t} d\Omega + \oint_\Gamma \rho(\vec{V} \cdot d\vec{S}) = 0 \\[4pt] \int_\Omega \frac{\partial (\rho\vec{V})}{\partial t} d\Omega + \oint_\Gamma \rho\vec{V}(\vec{V} \cdot d\vec{S}) + \oint_\Gamma p d\vec{S} = 0 \\[4pt] \int_\Omega \frac{\partial (\rho e)}{\partial t} d\Omega + \oint_\Gamma \rho e(\vec{V} \cdot d\vec{S}) + \oint_\Gamma p(\vec{V} \cdot d\vec{S}) = 0 \end{array}\right\} \quad (1)$$

$$p = \rho(k-1)(e - \frac{1}{2}V^2) \quad (2)$$

where Ω denotes a fixed volume with boundary Γ, $\vec{V}$ is total velocity vector, ρ, p, e are density, pressure and total energy per unit volume respectively. k is ratio of specific heats.

Discretization The above integral equations, Eq.(1), can be discretized into the following ordinary differential equations with summation around the six surfaces (numbered as l) of the hexahedral cell and the variables are defined at the cell center:

$$\left.\begin{array}{l} \frac{d\rho}{dt} = -\frac{1}{Vol}\sum \rho_l(\vec{V} \cdot \Delta \vec{S})_l \\[4pt] \frac{d(\rho V_x)}{dt} = -\frac{1}{Vol}\left[\sum (\rho V_x)_l(\vec{V} \cdot \Delta \vec{S})_l + \sum (p \cdot \Delta S_x)_l\right] \\[4pt] \frac{d(\rho V_y)}{dt} = -\frac{1}{Vol}\left[\sum (\rho V_y)_l(\vec{V} \cdot \Delta \vec{S})_l + \sum (p \cdot \Delta S_y)_l\right] \\[4pt] \frac{d(\rho V_z)}{dt} = -\frac{1}{Vol}\left[\sum (\rho V_z)_l(\vec{V} \cdot \Delta \vec{S})_l + \sum (p \cdot \Delta S_z)_l\right] \\[4pt] \frac{d(\rho e)}{dt} = -\frac{1}{Vol}\left[\sum (\rho e + p)_l(\vec{V} \cdot \Delta \vec{S})_l\right] \end{array}\right\} \quad (3)$$

where Vol is cell volume, $\Delta \vec{S}_l = (\Delta S_{xl}, \Delta S_{yl}, \Delta S_{zl})^T$ is area vector of the cell surface l, V_x, V_y, and V_z are velocity components in x, y, z direction respectively.

Since the variables are stored at the cell center, their values at cell volumes can be obtained by simply averaging the corresponding values from the adjacent cell centers. This kind of finite volume spatial discretization reduces to central difference scheme which is formally second order accurate for smooth grid. Eq.(3) are solved by using the modified multi-stage Runge-Kutta time-marching method as described in Chen, et al.[2] The accelerating convergence techniques such as implicit residual averaging are also used.

Grid   The computational grid used for Euler prediction is a non-orthogonal, equi-proportional H-type grid. It has 21 lines in the radial and circumferential directions, and 64 lines between the inlet and outlet boundaries, as shown in Fig.1.

Boundary Conditions   At inlet boundary, the total pressure, total temperature, velocity and its flow angles are given as the test conditions. At the outlet boundary, the static pressure is specified approximately as the local atmospheric pressure. On the solid walls are used the impermeable conditions, i.e. $\vec{V} \cdot \vec{S} = 0$. The periodic boundary conditions are satisfied for all the variables in the upstream and downstream domains of the cascade.

3 Pressure Correction Navier-Stokes Algorithm

Governing Equations   The three-dimensional Navier-Stokes equations coupled with $k - \epsilon$ turbulence model for steady, compressible turbulent flows in an arbitrary non-orthogonal curvilinear coordinate system are written in a generalized form as:

$$\frac{\partial}{\partial x^j}(\rho\sqrt{g}w^j \varphi) = \frac{\partial}{\partial x^j}\left(\Gamma_g^{ij} \frac{\partial \varphi}{\partial x^i}\right) + S_\varphi(x^1, x^2, x^3) \qquad (4)$$

$$p = \rho RT \qquad (5)$$

where:

$$\varphi = [1, \quad V_x, \quad V_y, \quad V_z, \quad I, \quad k, \quad g]^T \qquad (6)$$

$$\Gamma = [0, \quad \mu_{ef}\sqrt{g}, \quad \mu_{ef}\sqrt{g}, \quad \mu_{ef}\sqrt{g}, \quad \lambda_{ef}\sqrt{g}/C_p, \quad \mu_t\sqrt{g}/\sigma_k, \mu_t\sqrt{g}/\sigma_\epsilon]^T \qquad (7)$$

$$w^j = V_x \frac{\partial x^j}{\partial x} + V_y \frac{\partial x^j}{\partial y} + V_z \frac{\partial x^j}{\partial z} \qquad (j = 1, 2, 3) \qquad (8)$$

$$g^{ij} = \left(\frac{\partial x^i}{\partial x}\right)\left(\frac{\partial x^j}{\partial x}\right) + \left(\frac{\partial x^i}{\partial y}\right)\left(\frac{\partial x^j}{\partial y}\right) + \left(\frac{\partial x^i}{\partial z}\right)\left(\frac{\partial x^j}{\partial z}\right) \qquad (i,j = 1, 2, 3) \qquad (9)$$

in above equations φ denotes a general dependent variable, Γ is the diffusion coefficients, I is the stagnation rothalpy, k and ϵ are the kinetic energy and its dissipation rate respectively, μ_{ef} and μ_t are the effective and turbulent viscosity, λ_{ef} is the effective conductivity, w^1, w^2 and w^3 are the contravarient velocity components, g and g^{ij} are the metrical tensors, and $S_\varphi(x^1, x^2, x^3)$ represent the source terms.

Discretization   The discretized equations are obtained by integrating Eq.(4) over the appropriate control volumes and expressed in terms of neighbouring grid points. A hybrid difference form were adopted for improving the stability and convergence of the numerical solutions. A central difference approximation was used for all the diffusion terms, and an upwinding scheme for the convection terms. In order to avoid the associated pressure and velocity distribution oscillations effectively, a staggered grid system is used in the calculations. All the scalar variables are stored at each of the grid points. The three Cartesian velocity components and the contravariant components are stored in a staggered arrangement at each face of a control volume, i.e. the centre of adjacent grid points (details see Chen, et al.[3]). The discretized algebraic equations for all the scalar variables can be written in the general version as:

$$a_p^\varphi \varphi_p = \sum_i a_i^\varphi \varphi_i + S_u^\varphi \qquad (i = E, W, S, N, U, D) \tag{10}$$

and the discretized momentum equations can be written as:

$$a_p^\varphi \varphi_p = \sum_i a_i^\varphi \varphi_i + S_u^\varphi \qquad (i = e, w, s, n, u, d) \tag{11}$$

where φ_p denotes one of the variables p', k and ϵ at point P surrounded by neighbouring points E, W, S, N, U and D. φ_p denotes one of the variables ρV_x, ρV_y and ρV_z at point p surrounded by neighbouring points e, w, s, n, u and d. P and p are the centres of two neighbouring staggered grids. S_u^φ represent the linearized source terms.

Eq.(10) and Eq.(11) are solved by successive iterations with linear relaxation method. All the field variables iterate until each transport equation satisfies a prebuilt convergence criterion. A fully-converged solution is declared when the maximum normalized residuals of each governing equations at all the calculated grid points are no greater than 10^{-4}.

Grid Fig.2 shows the three-dimensional, H-type grid ($71 \times 21 \times 21$) used for Navier-Stokes prediction. The grid points are clustered near the solid walls by using a coordinate transformation formula[10] for improving the computational accuracy.

Boundary Conditions In addition to the same inlet, outlet and periodic boundary conditions as that given in the above-mentioned Euler algorithm, the inlet kinetic energy and its dissipation rate are also required. On the solid walls, no-slip conditions and zero values of k and ϵ are used, thermal-isolated walls, i.e. $(\partial T/\partial n)_w = 0$, are assumed, and the wall functions are applied to decrease the numbers of grid points near the walls.

4 Numerical Example

The high-turning annular turbine cascade presented in this paper is representative of modern highly loaded blade design philosophy, and a typical example for assessing the validation of the three-dimensional flow computational codes. Fig.3 shows the blade airfoil section. It has a constant profile along the blade height and untwisted. The blades are installed radially about point M. The inlet conditions are:

$p_{t0} = 8980.27 N/m^2$, $T_{t0} = 297.6K$, $Re_0 = 2.59 \times 10^5$, $Tu_\infty = 6.5\%$. The experiments were carried out in the annular cascade wind tunnel of the Institute of Engineering Thermophysics developed by the authors. Details of the experimentation are given in Wang's paper[11]. The test data can not only reveal the details of the flow fields, but also provide a tolerable justication for the use of the computations. In the following section, some of the experimental results are presented and compared with the numerical predictions.

5 Codes Validation

In Fig.4 and Fig.5, the predictions of static pressure contours on blade surfaces are compared with the measured data. On the blade pressure surface, the radial pressure gradients are small over most of the blade axial chord, but in the corner between the trailing edge nearby and inner wall, the pressure varies rapidly, and the axial and radial pressure gradients increased towards the low-pressure core. On the blade suction surface, a two-dimensional flow region exists between the leading edge and 30% axial chord, but the pressure falls rapidly towards a low-pressure zone located at about 65% axial chord near the inner wall, and a minimum pressure vortex core is formed there correspondingly. The comparisons show that, both numerical solutions show similar pressure contours, and agree well with the experimental results.

Fig.6 and Fig.7 show the comparisons between the measured and predicted spanwise distributions of pitchwise area-averaged static pressure coefficient and flow angle at the axial plane $x/C_{ax} = 1.06$. The time-marching Euler solutions overestimate C_{psm} and underestimate α_m, and the pressure correction Navier-Stokes solutions show more closely with the measured data. But the overall results of both numerical methods show good agreement with the test data.

6 Conclusions

From the above comparisons we can conclude that, for the typical numerical case of high turning annular turbine cascade, both of these two numerical computation codes can simulate the three-dimensional flows well to a certain degree, and have good reliability and accuracy for predicting the static pressure and velocity fields. Comparatively, the pressure correction Navier-Stokes solutions are more closely and reasonable with the measured data than the time-marching Euler solutions.

References

[1] Mcnally, W.D., et al., ASME J. Fluids Engineering, Vol.107, pp.6-21, 1985.

[2] Chen Naixing, et al., the 18th ICAS, Vol.1, pp.68-76, Beijing, 1992.

[3] Chen Naixing, et al., the 2nd ISAIF, pp.149-157, Prague, 1993.

[4] Miao Yongmiao, et al., ASME Paper No.91-GT-171, 1991.

[5] Denton, J.D., ASME J. Engineering for Power, Vol.105, pp.514-524, 1983.

[6] Hah, C., ASME J. Engg. for Gas Turbines and Power, Vol.106, pp.421-429, 1984.

[7] Jameson, A., et al., ASME Paper No.84-GT-76, 1984.

[8] McDonald, H., et al., ASME Paper No.85-GT-66, 1985.

[9] Yamamoto, S., et al., Paper 87-Tokyo-IGTC-39, 1987.

[10] Chen Naixing, et al., the 1st ISAIF, pp.3-12, Beijing, 1990.

[11] Wang Wensheng, Ph.D. Thesis, I.E.T. of CAS, Beijing, 1992.

Fig.1 Computational grid ($64 \times 21 \times 21$) for Euler prediction

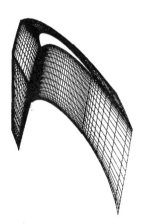

Fig.2 Computational grid ($71 \times 21 \times 21$) for Navier–Stokes prediction

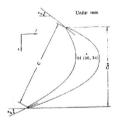

Fig.3 The blade profile

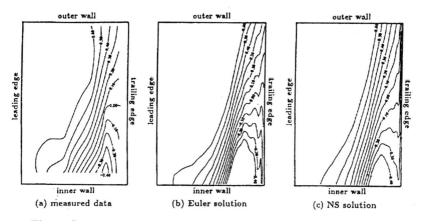

Fig.4 Comparison between the measured and predicted static pressure contours on blade pressure surface

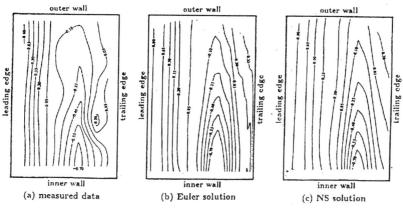

Fig.5 Comparison between the measured and predicted static pressure contours on blade suction surface

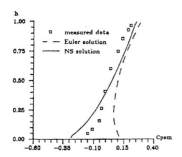

Fig.6 Comparison between the measured and predicted spanwise distributions of pitchwise aera-averaged static pressure coefficients at the axial plane $x / C_{ax} = 1.06$

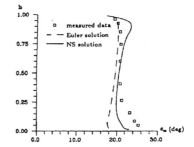

Fig.7 Comparison between the measured and predicted spanwise distributions of pitchwise aera-averaged flow angle at the axial plane $x / C_{ax} = 1.06$

5. Author Index

Andrews, J.G., 122
Arakeri, J.H., 550
Arina, R., 228
Arora, M., 379
Bai, L., 435
Bandyopadhyay, G., 489
Bassi, F., 234
Basu, A.J., 476
Behnia, M., 447
Biswas, R., 127
Bogolepov, V.V., 441
Brandes-Duncan, B., 337
Bruneau, Ch. -H., 320
Cambier, J.-L., 567
Chanteur, G., 86
Catalano, L.A., 241
Chalot, F., 133
Charrier, P., 251
Chatterjee, A., 139
Chattot, J.-J., 413
Chen, H., 501
Chen, P.Y.P., 447
Cherny, S.G., 280
Chollet, J.P., 348
Condra, T.J., 418
Croisille, J.P., 86
Dadone, A., 188
Daiguji, H., 314
Damodaran, M., 507
Dang-Vu, H., 385
Davidson, L., 372
De Palma, P., 241
Decker, W.J., 325
Dedesh, V.V., 246
Delcarte, C., 385
Deshpande, S.M., 101, 524
Deshpande, S.S., 91
Dey, P.K., 550
Dick, E., 353
Dijkstra, D., 358

Dorning, J.J., 325
Engel, K., 454
Estivalezes, J.L., 96
Eulitz, F., 454
Fabrie, P., 320
Faden, M., 454
Favini, B., 307
Fengxian, Z., 578
Fiebig, M., 435
Flandrin, L., 251
Fletcher, C.A.J., 495
Floryan, J.M., 287
Fontaine, J., 391
Fort, J., 461
Fortunato, B., 259
Friedrich, R., 330
Gao, Z., 396
Gazaix, M., 144
Ghia, K., 337
Ghia, U., 337
Ghosh, A.K., 101
Gregoire, J.P., 149
Groenner, J., 265
Grossman, B., 188
Guilmineau, E., 466
Hänel, D., 155
Hartmann, C., 472
Hilgenstock, M., 265
Holst, M.D., 418
Hunék, M., 461
Iollo, A., 274
Jameson, A., 71,106
Jia, W., 195
Karamyshev, V.B., 280
Karniadakis, G.E., 429
Kasbarian, C., 133
Keeling, S.L., 529
Keyes, D.E., 1
Khanfir, R., 86
Klopfer, G.H., 285

Koumoutsakos, P., 21
Kovenya, V.M., 280
Kozel, K., 461
Krishnamurthy, R., 524
Kulkarni, P.S., 524
Kwon, K.D., 539
Lain, J., 461
Laschefski, H, 368.
Leclercq, M.-P., 133
Leonard, A., 21
Li, Yuguo, 402
Lipatov, I.I., 441
Loc, Ta Phuoc, 391, 561
Lynn, J.F., 161
Magi, V., 259
Mallet M., 133
Martinelli, L., 106
Mathew, J., 476
Mavriplis, D.J., 178
Mejak, G., 481
Mellen, C.P., 201
Merkle, C.L., 573
Milton, B.E., 447
Mitra, N.K., 368, 435
Mizuta, Yo, 206
Mochimaru, Y., 485
Modi, V.J., 489
Mohammadi, B., 31
Morgret, C.H., 529
Morton, K.W., 122
Mukharjea, S.K., 534
Munshi, S.R., 489
Murali Krishna, K., 534
Muralirajan, K.S., 337
Nagarathinam, M., 524
Nagatomo, H., 314
Naixing, C., 578
Nakamura, Y., 195
Napolitano, M., 241
Nasuti, F., 407
Nishitani, T., 512
O'Shea, N., 495
Obayashi, S., 285

Olsson, E., 372
Onofri, M., 407
Oshima, K., 501
Pagano, A., 293
Pallis, J.M, 413
Parthasarthy, K.N, 529
Pascazio, G., 241
Paulsen, B.B., 418
Pironneau, O., 31
Pokorny, S., 454
Pot, G., 149
Prahlad, T.S., 39
Queutey, P., 466
Raghurama Rao, S.V., 112
Ramakrishnananda, B., 507
Ramamurthi, K., 343
Ramella, F., 228
Ravachol, M., 133
Ravi, K., 343
Ravi, K., 519
Raymond, P., 299
Rebay, S., 234
Rist, M.J., 529
Roe, P.L., 379
Roesner, K.G., 472
Rogier, F., 117
Rokicki, J., 287
Rudman, M., 402
Rusaas, J., 418
Sa, J.-Y., 423
Saha, S., 534
Salas, M.D., 274
Satofuka, N., 512
Saxena, S.K., 519
Schneider, J., 117
Schulze, D., 211
Šejna, M., 461
Sekar, S., 524
Sen, D., 216
Sherwin, S.J., 429
Shevare, G.R., 139
Shyy, W., 221
Si Ameur, M., 348

Sickles, W.L., 529
Singh, K.P., 534
Sokolov, L.A., 441
Song, D.J., 539
Soni, B.K., 166
Sreenivas, K.R., 550
Srinivas, K., 201
Srinivasan, G.R., 555
Srinivasan, J., 550
Steelant, J., 353
Stoufflet, B., 133
Strawn, R.C., 127
Szmelter, J., 293
Ta'asan, S., 173, 274
Tenaud, C., 561
Theofilis, V., 358
Thornburg, Hugh, 166
Tokarcik-Polsky, S., 567
Toumi, I., 299
Tsuei, H.-H., 573
Udaykumar, H.S., 221
Valorani, M., 307

van Leer, B., 161
Vavrincová, M., 461
Venkatakrishnan, V., 178
Venkateswaran, S., 363
Villedieu, P., 96
Vilsmeier, R., 155
von Lavante, E., 265
Wagner, C., 330
Weatherill, N.P., 54
Wensheng, W., 578
Winckelmans, G., 21
Winkelsträter, M., 368
Won, S.Y., 539
Wu, Zi-Niu, 183
Xu, K., 106
Yamamoto, S, 314
Yanji, Xu, 578
Yegorov, I.V., 246
Yokomizo, T., 489
Zandbergen, P.J., 358
Zhou, G., 372
Zhuang, F.G., 396

PHOTOGRAPH TAKEN ON THE OCCASION OF ICNMFD 14

Lecture Notes in Physics

For information about Vols. 1–419
please contact your bookseller or Springer-Verlag

Vol. 420: F. Ehlotzky (Ed.), Fundamentals of Quantum Optics III. Proceedings, 1993. XII, 346 pages. 1993.

Vol. 421: H.-J. Röser, K. Meisenheimer (Eds.), Jets in Extragalactic Radio Sources. XX, 301 pages. 1993.

Vol. 422: L. Päivärinta, E. Somersalo (Eds.), Inverse Problems in Mathematical Physics. Proceedings, 1992. XVIII, 256 pages. 1993.

Vol. 423: F. J. Chinea, L. M. González-Romero (Eds.), Rotating Objects and Relativistic Physics. Proceedings, 1992. XII, 304 pages. 1993.

Vol. 424: G. F. Helminck (Ed.), Geometric and Quantum Aspects of Integrable Systems. Proceedings, 1992. IX, 224 pages. 1993.

Vol. 425: M. Dienes, M. Month, B. Strasser, S. Turner (Eds.), Frontiers of Particle Beams: Factories with e^+ e^- Rings. Proceedings, 1992. IX, 414 pages. 1994.

Vol. 426: L. Mathelitsch, W. Plessas (Eds.), Substructures of Matter as Revealed with Electroweak Probes. Proceedings, 1993. XIV, 441 pages. 1994

Vol. 427: H. V. von Geramb (Ed.), Quantum Inversion Theory and Applications. Proceedings, 1993. VIII, 481 pages. 1994.

Vol. 428: U. G. Jørgensen (Ed.), Molecules in the Stellar Environment. Proceedings, 1993. VIII, 440 pages. 1994.

Vol. 429: J. L. Sanz, E. Martínez-González, L. Cayón (Eds.), Present and Future of the Cosmic Microwave Background. Proceedings, 1993. VIII, 233 pages. 1994.

Vol. 430: V. G. Gurzadyan, D. Pfenniger (Eds.), Ergodic Concepts in Stellar Dynamics. Proceedings, 1993. XVI, 302 pages. 1994.

Vol. 431: T. P. Ray, S. Beckwith (Eds.), Star Formation and Techniques in Infrared and mm-Wave Astronomy. Proceedings, 1992. XIV, 314 pages. 1994.

Vol. 432: G. Belvedere, M. Rodonò, G. M. Simnett (Eds.), Advances in Solar Physics. Proceedings, 1993. XVII, 335 pages. 1994.

Vol. 433: G. Contopoulos, N. Spyrou, L. Vlahos (Eds.), Galactic Dynamics and N-Body Simulations. Proceedings, 1993. XIV, 417 pages. 1994.

Vol. 434: J. Ehlers, H. Friedrich (Eds.), Canonical Gravity: From Classical to Quantum. Proceedings, 1993. X, 267 pages. 1994.

Vol. 435: E. Maruyama, H. Watanabe (Eds.), Physics and Industry. Proceedings, 1993. VII, 108 pages. 1994.

Vol. 436: A. Alekseev, A. Hietamäki, K. Huitu, A. Morozov, A. Niemi (Eds.), Integrable Models and Strings. Proceedings, 1993. VII, 280 pages. 1994.

Vol. 437: K. K. Bardhan, B. K. Chakrabarti, A. Hansen (Eds.), Non-Linearity and Breakdown in Soft Condensed Matter. Proceedings, 1993. XI, 340 pages. 1994.

Vol. 438: A. Pękalski (Ed.), Diffusion Processes: Experiment, Theory, Simulations. Proceedings, 1994. VIII, 312 pages. 1994.

Vol. 439: T. L. Wilson, K. J. Johnston (Eds.), The Structure and Content of Molecular Clouds. 25 Years of Molecular Radioastronomy. Proceedings, 1993. XIII, 308 pages. 1994.

Vol. 440: H. Latal, W. Schweiger (Eds.), Matter Under Extreme Conditions. Proceedings, 1994. IX, 243 pages. 1994.

Vol. 441: J. M. Arias, M. I. Gallardo, M. Lozano (Eds.), Response of the Nuclear System to External Forces. Proceedings, 1994, VIII. 293 pages. 1995.

Vol. 442: P. A. Bois, E. Dériat, R. Gatignol, A. Rigolot (Eds.), Asymptotic Modelling in Fluid Mechanics. Proceedings, 1994. XII, 307 pages. 1995.

Vol. 443: D. Koester, K. Werner (Eds.), White Dwarfs. Proceedings, 1994. XII, 348 pages. 1995.

Vol. 444: A. O. Benz, A. Krüger (Eds.), Coronal Magnetic Energy Releases. Proceedings, 1994. X, 293 pages. 1995.

Vol. 445: J. Brey, J. Marro, J. M. Rubí, M. San Miguel (Eds.), 25 Years of Non-Equilibrium Statistical Mechanics. Proceedings, 1994. XVII, 387 pages. 1995.

Vol. 446: V. Rivasseau (Ed.), Constructive Physics. Results in Field Theory, Statistical Mechanics and Condensed Matter Physics. Proceedings, 1994. X, 337 pages. 1995.

Vol. 447: G. Aktaş, C. Saçlıoğlu, M. Serdaroğlu (Eds.), Strings and Symmetries. Proceedings, 1994. XIV, 389 pages. 1995.

Vol. 448: P. L. Garrido, J. Marro (Eds.), Third Granada Lectures in Computational Physics. Proceedings, 1994. XIV, 346 pages. 1995.

Vol. 449: J. Buckmaster, T. Takeno (Eds.), Modeling in Combustion Science. Proceedings, 1994. X, 369 pages. 1995.

Vol. 450: M. F. Shlesinger, G. M. Zaslavsky, U. Frisch (Eds.), Lévy Flights and Related Topics in Physics. Proceedigs, 1994. XIV, 347 pages. 1995.

Vol. 452: A. M. Bernstein, B. R. Holstein (Eds.), Chiral Dynamics: Theory and Experiment. Proceedings, 1994. VIII, 351 pages. 1995.

Vol. 453: S. M. Deshpande, S. S. Desai, R. Narasimha (Eds.), Fourteenth International Conference on Numerical Methods in Fluid Dynamics. Proceedings, 1994. XIII, 589 pages. 1995.

Vol. 454: J. Greiner, H. W. Duerbeck, R. E. Gershberg (Eds.), Flares and Flashes, Germany 1994. XXII, 477 pages. 1995.

Vol. 455: F. Occhionero (Ed.), Birth of the Universe and Fundamental Physics. Proceedings, 1994. XV, 387 pages. 1995.

Vol. 456: H. B. Geyer (Ed.), Field Theory, Topology and Condensed Matter Physics. Proceedings, 1994. XII, 206 pages. 1995.

New Series m: Monographs

Vol. m 1: H. Hora, Plasmas at High Temperature and Density. VIII, 442 pages. 1991.

Vol. m 2: P. Busch, P. J. Lahti, P. Mittelstaedt, The Quantum Theory of Measurement. XIII, 165 pages. 1991.

Vol. m 3: A. Heck, J. M. Perdang (Eds.), Applying Fractals in Astronomy. IX, 210 pages. 1991.

Vol. m 4: R. K. Zeytounian, Mécanique des fluides fondamentale. XV, 615 pages, 1991.

Vol. m 5: R. K. Zeytounian, Meteorological Fluid Dynamics. XI, 346 pages. 1991.

Vol. m 6: N. M. J. Woodhouse, Special Relativity. VIII, 86 pages. 1992.

Vol. m 7: G. Morandi, The Role of Topology in Classical and Quantum Physics. XIII, 239 pages. 1992.

Vol. m 8: D. Funaro, Polynomial Approximation of Differential Equations. X, 305 pages. 1992.

Vol. m 9: M. Namiki, Stochastic Quantization. X, 217 pages. 1992.

Vol. m 10: J. Hoppe, Lectures on Integrable Systems. VII, 111 pages. 1992.

Vol. m 11: A. D. Yaghjian, Relativistic Dynamics of a Charged Sphere. XII, 115 pages. 1992.

Vol. m 12: G. Esposito, Quantum Gravity, Quantum Cosmology and Lorentzian Geometries. Second Corrected and Enlarged Edition. XVIII, 349 pages. 1994.

Vol. m 13: M. Klein, A. Knauf, Classical Planar Scattering by Coulombic Potentials. V, 142 pages. 1992.

Vol. m 14: A. Lerda, Anyons. XI, 138 pages. 1992.

Vol. m 15: N. Peters, B. Rogg (Eds.), Reduced Kinetic Mechanisms for Applications in Combustion Systems. X, 360 pages. 1993.

Vol. m 16: P. Christe, M. Henkel, Introduction to Conformal Invariance and Its Applications to Critical Phenomena. XV, 260 pages. 1993.

Vol. m 17: M. Schoen, Computer Simulation of Condensed Phases in Complex Geometries. X, 136 pages. 1993.

Vol. m 18: H. Carmichael, An Open Systems Approach to Quantum Optics. X, 179 pages. 1993.

Vol. m 19: S. D. Bogan, M. K. Hinders, Interface Effects in Elastic Wave Scattering. XII, 182 pages. 1994.

Vol. m 20: E. Abdalla, M. C. B. Abdalla, D. Dalmazi, A. Zadra, 2D-Gravity in Non-Critical Strings. IX, 319 pages. 1994.

Vol. m 21: G. P. Berman, E. N. Bulgakov, D. D. Holm, Crossover-Time in Quantum Boson and Spin Systems. XI, 268 pages. 1994.

Vol. m 22: M.-O. Hongler, Chaotic and Stochastic Behaviour in Automatic Production Lines. V, 85 pages. 1994.

Vol. m 23: V. S. Viswanath, G. Müller, The Recursion Method. X, 259 pages. 1994.

Vol. m 24: A. Ern, V. Giovangigli, Multicomponent Transport Algorithms. XIV, 427 pages. 1994.

Vol. m 25: A. V. Bogdanov, G. V. Dubrovskiy, M. P. Krutikov, D. V. Kulginov, V. M. Strelchenya, Interaction of Gases with Surfaces. XIV, 132 pages. 1995.

Vol. m 26: M. Dineykhan, G. V. Efimov, G. Ganbold, S. N. Nedelko, Oscillator Representation in Quantum Physics. IX, 279 pages. 1995.

Vol. m 27: J. T. Ottesen, Infinite Dimensional Groups and Algebras in Quantum Physics. IX, 218 pages. 1995.

Vol. m 28: O. Piguet, S. P. Sorella, Algebraic Renormalization. IX, 134 pages. 1995.

Vol. m 29: C. Bendjaballah, Introduction to Photon Communication. VII, 193 pages. 1995.

Vol. m 30: A. J. Greer, W. J. Kossler, Low Magnetic Fields in Anisotropic Superconductors. VII, 161 pages. 1995.

Vol. m 31: P. Busch, M. Grabowski, P. J. Lahti, Operational Quantum Physics. XI, 230 pages. 1995.

Vol. m 33: R. Alkofer, H. Reinhardt, Chiral Quark Dynamics. VIII, 115 pages. 1995.